More information about this series at http://www.springer.com/series/8385

Metal Ions in Life Sciences

Volume 15

Guest editors
Peter M.H. Kroneck
Martha E. Sosa Torres

Series editors
Astrid Sigel
Helmut Sigel
Roland K.O. Sigel

Astrid Sigel • Helmut Sigel • Roland K.O. Sigel
Series Editors

Peter M.H. Kroneck • Martha E. Sosa Torres
Guest Editors

Sustaining Life on Planet Earth: Metalloenzymes Mastering Dioxygen and Other Chewy Gases

Springer

Guest Editors
Peter M.H. Kroneck
Fachbereich Biologie
Universität Konstanz
Universitätsstrasse 10
D-78457 Konstanz, Germany
peter.kroneck@uni-konstanz.de

Martha E. Sosa Torres
Departamento de Química Inorganica y Nuclear
Facultad de Química
Universidad Nacional Autónoma de México
Ciudad Universitaria
México, D.F. 04510, México
mest@unam.mx

Series Editors
Astrid Sigel
Department of Chemistry
Inorganic Chemistry
University of Basel
Spitalstrasse 51
CH-4056 Basel, Switzerland
astrid.sigel@unibas.ch

Helmut Sigel
Department of Chemistry
Inorganic Chemistry
University of Basel
Spitalstrasse 51
CH-4056 Basel, Switzerland
helmut.sigel@unibas.ch

Roland K.O. Sigel
Department of Chemistry
University of Zürich
Winterthurerstrasse 190
CH-8057 Zürich, Switzerland
roland.sigel@chem.uzh.ch

ISSN 1559-0836 ISSN 1868-0402 (electronic)
Metal Ions in Life Sciences
ISBN 978-3-319-12414-8 ISBN 978-3-319-12415-5 (eBook)
DOI 10.1007/978-3-319-12415-5

Library of Congress Control Number: 2014958669

Springer Cham Heidelberg New York Dordrecht London

Cover illustration: Cover figure of the MILS series since Volume 11: RNA-protein interface of the Ile-tRNA synthetase complex held together by a string of Mg^{2+} ions, illustrating the importance of metal ions in both the protein and the nucleic acid world as well as connecting the two; hence, representing the role of *Metal Ions in Life Sciences*. tRNA synthetases are not only essential to life, but also serve as a target for novel classes of drugs making such RNA-protein complexes crucial also for the health sciences. The figure was prepared by Joachim Schnabl and Roland K. O. Sigel using the PDB coordinates 1FFY.

Printed on acid-free paper

Springer International Publishing AG Switzerland is part of Springer Science+Business Media (www.springer.com)

Historical Development and Perspectives of the Series

Metal Ions in Life Sciences*

It is an old wisdom that metals are indispensable for life. Indeed, several of them, like sodium, potassium, and calcium, are easily discovered in living matter. However, the role of metals and their impact on life remained largely hidden until inorganic chemistry and coordination chemistry experienced a pronounced revival in the 1950s. The experimental and theoretical tools created in this period and their application to biochemical problems led to the development of the field or discipline now known as *Bioinorganic Chemistry*, *Inorganic Biochemistry*, or more recently also often addressed as *Biological Inorganic Chemistry*.

By 1970 *Bioinorganic Chemistry* was established and further promoted by the book series *Metal Ions in Biological Systems* founded in 1973 (edited by H.S., who was soon joined by A.S.) and published by Marcel Dekker, Inc., New York, for more than 30 years. After this company ceased to be a family endeavor and its acquisition by another company, we decided, after having edited 44 volumes of the *MIBS* series (the last two together with R.K.O.S.) to launch a new and broader minded series to cover today's needs in the *Life Sciences*. Therefore, the Sigels new series is entitled

Metal Ions in Life Sciences.

After publication of the first four volumes (2006–2008) with John Wiley & Sons, Ltd., Chichester, UK, and the next five volumes (2009–2011) with the Royal Society of Chemistry, Cambridge, UK, we are happy to join forces now in this still new endeavor with Springer Science & Business Media B.V., Dordrecht, The Netherlands, a most experienced Publisher in the *Sciences*.

---

* Reproduced with some alterations by permission of John Wiley & Sons, Ltd., Chichester, UK (copyright 2006) from pages v and vi of Volume 1 of the series *Metal Ions in Life Sciences* (MILS-1).

The development of *Biological Inorganic Chemistry* during the past 40 years was and still is driven by several factors; among these are: (i) the attempts to reveal the interplay between metal ions and peptides, nucleotides, hormones, or vitamins, etc.; (ii) the efforts regarding the understanding of accumulation, transport, metabolism, and toxicity of metal ions; (iii) the development and application of metal-based drugs; (iv) biomimetic syntheses with the aim to understand biological processes as well as to create efficient catalysts; (v) the determination of high-resolution structures of proteins, nucleic acids, and other biomolecules; (vi) the utilization of powerful spectroscopic tools allowing studies of structures and dynamics; and (vii) more recently, the widespread use of macromolecular engineering to create new biologically relevant structures at will. All this and more is and will be reflected in the volumes of the series *Metal Ions in Life Sciences*.

The importance of metal ions to the vital functions of living organisms, hence, to their health and well-being, is nowadays well accepted. However, in spite of all the progress made, we are still only at the brink of understanding these processes. Therefore, the series *Metal Ions in Life Sciences* will endeavor to link coordination chemistry and biochemistry in their widest sense. Despite the evident expectation that a great deal of future outstanding discoveries will be made in the interdisciplinary areas of science, there are still "language" barriers between the historically separate spheres of chemistry, biology, medicine, and physics. Thus, it is one of the aims of this series to catalyze mutual "understanding".

It is our hope that *Metal Ions in Life Sciences* proves a stimulus for new activities in the fascinating "field" of *Biological Inorganic Chemistry*. If so, it will well serve its purpose and be a rewarding result for the efforts spent by the authors.

Astrid Sigel and Helmut Sigel
Department of Chemistry, Inorganic Chemistry,
University of Basel, CH-4056 Basel, Switzerland

Roland K.O. Sigel
Department of Chemistry,
University of Zürich, CH-8057 Zürich, Switzerland

October 2005,
October 2008,
and August 2011

Preface to Volume 15

Sustaining Life on Planet Earth: Metalloenzymes Mastering Dioxygen and Other Chewy Gases

In this volume of the *Metal Ions in Life Sciences* series the mastering of dioxygen (O_2), methane (CH_4), and ammonia (NH_3) by mainly manganese-, iron- and copper-dependent metalloenzymes and their biomimetic complexes is discussed. It is closely related to Volume 14, *The Metal-Driven Biogeochemistry of Gaseous Compounds in the Environment*, which deals with the biogeochemistry of gases including dihydrogen (H_2), carbon monoxide (CO), acetylene ($HC{\equiv}CH$), dinitrogen (N_2), nitrous oxide (N_2O), hydrogen sulfide (H_2S), and dimethylsulfide (CH_3-S-CH_3). The accumulation of O_2 in the atmosphere forever changed the surface chemistry of the Earth. Dioxygen, as electron acceptor, is used in the respiration of numerous different organisms that conduct a wide variety of chemically complex metabolisms. To produce O_2, and to conserve energy by activating and transforming O_2, CH_4, or NH_3, sophisticated metal-dependent enzymes had to be evolved by Nature. These catalysts can overcome unusually high activation barriers of kinetically inert molecules, still a tremendous challenge in the chemical laboratory today.

In the first chapter, the reader is shortly introduced to several aspects and properties of this special molecule "dioxygen" and its extraordinary impact on our current Earth. Just think of water (H_2O), perhaps the most important compound containing oxygen, a superb solvent for numerous biomolecules, and the main source of O_2 in the atmosphere. Carl Zimmer reports in his article entitled "The Mystery of Earth's Oxygen" (The New York Times, October 3, 2013) about the work of the geochemist D. E. Canfield from the University of Southern Denmark: "There's something astonishing in every breath we take. What is even more astonishing is that the Earth started out with an oxygen-free atmosphere, it took billions of years before there was enough of it to keep animals like us alive". Clearly, oxygen must be considered one of the most important elements on Earth, it means life for all aerobes. Eliminate O_2 and they cannot conserve enough energy to support an active lifestyle.

Chapter 2 deals with the light-driven production of dioxygen by photosynthetic organisms. O_2 is abundant in the atmosphere because of its constant regeneration by the photosynthetic oxidation of H_2O. This process is catalyzed by a unique Mn_4CaO_5 cluster located in photosystem II, a gigantic multi-subunit membrane protein complex. Results and interpretations, especially from state-of-the-art X-ray spectroscopy studies, are summarized. These studies focus on the geometric and electronic structure and the changes as the Mn_4CaO_5 site proceeds through the catalytic cycle.

The following Chapter 3 is devoted to O_2-generating reactions in the dark. These are rare in biology and difficult to mimic synthetically. Recently, perchlorate-respiring bacteria have been discovered which carry a heme-containing chlorite dismutase. Notably, the enzyme bears no structural or sequence relationships with known heme peroxidases or other heme proteins. These microorganisms detoxify chlorite (ClO_2^-), the end product of the perchlorate (ClO_4^-) respiratory pathway, by rapidly converting ClO_2^- to O_2 and chloride (Cl^-).

In Chapter 4 a long time embattled enzyme is reviewed: Cytochrome *c* oxidase, the terminal oxidase of cell respiration. This redox-driven proton pump reduces molecular oxygen to H_2O. Highly resolved three-dimensional structures of the bovine enzyme in various oxidation and ligand binding states have been obtained; they show that the O_2 reduction site – a dinuclear Fe (heme a_3), Cu (Cu_B) center – drives a non-sequential four-electron transfer for complete reduction of O_2 to H_2O without the release of toxic reaction intermediates like the superoxide anion ($O_2^{\bullet -}$), hydrogen peroxide (H_2O_2), or the hydroxyl radical ($OH^{\bullet}$). X-ray structural and mutational analyses of bovine cytochrome *c* oxidase, which hosts a sophisticated catalytic machinery for efficient proton and electron delivery, reveal three possible proton transfer pathways which can transfer pumped protons and water-forming protons.

Chapter 5 surveys recent important advances in the field of transition metal complexes and the activation of O_2. Studies of synthetic models of the diverse iron and copper active sites have led to fundamental chemical insights into how O_2 coordinates to mono- and multinuclear Fe and Cu centers and is reduced to superoxo, peroxo, hydroperoxo, and, after O-O bond scission, oxo species relevant to proposed intermediates in catalysis. The involvement of disparate metal ions, nuclearities, geometries, and supporting ligands provides a rich tapestry of reaction pathways by which O_2 is activated.

Chapters 6 and 7 focus on the functionalization of the gases methane (CH_4) and ammonia (NH_3), both in the presence and absence of dioxygen. In view of their fundamental importance, a remarkable set of tools appears to exist in Nature to convert CH_4 and NH_3. These are inert molecules and complex transition metal-dependent enzymes (methane and ammonia monooxygenases) isolated from aerobic microorganisms and have been reported to break up the N-H and C-H bonds. Two distinct methane monooxygenases, a copper-dependent membrane protein and an iron-dependent cytosolic protein, catalyze the conversion of CH_4 to methanol (CH_3OH), thus playing a significant role in the biogeochemistry of this potent greenhouse gas. The reaction of the reduced Fe (or Cu) centers with O_2 leads to intermediates that activate the relatively inert C-H bonds of hydrocarbons to yield oxidized products. Notably, there exist "impossible" microorganisms which use the oxidative power of nitric oxide (NO) by forging this molecule to ammonium (NH_4^+), thereby making hydrazine (N_2H_4). Others can disproportionate NO into N_2 and O_2. This intracellularly produced O_2 enables these "impossible" bacteria to adopt an aerobic mechanism for methane oxidation.

In summary, this volume, like the preceding volume 14 of the *Metal Ions in Life Sciences* series, offers a wealth of profound information about important processes in our current biosphere. The emphasis is on the fundamental role of molecular oxygen for all aerobically living organisms including humans, animals, and plants. The crucial role of transition metals, specifically of manganese, iron, and copper, is addressed in the activation, production, and transformation of molecular oxygen, but also in the functionalization of methane and ammonia and their impact on the environment.

Peter M.H. Kroneck
Fachbereich Biologie
Universität Konstanz
D-78457 Konstanz, Germany

Martha E. Sosa Torres
Departamento de Química Inorgánica y Nuclear
Facultad de Química
Universidad Nacional Autónoma de México
México, D.F. 04510, México

Contents

Contributors to Volume 15

Numbers in parentheses indicate the pages on which the authors' contributions begin.

Jennifer L. DuBois Department of Chemistry and Biochemistry, Montana State University, Bozeman, MT 59717, USA, jdubois@chemistry.montana.edu (45)

Mike S.M. Jetten Department of Microbiology, Institute of Wetland and Water Research (IWWR), Radboud University of Nijmegen, Heyendaalseweg 135, 6525AJ Nijmegen, The Netherlands, m.jetten@science.ru.nl (257)

Jan T. Keltjens Department of Microbiology, Institute of Wetland and Water Research (IWWR), Radboud University of Nijmegen, Heyendaalseweg 135, 6525AJ Nijmegen, The Netherlands, j.keltjens@science.ru.nl (257)

Jan Kern Physical Biosciences Division, Lawrence Berkeley National Laboratory, Berkeley, CA 94720, USA, jfkern@lbl.gov (13)

Sergey Koroidov Department of Chemistry, Chemistry Biology Centre (KBC), Umeå University, S-90187 Umeå, Sweden (13)

Peter M.H. Kroneck Fachbereich Biologie, Universität Konstanz, Universitätsstrasse 10, D-78457 Konstanz, Germany, peter.kroneck@uni-konstanz.de (1)

Stephen J. Lippard Department of Chemistry, Massachusetts Institute of Technology, Cambridge, MA 02139, USA, lippard@mit.edu (205)

Johannes Messinger Department of Chemistry, Chemistry Biology Centre (KBC), Umeå University, S-90187 Umeå, Sweden, johannes.messinger@chem.umu.se (13)

Håkan Nilsson Department of Chemistry, Chemistry Biology Centre (KBC), Umeå University, S-90187 Umeå, Sweden (13)

Sunil Ojha Department of Chemistry and Biochemistry, Montana State University, Bozeman, MT 59717, USA (45)

Joachim Reimann Department of Microbiology, Institute of Wetland and Water Research (IWWR), Radboud University of Nijmegen, Heyendaalseweg 135, 6525AJ Nijmegen, The Netherlands, j.reimann@science.ru.nl (257)

Juan P. Saucedo-Vázquez Departamento de Química Inorgánica y Nuclear, Facultad de Química, Universidad Nacional Autónoma de México, Ciudad Universitaria, México, D.F. 04510, México, jpablo_sa@unam.mx (1)

Matthew H. Sazinsky Department of Chemistry, Pomona College, Claremont, CA 91711, USA, matthew.sazinsky@pomona.edu (205)

Atsuhiro Shimada Picobiology Institute, Graduate School of Life Science, University of Hyogo, Kamigohri Akoh Hyogo, 678-1297, Japan, ashima@sci.u-hyogo.ac.jp (89)

Kyoko Shinzawa-Itoh Picobiology Institute, Graduate School of Life Science, University of Hyogo, Kamigohri Akoh Hyogo, 678-1297, Japan, shinzawa@sci.u-hyogo.ac.jp (89)

Martha E. Sosa Torres Departamento de Química Inorgánica y Nuclear, Facultad de Química, Universidad Nacional Autónoma de México, Ciudad Universitaria, México, D.F. 04510, México, mest@unam.mx (1)

William B. Tolman Department of Chemistry, University of Minnesota, 207 Pleasant St. SE, Minneapolis, MN 55455, USA, wtolman@umn.edu (131)

Vittal K. Yachandra Physical Biosciences Division, Lawrence Berkeley National Laboratory, Berkeley, CA 94720, USA, vkyachandra@lbl.gov (13)

Junko Yano Physical Biosciences Division, Lawrence Berkeley National Laboratory, Berkeley, CA 94720, USA, jyano@lbl.gov (13)

Gereon M. Yee Department of Chemistry, University of Minnesota, 207 Pleasant St. SE, Minneapolis, MN 55455, USA (131)

Shinya Yoshikawa Picobiology Institute, Graduate School of Life Science, University of Hyogo, Kamigohri Akoh Hyogo, 678-1297, Japan, yoshi@sci.u-hyogo.ac.jp (89)

Titles of Volumes 1–44 in the *Metal Ions in Biological Systems* Series

edited by the SIGELs
and published by Dekker/Taylor & Francis (1973–2005)

Volume 1:	**Simple Complexes**
Volume 2:	**Mixed-Ligand Complexes**
Volume 3:	**High Molecular Complexes**
Volume 4:	**Metal Ions as Probes**
Volume 5:	**Reactivity of Coordination Compounds**
Volume 6:	**Biological Action of Metal Ions**
Volume 7:	**Iron in Model and Natural Compounds**
Volume 8:	**Nucleotides and Derivatives: Their Ligating Ambivalency**
Volume 9:	**Amino Acids and Derivatives as Ambivalent Ligands**
Volume 10:	**Carcinogenicity and Metal Ions**
Volume 11:	**Metal Complexes as Anticancer Agents**
Volume 12:	**Properties of Copper**
Volume 13:	**Copper Proteins**
Volume 14:	**Inorganic Drugs in Deficiency and Disease**
Volume 15:	**Zinc and Its Role in Biology and Nutrition**
Volume 16:	**Methods Involving Metal Ions and Complexes in Clinical Chemistry**
Volume 17:	**Calcium and Its Role in Biology**
Volume 18:	**Circulation of Metals in the Environment**
Volume 19:	**Antibiotics and Their Complexes**
Volume 20:	**Concepts on Metal Ion Toxicity**
Volume 21:	**Applications of Nuclear Magnetic Resonance to Paramagnetic Species**
Volume 22:	**ENDOR, EPR, and Electron Spin Echo for Probing Coordination Spheres**
Volume 23:	**Nickel and Its Role in Biology**
Volume 24:	**Aluminum and Its Role in Biology**
Volume 25:	**Interrelations Among Metal Ions, Enzymes, and Gene Expression**

Contents of Volumes in the *Metal Ions in Life Sciences* Series

edited by the SIGELs

Volumes 1–4
published by John Wiley & Sons, Ltd., Chichester, UK (2006–2008)
<http://www.Wiley.com/go/mils>

Volume 5–9
by the Royal Society of Chemistry, Cambridge, UK (2009–2011)
<http://www.bioinorganic-chemistry.org/mils>

and from Volume 10 on
by Springer Science & Business Media BV, Dordrecht, The Netherlands (since 2012)
<http://www.mils-series.com>

Volume 1 Neurodegenerative Diseases and Metal Ions

Volume 3 The Ubiquitous Roles of Cytochrome P450 Proteins

Volume 9 Structural and Catalytic Roles of Metal Ions in RNA

Volume 10 Interplay between Metal Ions and Nucleic Acids

Volume 11 Cadmium: From Toxicity to Essentiality

Comments and suggestions with regard to contents, topics, and the like for future volumes of the series are welcome.

Chapter 1
The Magic of Dioxygen

Martha E. Sosa Torres, Juan P. Saucedo-Vázquez, and Peter M.H. Kroneck

Contents

Abstract Oxygen has to be considered one of the most important elements on Earth. Earlier, some dispute arose as to which of the three scientists, Carl Wilhelm Scheele (Sweden), Joseph Priestley (United Kingdom) or Antoine Lavoisier (France), should get credit for the *air of life*.

Today it is agreed that the Swede discovered it first, the *fire air* in 1772. The British chemist published it first, the *dephlogisticated air* in 1775, and the Frenchman understood it first, the *oxygen* in 1775–1778. Surely, there is credit enough for all three to split the "Nobel Prize" awarded by Carl Djerassi and Roald Hoffmann in their play *Oxygen*. Molecular oxygen means life. So-called aerobes – these include humans, animals, and plants – need O_2 to conserve the energy they have to gain from

M.E. Sosa Torres • J.P. Saucedo-Vázquez
Departamento de Química Inorgánica y Nuclear, Facultad de Química, Universidad Nacional Autónoma de México, Ciudad Universitaria, México, D.F. 04510, México
e-mail: mest@unam.mx; jpablo_sa@unam.mx

P.M.H. Kroneck (✉)
Fachbereich Biologie, Universität Konstanz, Universitätsstrasse 10, D-78457 Konstanz, Germany
e-mail: peter.kroneck@uni-konstanz.de

P.M.H. Kroneck, M.E. Sosa Torres (eds.), *Sustaining Life on Planet Earth: Metalloenzymes Mastering Dioxygen and Other Chewy Gases*, Metal Ions in Life Sciences 15, DOI 10.1007/978-3-319-12415-5_1

their environment. Eliminate O_2 and these organisms cannot support an active lifestyle. What makes dioxygen that special? It is a non-metal and oxidizing agent that readily reacts with most elements to form compounds, notably oxides. From a biological point of view, *the* most important compound of course is water, H_2O, which provides an excellent solvent for biomolecules. It influences the climate of the Earth, and it is the source of almost all of the molecular oxygen in the atmosphere.

Keywords Atmosphere • banded iron formation • dioxygen • energy conservation • evolution • reactive oxygen species

Please cite as: *Met. Ions Life Sci.* 15 (2015) 1–12

1 Introduction

The element oxygen, ${}^{16}_{8}O$, is a member of the chalcogen group of the periodic table. By mass, oxygen is the third-most abundant element in the universe, after hydrogen and helium. It is the most abundant element in the crust of the Earth, mostly tied up in carbonates, phosphates, and in a wider range of silicates, from clays to zeolites to quartz. Under standard conditions of pressure and temperature, two oxygen atoms combine to form the colorless gas dioxygen (O_2), the stable form of the element. Joseph Priestley (United Kingdom) and Carl Scheele (Sweden) independently discovered oxygen around 1770. It was the French scientist Antoine Lavoisier, who coined the name oxygen (from the Greek words *oxy genes* meaning *acid* and *forming*). His experiments were key to discredit the then-popular phlogiston theory of combustion and corrosion.

Dioxygen gas does not react with itself or the other main component of our atmosphere, dinitrogen (N_2), under normal conditions. However, the effect of ultraviolet light upon O_2 is to form the blue gas ozone, O_3, the second allotrope of oxygen. The decomposition of potassium chlorate ($KClO_3$) is one chemical route to O_2 and decomposition of potassium permanganate ($KMnO_4$) is another one (reactions 1 and 2) [1–5].

$$2\ KClO_3\ (400°C) \rightarrow 2\ KCl + 3\ O_2 \tag{1}$$

$$2\ KMnO_4\ (214°C) \rightarrow K_2MnO_4 + MnO_2 + O_2 \tag{2}$$

O_2 is kind of a strange molecule, for all its ubiquity. Oxygen's two least strongly held electrons, responsible for most of its chemistry, are in two anti-bonding π*orbitals. One electron goes into each of these HOMOs, thus the ground state of O_2 has two unpaired electrons and is paramagnetic (Figure 1). The ground state O_2 is a diradical (${}^3O_2$; triplet O_2) and can enter into many organic and inorganic reactions. Clearly, its appearance in the atmosphere changed the course of evolution on Earth. However, compared to other oxygen radicals, such as the superoxide anion ($O_2^{\bullet -}$), the hydroxyl radical (${}^{\bullet}OH$), and the peroxide radical (${}^{\bullet}OOH$), triplet

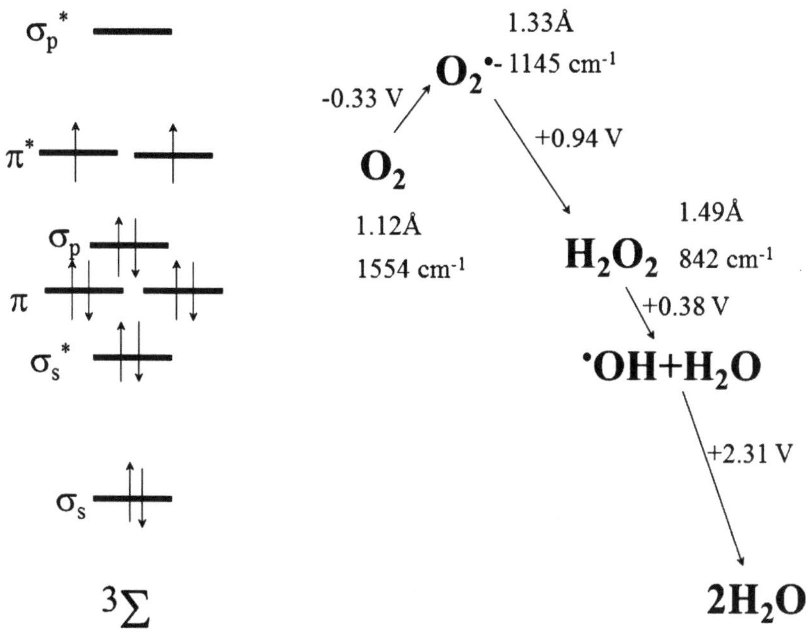

Figure 1 Key features of the oxygen molecule. **Left**: simplified energy diagram of molecular oxygen in the triplet ground state ($^3\Sigma$), showing two unpaired electrons in the anti-bonding π^* orbitals. **Right**: Properties of molecular oxygen and its reduced species superoxide ($O_2^{\bullet -}$), hydrogen peroxide (H_2O_2), hydroxyl radical ($^{\bullet}OH$), and water (H_2O). Redox potentials (V) are reported against the NHE (pH 7.25), O-O distances in Å, and O-O stretching frequencies in cm^{-1} [1–5].

O_2 is surprisingly unreactive. Along these lines, O_2 in its first excited state ($^1O_2$; singlet O_2) which is diamagnetic and lies 95 kJ above the triplet ground state, is much more reactive [5].

Oxygen has to be considered one of the most important elements on Earth. In short, oxygen means life. All aerobes, including humans, animals, and plants, need O_2 to unleash the energy they scavenge from their environment. Eliminate oxygen and organisms cannot keep up an active lifestyle, or even enable them to eat. However, oxygen has also a dark side. It can be toxic, usually the result from the harmful effects of breathing molecular oxygen at elevated partial pressures. The so-called central nervous system toxicity was first described by Paul Bert in 1878, referred to as the "Paul Bert effect" [6]. He showed that oxygen was toxic to many organisms including insects, earthworms, fungi, germinating seeds, birds, and other animals [7–9]. In 1899, pulmonary oxygen toxicity was described by J. Lorrain Smith when he noted central nervous system toxicity and discovered in experiments in mice and birds that 0.43 bar (43 kPa) had no effect but 0.75 bar (75 kPa) of oxygen was a pulmonary irritant [10, 11].

Not counting hydrocarbons, there exists a greater diversity of compounds with oxygen than with carbon. What makes oxygen so special, inspiring two famous chemists to write a play performed all over the world [12–14]? It is a non-metal and oxidizing agent that readily reacts with most elements to form compounds, notably oxides. Clearly, *the* most important compound of course is water, H_2O, which provides an excellent solvent for biomolecules, and moderates the climate of the Earth. Most important, it is the source of almost all the molecular oxygen in the atmosphere. One can easily picture a scenario with a planet without O_2. Let's speculate for a moment about Mars, our nearest planetary neighbor. Probably, it had once a water world too, ready to evolve life. However, a vital constituent was missing, namely a protective shield of ozone (O_3) derived from O_2. Without the ozone layer, the radiation from the Sun could slowly destroy the Martian water. The hydrogen floated off into space while the oxygen oxidized the iron-rich Martian surface soil responsible for its characteristic red color. Perhaps there is, or was, life on Mars, but if so, it never progressed beyond the bacterial stage [15]. Recently, beginning from two plausible starting points – an uninhabited or inhabited Mars – possible trajectories of martian habitability over time have been discussed. By identifying different trajectories of habitability, corresponding hypotheses could be described that allowed for the various trajectories to be disentangled and ultimately a determination of which trajectory Mars had taken and the changing relative abundance of its constituent environments [16].

In this chapter we will briefly discuss several aspects of this fascinating molecule O_2 and its impact on planet Earth, as described in depth in the following six chapters of this book. These cover the light-dependent production of O_2 by photosynthetic organisms, the production of O_2 in the dark by dismutation of oxyanions, the activation of O_2 by transition metal complexes and their role in catalysis, as well as the respiratory conservation of energy by the terminal oxidase cytochrome *c* oxidase. Following are two chapters on the functionalization of the greenhouse gas methane and on metal enzymes in "impossible" microorganisms catalyzing the anaerobic oxidation of ammonium and methane. In view of the remarkable amount of data accumulating in the field in recent years, we recommend several comprehensive reviews as primary references [5, 8, 9, 17–21].

2 The Rise of Dioxygen in the Atmosphere

The rapid increase of CO_2 concentration in Earth's modern atmosphere is a matter of major concern. But for the atmosphere of roughly two-and-half billion years ago, interest centers on a different gas: free oxygen (O_2) spawned by early biological production. The initial increase of O_2 in the atmosphere, its delayed build-up in the ocean, its increase to near-modern levels in the sea and air two billion years later, and its cause-and-effect relationship with life are among the most compelling and most heavily disputed stories in the history of the Earth [21].

We take our richly oxygenated world for granted and expect to find O_2 everywhere. After all, it makes up approximately 21 % of the current atmosphere. But remember, free O_2, at levels mostly less than 0.001 % of those present today, was anything but plentiful during the first half of the Earth's 4.5-billion-year history. Evidence for a permanent rise to appreciable concentrations of O_2 in the atmosphere between 2.4 and 2.1 billion years (Ga) ago began to accumulate in the 1960s. This increase, now popularly known as the Great Oxidation Event (GOE), left clear fingerprints in the rocks. For example, the first appearance of rusty red soils on land and the disappearance of easily oxidized minerals such as pyrite (FeS_2) from ancient stream beds both point to the presence of O_2 in the atmosphere. The notion of a GOE is now deeply entrenched in our understanding of the early Earth, with only a few researchers suggesting otherwise [22, 23].

After loss of the hydrogen, helium, and other hydrogen-containing gases from early Earth due to the Sun's radiation, primitive Earth was devoid of an atmosphere. Although it is estimated that oxygen is only the third most abundant element cosmically coming behind hydrogen and helium, in that order, it is the most abundant element on the Earth's crust [23, 24]. The first atmosphere was formed by outgassing of gases, such as H_2O, CO_2, SO_2, H_2S, and HCl, trapped in the interior of the early Earth. This still happens today in volcanoes. It is likely that the bulk of the atmosphere was derived from degassing early in the Earth's history [25–27]. According to geochemist D. E. Canfield from the University of Southern Denmark there is something astonishing in every breath we take. Even more astonishing is that the Earth started out with an oxygen-free atmosphere, it took billions of years before there was enough of it to keep animals like us alive [23, 28].

Early on, the biogeochemical cycling of the basic elements for life including carbon, nitrogen, oxygen, and sulfur, has attracted the interest of researchers in view of its importance for the Earth with a rapidly growing population, and in view of its impact on our environment and climate [29]. Atmospheric O_2 is controlled principally by the long-term (multi-million-year) geochemical cycles of carbon and sulfur. The effect of O_2 on the cycles of other elements that exhibit variable oxidation states has been shown to be far less significant quantitatively. The first documented description how the carbon and sulfur cycles affect O_2 was given in 1845 [30]. The following global reactions (3) and (4) were deduced.

$$CO_2 + H_2O \rightleftharpoons CH_2O + O_2 \tag{3}$$

$$2\ Fe_2O_3 + 8\ SO_4^{2-} + 16\ H^+ \rightleftharpoons 15\ O_2 + 4\ FeS_2 + 8\ H_2O \tag{4}$$

Reaction (3), from left to right, represents net photosynthesis (photosynthesis minus respiration) as represented by the burial of organic matter (CH_2O) in sediments. Going from right to left, reaction (3) represents two processes: (i) the oxidation of old sedimentary organic matter, and (ii) the sum of several reactions with the resulting reduced carbon-containing compounds released to the Earth's surface where they are oxidized to CO_2 by atmospheric or oceanic O_2. Reaction (4), from right to left, represents the oxidation of pyrite, FeS_2, during weathering of the continents (organic sulfur is included here with pyrite for simplification) and the sum of thermal pyrite decomposition and the oxidation of resulting reduced sulfur-containing gases produced by metamorphism and magmatism. Going from left to right, reaction (4) is the sum of several reactions. These are (i) photosynthesis and initial burial of organic matter (reaction 3), (ii) early diagenetic bacterial sulfate reduction to hydrogen sulfide, H_2S, with organic matter serving as the reducing agent, and (iii) the precipitation of pyrite via the reaction of H_2S with Fe_2O_3. The geochemical sulfur cycle also involves hydrothermal reactions between oceanic crust and seawater sulfate at mid-ocean rises, and these reactions have been advanced as major controls on atmospheric O_2 [30–32].

The evolution of oxygen-producing cyanobacteria was arguably *the* significant event in the history of life after the evolution of life itself [19, 21]. Molecular oxygen is a potent oxidizing agent, its accumulation into the atmosphere forever changed the surface chemistry of the Earth. It is also a favorable electron acceptor used in the respiration of countless different organisms that conduct a remarkable variety of different metabolisms. None of these could have prospered before the evolution of cyanobacteria. Thus, relationships between life, O_2, and the surface chemistry of the Earth are evident. The exact nature of these relationships requires an accurate understanding of the history of metabolic evolution, including knowledge of when cyanobacteria evolved, as well as a faithful reconstruction of the history of Earth surface oxidation and O_2 accumulation.

It is widely accepted that atmospheric O_2 concentrations remained persistently low for about the first two billion years of Earth's history. The first long-term oxygenation of the atmosphere was thought to have occurred approximately 2.3 billion years ago, during the GOE [33, 34]. However, high-resolution state-of-the-art techniques revealed an episode of enrichment of the redox-sensitive transition metals molybdenum and rhenium in the late Archean Mount McRae Shale in Western Australia. Correlations with organic carbon indicated that these metals were derived from contemporaneous seawater. Rhenium/osmium geochronology showed that the enrichment was a primary sedimentary feature dating to 2501 ± 8 million years ago. Molybdenum and rhenium were probably supplied to Archean oceans by oxidative weathering of crustal sulfide minerals.

These findings point to the presence of small amounts of O_2 in the environment more than 50 million years before the start of the GOE [35]. Furthermore, recent examinations of the distribution of chromium isotopes and redox-sensitive metals in the approximately 3-billion year-old Nsuze palaeosol and in the near-contemporaneous Ijzermyn iron formation from the Pongola Supergroup in South Africa revealed extensive mobilization of redox-sensitive elements through oxidative weathering. A best minimum estimate for atmospheric O_2 concentrations at that time of 3×10^{-4} times present levels were computed. In summary, the findings suggest the existence of appreciable levels of atmospheric O_2 about 3 billion years ago, more than 600 million years before the GOE and some 300–400 million years earlier than previous indications for Earth surface oxygenation [23].

Because oxygenic photosynthesis must be regarded as the only significant source of free O_2 on the Earth's surface, any evaluation of the oxygenation history must begin by asking when this metabolism evolved. Current estimates span well over a billion years representing almost one-third of the Earth's history. Part of the problem is related to the difficulties in differentiating between oxidation

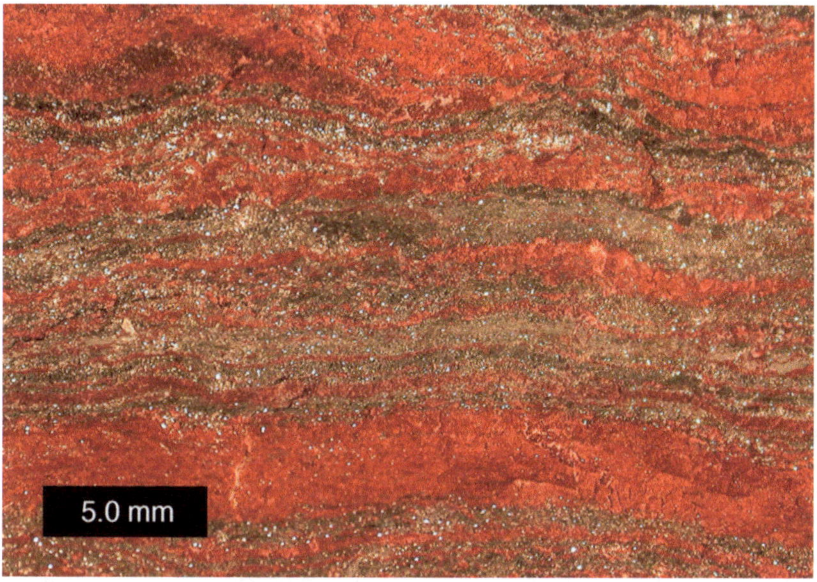

Figure 2 Photograph of a Michigan banded-iron formation (BIF). Image by courtesy of Professor M. A. Wilson, Department of Geology, The College of Wooster, Wooster, USA [36].

pathways that can be either biotic or abiotic and can occur with and without free O_2. Banded iron formations (BIFs), for example, are loaded with iron oxide minerals that often give these ancient deposits their spectacular red colors (Figure 2) [36].

The atmosphere has apparently been oxygenated since the GOE approximately 2.4 Ga ago, but when the photosynthetic oxygen production began, is still a matter of debate. However, geological and geochemical evidence from older sedimentary rocks indicates that oxygenic photosynthesis evolved well before this oxygenation event. Fluid-inclusion oils detected in approximately 2.45 Ga old sandstones contained hydrocarbon biomarkers evidently sourced from similarly ancient kerogen, preserved without subsequent contamination, and derived from organisms producing and requiring molecular oxygen. Molybdenum and rhenium abundances and sulfur isotope systematics of slightly older (2.5 Ga) kerogenous shales document a transient pulse of atmospheric O_2. As early as approximately 2.7 Ga, stromatolites and biomarkers from evaporative lake sediments deficient in exogenous reducing power strongly imply that oxygen-producing cyanobacteria had already evolved. Even at approximately 3.2 Ga, thick and widespread kerogenous shales are consistent with aerobic photoautotrophic marine plankton, and uranium-lead data obtained from approximately 3.8 Ga old sediments suggest that this metabolism could have arisen by the start of the geological record. Hence, the hypothesis that oxygenic photosynthesis evolved well before the atmosphere became permanently oxygenated seems well supported [23].

Pre-photosynthetic niches were meagre with a much lower productivity compared to modern photosynthesis. Serpentinization and volcanism reliably provided H_2, methanogens and acetogens reacted CO_2 with H_2 to conserve energy and to produce organic matter. These skills pre-adapted a bacterium for anoxygenic photosynthesis, probably starting with H_2 in lieu of an oxygen acceptor. Use of reduced iron(II) and sulfide followed as abundant O_2 acceptors, allowing productivity to approach modern levels. Cyanobacteria evolved O_2 production but they did not immediately dominate the Earth. Eventually, both anoxygenic and oxygenic photosynthesis oxidized much of the Earth's crust and supplied sulfate to the ocean. The process of anoxygenic photosynthesis remained important until there was enough O_2 in down welling seawater to quantitatively oxidize massive sulfides at mid-ocean ridge axes. The evolution of oxygenic photosynthesis is generally accepted as the ultimate cause of the GOE, but it has proven difficult to constrain the timing of this evolutionary innovation. The oxidation of manganese in the water column requires substantial free O_2 concentrations, and thus any indication that manganese oxides were present in ancient environments would imply that oxygenic photosynthesis was ongoing.

Manganese oxides are not commonly preserved in ancient rocks, but there is a large fractionation of molybdenum isotopes associated with the sorption of molybdenum onto the manganese oxides that would be retained. Recent investigations on Mo isotopes from rocks of the Sinqeni Formation, Pongola Supergroup in South Africa – these rocks formed no less than 2.95 Ga ago – showed that the Mo isotopic signature was consistent with an interaction of Mn oxides. These experimental results suggest that O_2 from oxygenic photosynthesis started to accumulate in shallow marine settings at least half a billion years before the accumulation of significant levels of atmospheric oxygen [19, 37]. Along these lines, in order to illuminate the history of manganese-dependent water splitting, the behavior of the ancient manganese cycle was examined using newly obtained scientific drill cores through an early Paleoproterozoic succession (2.415 Ga) preserved in South Africa [38].

Applying microscale X-ray spectroscopic techniques coupled to optical and electron microscopy and carbon isotope ratios, the researchers demonstrated that the manganese was hosted exclusively in carbonate mineral phases derived from reduction of Mn oxides during diagenesis of primary sediments. Additional observations revealed that the original Mn-oxide phases were not produced by reactions with O_2. These results point to a different high-potential oxidant and suggest that the oxidative branch of the manganese cycle predates the rise of O_2, and provide strong support for the hypothesis that the water-oxidizing complex of photosystem II evolved from an earlier transitional photosystem capable of single-electron oxidation reactions of manganese [38].

3 The Dark Side of Dioxygen

The cellular biochemistry of molecular oxygen is Janus-faced. Its good side includes numerous enzyme-catalyzed reactions of O_2 that drive respiration and normal metabolism. In 1956 the famous biochemist Otto Warburg pointed out that the respiration of cancer cells was irreversibly damaged, since it never returned to normal. If respiration was damaged it might be either that the O_2 consumption had been decreased or that, with undiminished O_2 consumption, the coupling between respiration and the formation of adenosine 5′-triphosphate had been broken [39]. Several examples of beneficial reactions are covered in this book, for example, reactions of O_2 catalyzed at transition metal sites in highly sophisticated enzymes such as cytochrome *c* oxidase and iron- or copper-dependent monooxygenases. However, the dark side of molecular oxygen encompasses deleterious reactions of species derived from O_2 that can lead to damage of cellular components. These reactive oxygen species (ROS) have historically been perceived almost exclusively as agents of the dark side, but it has more recently become clear that they play beneficial roles as well [40, 41].

As discussed earlier, the O_2 molecule is unusual in having two unpaired electrons in its most stable form (Figure 1). As a consequence, its direct reactions with other molecules are generally slow in the absence of catalysts or radical initiators and are therefore not the primary causes of so-called oxidative stress. ROS, i.e., superoxide, hydrogen peroxide, organic peroxides, hydroxyl radical, and other energetic molecules derived from their further reactions appear to cause the oxidative damage. Their ultimate source appears to be in the mitochondria, where side reactions of O_2 with components of the respiratory chain reduce it to the superoxide anion.

The threat by ROS is diminished by families of defensive enzymes which include the superoxide dismutases, the catalases, and the peroxidases. In addition, free radical chain reactions are controlled by antioxidants, such as L-ascorbate and the tocopherols. The oxidative damage, which occurs in spite of the action of these defenses, is largely repaired or is nullified by *de novo* biosynthesis. Yet some damage is sustained and it contributes to mutagenesis, to senescence and to numerous pathological processes [40, 41].

In a recent FASEB editorial entitled "Free Radicals Can Kill You: Lavoisier's Oxygen Revolution", editor-in-chief Gerald Weissmann expressed his thoughts about molecular oxygen [42]. We have known since Lavoisier that flames burn, metals rust, and we take breath, all thanks to O_2. But we hadn't learned how O_2 excess does harm until 1954, when Rebecca Gerschman and colleagues [43] worked out that molecular oxygen poisoning and radiation share the property of producing oxygen-derived free radicals (Figure 1). Add any of these to living tissues, and they do damage. Notably, over 15,000 articles about oxidative stress appeared in the literature just in 2009, a 16-fold increase in only 5 years [42]. Lavoisier had the chemistry right when he told us what happened to the animal oils of our cells when O_2 reacts: *It is evident that the oils, being composed of hydrogen and charcoal combined, are true carbonohydrous or hydrocarbonous radicals; and, indeed by adding oxygen, they are convertible to vegetable oxides and acids according to their degrees of oxidation* [44].

4 Outlook

Complex life on Earth crucially depends on strong redox disequilibria afforded by the almost ubiquitous presence of highly oxidizing molecular oxygen. Yet, as briefly discussed in this chapter, the unravelling of the history of O_2 levels in the atmosphere remains difficult, and prior to the GOE some billion years ago, the amount of O_2 in the biosphere is considered to have been extremely low as compared with nowadays values. Thus, the evolutionary histories of life and of O_2 levels are likely intricately intertwined.

The obvious biological proxy for inferring the impact of changing O_2 levels on life is the evolutionary history of enzymes allowing organisms to take advantage of the redox power of molecular oxygen. These are the O_2 reductases linked to bioenergetics, the groups of cytochrome and quinol oxidases. Consequently, molecular phylogenies reconstructed for this enzyme superfamily have been exploited over the last two decades in attempts to elucidate the interlocking between O_2 levels in the environment and the evolution of respiratory bioenergetic processes [45–47]. Although based on strictly identical data sets, the phylogenetic approaches produced diametrically opposite scenarios with respect to the history of both the enzyme superfamily and molecular oxygen on the Earth.

In an effort to overcome the deadlock of molecular phylogeny, the currently available structural, functional, palaeogeochemical and thermodynamic information pertinent to the evolution of the O_2 reductase superfamily, which notably also encompasses the subfamily of nitric oxide reductases, was analyzed. Many of these enzymes appear to be constructed from only a limited set of building blocks, for example the mixed-valence copper A center [Cu(1.5+)...Cu(1.5+)] discovered in cytochrome *c* oxidase and nitrous oxide reductase [47]. Based on these non-phylogenetic data Ducluzeau and colleagues proposed a scenario which sees the low-affinity O_2 reductases as the most recent evolutionary innovation and the high-affinity O_2 reductases as arising independently from NO-reducing precursor enzymes [46].

Addendum During the finalization of this chapter, leading expert D. E. Canfield published the book entitled "Oxygen: A Four Billion Year History" [48].

Abbreviations and Definitions

BIF	banded iron formation
Ga	billion years (giga years)
GOE	great oxidation event
HOMO	highest occupied molecular orbital
NHE	normal hydrogen electrode
ROS	reactive oxygen species
1 atmosphere (atm)	1.013 bars

Acknowledgments The authors are grateful for continuous financial support by Deutsche Forschungsgemeinschaft and Universität Konstanz (PK), and CONACYT and DGAPA-UNAM (MEST).

References

1. WebElements: the periodic table on the web – Oxygen; www.webelements.com
2. N. N. Greenwood, A. Earnshaw, *Chemistry of the Elements*, 2nd ed, Butterworth-Heinemann, Oxford, UK, 1997.
3. A. J. Bard, R. Parsons, J. Jordan, *Standard Potentials in Aqueous Solutions*, IUPAC. Marcel Dekker, New York, USA, 1985.
4. J. Emsley, *Oxygen. Nature's Building Blocks: An A-Z Guide to the Elements*, Oxford University Press, Oxford, UK, 2001, pp. 297–304.
5. R. Hoffmann, *American Scientist* **2004**, *92*, 23–26.
6. P. Bert (first published in French in 1878, translated by M. A. Hitchcock, F. A. Hitchcock), *Barometric Pressure: Researches in Experimental Physiology*, College Book Company, Columbus, OH, USA, 1943.
7. D. N Patel, A. Goel, S. B. Agarwal, P. Garg, K. K Lakhani, *J. Ind. Acad. Clin. Med.* **2003**, *4*, 234–237.
8. D. L. Gilbert, *Oxygen: An Overall Biological View*, in *Oxygen and Living Processes*, Ed D. L. Gilbert, Springer-Verlag, New York, USA, 1981.
9. S. G. Jenkinson, *New Horizons* **1993**, *1*, 504–511.
10. J. Lorrain Smith, *J. Physiol.* (London: The Physiological Society and Blackwell Publishing) **1899**, *24*, 19–35.
11. E. Meirovithz, J. Sonn, A. Mayevsky, *Brain Res. Rev.* **2007**, 294–304.
12. C. Djerassi, R. Hoffmann, *J. Chem. Educ.* **2001**, *78*, 283–284.
13. Z. M. Lerman, *Chem. Educ. Int.* **2005**, *6*, 1–5.
14. C. Djerassi, R. Hoffmann, *Oxygen*, Wiley-VCH, Weinheim, Germany, 2001.
15. R. A. Kerr, *Science* **2013**, *339*, 1373.
16. C. S. Cockell, *Astrobiology* **2014**, *14*, 182–203.
17. R. M. Hazen, *The Story of Earth. The First 4.5 Billion Years, from Stardust to Living Planet*, Viking, New York, USA, 2012.
18. H. D. Holland, *Phil. Trans. R. Soc. B* **2006**, *361*, 903–915.
19. N. H. Sleep, D. K. Bird, *Phil. Trans. R. Soc. B* **2008**, *363*, 2651–2664.
20. L. R. Kump, *Perspective* **2012**, 410–411; DOI: 10.2113/gselements. 8.6.410.
21. T W. Lyons, C. T. Reinhard, N. J. Planavsky, *Nature* **2014**, *506*, 307–315.
22. D. E. Canfield, *Annu. Rev. Earth Planet. Sci.* **2005**, *33*, 1–36.
23. S. A. Crowe, L. N. Døssing, N. J. Beukes, M. Bau, S. J. Kruger, R. Frei, D. E. Canfield, *Nature* **2013**, *501*, 535–538.
24. D. P. Mellor, *Chemistry* **1964**, *37*, 12–16.
25. M. Dole, *J. Gen. Physiol.* **1965**, *49*, 5–27.
26. www.globalchange.umich.edu/globalchange1/current/lectures/Perry_Samson_lectures/evolution_atm/
27. J. F. Kasting, M. T. Howard, *Phil. Trans. R. Soc. B* **2006**, *361*, 1733–1742.
28. C. Zimmer, *The Mystery of Earth's Oxygen*, in *The New York Times*, October 3, 2013.
29. V. Smil, *Cycles of Life: Civilization and the Biosphere*, Scientific American Library, W. H. Freeman and Company, New York, USA, 1997.
30. J. J. Ebelmen, *Ann. Rev. Mines* **1845**, *12*, 627–654.
31. R. A. Berner, D. J. Berling, R. Dudley, J. M. Robinson, R. A. Wildman, Jr., *Ann. Rev. Earth Planet. Sci.* **2003**, *31*, 105–134.

32. P. M. H. Kroneck, *Met. Ions Biol. Syst.* **2005**, *43*, 1–10.
33. H. D. Holland, *Geochim. Cosmochim. Acta* **2002**, *66*, 3811–3826.
34. Q. J. Guo, H. Strauss, A. J. Kaufman, S. Schröder, J. Gutzmer, B. Wing, M. A. Baker, A. Bekker, Q. S. Jin, S.-T. Kim, J. Farquhar, *Geology* **2009**, *37*, 399–402.
35. A. D. Anbar, Y. Duan, T. W. Lyons, G. L. Arnold, B. Kendall, R. A. Creaser, A. J. Kaufman, G. W. Gordon, C. Scott, J. Garvin, R. Buick, *Science* **2007**, *317*, 1903–1906.
36. M. A. Wilson, image freely available from Wikipedia for any purpose: http://commons.wikimedia.org/wiki/User:Wilson44691
37. N. J. Planavsky, D. Asael, A. Hofmann, C. T. Reinhard, S. V. Lalonde, A. Knudsen, X. Wang, F. Ossa Ossa, E. Pecoits, A. J. B. Smith, N. J. Beukes, A. Bekker, T. M. Johnson, K. O. Konhauser, T.W. Lyons, O. J. Rouxel, *Nature Geoscience* **2014**, *7*, 283–286.
38. J. E. Johnson, S. M. Webb, K. Thomas, S. Ono, J. L. Kirschvink, W. W. Fischer, *Proc. Natl. Acad. Sci. USA* **2013**, *110*, 11238–11243.
39. O. Warburg, *Science* **1956**, *123*, 309–314.
40. I. Fridovich, *J. Experimental Biology* **1998**, *201*, 1203–1209.
41. J. S. Valentine, D. L. Wertz, T. J. Lyons, L.-L. Liou, J. J. Goto, E. Butler Gralla, *Curr. Opin. Chem. Biol.* **1998**, *2*, 253–262.
42. G. Weissmann, *FASEB J.* **2010**, *24*, 649–652.
43. R. Gerschman, D. L. Gilbert, S. W. Nye, P. Dwyer, W. O. Fenn, *Science* **1954**, *119*, 623–626.
44. A.-L. Lavoisier, *Traité Elémentaire de Chimie* (1789), illustrated by Madame Lavoisier. in *Lavoisier, Fourier, Faraday, Great Books of the Western World Series*, Eds R. M. Hutchins, M. J. Adler, Encyclopædia Britannica, Inc., Chicago, USA, 1952, see A.-L. Lavoisier, 1–160.
45. F. Baymann, E. Lebrun, M. Brugna, B. Schoepp-Cothenet, M.-T. Giudici-Orticoni, W. Nitschke, *Phil. Trans. Roy. Soc. Lond. B* **2003**, *358*, 267–274.
46. A.-L. Ducluzeau, B. Schoepp-Cothenet, R. van Lis, F. Baymann, M. J. Russell, W. Nitschke, *J. Roy. Soc. Interface* **2014**, *11*, 20140196.
47. W. G. Zumft, P. M. H. Kroneck, *Adv. Microb. Physiol.* **2007**, *52*, 108–226.
48. D.E. Canfield, *Oxygen: A Four Billion Year History*, Princeton University Press, Princeton, USA, 2014. See book review by T.W. Lyons, *Nature Chemistry* **2014**, *6*, 655.

Chapter 2
Light-Dependent Production of Dioxygen in Photosynthesis

Junko Yano, Jan Kern, Vittal K. Yachandra, Håkan Nilsson, Sergey Koroidov, and Johannes Messinger

Contents

J. Yano • J. Kern • V.K. Yachandra (✉)
Physical Biosciences Division, Lawrence Berkeley National Laboratory, Berkeley, CA 94720, USA
e-mail: jyano@lbl.gov; jfkern@lbl.gov; vkyachandra@lbl.gov

H. Nilsson • S. Koroidov • J. Messinger
Department of Chemistry, Chemistry Biology Centre (KBC), Umeå University, S-90187 Umeå, Sweden
e-mail: johannes.messinger@chem.umu.se

P.M.H. Kroneck, M.E. Sosa Torres (eds.), *Sustaining Life on Planet Earth: Metalloenzymes Mastering Dioxygen and Other Chewy Gases*, Metal Ions in Life Sciences 15, DOI 10.1007/978-3-319-12415-5_2

Abstract Oxygen, that supports all aerobic life, is abundant in the atmosphere because of its constant regeneration by photosynthetic water oxidation, which is catalyzed by a Mn_4CaO_5 cluster in photosystem II (PS II), a multi subunit membrane protein complex. X-ray and other spectroscopy studies of the electronic and geometric structure of the Mn_4CaO_5 cluster as it advances through the intermediate states have been important for understanding the mechanism of water oxidation. The results and interpretations, especially from X-ray spectroscopy studies, regarding the geometric and electronic structure and the changes as the system proceeds through the catalytic cycle will be summarized in this review. This review will also include newer methodologies in time-resolved X-ray diffraction and spectroscopy that have become available since the commissioning of the X-ray free electron laser (XFEL) and are being applied to study the oxygen-evolving complex (OEC). The femtosecond X-ray pulses of the XFEL allows us to outrun X-ray damage at room temperature, and the time-evolution of the photo-induced reaction can be probed using a visible laser-pump followed by the X-ray-probe pulse. XFELs can be used to simultaneously determine the light-induced protein dynamics using crystallography and the local chemistry that occurs at the catalytic center using X-ray spectroscopy under functional conditions. Membrane inlet mass spectrometry has been important for providing direct information about the exchange of substrate water molecules, which has a direct bearing on the mechanism of water oxidation. Moreover, it has been indispensable for the time-resolved X-ray diffraction and spectroscopy studies and will be briefly reviewed in this chapter. Given the role of PS II in maintaining life in the biosphere and the future vision of a renewable energy economy, understanding the structure and mechanism of the photosynthetic water oxidation catalyst is an important goal for the future.

Keywords calcium • manganese • oxygen-evolving complex • membrane-inlet mass spectrometry • photosystem II • X-ray crystallography • X-ray emission spectroscopy • X-ray free electron laser

Please cite as: *Met. Ions Life Sci.* 15 (2015) 13–43

1 Introduction

Most of the dioxygen in the atmosphere that aerobic life depends on is generated by plants, algae, and cyanobacteria by the light-induced oxidation of water in photosystem II (PS II) (reviewed in [1]). PS II is a multi-peptide membrane protein complex embedded in the thylakoid membranes. Photosynthetic water oxidation and dioxygen evolution is one of the most important, life-sustaining chemical processes occurring in the biosphere. PS II is usually found in a dimeric form in nature with each monomer comprised of more than 20 peptides and about 100 cofactors, which include chlorophylls (Chl), quinones, carotenoids, and lipids. The primary light-driven charge separation takes place in the PS II reaction center, P_{680}, where 4 Chls, 2 pheophytins, and 2 quinone molecules are symmetrically arranged in the two branches of PS II. After the light-induced formation of P_{680}^{+}, a redox-active tyrosine (Y_Z) is oxidized, which in turn oxidizes the oxygen evolving complex (OEC) on the lumenal side of the protein complex.

The OEC in PS II contains a heteronuclear Mn_4CaO_5 cluster (Figure 1, inset) which catalyzes the water oxidation reaction (equation 1),

$$2\ H_2O \rightarrow O_2 + 4\ e^- + 4\ H^+ \tag{1}$$

that couples the four-electron oxidation of water with the one-electron photochemistry occurring at the PS II reaction center, P_{680}. The OEC cycles through five intermediate S states (S_0 to S_4, known as the Kok cycle) that corresponds to the abstraction of four successive electrons from the OEC (Figure 1) [2, 3]. The dark stable S_1 state is the first oxidized state and subsequent illumination leads to the formation of the S_2 and S_3 states. Once four oxidizing equivalents are accumulated (S_4 state), a spontaneous reaction occurs that results in the release of O_2 and the formation of the most reduced state, the S_0 state. Upon further light excitation, the initial S_1 state is formed once again, and the catalytic cycle is resumed.

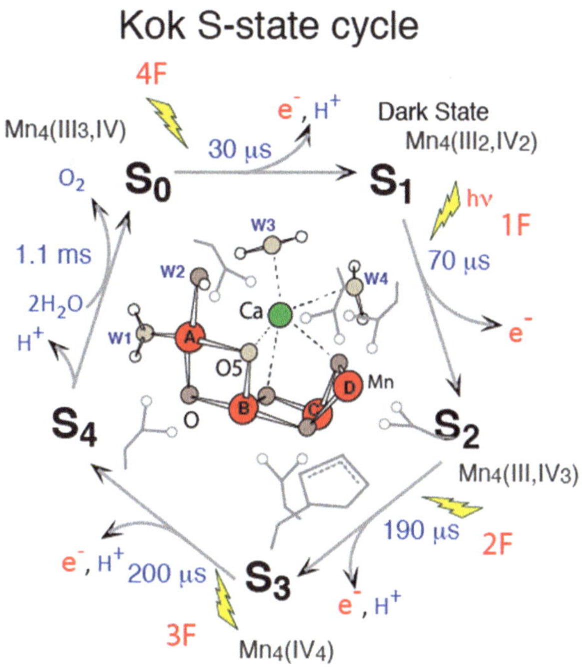

Figure 1 The Kok S state cycle for photosynthetic water oxidation and oxygen evolution. Proposed structure for the Mn_4CaO_5 cluster derived from the 1.9 Å X-ray structure modified using information from single-crystal polarized X-ray spectroscopy and EPR studies is in the inset. The proposed oxidation states for the Mn in the cluster in the various S state intermediates are shown.

Given the importance of PS II in maintaining life and the anticipated role of light-induced water-splitting for building a renewable energy economy, understanding the structure of the Mn_4CaO_5 catalyst and the mechanism of the water oxidation reaction is considered to be one of science's grand challenges [4]. Although details of the chemistry involved in water oxidation are slowly emerging, the mechanism of the reaction is not yet clear. In this chapter, we describe results from X-ray spectroscopy and diffraction studies, especially the use of time-resolved X-ray methods for room temperature studies using the recently introduced X-ray lasers. We will also describe the use of membrane inlet mass spectrometry for the elucidation of the mechanism of water-oxidation and its utility for time-resolved X-ray spectroscopy and diffraction measurements.

2 Geometric and Electronic Structure of the Mn_4CaO_5 Cluster

X-ray diffraction (XRD) [5–10], X-ray absorption spectroscopy (XAS) [11–13, 19], electron paramagnetic resonance (EPR) [14–16], infrared spectroscopy (IR) [17], and UV-vis [18] studies of PS II have provided valuable insights into the structure and mechanism of the Mn_4CaO_5 cluster (Figure 2).

Among them, the information regarding the geometric structural changes has come largely from extended X-ray absorption fine structure (EXAFS) studies (reviewed in [12, 13, 19]). EXAFS has provided Mn-Mn and Mn-Ca and Mn-O/N ligand distances with an accuracy of ~0.02 Å and resolution of ~0.1 Å. An important feature of EXAFS is that one can control the X-ray dose used by monitoring the Mn

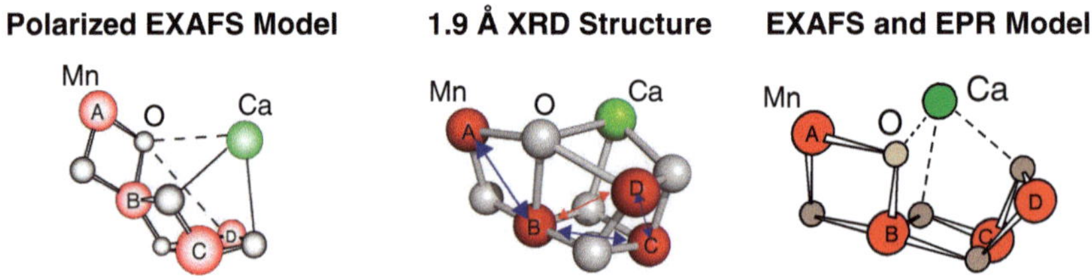

EXAFS		XRD
Mn-O/N Distances (average)	1.9 Å	2.2 Å
Mn-Mn Distances		
AB,BC,CD:	2.7-2.8 Å	2.8-3.0 Å
BD	3.3 Å	3.3 Å
Mn-Ca Distances		
2 to 3	3.4 Å	3.4 Å
2 to 1	3.9 Å	3.8 Å

Figure 2 The structural models for the Mn_4CaO_5 cluster from (**left**) the polarized EXAFS and Sr EXAFS studies, (**middle**) the 1.9 Å resolution XRD study, and (**right**) derived from EPR and EXAFS spectroscopic studies. The Mn-Mn and Mn-O/N ligand distances from the EXAFS and XRD studies are summarized below the respective structural model (adapted from [20, 23, 45]).

K-edge spectra, thus preventing the reduction of Mn and the concurrent disruption of the cluster, which occurs when PS II is exposed to the high X-ray dose used in protein crystallography (see below). EXAFS studies, including solution EXAFS [20], range-extended EXAFS [21, 22], and single crystal polarized EXAFS [23], have shown that in the S_1 state there are three short Mn-Mn distances at ~2.7 Å, one long Mn-Mn distance at ~3.3 Å, and three to four Mn-Ca distances [24, 25] at around 3.4–3.9 Å (based on Sr XAS studies [26–28]). The combination of polarized EXAFS data from single crystals of PS II with XRD led to three proposed models for the Mn_4CaO_5 cluster, one of which is shown in Figure 2 (left) [23].

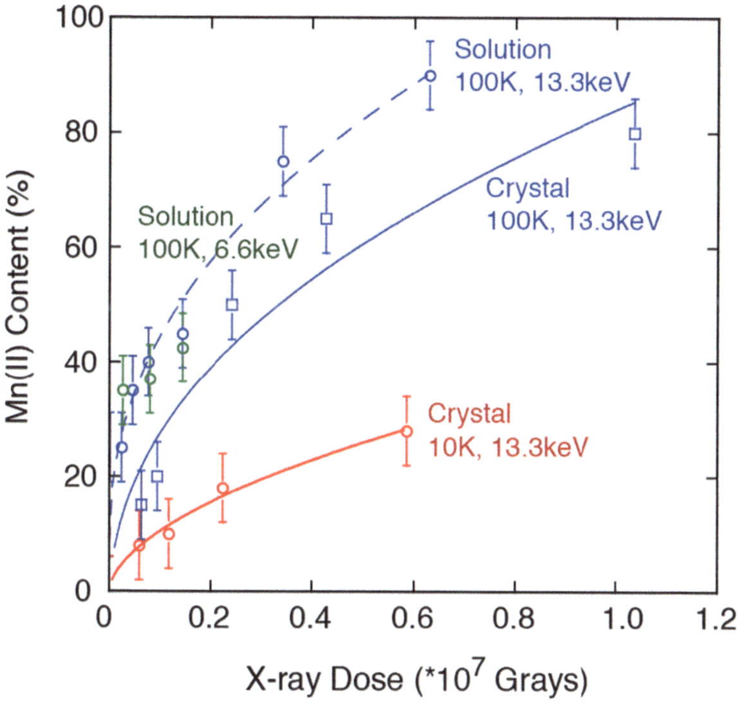

Figure 3 Radiation damage to PS II solutions and crystals as a function of X-ray dose. Dashed blue line is for PS II solutions illuminated at 100 K. The green circles are data from 6.6 keV illumination, while the blue squares are for 13.3 keV illumination. The blue solid line and blue circles are for crystals illuminated at 13.3 keV. The red solid line and red circles are for crystals illuminated at 10 K (adapted from [29]).

X-ray diffraction (XRD) studies using synchrotron radiation (SR) have been used over the last decade to study the structure of PS II [5–10]. However, the X-ray radiation damage to the redox active Mn_4CaO_5 cluster has been an issue, and the damage to the cluster precedes the loss of diffractivity of the crystals by about two orders of magnitude in dose in synchrotron-based crystallography (Figure 3) [29, 30]. This is accompanied by the disruption of the cluster as shown in Figure 4 [29, 30]. The recent study by Umena et al. at 1.9 Å [10], which has revealed the geometry of the Mn_4CaO_5

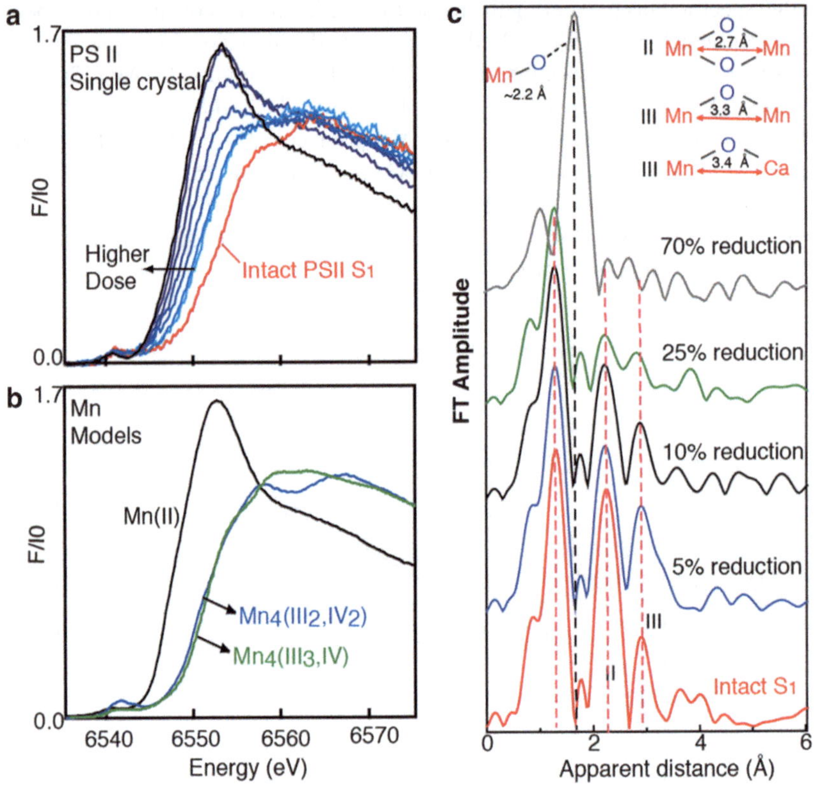

Figure 4 (**a**) Mn K-edge shift of PS II as a function of X-ray dose at 13.3 keV and 100 K. The spectrum at the highest inflection point energy is from an undamaged PS II crystal (in red). Light-blue to black lines are at increasing X-ray dose from 0.14, 0.21, 0.25, 0.54, 0.95, 2.3, and 5.0 × 10^{10} photons/μm^2. The exposure was at 100 K, and all XANES was collected at 10 K at low dose (1 × 10^7/μm^2). XANES shows that the increase in amplitude at ~6552 eV provides evidence for the photo-reduction to Mn^{II} in PS II crystals by exposure to X-rays. (**b**) A comparison of Mn K-edge spectra from two tetranuclear complexes, [$Mn_4O_3(OAc)_4(dbm)_3$] and [$(Mn_2O_2)_2(tphpn)_2$]$(ClO_4)_4$, in oxidation states Mn_3(III)Mn(IV) and Mn_2(III)Mn_2(IV) similar to those in intact PS II, and from Mn(II) in aqueous solution. (**c**) Fourier transforms of Mn EXAFS of PS II as a function of radiation dose. The Fourier transform of the EXAFS spectrum from an intact PS II solution sample is on the bottom (red). The three Fourier peaks are characteristic of a bridged Mn_4Ca complex, with peak I from bridging Mn-oxo and Mn-terminal ligand atoms, peak II is from Mn-Mn distances at 2.7 Å characteristic of di-μ-oxo bridged moieties and peak III is from mono-μ-oxo bridged Mn-Mn distances at 3.3 Å and Mn-Ca distances at 3.4 Å. The Fourier transforms from PS II samples exposed to radiation at 13.3 keV and 100 K, and containing 5 (blue), 10 (black), 25 (green), and 90 % (gray) photo-reduced Mn(II) centers. The Fourier peaks exhibit drastic changes, as the percent of Mn(II) increases, even at very low levels of X-ray dose. Peak II and peak III (vertical lines) that are characteristic of Mn-Mn distances at 2.7, 3.3, and 3.4 Å decrease and disappear along with peak I that is due to Mn-oxo bridging atoms. Peak I characteristic of bridging Mn-O distances is replaced by a longer Mn-O distance characteristic of Mn(II)-O terminal distances (adapted from [29]).

cluster (Figure 2, middle), however, used a significantly lower X-ray dose (X-ray dose of ~25 % Mn reduction level) and reported the atomic resolution structure of PS II (1.9 Å) in which four Mn and one Ca positions can be determined from the electron density map for the first time.

Although, the similarities in the number of Mn-Mn and Mn-Ca vectors is striking between the XRD and EXAFS structural models, noticeable changes still exist between the crystal structure and the EXAFS results in which the data was collected at much lower X-ray dose and the integrity of the sample was maintained (summarized in Figure 2). The Mn-ligand and Mn-Mn distances are longer in the 1.9 Å crystal structure than those determined using EXAFS by 0.1–0.2 Å for the Mn-Mn and ~0.3 Å for the Mn-O distances (Figure 2) [20]. These differences in distances could be caused by radiation damage during the XRD data collection and thus lead to some of the differences in the proposed models for the clusters.

2.1 *Geometric and Electronic Structural Changes During S State Transitions*

The 1.9 Å structure of PS II [10], nevertheless, serves as the current basis to relate spectroscopic data with the structural changes that occur during the catalytic S_i state transitions. Figure 2 (right) shows the structural model of the cluster modified from that shown in Figure 2 (left) taking into consideration the newest data from XRD and input from XAS [20] and EPR [14] studies. Based on this structure, possible changes that the Mn_4CaO_5 cluster undergoes during the S state transitions has been proposed as shown below in Figure 6. This model incorporates the ligands and basic structure of Mn found in the 1.9 Å crystal structure and builds upon this using EXAFS distances, FTIR, and EPR results and the changes in distances determined using Mn and Sr EXAFS of all the S states (Figure 5). The oxidation states assigned for the Mn atoms are based on both XANES and XES [31, 32] and EPR results [14, 33, 34].

The main topological difference between the structure proposed on the basis of EXAFS (in addition to the differences in the distances) [20, 23], EPR [14], and theoretical studies [35, 36] and the 1.9 Å XRD structure [10] is the position of the bridging oxygen atom O5 as shown in Figure 1 (inset), leading to a more open-cubane like structure. The open-cubane like structure for the S_1 and S_2 states is supported not only by polarized EXAFS of single crystals of PS II [20, 23], but has also been suggested by Siegbahn on the basis of theoretical studies [35], and by the Neese/Lubitz/Messinger groups on the basis of EPR studies for the S_2 state [14, 33, 34]. In the S_2 state a formal oxidation state distribution of (IV,IV,IV,III) for manganese atoms Mn1,2,3,4 (Figure 6) was assigned based on Mn K-edge XANES and Kβ emission spectroscopy [31], ^{55}Mn ENDOR measurements [34], and theoretical calculations [35], with one Mn being oxidized from Mn(III) to (IV) during the S_1 to S_2 transition.

Different nomenclatures have been used in the literature for identifying the Mn atoms, and they are all denoted in the caption for Figure 6. The shortening of one Mn-Mn interaction (~2.79 to ~2.74 Å) during the S_1 to S_2 transition is likely due to the change in oxidation state of one Mn (formally Mn(III) to Mn(IV)). FTIR studies indicate that the Mn3 atom ligated by Ala344 undergoes oxidation in the S_1 to S_2

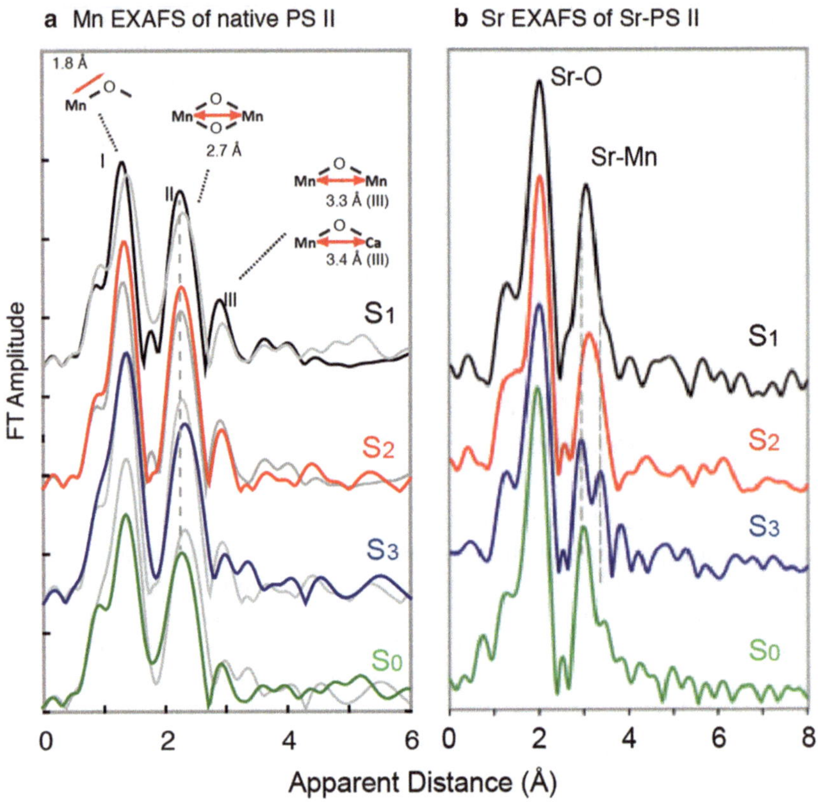

Figure 5 (**a**) The Mn EXAFS Fourier transforms from all the S states. The Mn-ligand, Mn-Mn, and Mn-Ca distances are characterized by the Fourier transforms peaks I, II, and III (adapted from [20]). (**b**) Sr EXAFS Fourier transforms of PS II from all the S states, showing the Sr-O and Sr-Mn distances (adapted from [28]). The significant changes in the distances during the S_2 and S_3 state, and the S_3 to S_0 transitions are seen. The vertical lines show the changes in the distances in the Mn-Mn and Mn-Sr distances.

state transition [37], however, it is possible that other Mn atoms could be oxidized. ENDOR studies [38, 39] suggest that Mn4 is the Mn(III) moiety in the S_2 state leaving open the possibility that either Mn3 or Mn1 is oxidized during the S_1 to S_2 transition.

The recent EPR/ENDOR studies support the formal oxidation state assignment of $Mn_4(III_3,IV)$ in the S_0 state and $Mn_4(III_2,IV_2)$ in the S_1 state [33, 40]. An oxidation state change of one Mn is also supported by Mn XANES and Kβ emission spectroscopy for the S_0 to S_1 state transition. The S_0 to S_1 state transition is also accompanied by the shortening of Mn-ligand distances as well as a Mn-Mn distance (~2.8 to ~2.7 Å) [41]. The shortening of the Mn-ligand and Mn-Mn distances could be due to the elimination of the Jahn-Teller effect at one Mn after its oxidation. An open cubane moiety in the S_0 state seems likely with either Mn2 or Mn1 being the Mn oxidized during the S_0 to S_1 state transition.

As indicated above, there is a consensus that Mn-centered oxidation occurs during the S_0 to S_1, and S_1 to S_2 transitions. However, there has been a long debate regarding the nature of the S_2 to S_3 transition [31, 32]. In the S_3 state, the question remains whether a Mn-centered oxidation occurs, $Mn_4(III,IV_3)$ to $Mn_4(IV_4)$, see above, or a ligand-centered oxidation takes place before O-O bond formation and release of O_2. Although Mn XANES studies have produced results that clearly indicate that in the S_0 to S_1 and S_1 to S_2 transitions Mn is oxidized, the S_2 to S_3 transition studies have not been conclusive, with one study suggesting that Mn may not be involved in the oxidation, while another study supports such a process [31, 32]. Mn Kβ emission spectra have indicated that the oxidizing equivalents may not be totally centered on the Mn. RIXS spectroscopy has also shown that formal oxidation states may be insufficient for describing the complex nature of the electronic structure in multinuclear clusters like the Mn_4CaO_5 cluster in PS II, and that the electrons are strongly delocalized in the Mn_4CaO_5 cluster, and that would mean that ligands may be involved in the redox chemistry [42, 43]. Mechanisms that rely on delocalization of charge on the ligands may be more relevant as discussed below.

EXAFS data show that Mn-Mn distances during the S_2 to S_3 state transition are elongated, compared to the S_0 to S_1 or S_1 to S_2 state transitions where contractions are observed. This suggests that the S_2 to S_3 step is not a simple one-oxidation state change of Mn, but is accompanied by changes in the geometry of the Mn_4CaO_5 cluster [20, 44]. Protonation of an oxo-bridge and the consequent elongation of Mn-Mn due is unlikely at the S_2 to S_3 state transition, unless protons from terminal water molecules are transferred to the bridging oxygens. A structural change that is caused by the shift of the oxygen O5 from the Mn1 side to the Mn4 side as illustrated in Figure 6 has been proposed. Such an O5 shuffling possibility has been suggested as a reason for the change from the S_2 low spin ($S = 1/2$) to S_2 high spin ($S = 7/2$), and on the basis of DFT calculations during the S_2 to S_3 state transition [45]. The motion of the oxygen O5 towards the Mn_3Ca open cubane site, generates a Mn_3CaO_4 closed cubane in the S_3 state, and such a structure is supported by Mn-Mn distances that have been seen in inorganic complexes [20]. A second possible structure for the S_3 state is also shown in Figure 6.

The 2.7 Å Mn-Mn distances on average are shown to be shortened during the S_3 to S_0 state transition via the S_4 state. This is counterintuitive as the Mn oxidation state changes from the most oxidized form in the S_3 state to the most reduced state form in the S_0 state. But such changes can be explained if the Mn_4CaO_5 geometry reverts back to a structure similar to the S_1 and S_2 states where the Mn_3Ca moiety shows open-cubane like structures, upon the S_3 to S_0 state transition.

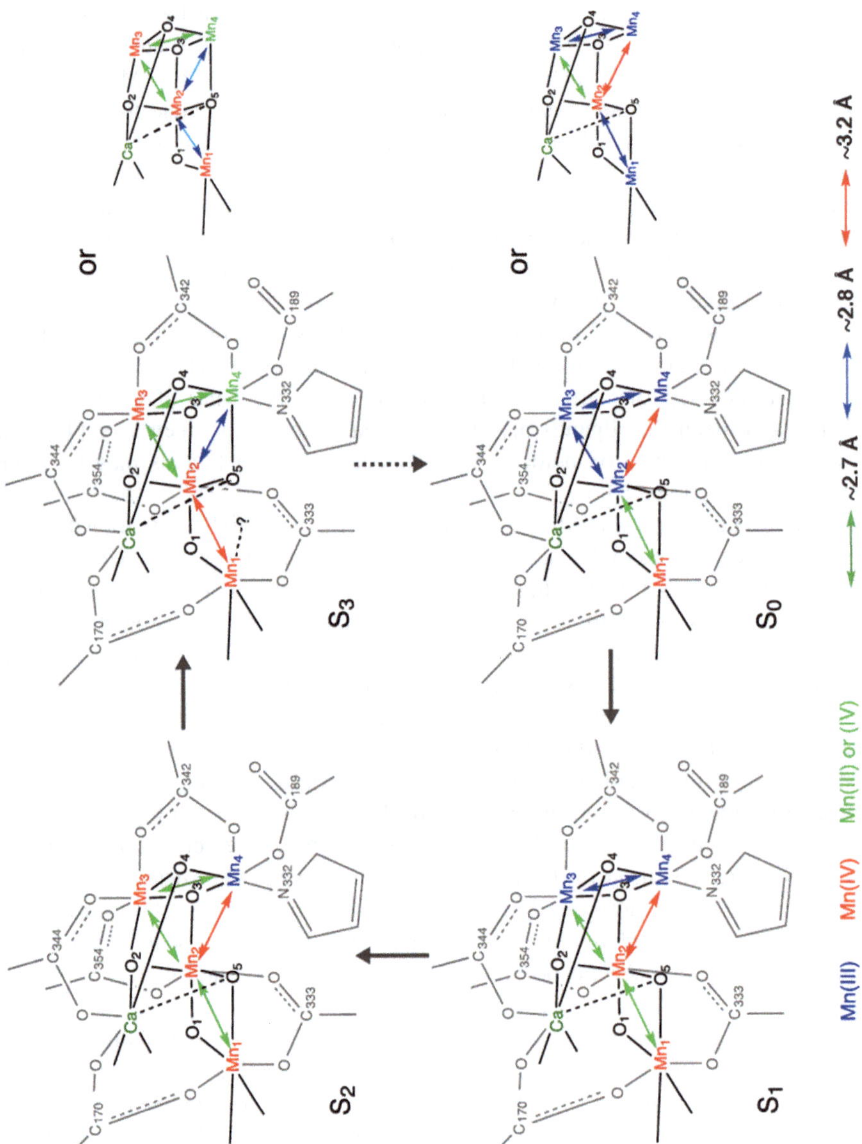
S2
S3
or
S1
S0
or
Mn(III)
Mn(IV)
Mn(III) or (IV)
~2.7 Å
~2.8 Å
~3.2 Å

Figure 6 Proposed structural changes during the S state transitions based on the EXAFS distance changes, and possible protonation states (at oxo-bridging and terminal water molecules) or changes in the ligand environment (type of ligands and ligation modes). The Mn-Mn distances at ~2.7 Å are indicated by green arrows, ~2.8 Å by blue arrows and ~3.2 Å by red arrows. The dashed line indicates that it may not be a bond. For the S_3 and the S_0 state two possible models are shown. Mn atoms are shown in blue (Mn(III)), red (Mn(IV)) or green (Mn(III) or Mn(IV)) possible), Ca in green and the surrounding ligand environment in grey (adapted from [20]). Several different nomenclatures have been used to refer to the Mn atoms. The numbering scheme shown here is from reference [20]. Mn (1,2,3,4) here corresponds to Mn(A,B,C,D) from polarized EXAFS studies and Mn(4,3,2,1) from the 1.9 Å XRD structure. The second N atom in the His ring and other atoms are not shown for simplicity.

Sr can functionally replace Ca [46, 47] in the OEC and, therefore, Sr XAS studies of Sr-substituted PS II (Mn_4SrO_5 cluster) have been used to study the structural changes of the Mn_4CaO_5 cluster [26–28]. The Mn-Sr (and by inference Mn-Ca) distance changes were observed during the S state transitions, with significant changes during the S_2 to S_3 state transition (Figure 5). These results, together with the Mn XAS data, have demonstrated that Ca (or Sr) plays an important role during the S_2 to S_3 state transition. This is in line with the fact that the OEC does not go beyond the $S_2Y_z^{\bullet}$ state when Ca (or Sr) is chemically depleted from PS II. Recent studies by Lohmiller et al. have shown that the depletion of Ca from the Mn_4CaO_5 core does not disturb the overall structure of the Mn_4 moiety or the spin states in the S_1 and S_2 states [48], as well as the geometry ([24] and unpublished data, T. Lohmiller). The fact that Ca can be removed more easily in the S_3 state (or that Ca can be more easily exchanged in the higher S states) compared to the S_1 and the S_2 states [49], together with the observations on the Ca-depleted system, implies that the Mn-Ca binding modes are changed upon the S_2 to S_3 state transition.

In addition to the changes to the core of the Mn_4CaO_5 cluster, terminal ligands from carboxylates, histidine, and water/hydroxo ligands could be involved in the catalytic reaction. It has been shown using site-directed mutagenesis studies that some ligands have critical roles in the OEC activity. The replacement of just one His terminal ligand by a glutamate residue (D1-His332Glu) resulted in a major change in the EXAFS and XANES spectra [50], illustrating the importance of the ligands in maintaining the active-site structure and how well-tuned the active site is by the residues surrounding the Mn_4CaO_5 cluster.

3 X-Ray Diffraction and Spectroscopy of Photosystem II at Room Temperature Using Femtosecond X-Ray Pulses

As described in the previous section, synchrotron-based XRD has been used over the last decade to study the structure of PS II using cryo-cooled crystals with improving resolution, with the most recent at a resolution of 1.9 Å. Detailed mechanistic studies are hampered by the intrinsic radiation-sensitivity of the Mn_4CaO_5 cluster to X-rays [29, 30], a problem that is common with other redox-active metalloenzymes [51–54]. The higher oxidation states of Mn(III)/Mn(IV) in PS II in the native state are rapidly reduced to Mn(II) upon exposure to X-rays, resulting in structural disruption with concomitant changes in the Mn-O-Mn bridging structure, changes in metal-metal and metal-ligand bond lengths. Thus, even at cryogenic temperatures synchrotron radiation (SR)-based XRD of the Mn_4CaO_5 structure in PS II is fundamentally limited by the radiation damage to the redox active metal site. While X-ray radiation damage makes it difficult to obtain structures of the stable intermediate states of PS II or other metallo-enzymes using frozen cryo-trapped states, it is almost impossible to study the transient intermediate states involved in the catalytic reaction, which can only be generated at ambient conditions, best at room temperature (RT). Thus, it is imperative to work at RT, if one needs to determine physiologically/or biologically relevant structures and changes in the catalytic cycle.

The geometric and electronic structure of the intact Mn_4CaO_5 cluster in the stable S_0 through S_3 states has been addressed especially by X-ray absorption and emission techniques at cryogenic temperatures using a well regulated low X-ray dose (see above). However, following the time course of the water-oxidation reaction at RT using X-ray absorption or emission spectroscopic features within the threshold of radiation damage has been unrealistic with SR sources. In particular studying the transient [S_4] state, when the O-O bond formation and the evolution of O_2 occurs, cannot be captured by traditional cryo-trapping methods and requires time-resolved detection at RT [17, 55–57]. Within ~1.3 milliseconds during the S_3-[S_4]-S_0 transition, which is initiated by the 3rd flash, the following sequence of events is proposed to occur [3]: (a) release of a proton from the $S_3Y_Z^{ox}$ state, (b) transfer of one electron to Y_Z^{ox} from the Mn_4CaO_5 cluster, (c) formation of the O-O bond (peroxo intermediate) coupled with a 2-electron reduction of the Mn_4CaO_5 cluster, (d) formation and release of O_2 together with a further 2-electron reduction of the Mn_4CaO_5 cluster, (e) binding of one or two water molecules to the cluster, and (f) release of another proton during steps (c) to (e). Determining the geometric and electronic structures of intermediates of this reaction (e.g., $S_3Y_Z^{ox}$ and peroxo state) is pivotal for testing the many hypotheses that have been proposed for the mechanism of the water-oxidation reaction (see below).

The recent introduction of X-ray free electron laser (XFEL) sources that produce intense and ultra-short X-ray pulses provide an opportunity to overcome the above described limitations of SR sources for both crystallography and spectroscopy of biological samples with the "collect before destroy" approach [58–60], which entails measuring the response of the system before the manifestation of radiation-induced changes. Unlike cryogenic conditions required at synchrotron sources, the experiments with XFELs can be carried out at RT, making it possible to obtain molecular movies of the catalyst at work by recording snapshots at different time points in the catalytic cycle. In such a study crystallography and spectroscopy can give complementary information: spectroscopy provides detailed information about changes in the Mn oxidation states and the chemical structure of the Mn_4CaO_5 cluster, and crystallography probes the structural changes of the Mn_4CaO_5 cluster and the overall protein. As the entire X-ray emission spectrum can be collected with one excitation energy, this is the method of choice, and as the same X-ray energy (7.1 keV) can be used for X-ray diffraction and excitation of Mn, both XRD and XES [61] methods can be applied simultaneously.

Figure 7 shows the design for an experimental setup for the simultaneous collection of both XRD and XES of the stable and transient intermediate states of PS II at RT using the femtosecond X-ray pulses from an XFEL.

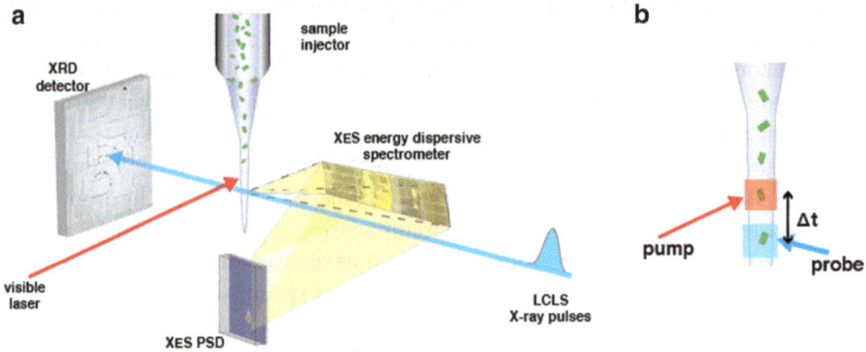

Figure 7 (**a**) XFEL setup for simultaneous collection of X-ray diffraction and emission spectra from PS II crystals in the various S state intermediates in a time-resolved manner. The crystals are injected using an electrofocusing jet to intersect the femtosecond X-ray pulses from the X-ray free electron laser. Downstream of the X-ray pulses is a multi-pixel multi-array detector for the collection of the X-ray diffraction data. Perpendicular to the direction of the X-ray beam is an energy-dispersive multi-crystal emission spectrometer, which focuses the emission spectrum onto a position sensitive detector. (**b**) Visible laser pulses advance the PS II into the various S states and the interval between the visible-laser pump and X-ray laser probe pulse gives the time dependence for the data collected.

3.1 *Simultaneous X-Ray Spectroscopy and Diffraction of Photosystem II*

Since the start of the first hard X-ray XFEL, the Linac Coherent Light Source (LCLS) at Stanford, the application of XFELs to important biological problems has evolved rapidly. Over the last three years, several pioneering "proof of principle" experiments both using crystallography and spectroscopy have been conducted, applying this concept to biological samples, both at LCLS and more recently at the SPring-8-Ångstrom Compact free electron LAser (SACLA), the XFEL at SPring-8 in Japan. The first application of both XES and XRD to PS II are described in this section.

3.1.1 X-Ray Emission Spectroscopy of Photosystem II at Room Temperature Using the X-Ray Free Electron Laser

In contrast to XAS, where the lowest unoccupied orbitals of the metal complexes are probed, XES probes the highest occupied orbitals of the metal complexes (Figure 8a, left) [61]. The highest occupied orbitals are of special interest as they are involved in the actual chemistry during a reaction. The excitation pulse used is at an energy higher than the binding energy of the electrons in 1s orbitals, and subsequent emission from the various levels can be examined using secondary optics. Kβ emission is from the 3p orbitals and is sensitive to the oxidation state and spin state of the metal. XES is well suited for experiments with XFELs because energy-dispersive detection schemes can be used, which allows for a full spectrum to be collected for each X-ray pulse. The scheme for such a dispersive spectrometer is shown in Figure 8a (right) [62]. It is comprised of cylindrically bent analyzer crystals arranged in a von Hamos geometry, where the X-ray emission signal is focused in the horizontal direction and dispersed in energy along the vertical direction onto a multi-pixel array detector. The signal is recorded for each individual shot and the single spectra from several shots can be added to obtain higher S/N ratios. Figure 8a (middle) illustrates the sensitivity of the XES technique to the oxidation states of Mn.

There have been concerns that the X-ray emission spectrum could be sensitive to changes in the electronic structure induced by the intense X-ray pulse, either because of the potential Coulomb explosion or inner-shell ionizations. XES data from solutions of Mn(II)Cl_2 and Mn_2(III,IV)Terpy are shown in Figure 8b and the spectra of both the complexes are in good agreement with the data collected at SR sources under cryogenic conditions [63]. This was especially relevant for the Mn(III,IV) complex, as this compound is highly redox sensitive and can only be measured in frozen solutions at 10 K using synchrotron X-rays. The absence of any deviation from cryogenic SR data indicated that undisturbed Kβ XES can be measured from aqueous solutions of transition metal compounds at the LCLS. More importantly, no change in the electronic structure of the Mn was observed, indicating that under the conditions normally used for hard X-ray protein crystallography at the LCLS, the "probe before destroy" approach is also feasible for spectroscopy of radiation-sensitive transition metals.

Figure 8c shows the Mn Kβ emission spectra of the S_1 state of PS II at RT using an XFEL. The RT XFEL spectrum is identical to that collected with SR at 8 K, and clearly different from the data from Mn(II)Cl_2 or that of PS II collected at RT at SR sources. The spectra of damaged PS II are similar to that obtained for Mn(II)Cl_2. The XFEL XES data are very encouraging as they clearly demonstrate that it is indeed possible to collect spectra from undamaged PS II using the very intense XFEL pulses even at RT. Moreover, these spectra from PS II can be collected simultaneously with the collection of the XRD data (see below), which shows that the crystallographic data originate from an intact PS II, and that XES can be used to determine the integrity of the PS II samples and to confirm the intermediate state of PS II (see below).

3.1.2 X-Ray Diffraction Studies of Photosystem II at Room Temperature Using the X-Ray Free Electron Laser

Single-crystal XRD experiments, when performed with conventional SR, generally use one or a few crystals that are rotated through a set of angles to collect a complete data set for deriving the electron density of the macromolecule. The XFEL destroys the sample with a single X-ray pulse, requiring the full data set to be assembled from a series of still diffraction shots of individual microcrystals [64, 65]. This technique is now becoming known as serial femtosecond crystallography (SFX). The ~40 femtosecond-duration XFEL pulse can deliver diffraction/spectroscopy information on time scales that outrun radiation damage, allowing PS II reaction dynamics to be studied under functional physiological conditions, while the small beam focus size permits the investigation of extremely small and weakly diffracting PS II microcrystals.

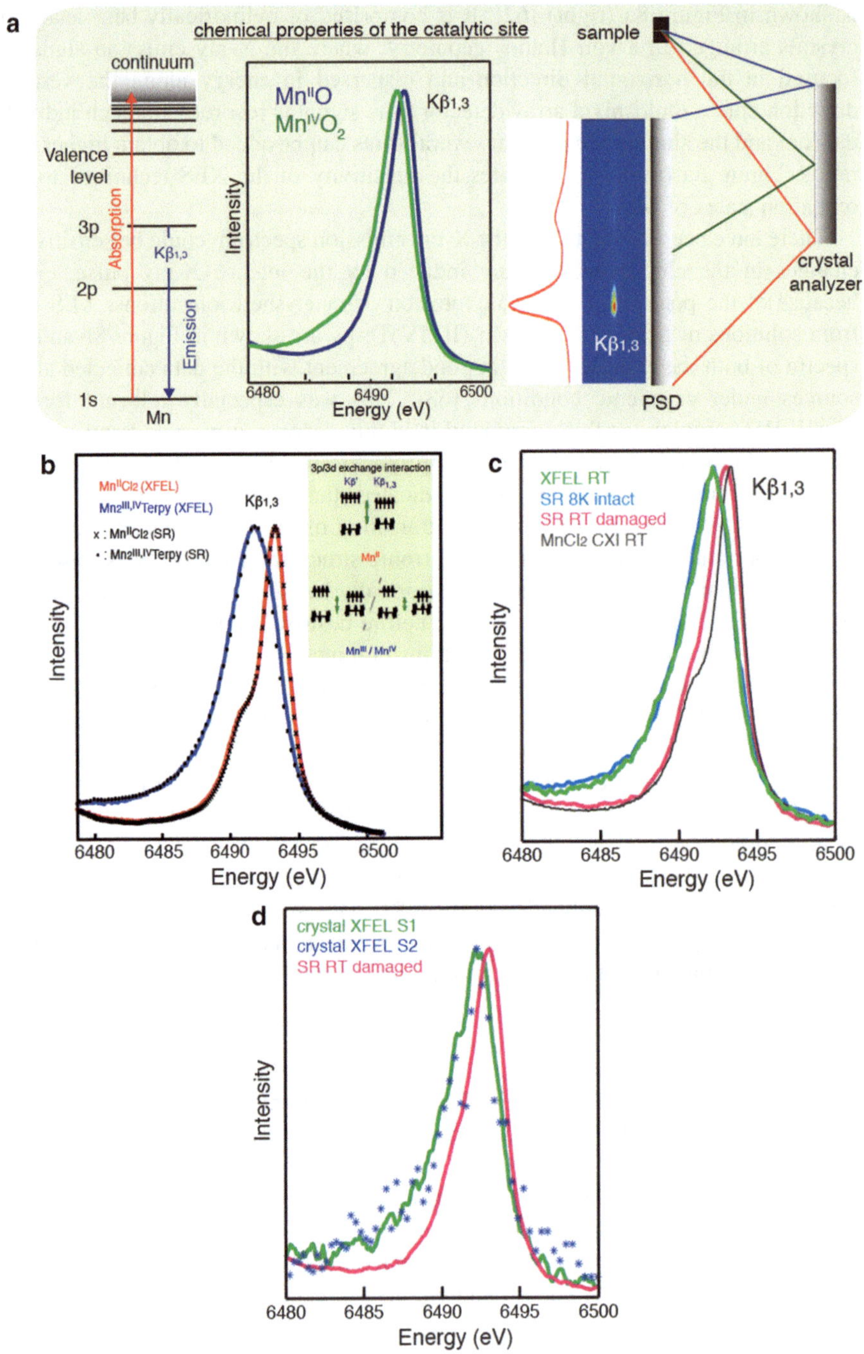

Figure 8 (**a**) The $K\beta_{1,3}$ emission spectra from Mn(II) and Mn(IV) are shown in the middle. $K\beta_{1,3}$ spectra are sensitive to the oxidation state of Mn. The energy level diagram is shown on the left for the $K\beta_{1,3}$ process and the cross-section of the energy dispersive emission spectrometer and the spectrum on the position sensitive detector, with the integrated emission spectrum at the right.

In the case of XRD measurements, using femtosecond pulses from XFELs it has been postulated that there is a self-gating mechanism where the signal contains contributions only from the initial undamaged sample. It is thought that the later part of the X-ray pulse will interact with a strongly distorted or a disrupted crystal lattice, contributing only to diffuse background scattering rather than the Bragg scattering. Hence, the signal that is collected would only reflect the mostly undamaged sample as it is seen during the initial part of the pulse. Thus, XRD data from XFELs is free of data resulting from damaged sites expected at the intensities that are common with XFELs. The initial XRD experiments of PS II at LCLS have largely confirmed these observations. Moreover, the simultaneous XES data obtained from the crystals has shown that under these conditions, the Mn_4CaO_5 cluster is also undamaged [66, 67].

The structure of PS II obtained using the XFEL is isomorphous to that from SR structures, showing that there are no specific large-scale differences either due to radiation damage or due to temperature differences (cryo-cooled crystals at SR and RT with XFEL) (Figure 9a). The maximum of the electron density map (Figure 9b) was found in the region of the Mn_4CaO_5 cluster. The lack of influence of model bias was confirmed by computing the various omit maps; omitting the Mn_4CaO_5 cluster, or the non-heme iron (located at the stromal side of the PS II complex), or the four central Chls from the phasing model. In each case, positive difference electron density for the omitted cofactors was visible in the omit map at the expected location and is an unbiased indication of the presence of these groups in the PS II microcrystals. The mF_o-DF_c difference peak in the Mn_4CaO_5 omit map is shown in Figure 9c.

The dose deposited on the PS II crystals by the XFEL for each individual shot was ~10^8 Grays (Gy or J kg^{-1}). This dose is an order of magnitude higher than the Henderson/Garman limit of 2–3 × 10^7 Gy [68, 69], generally considered to be the limit for loss of diffractivity in cryogenic SR-based XRD, and about 100 times higher than the dose used for SR XRD data collection of PS II at 100 K (~1 × 10^6 Gy for the 1.9 Å crystal structure). With PS II microcrystals, cryogenic SR measurements are not possible due to the extent of X-ray exposure and loss of diffraction; however, the same PS II microcrystals exhibit diffraction spots with the XFEL technique at RT. Despite applying this high dose at RT, there appears to be no loss of diffractivity or visible differences between the RT XFEL and cryogenic SR

Figure 8 (continued) (**b**) The $K\beta_{1,3}$ spectra from a Mn(II)Cl_2 solution and the Mn_2(III,IV)(Terpy) complex are given. The data collected using the XFEL at room temperature is shown in red and blue lines, and the data collected using synchrotron radiation at 10 K are shown as crosses and dots. The spectra at RT using the XFEL and SR at 10 K are identical. The inset shows the exchange coupling showing the sensitivity of the Kβ emission process to the number of unpaired electron in the 3d orbitals of Mn. (**c**) X-ray emission spectra of PS II solutions in the dark state collected using the XFEL at RT (green) or collected using SR under cryogenic conditions with low dose ("8 K intact", light blue) and using SR at RT under photo-reducing conditions ("RT damaged", pink). The spectrum from Mn(II)Cl_2 in aqueous solution collected at RT at the XFEL is shown (grey) for comparison. (**d**) The $K\beta_{1,3}$ XES data collected from PS II crystals using the XFEL in the S_2 state are shown in blue (✳). The XFEL spectrum of microcrystals of PS II in the S_1 state is shown as a green line. For comparison an X-ray emission spectrum of completely photo-reduced ("damaged") PS II collected at RT at a synchrotron is shown in pink (adapted from [63]).

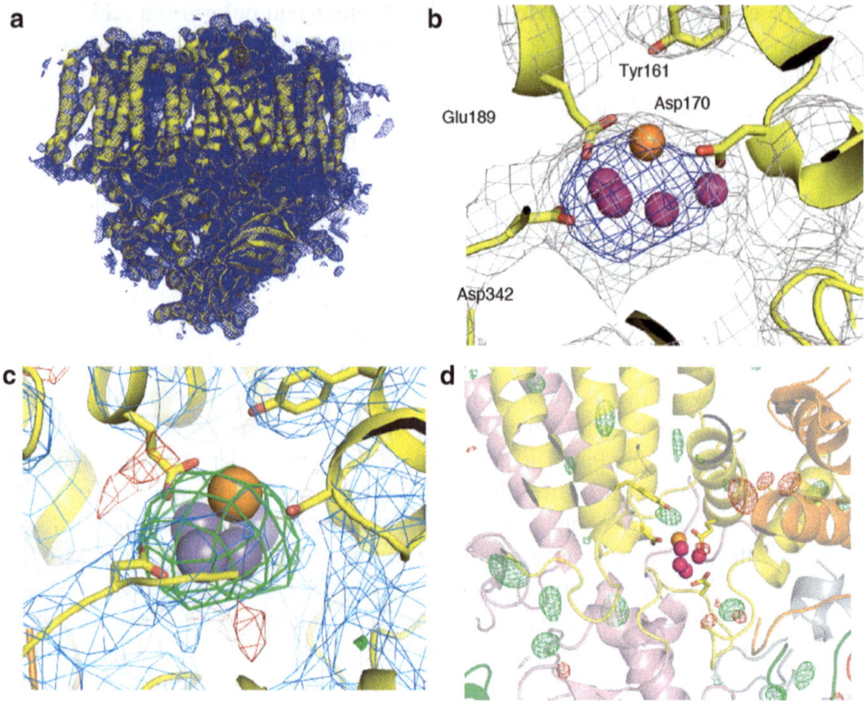

Figure 9 Electron density of PS II obtained from femtosecond XRD measured at the CXI instrument of LCLS. (**a**) Electron density of one monomer of the dimer is shown in blue with the protein shown in yellow, view is along the membrane plane with the lumenal side on bottom and the cytoplasmic side on top. The density is contoured at 1.2 σ. (**b**) Electron density in the vicinity of the OEC, Mn (magenta) and Ca (orange) ions are shown as spheres, the protein backbone in yellow and the electron density as grey (1.0 σ) and blue (4.0 σ) mesh. (**c**) Omit map obtained by excluding the Mn_4Ca cluster from the phasing model. Electron density is shown as in (c), view direction and coloring of Mn and Ca is similar to panel (b). (**d**) Isomorphous difference map between the XFEL-illuminated (S_2 state) and the XFEL-dark (S_1 state) XRD dataset in the region of the Mn_4CaO_5 cluster, with F_o-F_o difference contours shown at +3 σ (green) and −3 σ (red). The map indicates that there are no major changes between the S_1 and S_2 states at this resolution. Metal ions of the Mn_4CaO_5 cluster are shown for orientation as violet (Mn) and orange (Ca) spheres, subunits are indicated in yellow (D1), orange (D2), pink (CP43), and green (PsbO) (adapted from [66, 67]).

structures. This fact demonstrates that the femtosecond XFEL pulses (~45 fs) are short enough to outrun the damage processes present in conventional SR XRD [70].

It should be noted that, for several protein structures, there were recent reports on differences between structures obtained at RT and cryogenic temperatures [69]. PS II shows several large loop regions that are extrinsic to the membrane and are potentially flexible. In addition, only a small number of amino acid residues are involved in providing crystal contacts, raising the possibility that these loop regions could adopt different conformations depending on the crystal conditions. Interestingly, no deviations in position for these loop regions are found when comparing the SR cryogenic and the XFEL RT structures, indicating that there are no large-scale effects on the structure of PS II due to the freezing necessary for cryogenic XRD.

3.2 *Intermediate S State Transitions and Mechanism of Dioxygen Evolution*

A real advantage of XFEL-based XRD or spectroscopy is not only to study biological systems at RT under physiological conditions, but to also study the enzymatic reactions in real time so that one can understand the electronic and structural processes in play. In pursuit of this goal with PS II, and to advance PS II into the higher S states, an *in situ* illumination setup has been integrated into the sample delivery system (Figure 7). The illumination setup consists of three lasers directly coupled via fiber-optic cables to the silica capillary for sample delivery and a fourth laser intersecting the jet at the X-ray interaction point. The idea is to be able to advance PS II through the S state cycle and study the last light-induced S_3 to S_0 advance via the transient and other possible intermediate states in a time-resolved manner to understand the O-O bond formation step. In preparation for these studies, the very first illuminated state, the S_2 state, has been generated *in situ* and studies using both XRD and XES (see below).

3.2.1 X-Ray Free Electron Laser-Based X-Ray Diffraction and X-Ray Emission Spectroscopy of Photosystem II in the S_1 and S_2 States

Recently, XES and XRD data from PS II microcrystals highly enriched in the S_2 state have been successfully collected and analyzed. The XES data did not exhibit a significant difference from the S_1 data (Figure 9d), showing that there was no damage to the electronic structure of the Mn cluster in either state by the visible-laser pump/X-ray-probe method. Due to the limited sample amount, the quality of the S_2 data was poorer and the small shift expected between the S_1 and S_2 state could not be resolved within the signal/noise ratio of the spectra.

XRD data for PS II enriched in the S_2 state was collected using the XFEL with the setup described in Figure 7. As the S_2 state data are isomorphous to the S_1 state data, isomorphous difference map between the S_1 and S_2 states was calculated (Figure 9d). A detailed analysis of this map revealed no statistically relevant difference peaks. This indicates that within the limits of the currently available resolution there are no larger scale structural changes associated with the oxidation of the Mn_4CaO_5 cluster from the dark stable S_1 to the first illuminated S_2 state. These results represent the first step in determining the electronic and geometric structure changes during the water oxidation reaction in PS II in a time-resolved manner.

4 Membrane Inlet Mass Spectrometry and Photosystem II

Monitoring the isotopic composition of the product, O_2, by time-resolved isotope-ratio membrane-inlet mass spectrometry (TR-IR-MIMS) is a powerful method for kinetic and functional analyses in PS II research in particular, in combination and for the study of the mechanism of photosynthetic water-oxidation to O_2.

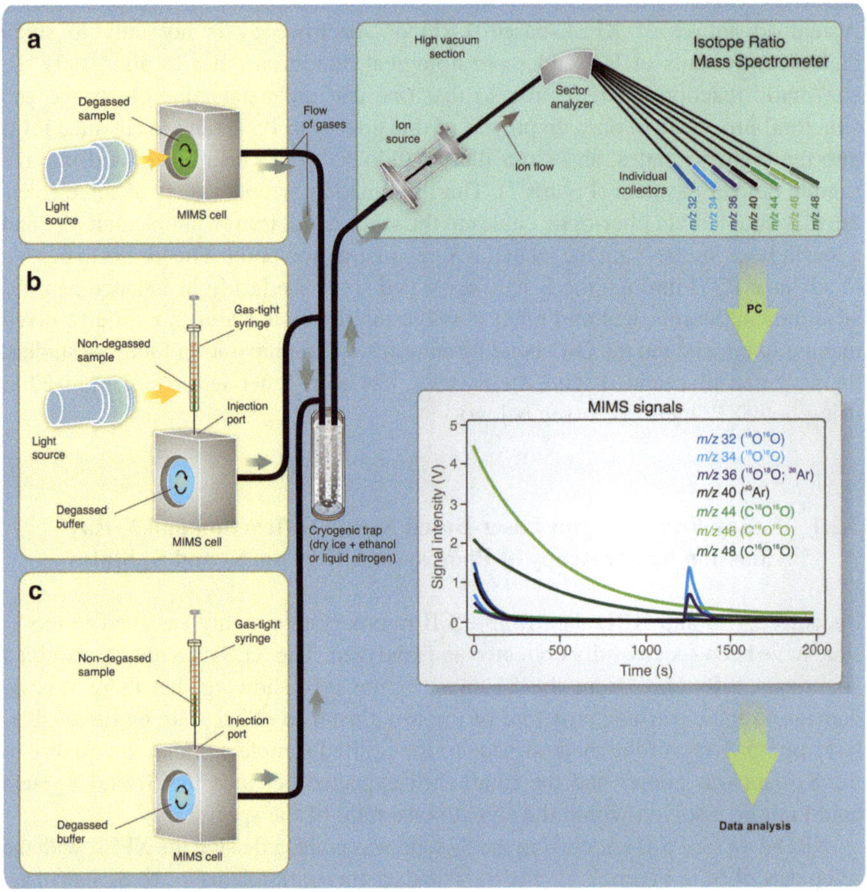

Figure 10 Simplified scheme of a MIMS set up, in which the production of gaseous analyte is initiated by (**a**) illumination of degassed photo-active samples, (**b**) illumination of non-degassed samples that were subsequently injected into a degassed buffer, and (**c**) the reaction of non-degassed chemicals injected into the degassed sample. The gaseous reaction products penetrate through the gas-permeable membrane and reach the high vacuum section where they are ionized by electron impact. The generated ions are separated by a magnetic field according to their mass-to-charge ratios and detected by Faraday cups. The resulting signals of the different isotopes are amplified and recorded online simultaneously.

Schematic views of a TR-MIMS set-up employing an isotope ratio mass spectrometer are shown in Figure 10. This type of mass spectrometer is normally equipped with an electron-impact ion source, magnetic sector field analyser, and individual detectors (Faraday cups) that provide simultaneous detection of several masses (ions) with high sensitivity and signal stability. For its ability to monitor and to selectively analyze all *isotopologues* (molecules that differ only in their isotopic composition) of gaseous products with one instrument, the TR-MIMS approach in combination with isotope enrichments became an indispensable tool for kinetic and functional analyses of photosynthetic enzymes [71–73]. The key part of the TR-MIMS instrument is a gas inlet system that is integrated within a MIMS cell. The design of MIMS cells may vary depending on the measuring purposes [71, 72], but all of them contain a gas-permeable membrane functioning as analyte inlet system into the vacuum of the mass spectrometer. The coupling of such a cell to various light sources (e.g., Xenon lamps or lasers) allows carrying out the measurements of light-induced O_2 evolution in photosynthetic samples or light-driven O_2-evolving artificial catalysts. Before entering the ion source of the mass spectrometer the analytes pass through a cryogenic trap, which freezes out water vapor that inadvertently pervaporate through the membrane in trace amounts. Enrichment of the aqueous sample suspension with oxygen's heavy isotope (^{18}O) for isotope ratio measurements of O_2 (and/or CO_2) isotopologues is a useful and commonly used tool in studies of water-splitting chemistry and/or related reactions. Therefore, most of the experiments are carried out in $H_2^{18}O$-labelled sample suspensions/ solutions.

4.1 *Membrane Inlet Mass Spectrometry and S State Turnover in X-Ray Free Electron Laser Studies*

In the simultaneous XRD and XES experiments using the XFEL, although XES provides an *in situ* check for the quality of the sample and the advancement of the S states in the Kok cycle, it is advantageous to have another independent method to confirm the turnover status of the PS II samples. MIMS has been the method of choice for these studies.

In order to achieve a high population of the illuminated states, the illumination parameters had to be optimized. For this purpose, MIMS has been critical. Using this method the conditions required for the best turnover under the conditions of the XFEL experiments have been optimized. These include the flow rate of PS II, the power and frequency of the laser illumination, and the delay conditions. The MIMS experiment has been used to achieve about 73 % of the sample in the first illuminated state (S_2, see Figure 11). Such independent assessments and optimizations of the turnover are critical for a success of the XFEL experiment.

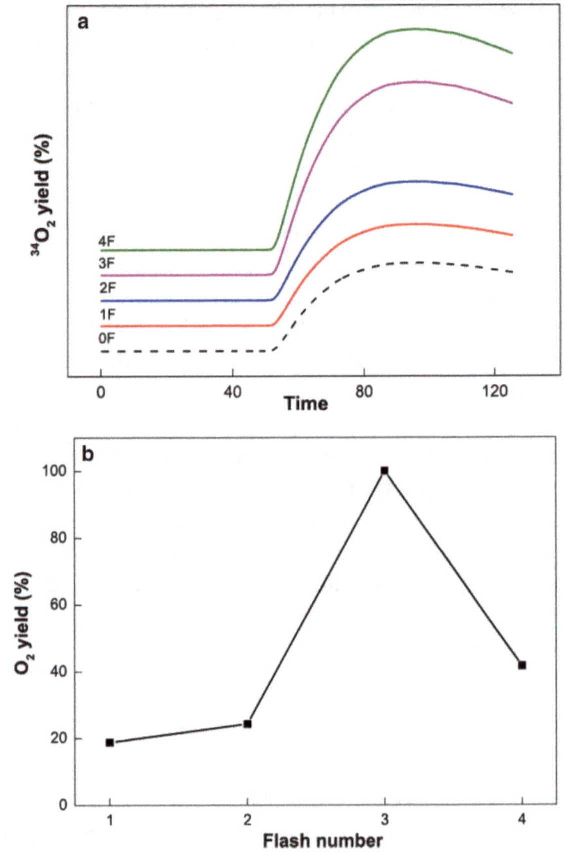

Figure 11 On-line MIMS measurements of light-induced O_2 yield detected as mixed labeled $^{16}O^{18}O$ species after illumination of photosystem II from *Thermosynechococcus elongatus* with 0, 1, 2, 3, and 4 flashes at pH 6.5 and 20 °C (yield *versus* time in seconds) (panel **a**). Panel (**b**) displays the flash pattern derived from (**a**) by subtracting the signal of x flashes from that obtained with x + 1 flashes (panel **b** adapted from [67]).

4.2 *Time-Resolved Membrane Inlet Mass Spectrometry and Insights into Oxygen Evolution*

The most significant contribution of the TR-MIMS in understanding of the water-oxidation mechanism has been its application for studying substrate H_2O binding in the different S states of the OEC. In these experiments the binding of water to the OEC was probed by the rapid injection of $H_2^{18}O$ into the PS II samples which were preset into the desired S state by pre-illumination with 0, 1, 2, or 3 flashes as illustrated in Figure 12. After the desired incubation time, O_2 evolution is induced by a sequence of additional flashes. The exchange rates of the two water molecules are calculated from the dependence of the $^{16}O^{18}O$ and $^{18}O^{18}O$ yields as a function of incubation time (Table 1). A short mixing time of the $H_2^{18}O$ with PS II samples after injection and a very low level of dissolved O_2 in the $H_2^{18}O$ are highly important for these experiments since they determine the time resolution of the TR-MIMS measurements. In the first $H_2^{16}O/H_2^{18}O$-exchange TR-MIMS experiments the water

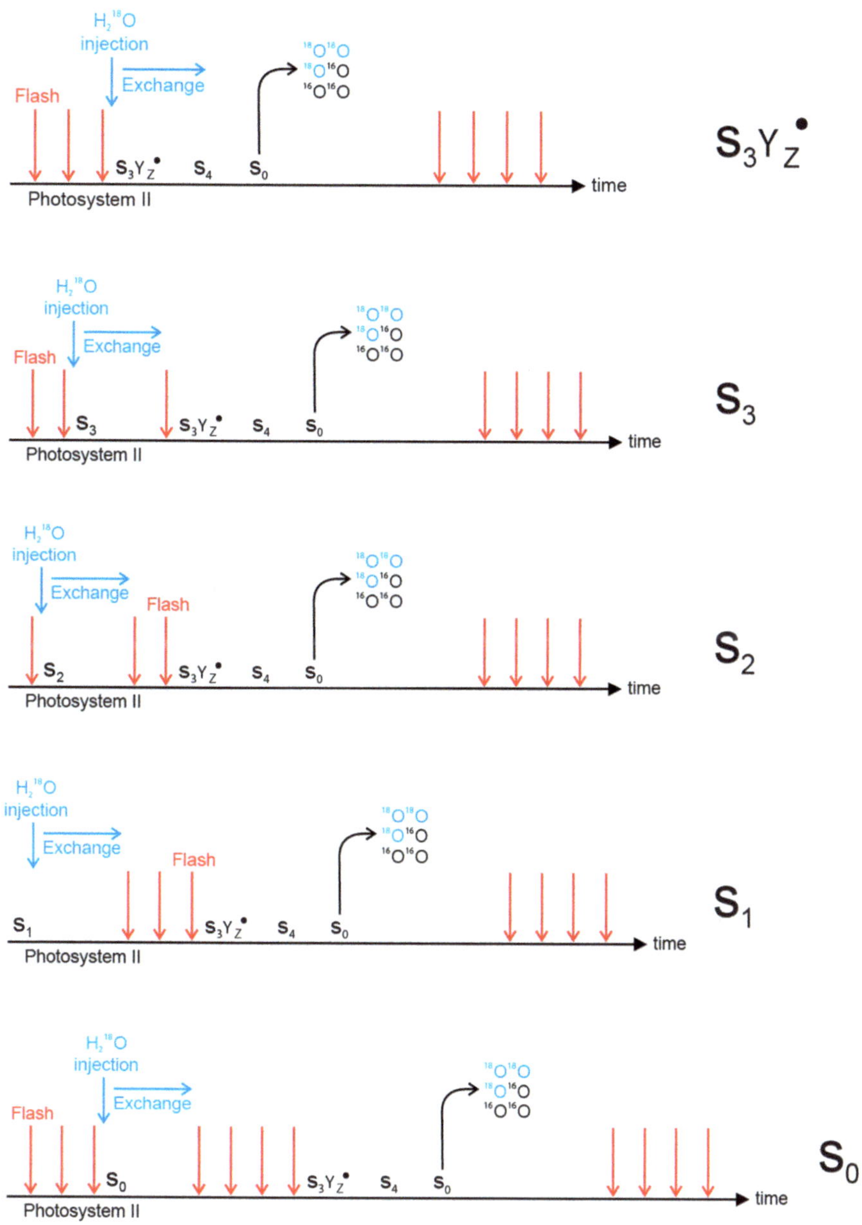

Figure 12 The protocol for the TR-MIMS measurements of substrate water exchange in the S states.

Table 1 S state dependence of substrate water exchange rates measured by TR-MIMS in spinach thylakoids at 10 °C, pH 6.8.[a]

S_i state	k_s, s^{-1}	k_f, s^{-1}
S_0	~10	–
S_1	~0.02	>120
S_2	~2.0	~120
S_3	~2.0	~40

[a]Table adapted from [80].

exchange kinetics could not be resolved [74, 75]. The development of the MIMS cell by Messinger, Badger, and Wydrzynski [76], which allowed for fast mixing of $H_2^{18}O$ with the sample and also implemented O_2 removal from the labeled water by the glucose – glucose oxidase – catalase method, greatly improved the time resolution down to the milliseconds scale and allowed measurements of substrate water exchange in all S states [76–78].

Figure 13 illustrates the characteristic water exchange kinetics in the S_3 state. The yields of the singly-labeled ($^{16}O^{18}O$) and doubly-labeled ($^{18}O^{18}O$) isotopologues of O_2 are plotted as a function of $H_2^{18}O$ incubation time in the S_3 state. Figure 13 (top) shows the result when only one of the two possible ^{18}O-water substrates is exchanged, while Figure 13 (bottom) is for the case when both ^{18}O-waters are exchanged. The biphasic behavior of the $^{16}O^{18}O$ rise (detected at $m/z = 34$) is known to represent the exchange rates of two independent *slowly* (W_s, see below) and *fast* exchanging substrate water molecules (W_f, see below) bound at separate sites within the OEC. In contrast, the $^{18}O^{18}O$ product (monitored at $m/z = 36$) exhibits a mono-exponential rise with a rate equal to that of the slow phase kinetics of the $^{16}O^{18}O$ data, thus-reflecting the exchange of the same 'slowly' exchanging substrate water as observed at $m/z = 34$. This finding clearly confirms that the two phases of the $^{16}O^{18}O$ data are an intrinsic feature of the OEC and do not originate from PS II heterogeneity [73, 76, 77].

Further TR-MIMS experiments also revealed that the 'slowly' exchanging water is bound to the OEC in all the S states, while the 'fast' exchanging water was detected only in the S_2 and S_3 states [77–80]. Thus, the TR-MIMS technique provides not only the most direct evidence for independent substrate water binding within the OEC, but also allows to monitor the change in their binding affinities throughout the reaction cycle. For a complete overview of the TR-MIMS findings in this field, we refer the readers to recent reviews [72, 81, 82].

The most likely mechanisms presently being considered are presented in Figure 14 below. The water-exchange results do not support the nucleophile attack mechanisms in which Ca-bound water attacks a terminal oxo or a μ-oxo bridge. Similarly, mechanisms involving two μ-oxo bridges, or two terminal waters seem unlikely. Two likely options are that O5 is W_s, where W_f is either W_2, the terminal hydroxo ligand to Mn_{A4}, or a water not seen in the crystal structure (see above, Figure 1, inset) [35, 82–84].

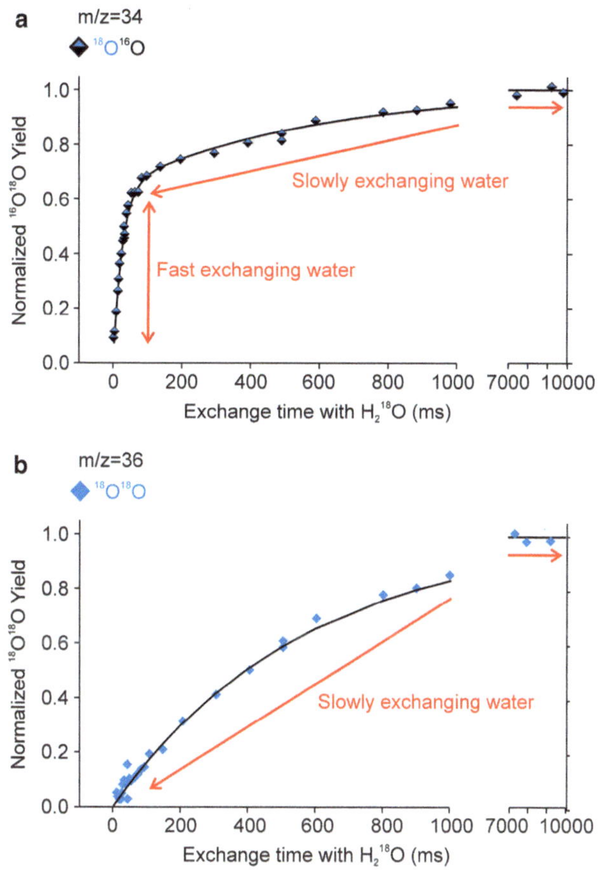

Figure 13 The characteristic water exchange kinetics in the S_3 state as measured in spinach thylakoids with the time resolution of 8 ms. The yields of the singly-labeled ($^{16}O^{18}O$) and doubly-labeled ($^{18}O^{18}O$) isotopologues of molecular oxygen are plotted as a function of $H_2^{18}O$ incubation time in the S_3 state. While the former plot reflects the result when only one of the two possible ^{18}O-water substrates is exchanged, the latter one is for the case when both ^{18}O-waters are exchanged. The biphasic behavior of the $^{16}O^{18}O$ rise (detected at $m/z = 34$) is known to represent the exchange rates of two independent *slowly* (W_s) and *fast* exchanging substrate water molecules (W_f) bound at separate sites within the OEC (see Table 1). The data in panel (**b**) are kinetically limited by the exchange of W_s and thus show a single slow phase identical to that in panel (**a**).

5 Concluding Remarks and Future Directions

Ultimately, understanding the water splitting mechanism requires the characterization of the last step (S_3-S_4-S_0), where the O-O bond is formed and O_2 is evolved. Unlike other intermediate S states (S_0 through S_3), the kinetically unstable S_4 state cannot be cryo-trapped by freeze-quenching techniques, and therefore requires time-resolved studies at ambient conditions [55]. The S_3-S_4-S_0 transition includes five sequential events: (a) extraction of one electron, (b) release of two protons, (c) release of molecular oxygen, (d) binding of one or two substrate water molecules to the Mn_4CaO_5 cluster, and (e) four-electron-reduction of the Mn_4CaO_5 cluster.

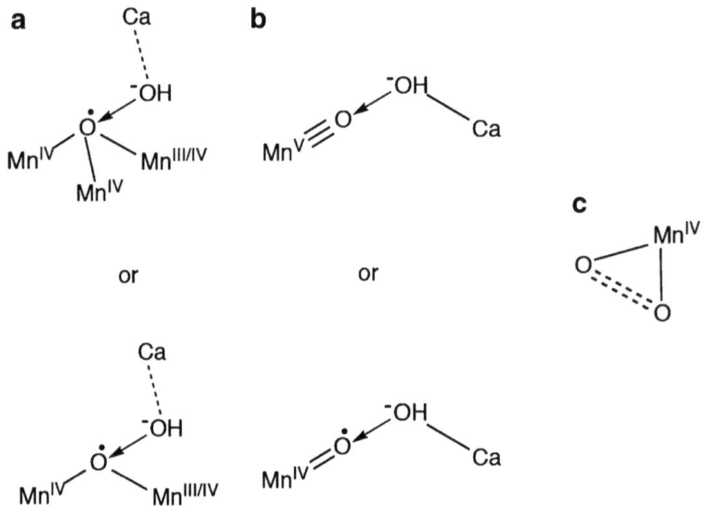

Figure 14 Schematic presentation of three mechanisms proposed for O-O bond formation during the S_3 - $[S_4]$ - S_0 transition. (**a**) involves a bridging O ligand and a radical and/or a delocalized charge on the Mn cluster. (**b**) involves the electrophilic mechanism with a high valent Mn(V) and terminal O ligation, and (**c**) involves an end-on peroxo ligand to a single Mn center. These schematics do not represent all the possible variations that have been proposed within the categories of radical-based and nucleophilic attack mechanisms.

Several O-O bond formation mechanisms have been proposed [83], and three representative ones are shown in Figure 14. Mechanism (a) invokes an O-O bond formation by a radical-type mechanism, mechanism (b) involves an electrophilic high-valent Mn(V), and mechanism (c) has an end-on peroxo ligand to a single Mn center. Fundamental differences in the chemistry of O-O bond formation and O_2 evolution exists between the three types of mechanisms. In the first case there is delocalization of the charge to the ligands, while in the second case the charge is mainly centered on the Mn. Not much is known about the electronic structure of the third case, although such intermediates have been observed for Cu and Fe metalloenzymes and Mn model compounds. Mn(V) has also been proposed as a transient intermediate in the catalytic cycle of epoxidation reactions, and the water-oxidation in PS II might involve a similar intermediate. The high reactivity of the S_4 state intermediates is not in line with the stability of synthetic Mn(V) compounds, which are all in a low-spin configuration. We would therefore rather expect a high-spin, reactive Mn(V) to be involved. However, so far there is no direct spectroscopic evidence for the existence of such a Mn(V) species in PS II. It is known that up to the S_3 state all four Mn centers are in a high-spin configuration, hence it is likely that if Mn(V) is involved it would also be in a high-spin configuration, and therefore reactive. The presence of Mn(V) or lack thereof and its spin state is particularly important for developing artificial inorganic water splitting catalysts, a central issue in solar fuel generation. These questions, along with the need to characterize the proposed transient S_4 state and other kinetic intermediates that may be present, offer challenges for the future and they need to be addressed before we can fully understand where and how plants oxidize water to dioxygen.

Abbreviations and Definitions

Chl	chlorophyll
CXI	coherent X-ray imaging
Dbm	dibenzoylmethane
DFT	density functional theory
ENDOR	electron nuclear double resonance
EPR	electron paramagnetic resonance
EXAFS	extended X-ray absorption fine structure
fs	femtosecond
FTIR	Fourier transform infrared spectroscopy
Htphpn	N,N,N',N'-tetra(2-methylpyridyl)-2-hydroxypropanediamine
IR	infrared spectroscopy
LCLS	Linac coherent light source
MIMS	membrane inlet mass spectrometry
OAc	acetate
OEC	oxygen-evolving complex
PS II	photosystem II
PSD	position sensitive detector
RIXS	resonant inelastic X-ray scattering
RT	room temperature
SACLA	Spring-8 Ångstrom compact free electron laser
SFX	serial femtosecond crystallography
SR	synchrotron radiation
Terpy	2,2':6',2"-terpyridine
TR-MIMS	time-resolved membrane inlet mass spectrometry
XANES	X-ray absorption near edge spectroscopy
XAS	X-ray absorption spectroscopy
XES	X-ray emission spectroscopy
XFEL	X-ray free electron laser
XRD	X-ray diffraction

Acknowledgments The research reviewed here was supported by the NIH Grant GM 55302 (V.K.Y.) and the Director, Office of Science, Office of Basic Energy Sciences (OBES), Division of Chemical Sciences, Geosciences, and Biosciences of the Department of Energy (DOE) under Contract DE-AC02-05CH11231 (J.Y. and V.K.Y.). The Solar Fuels Strong Research Environment (Umeå University), the Artificial Leaf Project (K&A Wallenberg Foundation), VR and Energimyndigheten (J.M.), and the Human Frontier Research Grant, RGP0063/2013 (J.Y.) are acknowledged for supporting this project. The LCLS, and the synchrotron facilities at Stanford Synchrotron Radiation Lightsource (SSRL), the Advanced Light Source (ALS), and the Advanced Photon Source (APS), used in the course of these studies are all supported by DOE OBES. We thank all our students, postdoctoral fellows, and collaborators for their very important contributions to the PS II studies at LCLS, and the staff at LCLS for their support of the XFEL experiments.

References

1. *Photosystem II: The Light-Driven Water:Plastoquinone Oxidoreductase*, Vol. 22 of *Advances in Photosynthesis and Respiration*, Eds T. Wydrzynski, S. Satoh, Springer, Dordrecht, 2005.
2. B. Kok, B. Forbush, M. McGloin, *Photochem. Photobiol.* **1970**, *11*, 457–475.
3. G. Renger, *J. Photochem. Photobiol. B Biol.* **2011**, *104*, 35–43.
4. N. S. Lewis, D. G. Nocera, *Proc. Natl. Acad. Sci. USA* **2006**, *103*, 15729–15735.
5. A. Zouni, H.-T. Witt, J. Kern, P. Fromme, N. Krauß, W. Saenger, P. Orth, *Nature* **2001**, *409*, 739–743.
6. N. Kamiya, J. R. Shen, *Proc. Natl. Acad. Sci. USA* **2003**, *100*, 98–103.
7. K. N. Ferreira, T. M. Iverson, K. Maghlaoui, J. Barber, S. Iwata, *Science* **2004**, *303*, 1831–1838.
8. B. Loll, J. Kern, W. Saenger, A. Zouni, J. Biesiadka, *Nature* **2005**, *438*, 1040–1044.
9. A. Guskov, J. Kern, A. Gabdulkhakov, M. Broser, A. Zouni, W. Saenger, *Nature Struct. Mol. Biol.* **2009**, *16*, 334–342.
10. Y. Umena, K. Kawakami, J.-R. Shen, N. Kamiya, *Nature* **2011**, *473*, 55–60.
11. V. K. Yachandra, K. Sauer, M. P. Klein, *Chem. Rev.* **1996**, *96*, 2927–2950.
12. J. Yano, V. K. Yachandra, *Inorg. Chem.* **2008**, *47*, 1711–1726.
13. J. Yano, V. K. Yachandra, *Chem. Rev.* **2014**, *114*, 4175–4205.
14. N. Cox, D. A. Pantazis, F. Neese, W. Lubitz, *Acc. Chem. Res.* **2013**, *46*, 1588–1596.
15. A. Haddy, *Photosynth. Res.* **2007**, *92*, 357–368.
16. G. W. Brudvig, *Phil. Trans. Roy. Soc. B-Biol. Sci.* **2008**, *363*, 1211–1218.
17. T. Noguchi, H. Suzuki, M. Tsuno, M. Sugiura, C. Kato, *Biochemistry* **2012**, *51*, 3205–3214.
18. F. Rappaport, N. Ishida, M. Sugiura, A. Boussac, *Energ. Environ. Sci.* **2011**, *4*, 2520–2524.
19. H. Dau, M. Haumann, *Coord. Chem. Rev.* **2008**, *252*, 273–295.
20. C. Glöckner, J. Kern, M. Broser, A. Zouni, V. Yachandra, J. Yano, *J. Biol. Chem.* **2013**, *288*, 22607–22620.
21. J. Yano, Y. Pushkar, P. Glatzel, A. Lewis, K. Sauer, J. Messinger, U. Bergmann, V. K. Yachandra, *J. Am. Chem. Soc.* **2005**, *127*, 14974–14975.
22. Y. Pushkar, J. Yano, P. Glatzel, J. Messinger, A. Lewis, K. Sauer, U. Bergmann, V. Yachandra, *J. Biol. Chem.* **2007**, *282*, 7198–7208.
23. J. Yano, J. Kern, K. Sauer, M. Latimer, Y. Pushkar, J. Biesiadka, B. Loll, W. Saenger, J. Messinger, A. Zouni, V. K. Yachandra, *Science* **2006**, *314*, 821–825.
24. M. J. Latimer, V. J. DeRose, I. Mukerji, V. K. Yachandra, K. Sauer, M. P. Klein, *Biochemistry* **1995**, *34*, 10898–10909.
25. R. M. Cinco, K. L. M. Holman, J. H. Robblee, J. Yano, S. A. Pizarro, E. Bellacchio, K. Sauer, V. K. Yachandra, *Biochemistry* **2002**, *41*, 12928–12933.

26. R. M. Cinco, J. H. Robblee, A. Rompel, C. Fernandez, V. K. Yachandra, K. Sauer, M. P. Klein, *J. Phys. Chem. B* **1998**, *102*, 8248–8256.
27. R. M. Cinco, J. H. Robblee, J. Messinger, C. Fernandez, K. L. M. Holman, K. Sauer, V. K. Yachandra, *Biochemistry* **2004**, *43*, 13271–13282.
28. Y. Pushkar, J. Yano, K. Sauer, A. Boussac, V. K. Yachandra, *Proc. Natl. Acad. Sci. USA* **2008**, *105*, 1879–1884.
29. J. Yano, J. Kern, K.-D. Irrgang, M. J. Latimer, U. Bergmann, P. Glatzel, Y. Pushkar, J. Biesiadka, B. Loll, K. Sauer, J. Messinger, A. Zouni, V. K. Yachandra, *Proc. Natl. Acad. Sci. USA* **2005**, *102*, 12047–12052.
30. M. Grabolle, M. Haumann, C. Muller, P. Liebisch, H. Dau, *J. Biol. Chem.* **2006**, *281*, 4580–4588.
31. J. Messinger, J. H. Robblee, U. Bergmann, C. Fernandez, P. Glatzel, H. Visser, R. M. Cinco, K. L. McFarlane, E. Bellacchio, S. A. Pizarro, S. P. Cramer, K. Sauer, M. P. Klein, V. K. Yachandra, *J. Am. Chem. Soc.* **2001**, *123*, 7804–7820.
32. M. Haumann, C. Muller, P. Liebisch, L. Iuzzolino, J. Dittmer, M. Grabolle, T. Neisius, W. Meyer-Klaucke, H. Dau, *Biochemistry* **2005**, *44*, 1894–1908.
33. L. V. Kulik, B. Epel, W. Lubitz, J. Messinger, *J. Am. Chem. Soc.* **2005**, *127*, 2392–2393.
34. L. V. Kulik, B. Epel, W. Lubitz, J. Messinger, *J. Am. Chem. Soc.* **2007**, *129*, 13421–13435.
35. P. E. M. Siegbahn, *Acc. Chem. Res.* **2009**, *42*, 1871–1880.
36. K. Yamaguchi, H. Isobe, S. Yamanaka, T. Saito, K. Kanda, M. Shoji, Y. Umena, K. Kawakami, J. R. Shen, N. Kamiya, M. Okumura, *Int. J. Quantum Chem.* **2013**, *113*, 525–541.
37. H. A. Chu, W. Hillier, R. J. Debus, *Biochemistry* **2004**, *43*, 3152–3166.
38. N. Cox, L. Rapatskiy, J. H. Su, D. A. Pantazis, M. Sugiura, L. Kulik, P. Dorlet, A. W. Rutherford, F. Neese, A. Boussac, W. Lubitz, J. Messinger, *J. Am. Chem. Soc.* **2011**, *133*, 3635–3648.
39. J. H. Su, N. Cox, W. Ames, D. A. Pantazis, L. Rapatskiy, T. Lohmiller, L. V. Kulik, P. Dorlet, A. W. Rutherford, F. Neese, A. Boussac, W. Lubitz, J. Messinger, *Biochim. Biophys. Acta* **2011**, *1807*, 829–840.
40. L. Kulik, B. Epel, J. Messinger, W. Lubitz, *Photosynth. Res.* **2005**, *84*, 347–353.
41. J. H. Robblee, J. Messinger, R. M. Cinco, K. L. McFarlane, C. Fernandez, S. A. Pizarro, K. Sauer, V. K. Yachandra, *J. Am. Chem. Soc.* **2002**, *124*, 7459–7471.
42. P. Glatzel, U. Bergmann, J. Yano, H. Visser, J. H. Robblee, W. W. Gu, F. M. F. de Groot, G. Christou, V. L. Pecoraro, S. P. Cramer, V. K. Yachandra, *J. Am. Chem. Soc.* **2004**, *126*, 9946–9959.
43. P. Glatzel, H. Schroeder, Y. Pushkar, T. Boron, III, S. Mukherjee, G. Christou, V. L. Pecoraro, J. Messinger, V. K. Yachandra, U. Bergmann, J. Yano, *Inorg. Chem.* **2013**, *52*, 5642–5644.
44. W. C. Liang, T. A. Roelofs, R. M. Cinco, A. Rompel, M. J. Latimer, W. O. Yu, K. Sauer, M. P. Klein, V. K. Yachandra, *J. Am. Chem. Soc.* **2000**, *122*, 3399–3412.
45. W. Ames, D. A. Pantazis, V. Krewald, N. Cox, J. Messinger, W. Lubitz, F. Neese, *J. Am. Chem. Soc.* **2011**, *133*, 19743–19757.
46. A. Boussac, A. W. Rutherford, *Biochemistry* **1988**, *27*, 3476–3483.
47. D. F. Ghanotakis, G. T. Babcock, C. F. Yocum, *FEBS Lett.* **1984**, *167*, 127–130.
48. T. Lohmiller, N. Cox, J.-H. Su, J. Messinger, W. Lubitz, *J. Biol. Chem.* **2012**, *287*, 24721–24733.
49. C. F. Yocum, *Coord. Chem. Rev.* **2008**, *252*, 296–305.
50. J. Yano, L. M. Walker, M. A. Strickler, R. J. Service, V. K. Yachandra, R. J. Debus, *J. Biol. Chem.* **2011**, *286*, 9257–9267.
51. K. G. V. Sigfridsson, P. Chernev, N. Leidel, A. Popovic-Bijelic, A. Gräslund, M. Haumann, *J. Biol. Chem.* **2013**, *288*, 9648–9661.
52. K. D. Daughtry, Y. Xiao, D. Stoner-Ma, E. Cho, A. M. Orville, P. Liu, K. N. Allen, *J. Am. Chem. Soc.* **2012**, *134*, 2823–2834.

53. S. V. Antonyuk, M. A. Hough, *Biochim. Biophys. Acta-Proteins and Proteomics* **2011**, *1814*, 778–784.
54. H.-P. Hersleth, K. K. Andersson, *Biochim. Biophys. Acta-Proteins and Proteomics* **2011**, *1814*, 785–796.
55. M. Haumann, P. Liebisch, C. Muller, M. Barra, M. Grabolle, H. Dau, *Science* **2005**, *310*, 1019–1021.
56. F. Rappaport, M. Blancharddesce, J. Lavergne, *Biochim. Biophys. Acta-Bioenerg.* **1994**, *1184*, 178–192.
57. M. R. Razeghifard, R. J. Pace, *Biochemistry* **1999**, *38*, 1252–1257.
58. R. Neutze, R. Wouts, D. van der Spoel, E. Weckert, J. Hajdu, *Nature* **2000**, *406*, 752–757.
59. H. N. Chapman, P. Fromme, A. Barty, T. A. White, R. A. Kirian, A. Aquila, M. S. Hunter, J. Schulz, D. P. DePonte, U. Weierstall, R. B. Doak, F. R. Maia, A. V. Martin, I. Schlichting, L. Lomb, N. Coppola, R. L. Shoeman, S. W. Epp, R. Hartmann, D. Rolles, A. Rudenko, L. Foucar, N. Kimmel, G. Weidenspointner, P. Holl, M. Liang, M. Barthelmess, C. Caleman, S. Boutet, M. J. Bogan, J. Krzywinski, C. Bostedt, S. Bajt, L. Gumprecht, B. Rudek, B. Erk, C. Schmidt, A. Homke, C. Reich, D. Pietschner, L. Struder, G. Hauser, H. Gorke, J. Ullrich, S. Herrmann, G. Schaller, F. Schopper, H. Soltau, K. U. Kuhnel, M. Messerschmidt, J. D. Bozek, S. P. Hau-Riege, M. Frank, C. Y. Hampton, R. G. Sierra, D. Starodub, G. J. Williams, J. Hajdu, N. Timneanu, M. M. Seibert, J. Andreasson, A. Rocker, O. Jonsson, M. Svenda, S. Stern, K. Nass, R. Andritschke, C. D. Schroter, F. Krasniqi, M. Bott, K. E. Schmidt, X. Wang, I. Grotjohann, J. M. Holton, T. R. Barends, R. Neutze, S. Marchesini, R. Fromme, S. Schorb, D. Rupp, M. Adolph, T. Gorkhover, I. Andersson, H. Hirsemann, G. Potdevin, H. Graafsma, B. Nilsson, J. C. H. Spence, *Nature* **2011**, *470*, 73–77.
60. S. Boutet, L. Lomb, G. J. Williams, T. R. M. Barends, A. Aquila, R. B. Doak, U. Weierstall, D. P. DePonte, J. Steinbrener, R. L. Shoeman, M. Messerschmidt, A. Barty, T. A. White, S. Kassemeyer, R. A. Kirian, M. M. Seibert, P. A. Montanez, C. Kenney, R. Herbst, P. Hart, J. Pines, G. Haller, S. M. Gruner, H. T. Philipp, M. W. Tate, M. Hromalik, L. J. Koerner, N. van Bakel, J. Morse, W. Ghonsalves, D. Arnlund, M. J. Bogan, C. Caleman, R. Fromme, C. Y. Hampton, M. S. Hunter, L. C. Johansson, G. Katona, C. Kupitz, M. Liang, A. V. Martin, K. Nass, L. Redecke, F. Stellato, N. Timneanu, D. Wang, N. A. Zatsepin, D. Schafer, J. Defever, R. Neutze, P. Fromme, J. C. H. Spence, H. N. Chapman, I. Schlichting, *Science* **2012**, *337*, 362–364.
61. P. Glatzel, U. Bergmann, *Coord. Chem. Rev.* **2005**, *249*, 65–95.
62. R. Alonso-Mori, J. Kern, D. Sokaras, T.-C. Weng, D. Nordlund, R. Tran, P. Montanez, J. Delor, V. K. Yachandra, J. Yano, U. Bergmann, *Rev. Sci. Instrum.* **2012**, *83*, 073114.
63. R. Alonso-Mori, J. Kern, R. J. Gildea, D. Sokaras, T.-C. Weng, B. Lassalle-Kaiser, T. Rosalie, J. Hattne, H. Laksmono, J. Hellmich, C. Gloeckner, N. Echols, R. G. Sierra, D. W. Schafer, J. Sellberg, C. Kenney, R. Herbst, J. Pines, P. Hart, S. Herrmann, R. W. Grosse-Kunstleve, M. J. Latimer, A. R. Fry, M. M. Messerschmidt, A. Miahnahri, M. M. Seibert, P. H. Zwart, W. E. White, P. D. Adams, M. J. Bogan, S. Boutet, G. J. Williams, A. Zouni, J. Messinger, P. Glatzel, N. K. Sauter, V. K. Yachandra, J. Yano, U. Bergmann, *Proc. Natl. Acad. Sci. USA* **2012**, *109*, 19103–19107.
64. J. Hattne, N. Echols, R. Tran, J. Kern, R. J. Gildea, A. S. Brewster, R. Alonso-Mori, C. Glöckner, J. Hellmich, H. Laksmono, R. G. Sierra, B. Lassalle-Kaiser, A. Lampe, G. Han, S. Gul, D. DiFiore, D. Milathianaki, A. R. Fry, A. Miahnahri, W. E. White, D. W. Schafer, M. M. Seibert, J. E. Koglin, D. Sokaras, T.-C. Weng, J. Sellberg, M. J. Latimer, P. Glatzel, P. H. Zwart, R. W. Grosse-Kunstleve, M. J. Bogan, M. Messerschmidt, G. J. Williams, S. Boutet, J. Messinger, A. Zouni, J. Yano, U. Bergmann, V. K. Yachandra, P. D. Adams, N. K. Sauter, *Nature Methods* **2014**, *11*, 545–548.
65. R. G. Sierra, H. Laksmono, J. Kern, R. Tran, J. Hattne, R. Alonso-Mori, B. Lassalle-Kaiser, C. Glockner, J. Hellmich, D. W. Schafer, N. Echols, R. J. Gildea, R. W. Grosse-Kunstleve, J. Sellberg, T. A. McQueen, A. R. Fry, M. M. Messerschmidt, A. Miahnahri, M. M. Seibert, C. Y. Hampton, D. Starodub, N. D. Loh, D. Sokaras, T.-C. Weng, P. H. Zwart, P. Glatzel,

D. Milathianaki, W. E. White, P. D. Adams, G. J. Williams, S. Boutet, A. Zouni, J. Messinger, N. K. Sauter, U. Bergmann, J. Yano, V. K. Yachandra, M. J. Bogan, *Acta Cryst. D* **2012**, *68*, 1584–1587.
66. J. Kern, R. Alonso-Mori, J. Hellmich, R. Tran, J. Hattne, H. Laksmono, C. Glöckner, N. Echols, R. G. Sierra, J. Sellberg, B. Lassalle-Kaiser, R. J. Gildea, P. Glatzel, R. W. Grosse-Kunstleve, M. J. Latimer, T. A. McQueen, D. DiFiore, A. R. Fry, M. Messerschmidt, A. Miahnahri, D. W. Schafer, M. M. Seibert, D. Sokaras, T. C. Weng, P. H. Zwart, W. E. White, P. D. Adams, M. J. Bogan, S. Boutet, G. J. Williams, J. Messinger, N. K. Sauter, A. Zouni, U. Bergmann, J. Yano, V. K. Yachandra, *Proc. Natl. Acad. Sci. USA* **2012**, *109*, 9721–9726.
67. J. Kern, R. Alonso-Mori, R. Tran, J. Hattne, R. J. Gildea, N. Echols, C. Glöckner, J. Hellmich, H. Laksmono, R. G. Sierra, B. Lassalle-Kaiser, S. Koroidov, A. Lampe, G. Han, S. Gul, D. DiFiore, D. Milathianaki, A. R. Fry, A. Miahnahri, D. W. Schafer, M. Messerschmidt, M. M. Seibert, J. E. Koglin, D. Sokaras, T.-C. Weng, J. Sellberg, M. J. Latimer, R. W. Grosse-Kunstleve, P. H. Zwart, W. E. White, P. Glatzel, P. D. Adams, M. J. Bogan, G. J. Williams, S. Boutet, J. Messinger, A. Zouni, N. K. Sauter, V. K. Yachandra, U. Bergmann, J. Yano, *Science* **2013**, *340*, 491–495.
68. R. Henderson, *Philos. Trans. R. Soc. Lond. Ser. B-Biol. Sci.* **1990**, *241*, 6–8.
69. R. L. Owen, E. Rudino-Pinera, E. F. Garman, *Proc. Natl. Acad. Sci. USA* **2006**, *103*, 4912–4917.
70. A. Barty, C. Caleman, A. Aquila, N. Timneanu, L. Lomb, T. A. White, J. Andreasson, D. Arnlund, S. Bajt, T. R. M. Barends, M. Barthelmess, M. J. Bogan, C. Bostedt, J. D. Bozek, R. Coffee, N. Coppola, J. Davidsson, D. P. DePonte, R. B. Doak, T. Ekeberg, V. Elser, S. W. Epp, B. Erk, H. Fleckenstein, L. Foucar, P. Fromme, H. Graafsma, L. Gumprecht, J. Hajdu, C. Y. Hampton, R. Hartmann, A. Hartmann, G. Hauser, H. Hirsemann, P. Holl, M. S. Hunter, L. Johansson, S. Kassemeyer, N. Kimmel, R. A. Kirian, M. N. Liang, F. Maia, E. Malmerberg, S. Marchesini, A. V. Martin, K. Nass, R. Neutze, C. Reich, D. Rolles, B. Rudek, A. Rudenko, H. Scott, I. Schlichting, J. Schulz, M. M. Seibert, R. L. Shoeman, R. G. Sierra, H. Soltau, J. C. H. Spence, F. Stellato, S. Stern, L. Struder, J. Ullrich, X. Wang, G. Weidenspointner, U. Weierstall, C. B. Wunderer, H. N. Chapman, *Nature Photonics* **2012**, *6*, 35–40.
71. L. Konermann, J. Messinger, W. Hillier, in *Biophysical Techniques in Photosynthesis*, Eds J. Amesz, A. J. Hoff, Springer, Dordrecht, 2008, pp. 167–190.
72. K. Beckmann, J. Messinger, M. R. Badger, T. Wydrzynski, W. Hillier, *Photosynth. Res.* **2009**, *102*, 511–522.
73. D. Shevela, J. Messinger, *Front. Plant Sci.* **2013**, *4*.
74. K. P. Bader, G. Renger, G. H. Schmid, *Photosynth. Res.* **1993**, *38*, 355–361.
75. R. Radmer, O. Ollinger, *FEBS Letts.* **1986**, *195*, 285–289.
76. J. Messinger, M. Badger, T. Wydrzynski, *Proc. Natl. Acad. Sci. USA* **1995**, *92*, 3209–3213.
77. W. Hillier, J. Messinger, T. Wydrzynski, *Biochemistry* **1998**, *37*, 16908–16914.
78. W. Hillier, T. Wydrzynski, *Biochemistry* **2000**, *39*, 4399–4405.
79. G. Hendry, T. Wydrzynski, *Biochemistry* **2003**, *42*, 6209–6217.
80. W. Hillier, T. Wydrzynski, *Phys. Chem. Chem. Phys.* **2004**, *6*, 4882–4889.
81. W. Hillier, T. Wydrzynski, *Coord. Chem. Rev.* **2008**, *252*, 306–317.
82. N. Cox, J. Messinger, *Biochim. Biophys. Acta* **2013**, *1827*, 1020–1030.
83. J. Messinger, *Phys. Chem. Chem. Phys.* **2004**, *6*, 4764–4771.
84. H. Nilsson, T. Krupnik, J. Kargul, J. Messinger, *Biochim. Biophys. Acta-Bioenerg.* **2014**, *1837*, 1257–1262.

Chapter 3
Production of Dioxygen in the Dark: Dismutases of Oxyanions

Jennifer L. DuBois and Sunil Ojha

Contents

J.L. DuBois (✉) • S. Ojha
Department of Chemistry and Biochemistry, Montana State University, Bozeman, MT 59717, USA
e-mail: jdubois@chemistry.montana.edu

P.M.H. Kroneck, M.E. Sosa Torres (eds.), *Sustaining Life on Planet Earth: Metalloenzymes Mastering Dioxygen and Other Chewy Gases*, Metal Ions in Life Sciences 15, DOI 10.1007/978-3-319-12415-5_3

Abstract O_2-generating reactions are exceedingly rare in biology and difficult to mimic synthetically. Perchlorate-respiring bacteria enzymatically detoxify chlorite (ClO_2^-), the end product of the perchlorate (ClO_4^-) respiratory pathway, by rapidly converting it to dioxygen (O_2) and chloride (Cl^-). This reaction is catalyzed by a heme-containing protein, called chlorite dismutase (Cld), which bears no structural or sequence relationships with known peroxidases or other heme proteins and is part of a large family of proteins with more than one biochemical function. The original assumptions from the 1990s that perchlorate is not a natural product and that perchlorate respiration might be confined to a taxonomically narrow group of species have been called into question, as have the roles of perchlorate respiration and Cld-mediated reactions in the global biogeochemical cycle of chlorine. In this chapter, the chemistry and biochemistry of Cld-mediated O_2 generation, as well as the biological and geochemical context of this extraordinary reaction, are described.

Keywords chlorite • dioxygen • heme • perchlorate • peroxidases

Please cite as: *Met. Ions Life Sci.* 15 (2015) 45–87

1 Introduction

The S4 state of photosystem II (PSII) – the catalytic species poised to make the O=O bond of O_2 – is one of the most sought-after, and most elusive, chemical structures in the history of biological chemistry [1, 2]. In spite of years of effort and a wide spectrum of approaches, the particulars of how PSII makes this deceptively simple bond are still controversial. At the same time, nature offers few other paradigms for reactions that couple two oxygen atoms. Currently, there is just one other known natural process: chlorite (ClO_2^-) is converted to chloride (Cl^-) and dioxygen (O_2) by an enzyme called chlorite dismutase (Cld).

These two systems make for an interesting comparison. PSII is an exceedingly complicated, membrane-associated molecular machine, comprised of around 20 protein subunits (depending on the organism) and a series of cofactors including 35 chlorophyll a molecules, 12 β-carotenes, two pheophytins, two plastoquinones, two hemes, one bicarbonate, 20 accessory lipids, and one non-heme Fe^{2+}, all in addition to the famous distorted cubane Mn_4CaO_5 cluster where O_2 evolution takes place [3]. This impressive apparatus harvests energy from light, using it to drive an exquisitely timed series of electron and proton transfers that result in the net splitting of water and emission of O_2 gas.

In sharp contrast to PSII, Cld is a relatively simple, water-soluble heme enzyme that, outside of making an O=O bond with exquisite efficiency, has little else to do [4]. As a consequence, Cld provides a window into one set of electronic and structural requirements for joining a pair of O atoms under biological conditions. Understanding the Cld reaction may therefore yield some analogies for PSII-S4 and perhaps synthetic, structurally still simpler water-splitting catalysts. At the same time, the Cld reaction extends the known repertoire of heme catalysis since, though many heme proteins and model complexes react with chlorite, the issuance of O_2 from any of them is an exceedingly rare outcome. The following chapter describes the structures and chemical mechanisms of the Clds, their diversity, and what is currently known about the geochemical and biological forces that gave rise to this type of catalysis. It concludes with a summary of some of the relevant small molecule and heme protein chemical reactivity with chlorite, asking effectively why they are *not* "Clds," and what limits O_2 generation in these cases.

2 Geochemistry of the Oxochlorates

Any thermodynamically well-matched combination of electron source and sink, no matter how exotic, can seemingly be found driving the metabolism of some microbial species. The oxyanions of main group elements such as nitrogen, sulfur, selenium, and arsenic, organic compounds of nitrogen and sulfur, and even solid mineral species (manganese, iron) can all serve as respiratory electron acceptors, provided that an electron donor with a larger reduction potential is available in the same environment. The redox tower diagram illustrates this point, listing several "edibles" (electron sources) and "breathables" (electron sinks) known to be used by various microbial species (Figure 1) [5]. Metabolism of these elements benefits the microbe, and at the same time strongly impacts global geochemical element cycles.

The ability to assume or toggle between different metabolic strategies depends on the presence of the appropriate enzymatic pathways in the microbe's genome. Some environmental microbes appear to be avid collectors of such pathways, duplicating and transferring them laterally from species to species and providing the raw materials from which new metabolic capabilities can emerge. Evolution, however, depends on the presence of environmental selection pressures to drive it. Microbial perchlorate respiration therefore was a surprise when it was first discovered in the late 1990s [6–11]. At that time, perchlorate was assumed to be almost exclusively a man-made chemical species with few known natural repositories. Wide-scale industrial production of perchlorate as a rocket propellant,

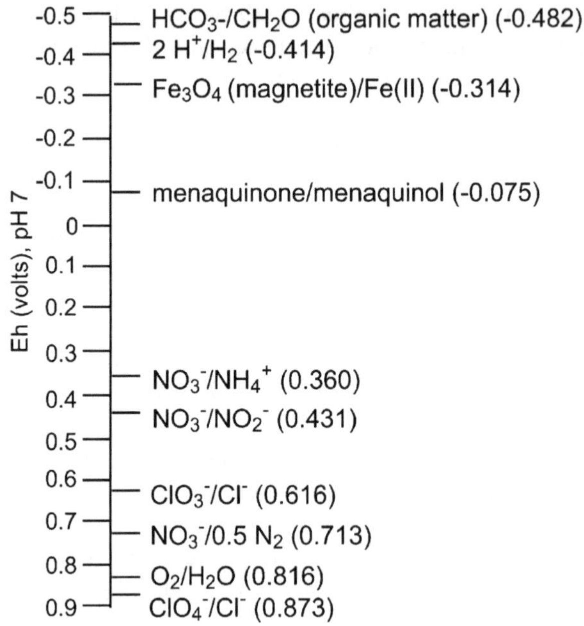

Figure 1 Redox tower, showing potential electron donors and acceptors used by microbes. The electron donor ("edible") and acceptor ("breathable") must together yield a process that is thermodynamically downhill. In terms of the tower, the edible must be higher than the chosen breathable. Tower data are adapted from reference [5].

constituent of fireworks and explosives, and chemical oxidant moreover only began in the U.S. in 1910 [12], and was increasingly identified as a freshwater pollutant at around the same time that perchlorate-respiring bacteria were discovered [13, 14]. The evolution of microbial perchlorate respiration therefore would have had to occur very quickly in response to man-made perchlorate, or in response to natural sources of perchlorate that had yet to be discovered. Which could it be?

Among other evidence, the identification of perchlorate respiration in diverse microbes from geographically far-flung locales, including places untainted by man-made oxochlorates, now suggests that perchlorate and bacteria that can eliminate it have indeed been with us for a long time. Perchlorate respiration in fact appears to play a pivotal and previously unappreciated role in the global cycling of chlorine. It is also potentially responsible for the dramatically different distribution of perchlorate on the surface of the Earth *versus* Mars, as described below.

2.1 *Oxochlorates as Respiratory Anions*

Using perchlorate as a terminal electron sink in respiratory pathways makes great chemical sense. It is highly water-soluble [14], a fact that also colors its ability to distribute in soils and travel in groundwaters [15]. Thermodynamically, it is a powerful electron acceptor. The standard reduction potential for the complete reduction of perchlorate to chloride is on par with the five-electron reduction of nitrate (a more biologically common respiratory anion), or the four-electron reduction of O_2 [14, 16, 17]:

$$ClO_4^- + 8\ H^+ + 8\ e^- \rightarrow Cl^- + 4\ H_2O \quad E^\circ = 1.287\ V \tag{1}$$

$$NO_3^- + 6\ H^+ + 5\ e^- \rightarrow ½\ N_2 + 3\ H_2O \quad E^\circ = 0.713\ V \tag{2}$$

$$O_2 + 4\ H^+ + 4\ e^- \rightarrow 2\ H_2O \quad E^\circ = 1.229\ V \tag{3}$$

The successive reductions of chlorate (Cl^{5+}) to chlorite (Cl^{3+}), chlorite to hypochlorite (Cl^{1+}), and hypochlorite to Cl^- are likewise highly thermodynamically favored [14, 18, 19]:

$$ClO_3^- + 2\ H^+ + 2\ e^- \rightarrow ClO_2^- + H_2O \quad E^\circ = 1.201\ V \tag{4}$$

$$ClO_2{}^- + 2\ H^+ + 2\ e^- \rightarrow ClO^- + H_2O \quad E^\circ = 1.645\ V \tag{5}$$

$$ClO^- + 2\ H^+ + 2\ e^- \rightarrow Cl^- + H_2O \quad E^\circ = 1.49\ V \tag{6}$$

Biologically, reduction of perchlorate to the product chlorate or chlorate to chlorite occurs with transfer of an oxygen atom, which in each case is protonated to form water. Facilitating proton transfer is consequently likely to be important for each of these reductions. Note that all of the reactants and products are written here as anions, though the standard potentials are reported (pH = 0, *versus* NHE) and the pK_as of their proton dissociation reactions ($HClO_x \rightleftharpoons ClO_x^- + H^+$) span a very broad range: −10 (ClO_4^-), −1 (ClO_3^-), 1.72 (ClO_2^-), 7.5 (ClO^-), and −8 (Cl^-) [14, 19, 20].

Finally, perchlorate is also a good bet from a kinetic perspective. Like O_2, its reactions with many potential reductants tend to be slow [14]. As a consequence, perchlorate is relatively stable under aqueous/biological conditions. Rather than simply "burning up" in a biological matrix, its thermodynamically favored reduction, like O_2's, would need to be catalyzed.

2.2 *Natural Abundance of Perchlorate on Earth*

From a geological perspective, perchlorate seems a decidedly unattractive electron acceptor because it has not accrued to nearly the same extent as O_2 or nitrate, either on land or at sea [21]. Naturally occurring perchlorate at its most abundant can be found comingled with deposits of sodium nitrate in the Atacama Desert of northern Chile (see Figure 2 in Section 2.4). These deposits have been heavily mined as sources for nitrate-based fertilizers. Chilean nitrate is in fact estimated to be the source of ~81×10^6 kg of perchlorate imported to the U.S. (1909–1997) [22–24]. The Atacama itself is a highly unusual, arid, and geologically exotic place. It is believed to have persisted as a nearly rainless climate for 10–15 million years, over which time its nitrate deposits have accumulated. The arid climate and general lack of hydrologic events or large scale biotic (including human) incursions may be responsible for the preservation of the nitrate deposits and for the co-concentration of ions like perchlorate and chromate, which are otherwise not found in large amounts anywhere on earth [25]. These deposits have been shown to have a distinct oxygen isotope pattern, making it possible to distinguish natural and synthetic perchlorate [26].

Less extreme arid and semi-arid North American environments have been surveyed for perchlorate. Many of these have also accumulated the ion in scarcely detectable, very small amounts: for example, on the order of 1–10 μg kg^{-1} dry soil, where chloride might also be present at 3 or more orders of magnitude greater concentrations [27–30]. Considerably higher (100-fold) concentrations of perchlorate have been found in the very arid soils of the Antarctic [30]. Accumulation of the anion there may have occurred for the same reasons as in the Atacama: that is, because the land has remained extremely dry and unperturbed by either human or microbial activity [31].

The detection of perchlorate in areas where it might have remained undisturbed over long time periods suggests that there are indeed natural mechanisms for perchlorate production. These mechanisms must have been at work for at least several thousand years. Perchlorate has been detected in ground waters deposited thousands to tens of thousands years ago and in glacial ice dating as far back as 2000 years [32–34]. Results from surveys of ground water, ice, and relatively unperturbed deserts have been used to estimate a $0.1–3 \times 10^9$ kg "global inventory" of natural perchlorate presently on Earth [22, 28, 33].

Anthropogenic contributions of perchlorate to the environment are relatively recent and far less subtle. The redistribution of natural, Chilean perchlorate stowed away with the nitrate fertilizers has already been mentioned. Industrial perchlorate production from 1951 to 1997 in the U.S. has been estimated as 5×10^9 kg [22]: a figure on the same order as the Earth's entire estimated inventory of natural perchlorate. Man-made perchlorate has been used in the production of rockets, missiles, fireworks, and the like, where perchlorate's oxidizing power allows it to be used as a propellant and explosive [12]. Environmental surveys in the early 2000s identified perchlorate as a widespread contaminant in drinking water, crops, and milk, particularly in areas of the southwest United States where perchlorate had been produced or used [16, 21, 35–41]. Microbial perchlorate respiration was documented at around the same time [42, 43], and its origins were then unclear. The identification of perchlorate-respiring bacteria in non-contaminated sites and of perchlorate reservoirs in places predating human activity both suggest that natural perchlorate supplied the initial selection pressure needed for the evolution of perchlorate respiration. This suggestion has gained further support as perchlorate-reducing enzymes have been identified in both taxonomically broad and temporally ancient reaches of the tree of life (see Section 3.1.)

2.3 *Atmospheric and Extraterrestrial Origins of Perchlorate*

Natural perchlorate may be accruing on the Earth's surface, but there remains no documented geological or geochemical means of producing it. It would appear as if perchlorate simply falls out of the sky. This suggestion is not so facetious, as the identification of perchlorate in pristine and geographically widespread ground waters, ice cores, rain, and snow indeed suggests atmospheric origins [34, 44]. Three atmospheric sources of perchlorate have been proposed:

First, laboratory experiments with simulated lightning have shown that perchlorate is produced, albeit less efficiently than nitrogen oxides, in oxygenic atmospheres in the presence of Cl^- [45]. Higher voltages and lower humidity, conditions found in high energy storms in dry locales, promote the greatest yields [45]. Second, reactions of aerosolized chlorine and oxygen species, energized by ultraviolet light, have long been known to occur. For example, perchlorate can be formed from reactions between ozone (O_3) and chloride. As a consequence, perchlorate has been proposed to be a stable sink for chloride in the stratosphere, where because of kinetic stability it is expected to resist decomposition by photolytic processes [46, 47]. Perchlorate's presence in the stratosphere has since been experimentally verified by single particle mass spectrometry on research flights at altitudes of up to 19 km [46]. Such atmospheric reactions would be a highly probable source of continuously deposited, natural, terrestrial perchlorate. Third and finally, in addition to these atmospheric sources, localized deposits of perchlorate have been proposed to have been brought to the Earth's surface by meteors [28].

2.4 Perchlorate on Mars

In 2008, the Phoenix Lander spacecraft, with the Wet Chemistry Laboratory (WCL) onboard, landed on the surface of Mars (Figure 2) [153]. The WCL analyzed three separate cubic centimeter soil samples for their water-soluble constituents. Quite stunningly, ClO_4^- was found to be their major anionic constituent (0.4–0.6 % by mass) [48].

Figure 2 Side-by-side images of the Martian surface (**left**) and Atacama Desert (**right**). The Martian surface is rich in perchlorate, as it lacks the action of bacteria which would otherwise eliminate it. The Atacama Desert's hyperarid interior is largely lifeless, relatively devoid of geochemical or hydrological activity, and a surrogate for Mars in NASA training missions. It is also the only place on Earth where perchlorate has accrued to a significant extent. The Martian image was taken by the Spirit rover and is courtesy of NASA/JPL-Caltech. The photograph of the Atacama Desert was freely provided by Dr. Mark Claire of the Blue Marble Space Institute of Science.

The hyperarid Martian surface environment is not unlike the Atacama Desert or the dry valleys of the Antarctic, places on earth which have served as Mars surrogates in preparation for missions and places where perchlorate has also accumulated. How Martian perchlorate is generated is not currently known. It could be produced by atmospheric processes similar to those on earth, though the source of the oxygen atoms (O_2, O_3, CO_2, H_2O) is not clear. Mineral-catalyzed photochemical processes, using chloride and water as chemical ingredients, have also been proposed and given some Earth-bound experimental support [49]. Such processes would take advantage of the intense UV light permitted by the thin ozone layer around Mars.

Even if perchlorate production on Mars occurs at a similar or even slower rate than on Earth, the complete lack of hydrological or biological forces for its removal would certainly allow for it to accrue on the Martian surface. Whether this perchlorate could or ever did serve to support Martian microbial life, whether it could serve as a chemical feedstock for O_2 production, and what it tells us about the chemistry of the Martian atmosphere are all open questions for future discussion and perhaps even experimental work.

3 Perchlorate Respiration

3.1 Diversity of Perchlorate-Respiring Microbes

The first dissimilatory (i.e., respiratory) perchlorate-reducing bacteria were isolated in 1996 [42, 43], touching off research both into natural sources of environmental perchlorate (above) and the origins of perchlorate metabolism. Since 1996, $\geq$50 strains using chlorate or perchlorate as respiratory electron acceptors [collectively, "(per)chlorate"] have been discovered (Figure 3) [50]. The overwhelming majority of these are β-Proteobacteria isolated from freshwater, mesophilic, neutral-pH environments [6, 11, 51, 52].

It has been debated whether these species encompass the real diversity of perchlorate respirers, or whether they merely reflect where the light of investigation has shone most brightly. There is some evidence that the latter phenomenon may be at work. Sampling in a marine environment, for example, yielded the first ε-Proteobacterial dissimilatory perchlorate-respiring bacterium, from the Arcobacter genus [53]. How far this metabolism extends beyond the Proteobacteria, into Gram-positive bacteria or Archaea, for example, is not clear, as the current evidence is incomplete. *Moorella perchloratireducens* and *Sporomusa* strain An4 (Gram-positives, phylum Firmicutes) were identified from an underground gas storage facility and shown to have perchlorate-reducing activity [54, 55]. Similarly, strains of *Bifidobacteria* (Gram-positives, phylum Actinobacteria) have been shown to reduce perchlorate. In each case, though reduction of perchlorate was observed, respiration of the anion has not been verified nor have the genetics of

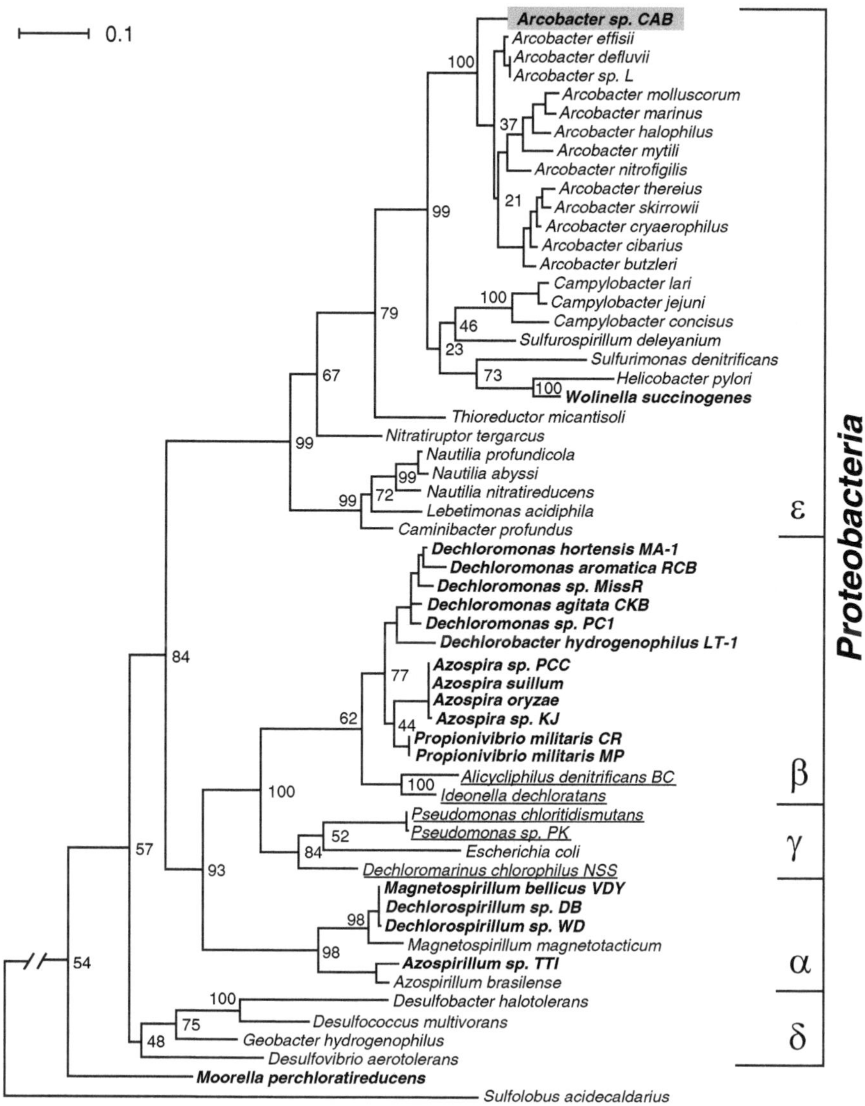

Figure 3 Diversity of characterized perchlorate- and chlorate-respiring bacteria. Maximum likelihood tree based on 16S rRNA sequences showing the phylogenetic relationships of known perchlorate (bold)- and chlorate (underlined)-reducing bacteria. Bootstrap values are based on 1,000 replications and are shown at the nodes of the tree. The scale bar represents 0.1 expected change per site [53].

reduction been probed [56]. Because the natural atmospheric events leading to perchlorate production would have existed perhaps deep in prehistory, evidence of perchlorate reduction was sought – and found – in an ancient organism. *Archaeoglobus fulgidis* is a hyperthermophilic Archaean from extreme subsurface sulfate-reducing environments that resemble conditions on early Earth. This organism appears to possess both genes encoding a perchlorate reductase enzyme (below) and the physiological capacity to use perchlorate as a respiratory anion [57].

3.2 Genetics and Genomics of Perchlorate-Respiring Microbes

Perchlorate-respiring bacteria convert perchlorate to O_2 and Cl^- via two enzymes: perchlorate reductase (Pcr) and chlorite dismutase (Cld). Pcr is encoded by a four-gene operon (*pcrABCD*) that is functionally similar and homologous to the operon that encodes respiratory nitrate reductase (*narGHJI*) in diverse bacteria. In some cases (examples include *Ideonella dechloratans* [58] and *Pseudomonas chloritidismutans* AW-1 [59]), bacteria may respire chlorate but not perchlorate. The responsible enzyme in these organisms is consequently called a chlorate reductase (Clr). It has been experimentally shown that Pcr enzymes can reduce chlorate; however, Clr appears to be unable to reduce perchlorate. These differences in substrate specificity may reflect the far greater kinetic lability of chlorate, which can in fact be reduced by several enzymes including Nar and selenite reductase (Ser), as well as abiotically by Fe(II) or Mn(II) [60, 61]. The ultimate end products of (per)chlorate respiration are water and chlorite. Though the latter anion could in principle be further reduced to provide yet more respiratory energy, it is not reduced. Instead, it is simply detoxified, via the enzyme Cld.

In spite of the similarity of the reactions they catalyze, the Pcr and Clr enzymes are encoded by distinct genes with separate evolutionary origins. The sequences of the PcrA and ClrA subunits, where the oxochlorate anions are reduced, place both enzymes into the dimethyl sulfoxide (DMSO) reductase class (Figure 4). This class can be further subdivided into types I, II, and III, corresponding to three distinct phylogenetic clades [62]. The type I enzymes include periplasmic nitrate reductase (NapA), assimilatory nitrate reductase (NasA), and arsenite oxidase (AsoA). PcrA and ClrA are both type II enzymes, along with selenate reductase (SerA) and the dissimilatory/respiratory nitrate reductase (NarG). However, within the type IIs, the PcrA subunits form their own monophyletic group that is most closely related to the NarGs. The ClrA subunits, by contrast, are more closely related to SerA [50]. Additionally, the four Clr- and Pcr-encoding genes occur in a different order relative to one another (*pcrABCD* versus *clrABDC*), where the gene ordering is conserved between Pcr/Nar and Clr/Ser, respectively.

Genomic analyses support the conclusion that perchlorate and chlorate metabolic pathways are genetically and etiologically different, as well as primed for horizontal transfer between species. The first perchlorate-reducing organism to have its genome sequenced was *Dechloromonas aromatica* strain RCB [63]. Genome sequences for *Azospira suillum* strain PS [64], the α-Proteobacterium *Magnetospirillum bellicus* strain VDYT [52], *Dechloromonas agitata* strain CKB [63],

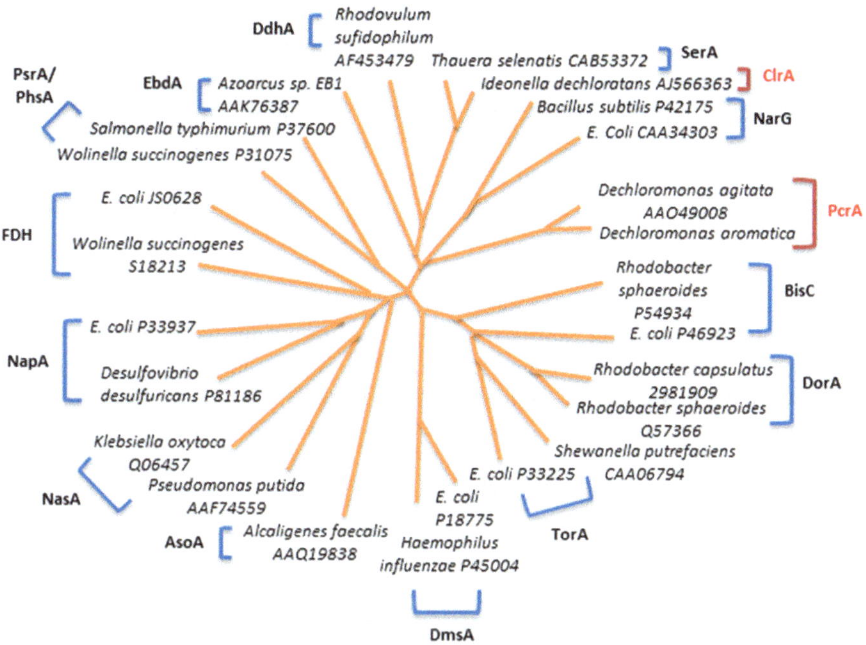

Figure 4 Unrooted neighbor-joining tree illustrating the phylogenetic relationships among type II DMSO-reductase superfamily members. GenBank accession numbers are given after the names. Based on information provided in [50].

and *Dechlorosoma suillum* PS [65] have since followed. Similarly, the genomes of five chlorate respiring bacteria (*Ideonella dechloratans*, *Pseudomonas* sp. strain PK, *Dechloromarinus chlorophilus* NSS, *Shewanella algae* ACDC [66], and *Alicycliphilus denitrificans* [67, 68]), have been sequenced [66]. Each set of genomes has been compared in order to better understand the evolutionary origins of perchlorate and chlorate metabolism, and the relationships between these two [65]. In perchlorate-respiring bacteria, the Pcr- and Cld-encoding genes are located on a genomic island. In chlorate respirers, the Clr and Cld encoding genes are on a "composite transposon," a region of the genome flanked by so-called insertion sequences. Insertion sequences are hot spots for bacterial evolution, serving as sites for homologous recombination and plasmid integration. Even among the relatively small available sample set, the transposon varies in its location (genome *versus* plasmid), copy number, and composition.

The *cld* gene in all of these organisms is, genomically speaking, something of a rogue. In perchlorate respirers, it appears to assume a number of possible orientations relative to the *pcrABCD* operon, and its transcription may be under separate regulation [69, 70]. Based on incongruence between protein and species phylogenies, *cld* genes have been predicted to have undergone horizontal transfer between species on several occasions [71, 72]. A functional chlorite-degrading chlorite dismutase protein was isolated, for example, from Candidatus *Nitrospira defluvii:* a nitrite-oxidizing bacterium without a (per)chlorate reductase and without known (per)chlorate-respiring capabilities. Genetic evidence suggests that the *cld* gene in this species was transferred from a taxonomically distant (per)chlorate respirer. The genomic analyses described above suggest that chlorite dismutase originated in the perchlorate respiratory islands, from which it has likely been uprooted and transferred (along with peripheral genetic sequences) to the composite transposons in the chlorate respirers [66]. The movement of a *cld* gene effectively completes the chlorate respiratory pathway. Such transfers appear to have happened on more than one occasion, again even in the relatively small pool of available genomic sequences.

Some chlorate respirers may have an additional *cld* gene that is not associated with the chlorate reductase transposon. These *clds* form a lineage, sometimes called "lineage II", that is closely related to but nonetheless evolutionarily distinct from the longer, respiration-associated lineage I proteins. Additionally, Proteobacteria with no known oxochlorate respiratory pathways – for either perchlorate or chlorate – may also contain a lineage II Cld. The chemical properties and biological roles of these "Clds" are not clear and are currently under investigation (below).

3.3 (Per)chlorate Reductases

Perchlorate reductases, like nitrate reductases, are molybdenum-dependent oxo-transferase enzymes. Molybdenum in the protein's α-subunit is coordinated to the thiolates of two pterin cofactors. Each pterin, in turn, is connected to a guanine dinucleotide by a pyrophosphate linkage, and the whole cofactor is known as bis(MGD)Mo [bis(molybdopterin guanine dinucleotide)-molybdenum]. The molybdenum may additionally have a variable number of hydroxide, carboxylate, and cysteine or selenocysteine ligands. In PcrA, the protein-derived ligand is likely a conserved aspartate. Generally speaking, proteins in this family catalyze the transfer of an oxo-atom to or from a substrate. The molybdenum is the holder of two electron equivalents in this transaction, cycling between Mo^{IV} and Mo^{VI}. Hence, the complete perchlorate reductase reaction cycle is:

$$\text{bis(MGD)Mo}^{IV} + ClO_4^- \rightleftharpoons \text{bis(MGD)Mo}^{VI}{=}O + ClO_3^- \qquad (7)$$

$$\text{bis(MGD)Mo}^{VI}{=}O + 2\ e^- + 2\ H^+ \rightleftharpoons \text{bis(MGD)Mo}^{IV} + H_2O \qquad (8)$$

$$\text{bis(MGD)Mo}^{IV} + ClO_3^- \rightleftharpoons \text{bis(MGD)Mo}^{VI}{=}O + ClO_2^- \qquad (9)$$

$$\text{bis(MGD)Mo}^{VI}{=}O + 2\ e^- + 2\ H^+ \rightleftharpoons \text{bis(MGD)Mo}^{IV} + H_2O \qquad (10)$$

In respiratory pathways, reduction reaction occurs at the end of an electron transport chain. Passage of electrons to the substrate is used to generate a proton gradient and, in turn, ATP. Here, perchlorate is initially reduced in a two-electron, water-producing step to chlorate (eqs 7 and 8). A similar Pcr- or Clr-mediated two-electron step reduces chlorate to chlorite (eqs 9 and 10). The electrons in each case come from the electron transport chain embedded in the bacterial inner membrane. These reducing equivalents ultimately originate from a variety of carbon sources that the (per)chlorate-respiring bacteria use as food.

The molybdopterin active site is housed in the α-subunit of Pcr, Clr, or Nar, encoded by PcrA, ClrA or NarG, respectively. Though the subunits are homologs, PcrA or ClrA are considerably smaller (60–104 kDa *versus* 104–150 kDa) and hence distinct from NarG, making it possible to predict from the sequence (even without computational analysis) that an operon could be involved in oxochlorate metabolism. PcrA- and ClrA-encoding genes moreover begin with a twin-arginine signal peptide, which target the Pcr and Clr enzymes for transport to the periplasmic space in their folded form.

The PcrA, ClrA, and NarG subunits interface with a second, iron-sulfur cluster containing a β-subunit, encoded by *pcrB*, *clrB*, or *narH*. The purpose of the β-subunit is to relay electrons from the respiratory electron transport chain to the α-subunit. Clr and Nar have a further, heme-containing γ-subunit (*clrC, narC*) that anchors the whole complex to the inner membrane and serves as an electron conduit (Table 1). Pcr, by contrast, appears to be a freely diffusing periplasmic enzyme. Electron equivalents originating from the membrane are predicted to reach the FeS clusters through a soluble, cytochrome *b*-containing γ-subunit (*pcrC*) and electron transporting mediators [73].

Table 1 Characterized chlorate and perchlorate reductases.

Organism	Known electron acceptors	Pcr or Clr subunits	Cofactors	Ref.
Azospira oryzae GR-1	ClO_4^-, ClO_3^-, NO_3^-, IO_3^-, BrO_3^-, Mn(IV), O_2	$\alpha_3\beta_3$	Mo, FeS	[43, 151]
Strain Perc1ace	ClO_4^-, NO_3^-	$\alpha_3\beta_3$		[146]
Azospira sp. KJ	ClO_4^-, ClO_3^-, O_2	αβ		[147]
Ideonella dechloratans	ClO_4^-, NO_3^-, IO_3^-, BrO_3^-, O_2	αβγ		[153, 154]
Pseudomonas sp. PDA	ClO_3^-, O_2	αβγ	Mo, FeS, heme *b*	[147]
Pseudomonas chloritidismutans	ClO_3^-, O_2	αβγ	Mo, FeS	[148]

Adapted from [70].

Finally, the δ-subunit of each of these protein complexes (encoded by *pcrD*/*clrD*) is a chaperone involved in maturation of the enzyme and possibly in insertion of the bis(MGD)Mo cofactor.

3.4 Chlorite Dismutases and Perchlorate Respiration

All bacteria with a functional Pcr or Clr have a *cld* gene that encodes a *bona fide*, functioning chlorite dismutase: that is, a protein that detoxifies chlorite by converting it to Cl^- and O_2. In the absence of this reaction or some other means of eliminating chlorite, this end product would accrue to a toxic level. Interestingly, the only known perchlorate-respiring Archaean, *A. fulgidis*, is able to overcome chlorite toxicity even though it lacks a *cld* gene. How *A. fulgidis* degrades chlorite is not known, but abiotic means have been proposed.

The gene encoding the respiration-associated Cld, as described above, is near to the *pcrABCD* or *clrABDC* operon but not necessarily in a static location [66]. The properties of these O_2-generating proteins are described in detail in Section 4. Notably, many species with no known (per)chlorate metabolism possess a *cld* gene. Current evidence suggests that these non-respiratory Clds are not efficient detoxifiers of chlorite and that they serve other cellular roles, as discussed below.

4 Oxygen Generation by Chlorite Dismutases

4.1 Structures

4.1.1 Primary Structures: Diversity and Hallmarks of O_2 Generation

For an organism to respire chlorate/perchlorate, some means of chlorite degradation is necessary because chlorite is toxic. However, the presence of a *cld* gene does not correlate with either the possession of (per)chlorate reductase encoding genes, or with the ability to reduce oxochlorates. The "lineage I" and "lineage II" Clds were described above as two closely related but distinct groups of *cld* genes found thus far primarily in Proteobacteria. When a rainbow of microbial genome sequences began to emerge in earnest in the early 2000s, *cld* genes could be found in a great number of them. In fact, *cld*-family genes have been identified in highly diverse microbes from at least 9 bacterial and 2 archaeal phyla (Figure 5) [71]. They cluster into at least three additional monophyletic sequence groups beyond lineages I and II.

The respiration-associated Clds come from lineage I, and these proteins all rapidly emit O_2 when presented with chlorite (Figure 6, Table 2). The well-studied Cld from *Dechloromonas aromatica* (*Da*Cld) survives an impressive 20,000 turnovers with this often-punishing oxidant before its heme irreversibly lyses. By contrast, Clds from diverse bacteria beyond the Proteobacterial phylum – for example, *Staphylococcus aureus* (Gram-positive pathogen, phylum Firmicutes) and *Thermus thermophilus* HB8 (Gram-negative hyperthermophile, phylum Deinococcus-Thermus) have no and very little chlorite-degrading capacity, respectively, and their parent organisms do not respire oxochlorates [74, 75]. So, do all Clds from non-respirers fail to produce O_2? Not exactly. Clds from two nitrite-oxidizing, non-(per)chlorate-respiring bacteria were shown to convert chlorite to dioxygen [71, 76], and they illustrate two known types of exceptions. The first of these, the Cld from Candidatus *Nitrospira defluvii,* is clearly a lineage I Cld that appears to have arrived in the *N. defluvii* genome via horizontal transfer from a perchlorate respiring organism. The environmental selection pressures promoting its transfer and retention in its new host are not yet understood. The second of these is the Cld from *Nitrobacter winogradskyi*, which is representative of the lineage II

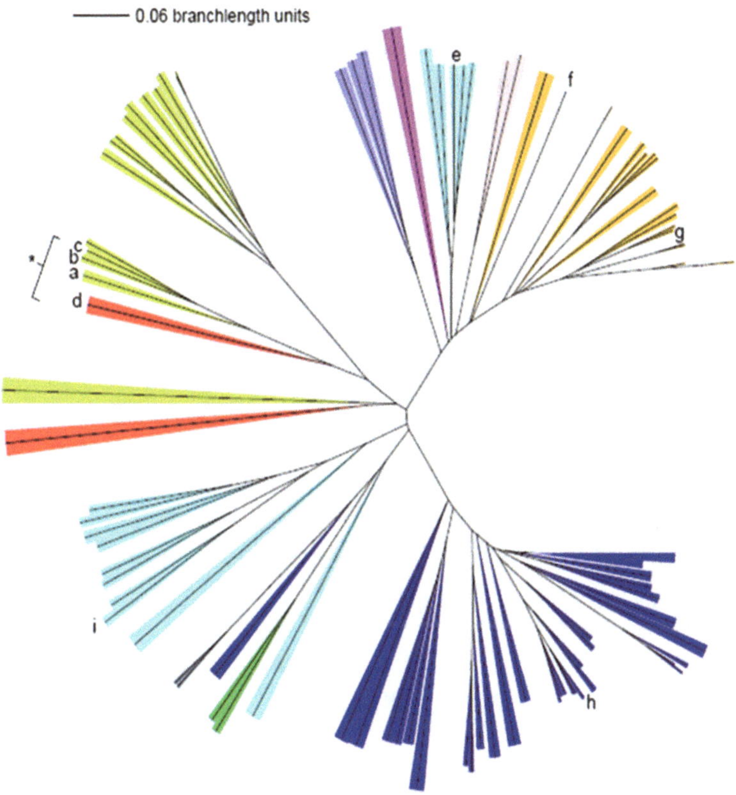

Figure 5 Phylogenetic trees illustrating the diversity of Cld sequences. The phylum/kingdom affiliation of each species is indicated by color: Proteobacteria (yellow), Firmicutes (orange), Nitrospirae (red), Actinobacteria (blue), Archaea (light blue), Deinococcus-Thermus (grey), Chloroflexi (green), Planctomycetes (dark purple), Veruccomicrobia (light purple), and Acidobacteria (pink). The Halobacteriacea, pictured near the bottom of the tree, form their own group distinct from the other archea. Species known to carry out chlorite detoxification are indicated with a bracket/asterisk. Clds from several species are designated with letters: (a) *Dechloromonas aromatica*; (b) *Dechloromonas agitata*; (c) *Ideonella dechloratans*; (d) *Nitrospira defluvii*; (e) *Halobacterium* sp. NRC-1; (f) *Thermus thermophilus* HB8; (g) *Geobacillus stearothermophilus*; (h) *Mycobacterium tuberculosis*; (i) *Thermoplasma acidophilum*. A three-iteration protein similarity (PSI-BLAST) search was performed using *Da*Cld as the bait sequence. The top 500 result sequences were aligned by ClustalX, and a phylogenetic tree was constructed. Representative sequences from each phylum were chosen for the above display. Settings used for the tree building were: random number = 111 and bootstrap maximum = 1000. The iTOL (Interactive Tree of Life) program was used for branch coloring and figure generation (http://itol.embl.de/). Reproduced with permission from [83]; copyright 2011 Elsevier.

Clds that are found predominantly or perhaps exclusively in Proteobacteria. Though the lineage II *clds* occur in genomes independently of (per)chlorate reductase genes, the *N. winogradskyi* Cld and potentially other members of this group (including the Cld from the pathogen *Klebsiella pneumoniae*) can produce O_2 from chlorite. The number of turnovers sustained before the heme is irreversibly destroyed, however, is sharply limited. Hence, while chlorite catalysis is observable, the catalyst is not robust, suggesting that chlorite decomposition may not be the biological function of the lineage II proteins.

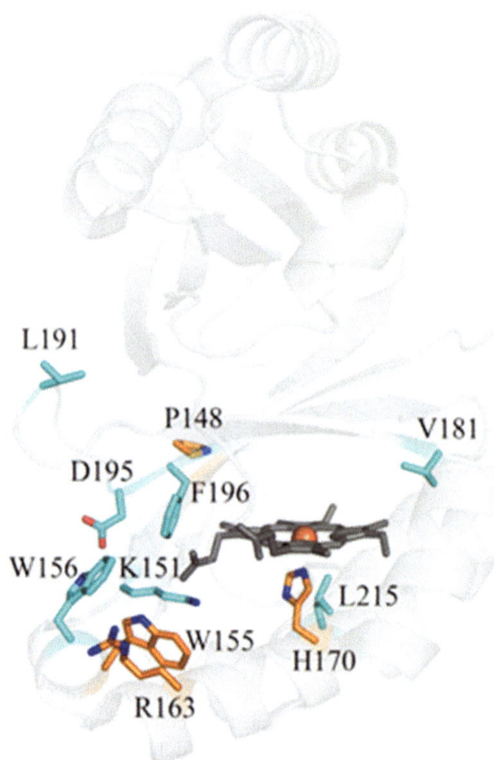

Figure 6 *Da*Cld monomer structure showing the locations of residues conserved among Clds from taxonomically diverse sources. The *Da*Cld monomer is shown as a faded gray cartoon. Strictly (carbon orange) and strongly (carbon cyan) conserved residues are shown as sticks and colored by atom (carbon orange and cyan, respectively). Heme is drawn in gray stick, and the iron is drawn as an orange sphere. All conserved residues fall within the C-terminal domain. This figure was generated using PyMOL (http://www.pymol.org/). Figure reproduced with permission from [83]; copyright 2011 Elsevier.

Table 2 Kinetic constants for the steady state reaction of chlorite dismutases.

Organism	k_{cat} (min^{-1})	K_M (μM)	k_{cat}/K_M ($M^{-1}\,s^{-1}$)	Temp. (°C)	pH[a]	Ref.
Dechloromonas aromatica	4.5×10^5	210	3.5×10^7	4	6.8	[98]
Ideonella dechloratans[b]	1.6×10^5	150	1.8×10^7	25	7.0	[149]
Ideonella dechloratans[c]	1.1×10^5	260	7.1×10^6	25	7.0	[150]
Strain GR-1	7.2×10^5	170	7.1×10^7	30	7.2	[145, 151]
Pseudomonas chloritidismutans AW-1	1.4×10^4	80	2.7×10^6	25	6.0	[152]
Thermus thermophilus HB8	4.6×10^1	12,000	5.9×10^1	25	7.0	[74]
Candidatus Nitrospira defluvii	2.1×10^3	58	6.0×10^5	30	7.0	[82]
Nitrobacter winogradskyi	1.1×10^4	90	2.1×10^6	30	7.0	[76]

[a]All reported measurements made in varying concentrations (10–100 mM) of phosphate buffers.
[b]Heterologously expressed in *E. coli*.
[c]Purified from native organism.

This conclusion is consistent with the lack of a known biological imperative for chlorite detoxification by their parent organisms, although it is possible that these Clds catalyze detoxification of chlorite produced by Nar-mediated chlorate (ClO_3^-) reduction. In short, the available evidence suggests that only the respiration-associated Clds – encoded by genes on perchlorate reductase islands or composite transposons near *pcrABCD*/*clrABDC* operons, or recently mobilized from such – efficiently convert chlorite to O_2 and Cl^-. Cld homologs from outside this relatively small group are biochemically distinct and appear to serve other biological roles.

Sequence alignments including lineage I, II, and the more taxonomically broad Clds show that the family as a whole is quite sequence-diverse (Figure 7, Table 3). Only three amino acids are strictly conserved family-wide. These include the proximal histidine ligand to the heme iron (H171 in *Da*Cld, with residue numbering reflecting the structurally characterized mature form of the protein lacking the SecB tag earmarking the protein for secretion to the periplasm) [4]. They also include a tryptophan near the porphyrin binding site (Trp155) and a proline residue important for maintaining the protein's fold (Pro148). This sequence diversity, the taxonomic breadth of organisms containing the gene, and the presence of several monophyletic groups of Clds suggests an ancient origin for *cld* in which divergence of the sequence occurs at pace with speciation events. The exception to this pattern comes, of course, for the heavily traveled *cld* genes associated with oxochlorate respiration [65, 66, 71, 77].

What sequence features distinguish the O_2-generating Clds from their non-O_2-generating counterparts? There are at least three (Figure 7, Table 3). First, in oxochlorate-respiring bacteria, the Cld-encoding sequence is preceded by a SecB peptide, signaling transport of the protein to the periplasm in its unfolded, heme-free form. Clds from the non-respirers, by contrast, are cytoplasmic. Second, both the longer lineage I Clds from oxochlorate respirers and the shorter lineage II Clds share a conserved arginine residue in the pocket above the heme plane. Proteins from each of these groups have been experimentally confirmed to produce O_2 rapidly, even if the lineage I proteins are considerably more robust (i.e., higher numbers of turnovers). The distal Arg residue is not conserved in the various non-O_2-producing branches of the phylogeny depicted in Figure 5, changing for example to serine (Euryarchaeota), glutamine (Firmicutes), or asparagine (Actinobacteria). Surprisingly, the arginine can be mutagenically substituted with an asparagine in *Da*Cld without completely abrogating activity with chlorite [78]. However, the Cld from *Staphylococcus aureus* (phylum Firmicutes), which has a distal glutamine residue in the analogous position, has no measurable chlorite-degrading activity. Finally, a second tryptophan near the heme's periphery (Trp156) and two residues from a triad of hydrophobic amino acids above the heme plane (Leu185 and Phe200) are strictly conserved among all Clds. The tryptophan appears to be critical for maintaining the bound heme as well as the protein's oligomeric state [79], and hydrophobicity in the distal pocket may be essential for stabilizing reaction intermediates (see Section 4.2.)

```
Brad   M--------- ---------- ---------- ---------- ---------- ---------F 2
Nvin   M--------- ---------- ---------- ---------- ---------- --------TF 3
Limn   M--------- ---------- ---------- ---------- ---------- -------NNH 4
Paer   M--------- ---------- ---------- ---------- ---------- -------NNH 4
Nmob   M--------- ---------- ---------- ---------- ---------- ------NARL 5
Kpne   M--------- ---------- ---------- ---------- ---------- ------NTRL 5
Ndef   M--------- ---NFRSAGR IAVLAGLLVL VAVWPAP--- -A----ADRE KLL-TESGVY 39
*Aory  MTNISIRSFK LSLIATTIGA AIAMVSSPVV AQQAMQPMQP MK----IERG TIL-TQPGVF 55
*Pchl  MTNMSIRSFK LSLAATAIGA AIAMASSPVV AQQAMQPMQP MK----IERG TIL-TQPGVF 55
*Dhor  MTNMSIRSFK LSLAATAIGA AIAMASSPVV AQQAMQPMQP MK----IERG TIL-TQPGVF 55
*Daro  MTNLSIHNFK LSLVAAVIGS AMVMTSSPVA AQQAMQPMQS MK----IERG TIL-TQPGVF(21)
*Dagi  MRKSTGLLLT FMALLSV--- -----GSSQA QQANMDAKPP MA---MPDMT KIL-TAPGVF 48
*Idec  M-KVRCVSLV AAGLLTIAGS AIGQPAPAPM PAMAPAAKPA MN--TPVDRA KIL-SAPGVF 56
*Mmag  MLKNRARTAI AAAITAAAIF AVSGAAQAQQ APMAGDMKPA MAAPMMADRA KLL-TSPGVF 59

Brad   RTFRGG-QSG GW-RVISIS- ---------- ---------- ---------- -PVTGEPLPF 28
Nvin   TVFTGG-DSG AW-SILSVA- ---------- ---------- ---------- -PVIGESLMA 29
Limn   YSFIGG-QEG QW-RVTRCD- ---------- ---------- ---------- -TVVGAPIEA 30
Paer   YSFIGG-SEG SW-RVTSCE- ---------- ---------- ---------- -TLIGIPLEI 30
Nmob   FAFVGG-EIG SW-RVIETK- ---------- ---------- ---------- -TVAGEGLAE 31
Kpne   FTFAGG-ETG VW-RVVRMD- ---------- ---------- ---------- -AVAGAPLPG 31
Ndef   GTFATFQMDH DWWDLPGESR VISVAEVKGL VEQWSGK--- ILVESYLLRG LSDHADLMFR 96
Aory   GVFTMFKLRP DWNKVPAMER KGAAEEVKKL IEKHKDN--- VLVDLYLTRG LETNSDFFFR 112
Pchl   GVFTMFKLRP DWSKVPAMER KSAAEEVKKL IEKHKDN--- VLVDLYLTRG LETNSDFFFR 112
Dhor   GVFTMFKLRP DWSKVPAMER KSAAEEVKKL IEKHKDN--- VLVDLYLTRG LETNSDFFFR 112
Daro   GVFTMFKLRP DWNKVPVAER KGAAEEVKKL IEKHKDN--- VLVDLYLTRG LETNSDFFFR (78)
Dagi   GNFSTYKVRP DYYKLSMAER KGAAAEVVAV VEKYKDK--- VKAEAYLTRG FEAQSDFFLR 105
Idec   VAFSTYKIRP DYFKVALAER KGAADEVMAV LEKHKEK--- VIVDAYLTRG YEAKSDYFLR 113
Mmag   GNFSTYKLRS DYYKLSAAER KGAAAEVMAV VEKHKAN--- IIADAYLTRG FEAQSDYFLR 116

Brad   MPALSVTDSE AISLPLVPSR -------NAW RLVGAPSSLR YTERAEKQQ- LTAV--QAGL 78
Nwin   ASHLAIAPSL SLGDTSAT-- -------TPW QLRGVASHAR YVERAEKIA- LTSV--QAGL 77
Limn   VPRLNVVNTA ASQLSQR--- -------GTW MLQGFTSNVR YAERHEINQ- LRAK--QEGL 77
Paro   VERVNVVNMP STNLIER--- -------GTW VLQGFTSNVR YAERHEINQ- LRAK--QEEL 77
Nmob   VKRLNVVNAA VPLLPDD--- -------AQW LLRGVTSNER YVTRSERAQ- LTAK--QPVL 78
Kpne   IPRLDVAAGS VSPQPLG--- -------TKW LLRGITSNER YVVREEKDR- LVAK--QPSL 78
Ndef   VHARTLSDTQ QFLSAFMGTR -LGRHLTSGG LLHGVSKKPT YV--AGFPES MKTEL-QVNG 152
Aory   INAYDLAKAQ TFMREFRSTT -VGKNADVFE TLVGVTKPLN YISKDKSPG- LNAGLSSATY 170
Pchl   IHAYDLAKAQ TFMREFRSTT -VGKNADVFE TLVGVTKPLN YISKDKSPE- LNAGLTSATY 170
Dhor   INAYDLAKAQ TFMREFRSTT -VGKNADVFE TLVGVTKPLN YISKDKSPG- LNAGLSSATY 170
Daro   INAYDLAKAQ TFMREFRSTT -VGKNADVFE TLVGVTKPLN YISKDKSPG- LNAGLSSATY(136)
Dagi   IHSYDMAATQ AFLVDFRATR -FGMNAEVTE NLVGMTKDLN YITKDKSPN- LNAGLTGATY 163
Idec   VHAYDAVAAQ AFLVDFRATR -FGMYSDVTE SLVGITKALN YISKDKSPD- LNKGLSGATY 171
Mmag   VHSMDMAATQ AFLVDFRATR -FGMYSDVTE NLVGITKALN YISKDKSPD- LNSGLSSASY 174

Brad   GRLEATSAAL IPIRKSQAWW ELTQEERRRI FEDKSHHIAS SLRFLPAIAR QLY--HSRDL 136
Nvin   GRNEATRAAL IPIRKSAAWW EMTQDERRAI FEDKSHHIAA SLKYLPAIAR QLY--HCRDI 135
Limn   SRPASTCAAL IPIKKNAQWW ALSQDERRAI FEAQSHHTEI GLAYLPEIAR QLH--HSRDL 135
Paer   NRPTSSCAAL IPIKKSPEWW AMSQEERREI FEAKSHHTEI GLAYLPEIAR QLH--HSRDL 135
Nmob   GRRQATCAAF IPIRKTASWW NLAQDERRMI LEESSNHIKT GLKYLPAVAR RLH--HCRDL 136
Kpne   GRAEATCAAL IPIRKNPSWW GLAQDERRKI FEEQSRHIHI GLQYLPAVAR RLH--HCRDL 136
Ndef   ESGSRPYAIV IPIKKDAEWW ALDQEARTAL MQE---HTQA ALPYLKTVKR KLY--HSTGL 207
Aory   SGPAPRYVIV IPVKKNAEWW NMSPEDRLKE MEV---HTAP TLAYLVNVKR KLY--HSTGL 225
Pchl   SGPAPRYVIV IPVKKNAEWW NMSPEERLKE MEV---HTTP TLAYLVNVKR KLY--HSTGL 225
Dhor   SGPAPRYVIV IPVKKNAEWW NMSPEERLKE MEV---HTTP TLAYLVNVKR KLY--HSTGL 225
Daro   SGPAPRYVIV IPVKKNAEWW NMSPEERLKE MEV---HTTP TLAYLVNVKR KLY—HSTGL(191)
Dagi   RDATPRYAFV IPVKKNADWW NLTDEQRLKE MET---HTLP TLANLVNVKR KLY--HSTGL 218
Idec   AGDAPRFAFM IPVKKNADWW NLTDEQRLKE MET---HTLP TLPFLVNVKR KLY--HSTGL 226
Mmag   TGDAPRYAIM IPVKKDAAWW NRSDEERLKE MEV---HTQP TLQYLVNVKR KLY—HSTGL 229
```

Figure 7 Structure-based sequence alignment of O_2-emitting Cld sequences, from perchlorate respirers and short Proteobacterial Clds with confirmed chlorite decomposition activity. Experimentally confirmed respiratory Clds are indicated with an asterisk by the species name. The proximal histidine-glutamic acid pair is highlighted in green, the conserved distal arginine in yellow, and the distal hydrophobic triad in purple, and the conserved pair of tryptophans and a proline residue are in blue.

```
Brad   G---EPFDFL TWFEFAPAHA SLFEELVGML RRTEEWTYVE R-EV------ ---DVRAVKE 183
Nwin   G---EPFDFL TWFEYAPEHA TMFEDLVGVL RATEEWTYVE R-EV------ ---DIRLARA 182
Limn   G---EPFDFL TWFEFAPEHT DAFDELLVKL RTSEEWKYVE R-EV------ ---DIRLVKD 182
Paer   G---EPFDFL TWFEFAPEHT AIFNKLLAQL RSSKEWEYVE R-EI------ ---DIRLVKN 182
Nmob   GGDAEPFDFL TWFEYAPSDS AAFDELVAEL RASQEWTYVD R-EI------ ---DMRLARD 186
Kpne   G-ESEPFDFL TWFEYSPSDE PGFNRLLAEL RASVEWKYVD R-EI------ ---DIRLVHE 185
Ndef   D---DV-DFI TYFE--TERL EDFHNLVRAL QQVKEFRHNR RFGHPTLLGT MSPLDEILEK 261
Aory   D---DT-DFI TYFE--TDDL TAFNNLMLSL AQVKENKFHV RWGSPTTLGT IHSPEDVIKA 279
Pchl   D---DT-DFI TYFE--TDDL TAFNNVVLSL AQVKENKFHV RWGSPTTLGT IHSPEDVIKA 279
Dhor   D---DT-DFI TYFE--TDDL TAFNNLMLSL AQVKENKFHV RWGSPTTLGT IHSPEDVMKA 279
Daro   D---DT-DFI TYFE--TDDL TAFNNLMLSL AQVKENKFHV RWGSPTTLGT IHSPEDVIKA(245)
Dagi   D---DT-DFI TYFE--TADL GAFNNLMLAL AKVPENKYHV RWGSPTVLGT IQSFDSVVNT 272
Idec   D---DT-DFI TYFE--TNDL GAFNNLMLSL AKVPENKYHV RWGNPTVLGT IQPIENLVKT 280
Mmag   D---DT-DFI TYFE--TNDL GAFNNLLISL AKVPENKHHV RWGSPTVLGT IQTFETVVKT 283
```

Fig. 7 (continued) The SecB-dependent secretion peptide sequence on the Cld sequence from *D. aromatica* (*Daro*) is shown in red. Sequence numbering with this peptide removed is shown in parentheses. The species names are abbreviated as: *Bradyrhizobium* sp. S23321 (*Brad*), *Nitrobacter winogradskyi* (*Nwin*), *Limnobacter* sp. MED105 (*Limn*), *Pseudomonas aeruginosa* (*Paer*), *Nitrococcus mobilis* (*Nmob*), *Klebsiella pneumoniae* (*Kpne*), Candidatus *Nitrospira defluvii* (*Ndef*), *Azospira oryzae* (*Aory*), *Pseudomonas chloritidismutans* (*Pchl*), *Dechloromonas hortensis* (*Dhor*), *Dechloromonas aromatica* (*Daro*), *Dechloromonas agitata* (*Dagi*), *Ideonella dechloratans* (*Idec*), *Magnetospirillum magnetotactica* (*Mmag*).

Table 3 Conservation of residues of likely functional relevance among the Clds.

Amino acid[a]	Proposed role	Oxochlorate-respiring bacteria[b]	Firmicutes	Euryarchaeota	Actinobacteria
H170	Proximal heme ligand	conserved	conserved	conserved	conserved
Elu220	H-bonding to axial His	conserved	conserved	K, R, M, L, E	variable
R183	distal polar residue	conserved	Q	S	A, Q[c]
L185	distal hydrophobic triad	conserved	I	T	L, I, or T
T198	distal hydrophobic triad	conserved	V, T	V	T, V, or L
F200	distal hydrophobic triad	conserved	L, F	Y	variable
W155	radical pathway	conserved	conserved	conserved	conserved
W156	radical pathway	conserved	Y	Y	W, T
H224	radical pathway	mostly conserved	R	W	variable
W227	radical pathway	mostly conserved	E	variable	variable
Dsp192	Ca^{2+}ligand	conserved	G	variable	D, G
T231	Ca^{2+}ligand	conserved	F	I, L, V	variable

[a]The residue numbering shown here refers to the structurally characterized, mature *Da*Cld with its SecB signal peptide removed.
[b]Includes sequences from all experimentally verified oxochlorate respirers.
[c]The residue at this position in alignments is A, but is adjacent to a conserved Q.

4.1.2 Heme-Binding Domain

The heme-binding domain of Clds consists of a series of α-helices lining the proximal side of the heme plane and a row of β-sheets over the distal, open coordination position. The iron is ligated by a histidine residue that is strictly conserved throughout the Cld family (Figure 6). In the full length Clds – that is, all Clds except those in lineage II – the protein monomer contains two mutually homologous domains, each about 140 amino acids in length. These resemble one another in secondary structure, with the exception of an added α-helix in the N-terminal domain, and are linked by a pseudo twofold axis of symmetry. The extra helix, the lack of a strictly conserved histidine, and a slightly more compact structure may be responsible for the failure of heme to bind in the N-terminal domain [4, 76, 80–82].

Heme is typically found in predominantly α-helical environments, such as in the globins, cytochrome P450, catalases, and peroxidases. Hence, the Cld heme-binding domain at first seemed somewhat unusual. However, it is now clear that the mixed α-helical/β-sheet domain of Clds is shared by a very large structural superfamily of proteins, many of which derive from microbes and only a small fraction of which associate with heme [75, 83]. Subfamilies range from quite close to Clds in sequence (and therefore ancestry) to relatively distant. At least two subfamilies in addition to the Clds are known to interact with heme. They are catalytically dissimilar to Clds, in spite of their identical heme-binding architecture and mutually conserved residues in the active site. The first subfamily contains the bacterial IsdG-type heme oxygenases that oxidatively cleave heme as a substrate, releasing the iron for cellular use [84]. Unlike Clds, the IsdG-family proteins have just one domain per monomer. Two heme-binding monomers come together to form a dimer that resembles the *Da*Cld bidomain monomer [85]. The heme in IsdI (from *Staphylococcus aureus*) also assumes a different orientation than in the Clds and is significantly distorted from planarity. Structural distortion of the tetrapyrrole is known to be important for catalysis [86].

The dye-decoloring peroxidases (DyPs) from bacteria and fungi form a second, diverse subfamily. These proteins are evolutionarily quite distant from the Clds, sharing as low as 6 % identity [83]. However, they are structurally very similar, having the same bidomain monomer in which only the C-terminal domain binds heme [87]. In the DyPs, however, the heme is flipped relative to its orientation in the Clds, placing its propionic acid side chains in a very different chemical environment. This again illustrates the remarkable versatility with which this protein domain is known to interact with heme [83]. One of the propionates is able to form a hydrogen bond to the conserved distal arginine. These features and the presence of a distal aspartic acid distinguish the DyPs from the Clds and may be important for their distinct chemistry [87, 88]. DyPs efficiently activate H_2O_2 to form high-valent iron-oxo intermediates which catalyze the one-electron oxidation of substrates. Clds do not, likely because they lack a base for deprotonating H_2O_2 and thereby activating it for binding to Fe(III) [89].

4.1.3 Active Site

Like heme proteins that mediate O_2 transport [90], gas sensing [91–95], and peroxidase catalysis, Clds have a proximal histidine ligand to the heme (Figure 8, left). A glutamic or aspartic acid forms a hydrogen bond to the proximal histidine's other, unligated nitrogen. Such hydrogen bonding is expected to increase the anionic (histidinate) character of the ligand in peroxidases, making the Fe(III) iron more electron-rich and therefore capable of donating significant electron density toward scission of a peroxide ligand in *trans* [96, 97]. However, resonance Raman shows that the histidine ligand in Clds, like in the gas-binding proteins, is neutral in character [98]. Also on the proximal side of the heme plane are two tryptophan residues, one strictly conserved throughout the entire Cld family (W155, *Da*Cld numbering) and the other conserved among O_2-producing lineage I Clds and related Clds from lineage II (W156). A third tryptophan (Trp227) is not strictly conserved, but is connected to the proximal histidine through a hydrogen bonding network. Finally, two tyrosine residues (Tyr118, Tyr177) are conserved among the lineage I and II Clds. Tyr118 is connected to a propionate side chain of the *Da*Cld heme by a hydrogen bond. From a structural perspective, either Tyr118 or Trp227 have a direct conduit to the heme and could therefore in principle stabilize radical intermediates, as described below.

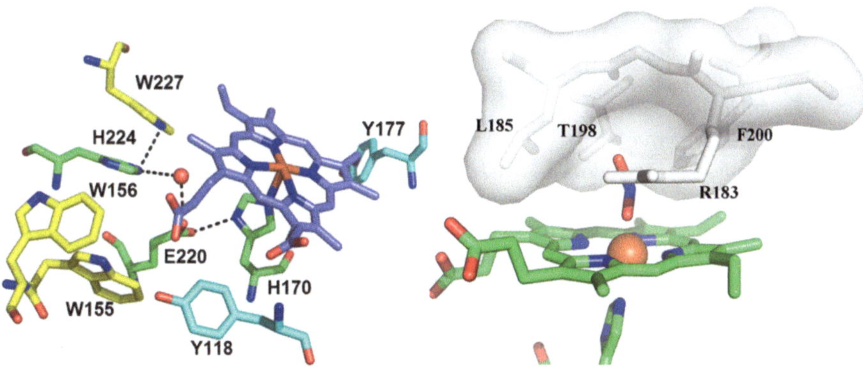

Figure 8 Structure of the proximal and distal sides of the heme in *Da*Cld. (**Left**) Proximal pocket with hydrogen bonding interactions indicated by dashed lines. A water molecule is clearly discernable even at low resolution and is represented as a red sphere. (**Right**) Residues lining the distal pocket are shown as white sticks with a transparent surface, except for Arg-183. Figures were originally generated using PyMOL (http://www.pymol.org/) from PDB 3Q08 using PyMOL and are adapted from [78] and [4].

The distal side of the heme plane (Figure 8, right) has several unique and likely functionally important features. Notably, there are no distal basic residues (histidine, aspartate, glutamate) in any of the Clds. A distal base is essential for rapid H_2O_2 activation in the well-studied peroxidases [99–101]. As mentioned above, there is instead an arginine directly above the heme plane of all lineage I and II Clds, changing to a serine or glutamine in most other Clds. This residue has no amino acid hydrogen-bonding partner in *Da*Cld or in any available crystal structures of O_2-evolving Clds, and is instead found hydrogen-bonded to the Fe(III) heme ligand (water/hydroxide, nitrite, thiocyanate). Adjacent to the distal arginine are three hydrophobic residues: Leu185, Thr198, Phe200. Although the exact identity of the residues can vary, a hydrophobic triad is strictly conserved at these positions. They form a low, hydrophobic ceiling above the porphyrin plane in what is overall a fairly hydrophobic pocket.

While all chlorite-degrading Clds that have been crystallographically characterized also have a heme bound, a few from non-chlorite-degrading species have been characterized in their heme-free, apo forms (Table 4). These were prepared mainly by structural genomics consortia, and hence the protein expression/purification conditions may not have supplied sufficient heme. However, it has also been observed that the Cld from *Staphylococcus aureus*, which does not convert chlorite to O_2, binds heme with very low affinity and is actually somewhat difficult to prepare in a heme-bound state [75]. Intrinsically low affinity may be responsible for the failure to capture the Cld structures in the heme-bound form. By the same token, it is notable that these proteins are stable enough to form crystals in the apo/heme-free form, though solution measurements have indicated that the apo-Clds from *N. winogradskyi* and Candidatus *N. defluvii* are more prone to denaturation than their holo-counterparts [102, 103].

Table 4 Some available X-ray crystal structures for chlorite dismutases.

PDB ID	Source	Description	Ref.
3Q08	*Dechloromonas aromatica*	Nitrite-bound Cld from a PRB[a]	[4]
2VXH	*Azospira oryzae*	Thiocyanate-bound Cld from a PRB	[8]
3NN2	Candidatus *Nitrospira defluvii*	Cyanide-bound chlorite-reactive Cld from an evolutionarily distant NPRB[a]	[82]
3QPI	*Nitrobacter winogradskyi*	Water-bound short dimeric Cld from a proteobacterial NPRB	[76]
1T0T	*Geobacillus stearothermophilus*	Heme-free Cld from NPRB (Firmicutes)	[b]
1VDH	*Thermus thermophilus HB8*	Heme-free Cld from NPRB (Firmicutes)	[74]
3DTZ	*Thermo acidophilum*	Heme-free Cld from NPR-Archaeon (Euryarchaeota)	[b]

[a]PRB = perchlorate respiring bacterium, NPRB = non-perchlorate-respiring bacterium; [b]Unpublished structure deposited in the Protein Data Bank.

4.1.4 Tertiary Structures and Oligomerization States

We conclude the discussion of Cld structures by noting the diversity of oligomerization states in which the Cld domain can be found in X-ray crystal structures. All full-length Clds (whether or not they catalyze the chlorite decomposition reaction) form a biochemically unusual homopentameric structure in which the heme-binding domains associate with each other along one face of the pentamer (Figure 9a). Uniquely in *Da*Cld, each monomer-monomer interface has a Ca^{2+} ion ligated by a strictly conserved aspartic acid (Asp192). The pentameric shape of the complex leaves an open, solvent-filled channel in the middle of the protein, the function of which (if any) is not clear. It is also unknown whether the pentameric form is physiologically relevant, although it represents the stable solution state of the pure protein. Full-length Clds in the pentameric state with heme bound and a conserved arginine in the distal pocket are moreover the most efficient catalysts in the conversion of chlorite to O_2 and Cl^- [103].

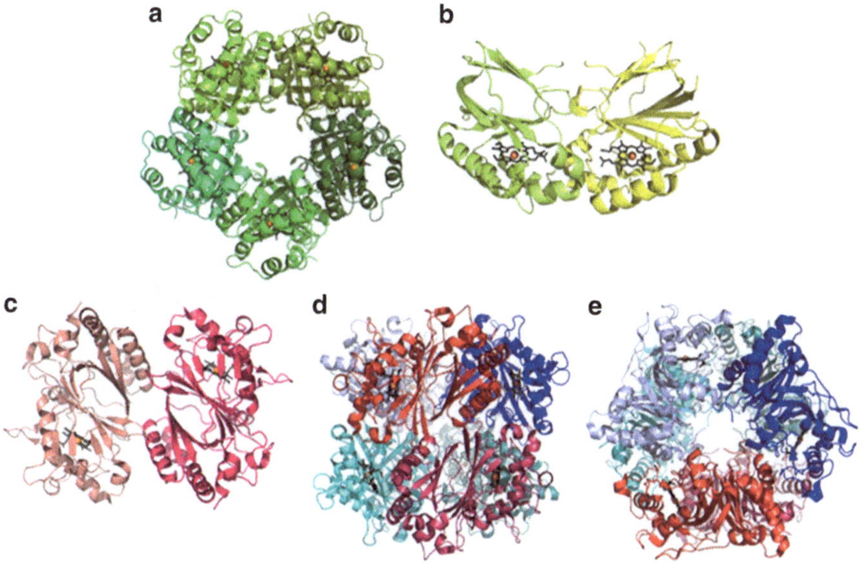

Figure 9 Biological oligomers of (**a**) *Da*Cld, (**b**) *Nitrospira winogradskyi* Cld, (**c**) TyrA from *Shewanella oneidensis*, and (**d**, **e**) DyP from *Rhodococcus jostii*, illustrating the diversity of multimeric structures assumed by structurally similar monomers in this superfamily. The monomers are individually colored, and the hemes are drawn in grey stick, with the iron as an orange sphere. Panel (**e**) is a 90° rotation of (**d**) towards the viewer about a horizontal axis. This figure was generated using PyMOL and reproduced with permission from [156]; copyright 2013 World Scientific Publishing Co.

By contrast, the short-length Cld from *N. winogradskyi* crystallizes as a dimer with a completely different monomer interface (Figure 9b) [76, 82]. This leaves the heme significantly more solvent exposed than in the pentameric structures (Figure 9a). The more open active site may be responsible for the significantly lower stability of the protein's heme. Though *Nw*Cld is capable of producing O_2 from chlorite, catalysis is limited by chlorite-dependent degradation of the heme, which appears to occur at much smaller concentrations of chlorite than in *Da*Cld.

The bacterial heme oxygenases (IsdG, IsdI) and DyPs share the heme binding domain structure of Clds but, as described above, are significantly different in sequence. The bidomain DyPs, unlike Clds, furthermore form head-to-tail dimers, where the C-terminal heme domain of one monomer interacts structurally with the N-terminal heme-free domain of the other. DyP structures may form independent dimers (Figure 9c) or a crystallographic trimer-of-dimers (Figure 9d and e).

By contrast, the monodomain IsdG-like proteins form independent dimers in which both monomers bind heme.

This diversity of oligomeric structures demonstrates that the heme-binding domain is structurally capable not only of accommodating heme in multiple orientations, but of interacting in a variety of different ways with partner proteins. Such intrinsic flexibility in their protein-protein interactions could be important for the biological functions or technological applications of diverse Clds.

4.2 Reactivity and Mechanism

4.2.1 Diversity of Reactions Catalyzed by Chlorite Dismutases

Clds catalyze at least three reactions: O–O bond formation, one-electron oxidations (peroxidase chemistry), and heme oxidation.

O–O bond formation from chlorite is the principal reaction catalyzed by full length Clds from (per)chlorate-respiring bacteria. As described above, these "*bona fide*" Clds are characterized by their pentameric oligomerization state, bidomain monomer structure, and possession of a distal pocket arginine in the heme-binding domain. They constitute a relatively small minority of all the sequenced Clds from taxonomically diverse sources. The small dimeric Clds have nearly the same active site structure as their pentameric homologs, including the distal arginine and the surrounding hydrophobic residues. These Clds also catalyze rapid O–O joining in spite of the lack of a known biological imperative by their host organisms for eliminating chlorite. However, the limited number of turnovers their hemes can sustain suggests that chlorite may not be their natural substrate [76, 104].

Whether short or long, the O_2-producing proteins also appear to be capable of catalyzing peroxidase-type one-electron oxidation reactions when either H_2O_2 or a peracid is the oxidant [89, 104]. A complete peroxidase reaction cycle has been documented for *Da*Cld, as described below. Catalysis of this reaction is very slow and the number of achievable turnovers per heme limited, consistent with the assumption that peroxidase chemistry is not the biological function of these Clds. Studies of the peroxidase reaction nonetheless have shed light on the basic chemical and biological capabilities of Clds and are therefore of interest [89].

Finally, Clds, like many heme-dependent oxidation catalysts, facilitate the oxidant-dependent degradation of their own hemes. The redox-stability of the heme in Clds is extraordinarily variable. *Da*Cld sustains ~20,000 turnovers with chlorite, per heme [105]. The dimeric Cld from *Klebsiella pneumoniae* sustains only ~2–3,000 turnovers under similar conditions [104]. The heme in the Cld from *Staphylococcus aureus* is completely destroyed by less than five equivalents of chlorite, without measurable production of O_2 [75]. Whether heme destruction serves a biological function in any of these proteins is not clear. However, it does seem to be the principal catalytic pathway followed by the *S. aureus* Cld (and likely by related pentameric Clds from non-oxochlorate-respiring bacteria) in the presence of a series of oxidants (chlorite, H_2O_2, peracids).

4.2.2 Possible Pathways for O–O Formation

There are a number of possible pathways for O_2 production from chlorite. Early efforts with *Da*Cld truncated the list by demonstrating that (i) O_2 is produced stoichiometrically with ClO_2^- degraded [105]; (ii) all of the emitted O_2 derives from chlorite, even in the presence of ^{18}O–labeled water [106]. These results strongly favored mechanisms where the two oxygen atoms of O_2 derive from the same molecule of ClO_2^-. Alternative mechanisms where chlorite acts as an oxygen atom or one-electron donor to the Fe(III) heme produce highly protein-reactive leaving groups (ClO^-, ClO_2) [107, 108] that would have to re-react quantitatively at the heme site to maintain the observed $1ClO_2^-:1O_2$ stoichiometry. They would moreover have to do so without undergoing exchange of their oxygen atoms with $^{18}OH_2$. Hence, the four possible mechanisms by which a single molecule of chlorite can be effectively rearranged to generate Cl^- and O_2, depicted in Figure 10, are preferred.

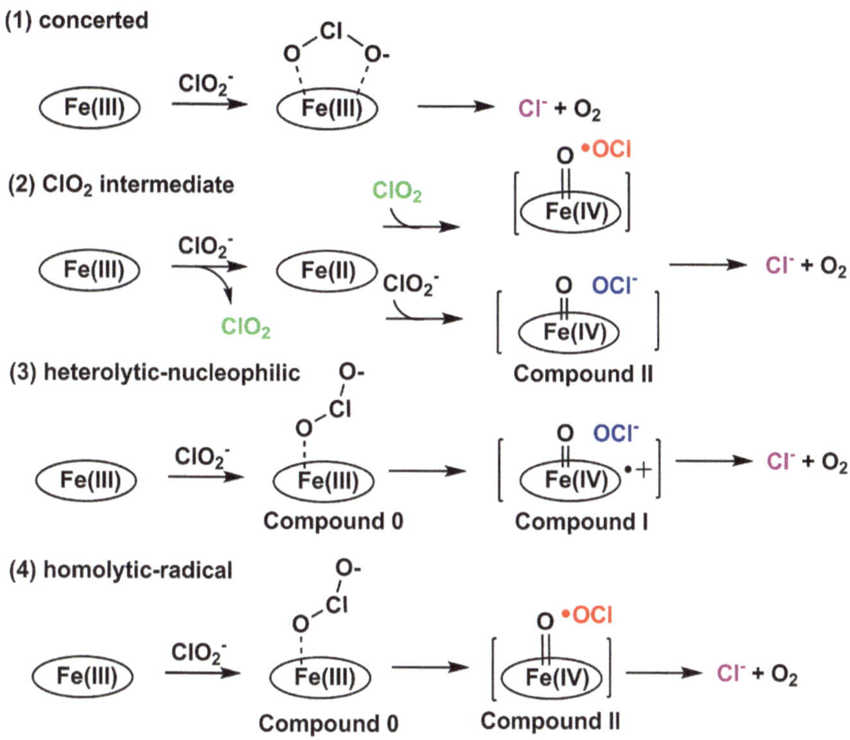

Figure 10 Possible mechanisms for O_2 formation from chlorite: (1) concerted, (2) chlorine dioxide intermediate, (3) heterolytic Cl–O cleavage/nucleophilic O–O formation, (4) homolytic Cl–O cleavage/radical O–O formation.

In pathway (1), a single molecule of chlorite binds Fe(III) in a bidentate, η^2 mode, and formation of O_2 occurs in a concerted fashion without any intermediate transfer of electrons or atoms to the iron. In pathway (2), electron transfer from chlorite to the Fe(III) heme, depicted here as an inner-sphere process, yields the Fe(II) porphyrin and chlorine dioxide (ClO_2). This reaction (eqn. 11) is thermodynamically favorable [109] and known to be catalyzed by certain synthetic metalloporphyrins (described below):

$$ClO_2^- \rightarrow ClO_2 + e^- \qquad E^\circ = -0.934\ V \tag{11}$$

Oxygen atom transfer from chlorine dioxide to the Fe(II) iron would yield a ferryl ($Fe^{IV}{=}O$) heme. This could recombine with the released chlorine oxide to make a formally Fe(III)-$OOCl^-$ species, spontaneously breaking down to yield Cl^- and O_2. The Fe(II) Cld may also react with chlorite to give O_2, which is also depicted in Figure 10.

Pathways (3) and (4) both begin with the formation of a Fe(III) heme/ chlorite complex. This complex is analogous to the Fe(III)-peroxy species (Fe(III)-OOH) formed as the initial intermediate in peroxidase reactions, known as Compound 0 [110]. The distal histidine of classical peroxidases removes the proton from H_2O_2 ($pK_a = 12$) to yield the peroxy anion, which binds avidly to Fe(III). Because the hypochlorous acid/chlorite pK_a is quite low (1.72), base catalysis would not be needed to form the analogous Fe(III)-chlorito complex.

Donation of two electrons from the Fe(III) heme leads to heterolytic scission of the coordinated O–Cl bond (pathway 3) and formation of a formally Fe(V)=O intermediate. This intermediate, observed pervasively in biological heme chemistry, is known as Compound I. Electronically, it is described as a ferryl (Fe(IV)=O) that is exchange-coupled to a radical on the non-innocent porphyrin ligand. Because loss of an electron occurs without concomitant loss of a proton, the resulting porphyrin radical is a cation (porphyrin$^{\bullet+}$). Unlike the parent tetrapyrrole, porphyrin$^{\bullet+}$ is not aromatic. Hence, the porphyrin's visible Soret absorbance band is significantly diminished in intensity in porphyrin$^{\bullet+}$, providing a simple indicator of the possible presence of Compound I. The oxygen atom in Compound I is capable of removing electrons or an electron/proton pair (hydrogen atom) from a substrate, thereby becoming reduced to the ferryl-porphyrin intermediate (Fe(IV)=O or Fe(III)-OH), also known as Compound II [111]. A good nucleophile may also attack the electron-deficient oxygen of Compound I. The oxygen atom of hypochlorite, particularly in its anionic/deprotonated form, is nucleophilic and could attack Compound I to yield the same Fe(III)-peroxychlorite intermediate proposed in pathway (1).

Alternatively, the Fe(III)-chlorite complex could react in a one-electron fashion to give the products of heterolytic O–Cl bond cleavage: Fe(IV)=O and the $ClO^\bullet$ (chlorine monoxide) radical. These two would need to re-react, once again to form a Fe(III) species with an O–O(Cl) bond (pathway 4). Chlorine monoxide is presumably a highly reactive species, perhaps not dissimilar to the hydroxyl radical. It has been described in atmospheric chemistry models but is not a known biological molecule.

There is no precedent for pathway (1) using either synthetic or biological catalysts. In fact, there are very few reported crystal structures for metals bound to chlorite, and none in either the through chloride or bidentate O,O-η^2 modes. While not a strict necessity, either of these binding modes and particularly the latter would seem to be the most logical for catalyzing the concerted formation of O_2 and Cl^-. By contrast, there are several direct precedents for parts of both mechanisms (2) and (3). The two long-proposed mechanisms for O–O joining in PSII and in synthetic water-splitting catalysts, for example, involve two oxygen atoms poised to react via either nucleophilic or radical processes [1, 2, 112, 113]. Mechanisms that are chemically analogous to pathways (3) and (4) have also been proposed for Fe(III) porphyrin-mediated isomerization of peroxynitrite ($ONOO^-$) to nitrate [114–117]. An Fe(III)-coordinated $ONOO^-$ anion could undergo heterolytic O–N bond cleavage to yield Compound I and the anionic leaving group (nitrite). Nitrite would then act as the nucleophile, attacking Compound I to form nitrate. Alternatively, homolytic O–N bond cleavage would yield Compound II and the neutral, radical leaving group (NO_2). Radical joining of NO_2 and the ferryl oxygen then lead to the Fe(III) porphyrin and NO_3^-.

4.2.3 Catalytic Efficiency

Clds that produce O_2 tend to do so very efficiently. Values for k_{cat}/K_M for several Clds are listed in Table 2, and many are near the diffusion limit. The rapidity with which O_2 is produced has made Clds useful generators of O_2 for stopped flow studies, where oxygen gas is produced within a few milliseconds in an initial mixing experiment between Cld in one syringe and chlorite in another [118]. The oxygenated solution is mixed with a third, concentrated solution of the reduced/anaerobic enzyme species of interest, and the subsequent reaction is monitored. Dioxygen is produced in this way at concentrations of up to 9 mM (25 °C, neutral pH) without observed visible light scattering in a stopped flow reaction chamber, which would be indicative of bubbling. It has been postulated that the solution remains transiently supersaturated over the very short time scale of the initial, O_2-producing mix, although the process of bubble growth has not been explicitly monitored over time.

Having the means to produce very large concentrations of O_2 with rapidity has allowed an important barrier to be surmounted. Namely, many intermediates in reactions between reduced biomolecules and O_2 cannot be observed because they decay more rapidly than they form. However, the second-order rate of their production can be enhanced by increasing [O_2]. Using *Da*Cld in this way, that is, as a reagent, has indeed yielded impressive results in cases previously limited by the rapid decay of the intermediate [118]. Other biotechnological applications of the reaction are currently under development.

4.2.4 Reaction Intermediates

While efficiency is critical for a detoxification enzyme and for the experiments described above, it can be the enemy of mechanistic studies. It has thus far not been possible to detect intermediates in the Cld/ClO_2^- reaction (e.g., Compounds 0, I, II) via stopped flow spectrophotometry. Hence, the Cld mechanism has not been monitored directly. By the same token, addition of a variety of colorimetric reagents in the presence of *Da*Cld during turnover with chlorite – small peroxidase substrates such as guaiacol or ascorbate, or the chlorination substrate monochlorodimedone (MCD) – has not resulted in observable side reactivity [105, 119]. This suggests that potential intermediates such as ClO_2 ($\varepsilon_{359\ nm} = 1.23\ mM^{-1}\ cm^{-1}$) [120], ClO^-, or possibly $ClO^\bullet$, any of which would react with MCD, either are not generated or cannot be accessed by MCD. The oxidation of molecules like guaiacol or ascorbate by potential intermediates Compound I or II is likewise not observed.

However, *Da*Cld can oxidize the peroxidase substrates guaiacol, ascorbate, and 2,2'-azino-bis(3-ethylbenzothiazoline)-6-sulfonic acid if either peroxide or peracetic acid is the oxidant. Moreover, the transient reactions of *Da*Cld with these oxidants have been examined [89]. These experiments showed that oxygen atom transfer from the peracid donor is very rapid and leads to the formation of Compound I, even in the presence of just one equivalent of oxidant. Hence, while it remains uncertain whether Compound I is an intermediate on the regular reaction pathway with chlorite, it is clear that the enzyme readily forms this species in the presence of another anionic oxygen atom donor that is thermodynamically predisposed toward oxygen atom transfer. These two donors are not identical, however. Unlike chlorite, for which the leaving group is relatively basic and therefore nucleophilic (hypochlorite, $pK_a = 7.5$), peracetic acid leaves behind a much less nucleophilic acetate ion ($pK_a = 4.5$). The reaction between the Fe(III) heme and chlorite would be expected to be catalyzed by a proton donor or positive charge that could stabilize the leaving group. The distal arginine has been proposed to fulfill this role, stabilizing hypochlorite in its anionic form and priming it for re-reaction with the oxygen atom of a hypothetical Compound I.

Interestingly, the *actual* Compound I formed in the reaction between peracetic acid and Fe(III) *Da*Cld reacts avidly with peroxidase substrates [89]. Under stopped flow conditions, a full peroxidase catalytic cycle is observed, with one equivalent of a one-electron donor (ascorbate) forming Compound II and the second returning the intermediate to the Fe(III) form:

$$\mathrm{Fe(III)(porphyrin)} + \mathrm{CH_3(C{=}O)COO^-} \rightarrow \mathrm{Fe(IV){=}O(porphyrin^{\bullet+})} + \mathrm{CH_3COO^-} \quad (12)$$

$$\mathrm{Fe(IV){=}O(porphyrin^{\bullet+})} + \mathrm{ascorbate} \rightarrow \mathrm{Fe(IV){-}OH(porphyrin)} + \mathrm{ascorbate^\bullet} \quad (13)$$

$$\mathrm{Fe(IV){-}OH(porphyrin)} + \mathrm{ascorbate} \rightarrow \mathrm{Fe(III)(porphyrin)} + \mathrm{ascorbate^\bullet} + \mathrm{H_2O} \quad (14)$$

These results demonstrate that *Da*Cld's Compound I can access peroxidase-like substrates, which in real peroxidase enzymes are known to bind in different locations above and adjacent to the heme plane [121].

By contrast, H_2O_2 reacts much more slowly with Fe(III) *Da*Cld than peracetic acid and, rather than producing Compound I, the only observed spectroscopic changes following the reaction correspond to heme oxidation. This observation is consistent with the Clds' lack of a distal pocket base. A distal histidine is absolutely required for rapid Compound I formation in the well-studied plant peroxidases [122–125]. When this residue is mutated to an aliphatic amino acid, the rate of Compound I formation slows by five orders of magnitude. In short, *Da*Cld can be a very effective peroxidase, but only in the presence of an acidic oxygen atom donor to the Fe(III) heme, for which a distal base is not required.

The *Da*Cld Compound I intermediate is not long-lived. It spontaneously converts to a ferryl without a coupled porphyrin cation radical. This conversion becomes more rapid at higher pH. The relative instability of Compound I is consistent with the generation of O_2 from a single molecule of chlorite. Namely, if Compound I is an intermediate on the O_2 production pathway, it may be too short-lived to react with oxygen-atom donors that are not already in the vicinity of the ferryl. Indeed, a Compound I species prepared in the stopped flow did not react with exogenously supplied ClO^-/HClO to form O_2, as the reaction appears to be kinetically out-competed by decay of the Compound I intermediate [89]. Though the pathway has not been studied, Compound I's breakdown has been presumed to occur via the conserved tyrosines that ring (and in one case form hydrogen bonds with) the heme [79].

Finally, although oxidation of guaiacol/ascorbate or chlorination of MCD is not observed during *Da*Cld/chlorite turnover, the peroxidase substrates increase (~10-fold) the number of molecules of chlorite that *Da*Cld can decompose before it irreversibly inactivates. The high valent intermediates are returned to the Fe(III) state and hence to activity with chlorite by these substrates. A similar type of "rescue" is likewise observed in other heme proteins [126–128]. This suggests that Compounds I, II, or both are intermediates on the pathway to whatever limits the catalytic lifetime of Clds (see below.) By contrast, MCD has no such rescuing effect, suggesting either that diffusible O/Cl species are not produced or are not appreciably responsible for substrate-dependent inactivation of *Da*Cld.

4.2.5 Structure-Activity Relationships: Highlights

The active site of the typical *bona fide,* chlorite-degrading Cld was discussed above in terms of what it is; it is also worthwhile to consider what it *is not.* Unlike peroxidases, which characteristically have a histidine-arginine pair in their distal pockets, Clds possess an arginine as the sole conserved, polar residue. The distal histidine is critical for peroxidase function as described above, acting as a base to deprotonate H_2O_2 and then as an acid toward the resulting Fe(III)-OOH (Compound 0) intermediate to generate Compound I and water. Such proton motions are critical for the rapid formation of both Compound 0 and Compound I in other heme proteins that activate O_2 or H_2O_2. Fe(III) Clds lack both a distal basic residue and a recognizable proton conduit to the heme iron. Perhaps not surprisingly, Clds bind anionic CN^- and $CH_3(C{=}O)COO^-$ (surrogates for OOH^-) much faster than protonated HCN or H_2O_2 [104]. *Da*Cld also rapidly forms Compound I following reaction with peracetic acid but not H_2O_2 [89]. In short, Clds lack the necessary machinery for efficiently binding and activating peroxide, and this colors what they can and cannot efficiently do with biological oxidants.

By contrast, Clds readily bind anions, including the substrate chlorite. Here, the distal arginine plays a key role. In multiple crystal structures, it forms a hydrogen bond to the axial ligand bound to the Fe(III) iron, from which it is separated by ~3 Å [4, 76, 80, 82]. Notably, the arginine has no identifiable hydrogen bonding partners among either the active site amino acids or the heme propionates, though an Arg-propionate interaction is present in some DyPs [83]. *Da*Cld, in fact, does not crystalize in the absence of a heme ligand, possibly in part because the motion of this residue would otherwise remain unrestrained. Resonance Raman experiments support a model where the arginine side chain interacts with axially-bound distal ligands and assumes different orientations ("in" or "out," relative to the heme plane) as a function of pH [78, 98]. In solution, substitution of the arginine with neutral glutamine by mutagenesis results in a strongly diminished k_{cat}/K_M(chlorite) and in a loss of affinity for anions [78]. This indicates that arginine is indeed critical for stabilizing the Michaelis complex, presumed to be a Compound 0-like Fe(III)-chlorito species [78].

The distal arginine is also sufficient for promoting rapid, heterolytic cleavage of the peracid O–O bond and stabilizing the resulting Compound I intermediate [89]. Similar functions have been attributed to the distal arginine in classical peroxidases, though in conjunction with an anionic proximal ligand to the heme iron. In peroxidases, anionic or histidinate character in this ligand is understood to add electron density to the Fe(III) and to consequently support heterolytic cleavage of a bond coordinated in *trans*. As described by the classic "push-pull" model of peroxidase catalysis, the electron-rich proximal ligand supplies a "push" toward cleaving a coordinated O–O bond, while the distal arginine supplies a concurrent "pull" [96, 124, 129]. The imidazole ring of the histidine gains anionic character via hydrogen bonding to an aspartate or glutamate. Though *Da*Cld has such a His-Glu hydrogen bond, resonance Raman showed the proximal ligand to be neutral in character [98]. O–O bond cleavage is relatively rapid in *Da*Cld in spite of the nature of the protein-derived ligand on the proximal side of the heme iron, suggesting that the distal pocket's "pull" alone is enough. The fact that *Da*Cld readily cleaves the O–O bond of coordinated peracetate ion also suggests that, in principle, it should be able to heterolytically cleave the O–Cl bond of a bound chlorite with similar avidity.

The mobility of the arginine could be important for an additional, important function: stabilizing the hypochlorite/hypochlorous acid (aqueous $pK_a = 7.5$) formed when the O–Cl bond of chlorite is broken. Nucleophilic attack of the leaving group on Compound I should be significantly more avid for the deprotonated hypochlorite than for hypochlorous acid, as described above. The positive charge of the arginine could locally lower the pK_a for the anion/acid pair, stabilizing the hypochlorite. It could also steer the hypochlorite into an optimal position for nucleophilic attack on Compound I.

Finally, the distal arginine could not manage to be such a key player were it not for a supporting cast of conserved, hydrophobic residues surrounding it. These residues constitute a sterically confined, chemically unreactive low ceiling over the heme plane [4]. Such an enclosed environment seems ideal for enforcing the recombination of a hypochlorite leaving group with Compound I.

4.2.6 Heme and Protein Stability in Diverse Chlorite Dismutase Family Proteins

The pentameric O_2-evolving *Da*Cld catalyzes ~20,000 turnovers (per heme) before irreversibly inactivating. The same number of chlorite equivalents completely eliminates the heme chromophore [105]. This suggests that oxidative destruction of the heme is responsible for the observed irreversible inactivation. Indeed, heme scission is a common side effect of oxidative catalysis: playing with matches will eventually cause even a robustly designed protein to get burned.

The mechanism of heme decomposition has not been studied. It has been observed, however, that far fewer turnovers are sustained by structurally distinct Clds [76]. The dimeric Cld from *Klebsiella pneumoniae*, for example, contains a distal arginine and catalyzes the chlorite to O_2 conversion with steady state parameters similar to those measured for *Da*Cld. However, the lifetime of the enzyme is clearly shorter; the measured turnover number is only 2,000–3,000 per heme [104]. The pentameric Cld from *Staphylococcus aureus* has a glutamine in place of the distal arginine. Its heme disappears following incubation with fewer than five equivalents of chlorite, and no O_2 evolution from this protein is observed [75]. The structural features leading to the great disparity in heme stability across the Cld family are not understood.

5 Synthetic and Biochemical Models

5.1 Chlorite as Reagent with Related Synthetic Metalloporphyrins

Quite apart from Clds and O_2 generation, NaOCl and $NaClO_2$ have an illustrious history as reagents in metalloporphyrin chemistry. Like iodosylbenzene, organic peracids (RCO_3H), and $KHSO_5$, the oxochlorates have been used as so-called "shunt" reagents: oxygen atom donors that, unlike O_2, do not require the input of a pair of activating electrons from the catalyst. Such shunt reagents have been used extensively in studies of cytochrome P450 monooxygenases and their synthetic mimics [130–132]. They have also been used in studies of peroxidases and heme oxygenases, where the oxygen-atom transfers to a Fe(III) heme center to yield Compound I. The same intermediate can equivalently be reached by the addition of O_2, three electrons, and one proton to a Fe(II) heme (P450), or peroxide minus one proton to the Fe(III) heme (peroxidases), with water as the leaving group.

Motivated by the goals of understanding and possibly mimicking P450 chemistry, Collman et al. were the first to describe O_2 evolution from synthetic Mn(III)porphyrin catalysts and ClO_2^- in non-aqueous media [133]. O_2 production occurred as a side reaction to a more P450-like process: the oxidation of unactivated alkanes (e.g., cyclohexane), which were converted in a two-electron process to alcohols and in a four-electron process to ketones. Notably, the turnover numbers using chlorite were >40-fold higher than if other oxygen-atom donors were used, and the ketone/alcohol ratio was at least double. These results suggested that chlorite was acting as something other than a simple oxygen-atom donor to the manganese. Further investigation then showed that radical trapping agents inhibited turnover, indicating that the Mn(III)/chlorite system acted as a radical initiator.

A two-pathway scheme was proposed to explain the observed oxidations of cyclohexane to cyclohexanol and cyclohexanone (Figure 11) [132]. In the first pathway, a Mn(III)/chlorite-derived species was proposed to generate a cyclohexyl radical by hydrogen-atom ($H^•$) abstraction. The organic radical then combined with O_2 to form a peroxyl radical adduct. The reaction of two such C_6H_{11}-$OO^•$ species would then lead to the formation of cyclohexanol, cyclohexanone, and, notably, O_2. A radical process helped explain the observation of ketone products, which were shown not to arise from the oxidation of alcohols formed initially as precursors. In the second proposed pathway, oxygen atom transfer from chlorite to the manganese(III)porphyrin catalyst (via heterolytic Cl–O bond cleavage) was proposed to yield a formal Mn(V)=O species and ClO^- as the leaving group. The high valent Mn(V)=O could then react via a more conventional, P450-like rebound mechanism to form the hydroxylation product:

$$\mathrm{Mn(V){=}O} + \mathrm{C_6H_6} \rightarrow \mathrm{Mn(IV){-}OH} + \mathrm{C_6H_5^{\bullet}} \rightarrow \mathrm{C_6H_5{-}OH} + \mathrm{Mn(III)} \quad (15)$$

In addition to the two alkane-oxidizing pathways, simple, metal-catalyzed decomposition of chlorite by two routes was also observed. In the first, Cl^- and O_2 were produced. Though a mechanism was not explicitly offered, it is possible that Mn(V)=O and Cl^- combined in the same manner proposed for the enzymatic/Cld reaction. The second pathway led to conversion of chlorite to chlorate and chloride in a 3:2:1 ratio:

$$3\ \mathrm{ClO_2^-} \rightarrow 2\ \mathrm{ClO_3^-} + \mathrm{Cl^-} \quad (16)$$

Because the manganese catalysts are inert toward chlorate, it persisted here as a dead-end product.

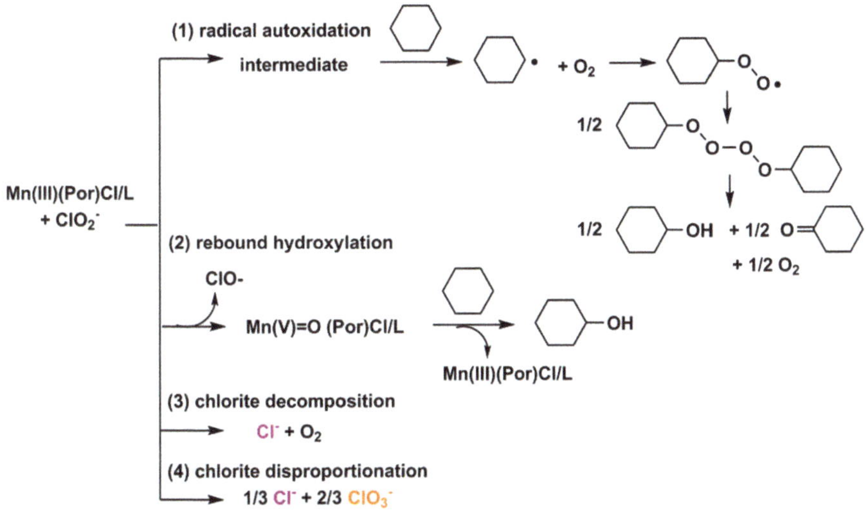

Figure 11 Reactions of Mn(III)porphyrin complexes (Por = porphyrin and L = *t*-butyl-pyridine) with chlorite and cyclohexane, where the use of chlorite as a P450-like "shunt" oxidant was tested. In pathway (1), an unknown Mn/chlorite intermediate forms a cyclohexyl radical. This reacts with O_2, either from air or produced from chlorite via pathway (3). Coupling of two peroxyl radicals ultimately leads to equimolar cyclohexanone, cyclohexanol, and O_2. Pathway (2) begins with the formation of a Mn(V)=O species via oxygen atom transfer from chlorite. This intermediate extracts a hydrogen atom from cyclohexane, forming the cyclohexyl radical and Mn(IV)–OH. Mn(IV)–OH and the radical recombine to yield the Mn(III)(Por)Cl starting material and cyclohexanol product. Finally, pathway (3) illustrates direct decomposition of chlorite to yield chloride and dioxygen, and pathway (4) shows its disproportionation. Scheme adapted from [133].

The laboratory of Abu-Omar subsequently investigated O_2 generation from metalloporphyrin/chlorite systems as the target reaction rather than as a side process [134–136]. Water-soluble iron porphyrin complexes with different aryl groups at the four *meso* positions were studied (Figure 12) as models for Clds. The *p*-sulfonatophenyl- and *N,N,N*-trimethylanilinium-substituted Fe(III) porphyrins led primarily to the same kind of chlorite-to-chlorate disproportionation observed by Collman et al. (eqn. 16) [133]. This mechanism of the reaction was studied in some detail, and shown to occur via the initial formation of an Fe(IV)=O(porphyrin$^{\bullet+}$) (Compound I) intermediate. Compound I was proposed to comproportionate with an aqua-Fe(III)porphyrin, resulting in a pair of Fe(IV)–OH(porphyrin) (Compound II) species. Compound II would then transfer its oxygen atom to chlorite, generating chlorate and the two-electron reduction product of the catalyst, Fe(II)porphyrin. The Fe(II)porphyrin thus formed can react with both chlorite and hypochlorite, heterolytically cleaving their Cl–O bonds and generating further reactive Fe(IV)–OH(porphyrin).

Though the first two iron porphyrins did not produce O_2 from chlorite, the fluorinated analog of one of them, [Fe(III)(TF_4TMAP)]$(OTf)_5$, did in 18 % yield

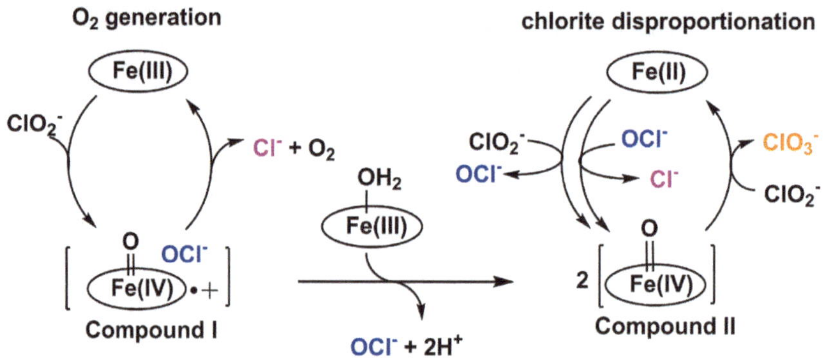

Figure 12 Reactions of $[Fe(III)TF_4TMAP]^{5+}$, yielding O_2, Cl^-, ClO_3^- as ultimate end products. The O_2-generating mechanism on the left resembles the heterolytic/nucleophilic pathway proposed for Clds in Figure 11. The Compound I-like intermediate can react with another molecule of the aquated, Fe(III) catalyst, yielding two equivalents of Compound II (ferryl). This intermediate reacts with further equivalents of chlorite to produce chlorate and the Fe(II) catalyst. Adapted from [135].

[OTf = triflate, TF_4TMAP = 5,10,15,20-tetrakis(tetrafluro-*N,N,N*-trimethylanilinium) porphyrinato]. The remaining chlorite followed the more customary pathway of disproportionation to chlorate and chloride. Production of O_2 was proposed to begin with oxygen atom transfer from chlorite to the Fe(III)porphyrin, forming Compound I. The same step was invoked as in the initiating step for chlorate generation. However, circa 1/5 of the time, the Compound I intermediate reacts with ClO^- instead of another Fe(III)porphyrin, generating O_2 and Cl^-. Notably, enclosure of the ClO^- leaving group and Fe(IV)=O(porphyrin$^{+\bullet}$) species, as well as site-isolation of the reactive metal, would both in principle improve the O_2 yield from the otherwise catalytically competent Fe(III)porphyrin. Isolation of the reactive intermediates is an obvious role served by the Cld protein environment.

In addition to the disoporoportionation and O_2 generation reactions described above, some manganese(III)porphyrins catalyze yet a further transformation of chlorite: its one electron oxidation to the chlorine dioxide (ClO_2) radical (Figure 13) [137, 138]. This reaction is preferentially catalyzed by the water-soluble Mn(III) analogs of $[Fe(III)(TF_4TMAP)]^{5+}$ and by tetrakis-5,10,15,20-(N,N-dimethylimidazolium) porphyrinatomanganese(III), $[Mn(TDMImP)]^{5+}$. In each case, the reaction was proposed to begin with oxygen atom transfer to the Mn(III), generating the Mn(V)=O product of heterolytic Cl–O bond cleavage in the $[Mn(TDMImP)]^{5+}$ system and a mixture of the heterolytic and homolytic (Mn(IV)=O) products in the presence of $[Mn(TF_4TMAP)]^{5+}$. The ClO^- leaving group was able to re-enter the catalytic cycle as an oxygen-atom donor to Mn(III) porphyrin. The $ClO^\bullet$ (product of homolytic Cl–O cleavage) was proposed to react in an uncatalyzed fashion with additional molecules of ClO_2^-. The high-valent metal-oxo species produced in either case served as one-electron oxidants toward chlorite [109], generating neutral chlorine dioxide gas (eqn. 11).

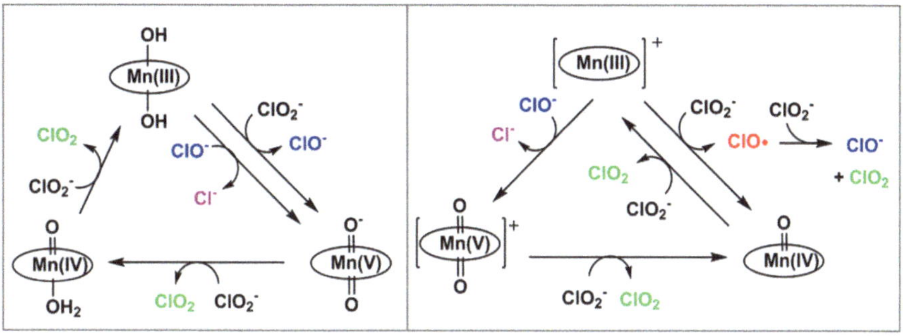

Figure 13 Reactions of Mn(III)porphyrins with chlorite to produce ClO_2. **Left**: Oxygen-atom transfer from chlorite yields a Mn(V)=O species that is then capable of two, sequential one-electron oxidations of a pair of chlorite anions in a peroxidase-like catalytic cycle (products = $2ClO_2$, green; manganese ligand = TDMImP). **Right**: A slightly more complicated proposed mechanism begins with electron transfer followed by atom transfer to the Mn(III)porphyrin (L = TF_4TMAP), forming an Mn(IV)=O species and $ClO^•$. The latter reacts with chlorite to form ClO_2 and ClO^-. Hypochlorite reacts with Mn(III)porphyrin to form a Mn(V)-dioxo species analogous to that in the left-hand panel, which oxidizes two molecules of chlorite to yield two ClO_2. Figure adapted from [137] and [138].

This reaction is interesting in itself from a technological point of view, because ClO_2 is a reagent used in water purification that must be produced near its point of use because it cannot be easily compressed and stored. It also provides a potential mechanistic contrast with Clds and the porphyrin complexes which generate Compound I but appear to use it to transfer an oxygen atom to hypochlorite (a two-electron process) rather than for the one-electron oxidation of chlorite. Presumably, recombination between the Compound I and hypochlorite generated in the confined space above the Cld heme kinetically out-competes the reaction between Compound I and another molecule of chlorite. This is a hypothesis that awaits further experimental investigation.

5.2 *Reactions of Chlorite with Horseradish Peroxidase: Implications for Chlorite Dismutases*

The use of chlorite as both an oxidant (oxygen-atom donor) and one-electron reductant by the Mn(III)porphyrins mirrors the results of a careful, systematic study of the reactions of horseradish peroxidase (HRP) and chlorite [120]. The complexity of this reaction had been previously noted [139, 140]. In strong contrast to Clds, Fe(III) HRP reacted with both chlorite and hypochlorite as oxygen-atom donors (generating Compound I), followed by two molecules of chlorite serving as one-electron donors to return the intermediate to the Fe(III) state (i.e., a complete peroxidase-like catalytic cycle, eqs 12–14). The chlorine dioxide products were also capable of donating an oxygen atom to the Fe(III) HRP, yielding Compound I and chlorine oxide (Figure 14). The ultimate fate of the highly-reactive $ClO^•$ radical was not clear.

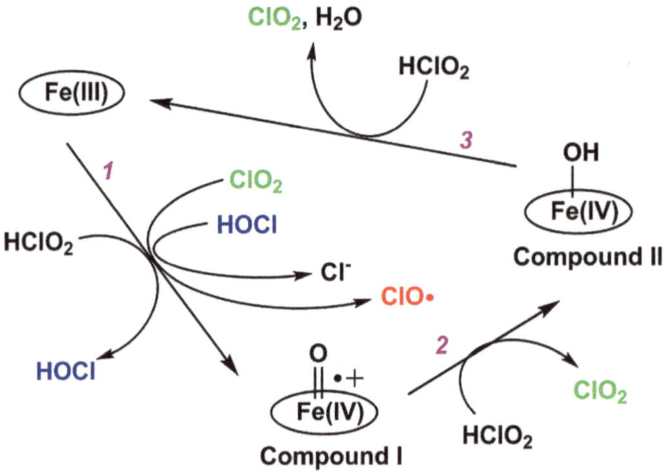

Figure 14 Reactions of horseradish peroxidase with chlorite. In the first step of the reaction (1), charge-neutral chlorous acid transfers an oxygen atom to Fe(III) HRP to form Compound I and hypochlorite/hypochlorous acid. Compound I then catalyzes two sequential one-electron oxidations of chlorous acid molecules (steps 2 and 3), producing two molecules of chlorine dioxide and returning the heme to the Fe(III) state. The inside of the diagram shows how both HClO and ClO_2 are able to re-enter the catalytic cycle as oxygen atom donors, producing Compound I and either Cl^-or $ClO^•$ as the leaving groups, respectively. Figure adapted from [120].

In spite of having a relatively well-enclosed active site that could in principle contain and steer $ClO^-/ClO^•$ leaving groups, neither appeared to recombine with the HRP-ferryl to make O_2. The reasons for this are unknown; however, the pH dependence of the HRP reactions with chlorite may lend clues. Compound I formation from Fe(III) HRP and ClO^- or ClO_2^-, but not ClO_2, was highly dependent on pH. In keeping with HRP's known preference for neutral ligands, including peroxide [141], the neutral/protonated species (HClO, $HClO_2$) reacted preferentially with the Fe(III) iron. The distal histidine in HRP and all the well-studied plant peroxidases is critical for H_2O_2 deprotonation to form the Fe(III)-OOH (Compound 0) complex. It may similarly be involved in deprotonating chlorous acid ($HClO_2$, $pK_a = 1.72$) [20, 123]. By contrast, Clds lack a distal histidine and strongly prefer to bind cyanide and peracids in their anionic forms, with the distal arginine helping to secure the bound anion [78, 89]. The subsequent reactions of HRP-Compounds I and II with chlorite were also highly pH-dependent. The plotted dependencies of log *k versus* pH were once again linear with a slope of −1, indicating that Compounds I and II preferentially react with the $HClO_2$ acid rather than ClO_2^-, in spite of the relative scarcity of the acid form (pK_a 1.72) in biochemically obtainable pH ranges. Again in contrast, it has been proposed that Compound I in Clds reacts with hypochlorite in its anionic form, in part because ClO^- is a better nucleophile than the HClO acid (pK_a 7.5), and in part because the steady state reaction is faster at low pH, where the distal arginine is expected to point toward the heme plane and to stabilize the ClO^- leaving group [78, 98]. The Clds' ability to stabilize the hypochlorite leaving group in its anionic form may be critical for promoting O–O bond formation, and for distinguishing Clds from (O)Cl–O^- bond-cleaving, Compound I-forming peroxidases that nonetheless do not make O_2, such as HRP.

Alternatively or in addition, peroxidases are conventionally understood to react with their substrates on the periphery of the heme rather than at the apex of the distal pocket [121, 142–144]. Direct reaction of the porphyrin cation radical and substrate prevents cytochrome-P450-like oxygen-atom transfer from the Compound I intermediate to the substrate, which occurs when the substrate is able to approach the ferryl oxygen from an optimal angle. In Clds, reaction of hypochlorite with Compound I requires access to the same ferryl oxygen atom, while peroxidase-style one-electron reductions can just as well occur at the heme edge. It is possible that, in HRP, the distal pocket is not sterically configured to promote attack of the ClO^-/HClO leaving group on the ferryl oxygen. Hence, O_2 production is not observed.

6 General Conclusions

As work with Clds illustrates, there is seemingly nothing that nature can't do with a protein and a heme. The unusual transformation of ClO_2^- to O_2 and Cl^- is greatly facilitated by enclosing heme in the relatively hydrophobic environment of chlorite dismutase, with a well-positioned arginine needed to recruit and steer the anionic substrate and intermediates. The arginine is a key innovation in an active site pocket that seems otherwise well-suited for little more than heme decomposition. The etiology of chlorite decomposition by Clds, the history of this fascinating protein family, and the biogeochemical consequences of natural perchlorate respiration all remain to be discovered by ongoing and future work.

Abbreviations and Definitions

Cld	chlorite dismutase
ClO^-	hypochlorite
ClO_2	chlorine dioxide
ClO_2^-	chlorite
ClO_3^-	chlorate
ClO_4^-	perchlorate
Clr	chlorate reductase
Compound 0	Fe(III)-anion complex
Compound I	Fe(IV)porphyrin cation radical
Compound II	Fe(IV)=O or Fe(IV)–OH
*Da*Cld	*Dechloromonas aromatica* chlorite dismutase
DMSO	dimethylsulfoxide

DyP	dye-decoloring peroxidase
HClO	hypochlorous acid
HRP	horseradish peroxidase
MCD	monochlorodimedone
MGD	molybdopterin guanine dinucleotide
Nar	nitrate reductase
NHE	normal hydrogen electrode
NPRB	non-perchlorate-respiring bacteria
*Nw*Cld	*Nitrospira winogradskyi* chlorite dismutase
$ONOO^-$	peroxynitrite
OTf	triflate
Pcr	perchlorate reductase
(per)chlorate	chlorate and/or perchlorate
PRB	perchlorate-respiring bacteria
PSII	Photosystem II
Ser	selenate reductase
TDMImP	tetrakis-5,10,15,20-(N,N-dimethylimidazolium)porphyrinato
TF_4TMAP	5,10,15,20-tetrakis(tetrafluro-*N*,*N*,*N*-trimethylanilinium) porphyrinato
WCL	wet chemistry laboratory

Acknowledgments Support for this work from the National Institutes of Health, National Institute for General Medical Sciences is gratefully acknowledged (GM090260), as are the scientific and intellectual contributions of laboratory members past and present.

References

1. C. S. Mullins, V. L. Pecoraro, *Coord. Chem. Rev.* **2008**, *252*, 416–443.
2. J. P. McEvoy, G. W. Brudvig, *Chem. Rev.* **2006**, *106*, 4455–4483.
3. K. N. Ferreira, T. M. Iverson, K. Maghlaoui, J. Barber, S. Iwata, *Science* **2004**, *303*, 1831–1838.
4. B. R. Goblirsch, B. R. Streit, J. L. DuBois, C. M. Wilmot, *J. Biol. Inorg. Chem.* **2010**, *15*, 879–888.
5. K. A. Weber, L. A. Achenbach, J. D. Coates, *Nature Rev. Microbiol.* **2006**, *4*, 752–764.
6. J. D. Coates, U. Michaelidou, R. A. Bruce, S. M. O'Connor, J. N. Crespi, L. A. Achenbach, *Appl. Environ. Microbiol.* **1999**, *65*, 5234–5241.
7. K. Kim, B. E. Logan, *Water Res.* **2001**, *35*, 3071–3076.
8. B. E. Logan, J. Wu, R. F. Unz, *Water Res.* **2001**, *35*, 3034–3038.
9. B. E. Logan, H. S. Zhang, P. Mulvaney, M. G. Milner, I. M. Head, R. F. Unz, *Appl. Environ. Microbiol.* **2001**, *67*, 2499–2506.
10. H. S. Zhang, M. A. Bruns, B. E. Logan, *Environ. Microbiol.* **2002**, *4*, 570–576.
11. R. A. Bruce, L. A. Achenbach, J. D. Coates, *Environ. Microbiol.* **1999**, *1*, 319–329.
12. C. W. Trumpolt, M. Crain, G. D. Cullison, S. J. P. Flanagan, L. Siegel, S. Lathrop, *Remediation* **2005**, *Winter*, 65–89.

13. R. Renner, *Environ. Sci. & Tech. News* **1998**, *32*, 210A.
14. E. T. Urbansky, *Bioremediation J.* **1998**, *2*, 81–95.
15. E. T. Urbansky, S. K. Brown, *J. Environ. Monitor.* **2003**, *5*, 455–462.
16. E. T. Urbansky, M. R. Schock, *J. Environ. Manage.***1999**, *56*, 79–95.
17. J. S. Valentine, C. S. Foote, A. Greenberg, J. F. Liebman, *Active Oxygen in Biochemistry*, Eds J. S. Valentine, C. S. Foote, A. Greenberg, J. F. Lieberman, Springer, Dordrecht, **1995**, pp. 481.
18. I. R. Epstein, K. Kustin, *J. Phys. Chem.* **1985**, *89*, 2275–2282.
19. J. Arnhold, E. Monzani, P. G. Fürtmuller, M. Zederbauer, L. Casella, C. Obinger, *Eur. J. Inorg. Chem.* **2006**, 3801–3811.
20. I. Fabian, G. Gordon, *Inorg. Chem.* **1991**, *30*, 3785–3787.
21. E. T. Urbansky, *Environ. Sci. Pollut. Res.* **2002**, *9*, 187–192.
22. P. K. DasGupta, J. V. Dyke, A. B. Kirk, W. A. Jackson, *Environ. Sci. Tech.* **2006**, *40*, 6608–6614.
23. E. T. Urbansky, *Environmental Impact of Fertilizer on Soil and Water* **2004**, *872*, 16–31.
24. E. T. Urbansky, S. K. Brown, M. L. Magnuson, C. A. Kelty, *Environ. Pollut.* **2001**, *112*, 299–302.
25. G. E. Ericksen, *Amer. Sci.* **1983**, *71*, 366–374.
26. H. M. Bao, B. H. Gu, *Environ. Sci. Tech.* **2004**, *38*, 5073–5077.
27. B. R. Scanlon, R. C. Reedy, W. A. Jackson, B. Rao, *Environ. Sci. Tech.* **2008**, *42*, 8648–8653.
28. B. Rao, T. A. Anderson, G. J. Orris, K. A. Rainwater, S. Rajagopalan, R. M. Sandvig, B. R. Scanlon, D. A. Stonestrom, M. A. Walvoord, W. A. Jackson, *Environ. Sci. Tech.* **2007**, *41*, 4522–4528.
29. S. Rajagopalan, T. A. Anderson, L. Fahlquist, K. A. Rainwater, M. Ridley, W. A. Jackson, *Environ. Sci. Tech.* **2006**, *40*, 3156–3162.
30. S. P. Kounaves, S. T. Stroble, R. M. Anderson, Q. Moore, D. C. Catling, S. Douglas, C. P. McKay, D. W. Ming, P. H. Smith, L. K. Tamppari, A. P. Zent, *Environ. Sci. Tech.* **2010**, *44*, 2360–2364.
31. D. K. Tipton, D. E. Rolston, K. M. Scow, *J. Environ. Quality* **2003**, *32*, 40–46.
32. L. N. Plummer, J. K. Bohlke, M. W. Doughten, *Environ. Sci. Tech.* **2006**, *40*, 1757–1763.
33. B. A. Rao, C. P. Wake, T. Anderson, W. A. Jackson, *Water, Air, Soil Pollut.* **2012**, *223*, 181–188.
34. V. I. Furdui, F. Tomassini, *Environ. Sci. Tech.* **2010**, *44*, 588–592.
35. G. Bordeleau, R. Martel, G. Ampleman, S. Thiboutot, *J. Environ. Qual.* **2008**, *37*, 308–317.
36. N. C. Sturchio, J. R. Hoaglund, III, R. J. Marroquin, A. D. Beloso, Jr., L. J. Heraty, S. E. Bortz, T. L. Patterson, *Ground Water* **2012**, *50*, 94–102.
37. P. N. Smith, C. W. Theodorakis, T. A. Anderson, R. J. Kendall, *Ecotoxicology* **2001**, *10*, 305–313.
38. M. L. Magnuson, E. T. Urbansky, C. A. Kelty, *Analyt. Chem.* **2000**, *72*, 25–29.
39. P. K. Dasgupta, A. B. Kirk, J. V. Dyke, S.-I. Ohira, *Environ. Sci. Tech.* **2008**, *42*, 8115–8121.
40. J. V. Dyke, K. Ito, T. Obitsu, Y. Hisamatsu, P. K. Dasgupta, B. C. Blount, *Environ. Sci. Tech.* **2007**, *41*, 88–92.
41. A. B. Kirk, M. Kroll, J. V. Dyke, S.-I. Ohira, R. A. Dias, P. K. Dasgupta, *Sci. Tot. Environ.* **2012**, *420*, 73–78.
42. W. Wallace, T. Ward, A. Breen, H. Attaway, *J. Indust. Microbiol.* **1996**, *16*, 68–72.
43. G. Rikken, A. Kroon, C. van Ginkel, *Appl. Microbiol. Biotech.* **1996**, *45*, 420–426.
44. P. K. Dasgupta, P. K. Martinelango, W. A. Jackson, T. A. Anderson, K. Tian, R. W. Tock, S. Rajagopalan, *Environ. Sci. Tech.* **2005**, *39*, 1569–1575.
45. B. Rao, S. Mohan, A. Neuber, W. A. Jackson, *Water, Air, Soil Pollut.* **2012**, *223*, 275–287.
46. L. Jaegle, Y. L. Yung, G. C. Toon, B. Sen, J. F. Blavier, *Geophys. Res. Lett.* **1996**, *23*, 1749–1752.
47. R. Simonaitis, J. Heicklen, *Planet. Space Sci.* **1975**, *23*, 1567–1569.
48. M. H. Hecht, S. P. Kounaves, R. C. Quinn, S. J. West, S. M. M. Young, D. W. Ming, D. C. Catling, B. C. Clark, W. V. Boynton, J. Hoffman, L. P. DeFlores, K. Gospodinova, J. Kapit, P. H. Smith, *Science* **2009**, *325*, 64–67.

49. J. D. Schuttlefield, J. B. Sambur, M. Gelwicks, C. M. Eggleston, B. A. Parkinson, *J. Am. Chem. Soc.* **2011**, *133*, 17521–17523.
50. K. S. Bender, C. Shang, R. Chakraborty, S. M. Belchik, J. D. Coates, L. A. Achenbach, *J. Bacteriol.* **2005**, *187*, 5090–5096.
51. J. C. Thrash, J. Pollock, T. Torok, J. D. Coates, *Appl. Microbiol. Biotech.* **2010**, *86*, 335–343.
52. J. C. Thrash, S. Ahmadi, T. Torok, J. D. Coates, *Appl. Microbiol. Biotech.* **2010**, *76*, 4730–4737.
53. C. I. Carlstrom, O. Wang, R. A. Melnyk, S. Bauer, J. Lee, A. Engelbrektson, J. D. Coates, *MBio* **2013**, *4*, 00217–13.
54. M. Balk, T. van Gelder, S. A. Weelink, A. J. A. Stams, *Appl. Environ. Microbiol.* **2008**, *74*, 403–409.
55. M. Balk, F. Mehboob, A. H. van Gelder, W. I. C. Rijpstra, J. S. S. Damste, A. J. M. Stams, *Appl. Microbiol. Biotech.* **2010**, *88*, 595–603.
56. C. P. Shelor, A. B. Kirk, P. K. Dasgupta, M. Kroll, C. A. Campbell, P. K. Choudhary, *Environ. Sci. Tech.* **2012**, *46*, 5151–5159.
57. M. G. Liebensteiner, M. W. H. Pinkse, P. J. Schaap, A. J. M. Stams, B. P. Lomans, *Science* **2013**, *340*, 85–87.
58. H. D. Thorell, K. Stenklo, J. Karlsson, T. Nilsson, *Appl. Environ. Microbiol.* **2003**, *69*, 5585–5592.
59. A. Wolterink, A. B. Jonker, S. W. M. Kengen, A. J. M. Stams, *Int. J. Syst. Evol. Microbiol.* **2002**, *52*, 2183–2190.
60. K. Yoshimatsu, T. Sakurai, T. Fujiwara, *FEBS Lett.* **2000**, *470*, 216–220.
61. R. M. Martinez-Espinosa, E. J. Dridge, M. J. Bonete, J. N. Butt, C. S. Butler, F. Sargent, D. J. Richardson, *FEMS Microbiol.* **2007**, *276*, 129–139.
62. A. McEwan, J. Ridge, C. McDevitt, P. Hugenholtz, *Geomicrobiol. J.* **2002**, *19*, 3–21.
63. J. D. Coates, R. Chakraborty, J. G. Lack, S. M. O'Connor, K. A. Cole, K. S. Bender, L. A. Achenbach, *Nature* **2001**, *411*, 1039–1043.
64. K. G. Byrne-Bailey, J. D. Coates, *J. Biotech.* **2012**, *194*, 2767–2768.
65. R. A. Melnyk, A. Engelbrektson, I. C. Clark, H. K. Carlson, K. Byrne-Bailey, J. D. Coates, *Appl. Environ. Microbiol.* **2011**, *77*, 7401–7404.
66. I. C. Clark, R. A. Melnyk, A. Engelbrektson, J. D. Coates, *mBio* **2013**, *4*, 00379–13.
67. S. Weelink, N. Tan, H. ten Broeke, C. van den Kieboom, W. van Doesburg, A. Langenhoff, J. Gerritse, H. Junca, A. Stams, *Appl. Environ. Microbiol.* **2008**, *74*, 6672–6681.
68. M. Oosterkamp, T. Veuskens, C. Plugge, A. Langenhoff, J. Gerritse, W. van Berkel, D. Pieper, H. Junca, L. Goodwin, H. Daligault, D. Bruce, J. Detter, R. Tapia, C. Han, M. Land, L. Hauser, H. Smidt, A. Stams, *J. Bacteriol.* **2011**, *193*, 5028–5029.
69. J. D. Coates, L. A. Achenbach, *Nature Rev. Microbiol.* **2004**, *2*, 569–580.
70. T. Nilsson, M. Rova, A. S. Backlund, *Biochim. Biophys. Acta* **2013**, *1827*, 189–197.
71. F. Maixner, M. Wagner, S. Lucker, E. Pelletier, S. Schmitz-Esser, K. Hace, E. Spieck, R. Konrat, D. Le Paslier, H. Daims, *Environ. Microbiol.* **2008**, *10*, 3043–3056.
72. K. S. Bender, M. R. Rice, W. H. Fugate, J. D. Coates, L. A. Achenbach, *Appl. Environ. Microbiol.* **2004**, *70*, 5651–5658.
73. A. S. Backlund, J. Bohlin, N. Gustavsson, T. Nilsson, *Appl. Environ. Microbiol.* **2009**, *75*, 2439–2445.
74. A. Ebihara, A. Okamoto, Y. Kousumi, H. Yamamoto, R. Masui, N. Ueyama, S. Yokoyama, S. Kuramitsu, *J. Struct. Funct. Gen.* **2005**, *6*, 21–32.
75. J. A. Mayfield, N. D. Hammer, R. C. Kurker, T. K. Chen, S. Ojha, E. P. Skaar, J. L. DuBois, *J. Biol. Chem.* **2013**, *288*, 23488–23504.
76. G. Mlynek, B. Sjoeblom, J. Kostan, S. Fuereder, F. Maixner, K. Gysel, P. G. Fürtmueller, C. Obinger, M. Wagner, H. Daims, K. Djinovic-Carugo, *J. Bacteriol.* **2011**, *193*, 2408–2417.
77. K. S. Bender, S. A. O'Connor, R. Chakraborty, J. D. Coates, L. A. Achenbach, *Appl. Environ. Microbiol.* **2002**, *68*, 4820–4826.

78. B. Blanc, J. A. Mayfield, C. A. McDonald, G. S. Lukat-Rodgers, K. R. Rodgers, J. L. DuBois, *Biochemistry* **2012**, *51*, 1895–1910.
79. B. Blanc, K. R. Rodgers, G. S. Lukat-Rodgers, J. L. DuBois, *Dalton Trans.* **2013**, *42*, 3156–3169.
80. D. C. de Geus, E. A. J. Thomassen, P.-L. Hagedoorn, N. S. Pannu, E. van Duijn, J. P. Abrahams, *J. Mol. Biol.* **2009**, *387*, 192–206.
81. A. Ebihara, A. Okamoto, Y. Kousumi, H. Yamamoto, R. Masui, N. Ueyama, S. Yokoyama, S. Kuramitsu, *J. Struct. Funct. Gen.* 2005, *6*, 21–32.
82. J. Kostan, B. Sjoeblom, F. Maixner, G. Mlynek, P. G. Fürtmueller, C. Obinger, M. Wagner, H. Daims, K. Djinovic-Carugo, *J. Struct. Biol.* **2010**, *172*, 331–342.
83. B. Goblirsch, R. C. Kurker, B. R. Streit, C. M. Wilmot, J. L. DuBois, *J. Mol. Biol.* **2011**, *408*, 379–398.
84. E. P. Skaar, A. H. Gaspar, O. Schneewind, *J. Biol. Chem.* **2004**, *279*, 436–443.
85. R. Y. Wu, E. P. Skaar, R. G. Zhang, G. Joachimiak, P. Gornicki, O. Schneewind, A. Joachimiak, *J. Biol. Chem.* **2005**, *280*, 2840–2846.
86. W. C. Lee, M. L. Reniere, E. P. Skaar, M. E. Murphy, *J. Biol. Chem.* **2008**, *283*, 30957–30963.
87. Y. Sugano, R. Muramatsu, A. Ichiyanagi, T. Sato, M. Shoda, *J. Biol. Chem.* **2007**, *282*, 36652–36658.
88. M. Ahmad, J. N. Roberts, E. M. Hardiman, R. Singh, L. D. Eltis, T. D. H. Bugg, *Biochemistry* **2011**, *50*, 5096–5107.
89. J. A. Mayfield, B. Blanc, K. R. Rodgers, G. S. Lukat-Rodgers, J. L. DuBois, *Biochemistry* **2013**, *52*, 6982–6994.
90. S. Adachi, S. Nagano, K. Ishimori, Y. Watanabe, I. Morishima, T. Egawa, T. Kitagawa, R. Makino, *Biochemistry* **1993**, *32*, 241–252.
91. A. Farhana, V. Saini, A. Kumar, J. R. Lancaster, Jr., A. J. C. Steyn, *Antiox. Redox Signal.* 2012, *17*, 1232–1245.
92. G. S. Lukat-Rodgers, K. R. Rodgers, *J. Biol. Inorg. Chem.* 1998, *3*, 274–281.
93. W. Gong, B. Hao, M. K. Chan, *Biochemistry* **2000**, *39*, 3955–3962.
94. S. Aono, H. Nakajima, *Coord. Chem. Rev.* **1999**, *192*, 267–282.
95. T. L. Poulos, *Curr. Opin. Struct. Biol.* **2006**, *16*, 736–743.
96. K. Choudhury, M. Sundaramoorthy, A. Hickman, T. Yonetani, E. Woehl, M. F. Dunn, T. L. Poulos, *J. Biol. Chem.* **1994**, *269*, 20239–20249.
97. T. L. Poulos, R. E. Fenna, in *Metal Ions in Biological Systems,* Vol. 30, Eds H. Sigel, A. Sigel, Marcel Dekker, Inc., New York, 1994, pp. 25–75.
98. B. R. Streit, B. Blanc, G. S. Lukat-Rodgers, K. R. Rodgers, J. L. DuBois, *J. Am. Chem. Soc.* **2010**, *132*, 5711–5724.
99. D. M. Davies, P. Jones, D. Mantle, *Biochem. J.* **1976**, *157*, 247–253.
100. P. Jones, H. B. Dunford, *J. Theor. Biol.* **1977**, *69*, 457–470.
101. J. E. Erman, L. B. Vitello, M. A. Miller, J. Kraut, *J. Am. Chem. Soc.* **1992**, *114*, 6592–6593.
102. S. Hofbauer, M. Bellei, A. Suendermann, K. F. Pirker, A. Hagmueller, G. Mlynek, J. Kostan, H. Daims, P. G. Fürtmueller, K. Djinovic-Carugo, C. Oostenbrink, G. Battistuzzi, C. Obinger, *Biochemistry* **2012**, *51*, 9501–9512.
103. S. Hofbauer, K. Gysel, G. Mlynek, J. Kostan, A. Hagmueller, H. Daims, P. G. Furtmueller, K. Djinovic-Carugo, C. Obinger, *Biochim. Biophys. Acta* **2012**, *1824*, 1031–1038.
104. DuBois lab, unpublished results.
105. B. R. Streit, J. L. DuBois, *Biochemistry* **2008**, *47*, 5271–5280.
106. A. Q. Lee, B. R. Streit, M. Zdilla, M. A. Abu-Omar, J. L. DuBois, *Proc. Natl. Acad. Sci. USA* **2008**, *105*, 15654–15659.
107. Y. Patel, D. Wong, L. Ingerman, P. McGinnis, M. Osier, Environmental Protection Agency report: "Toxicological Review of Chlorine Dioxide and Chlorite", **2000**; available for download from the world wide web.
108. R. A. Miller, B. E. Britigan, *Clin. Microbiol. Rev.* **1997**, *10*, 1–18.

109. U. K. Klaning, K. Sehested, J. Holcman, *J. Phys. Chem.* **1985**, *89*, 760–763.
110. H. B. Dunford, *Heme Peroxidases*, Wiley-VCH, New York, USA, 1999, pp. 528.
111. A. Gumiero, C. L. Metcalfe, A. R. Pearson, E. L. Raven, P. C. Moody, *J. Biol. Chem.* **2011**, *286*, 1260–1268.
112. T. A. Betley, Q. Wu, T. Van Voorhis, D. G. Nocera, *Inorg. Chem.* **2008**, *47*, 1849–1861.
113. I. Rivalta, G. W. Brudvig, V. S. Batista, *Curr. Opin. Chem. Biol.* **2012**, *16*, 11–18.
114. J. B. Lee, J. A. Hunt, J. T. Groves, *J. Am. Chem. Soc.* **1998**, *120*, 7493–7501.
115. J. Groves, J. Lee, J. Hunt, R. Shimanovich, N. Jin, *J. Inorg. Biochem.* **1999**, *74*, 28–28.
116. J. Su, J. Groves, *J. Am. Chem. Soc.* **2009**, *131*, 12979–12988.
117. J. Su, J. Groves, *Inorg. Chem.* **2010**, *49*, 6317–6329.
118. L. M. K. Dassama, T. H. Yosca, D. A. Conner, M. H. Lee, B. Blanc, B. R. Streit, M. T. Green, J. L. DuBois, C. Krebs, J. M. Bollinger, Jr., *Biochemistry* **2012**, *51*, 1607–1616.
119. J. L. DuBois, J. M. Mayfield, "Dioxygen-Generating Chlorite Dismutases and the CDE Protein Superfamily", Chapter 90 in *Handbook of Porphyrin Science*, Vol. 19, Eds K. M. Kadish, K. M. Smith, and R. Guilard, World Scientific, Singapore, 2012, pages 231–283.
120. C. Jakopitsch, H. Spalteholz, P. G. Fürtmuller, J. Arnhold, C. Obinger, *J. Inorg. Biochem.* **2008**, *102*, 293–302.
121. A. Gumiero, E. J. Murphy, C. L. Metcalfe, P. C. E. Moody, E. L. Raven, *Arch. Biochem. Biophys.* **2010**, *500*, 13–20.
122. A. N. Hiner, E. L. Raven, R. N. Thorneley, F. García-Cánovas, J. N. Rodríguez-López, *J Inorg. Biochem.* **2002**, *91*, 27–34.
123. J. E. Erman, L. B. Vitello, M. A. Miller, A. Shaw, K. A. Brown, J. Kraut, *Biochemistry* **1993**, *32*, 9798–9806.
124. B. C. Finzel, T. L. Poulos, J. Kraut, *J. Biol. Chem.* **1984**, *259*, 3027–3036.
125. S. L. Edwards, N. H. Xuong, R. C. Hamlin, J. Kraut, *Biochemistry* **1987**, *26*, 1503–1511.
126. J. Hernandez-Ruiz, M. B. Arnao, A. N. P. Hiner, F. Garcia-Canovas, M. Acosta, *Biochem. J.* **2001**, *354*, 107–114.
127. A. N. P. Hiner, J. N. Rodriguez-Lopez, M. B. Arnao, E. L. Raven, F. Garcia-Canovas, M. Acosta, *Biochem. J.* **2000**, *348*, 321–328.
128. J. N. Rodriguez-Lopez, J. Hernandez-Ruiz, F. Garcia-Canovas, R. N. F. Thorneley, M. Acosta, M. B. Arnao, *J. Biol. Chem.* **1997**, *272*, 5469–5476.
129. S. L. Edwards, T. L. Poulos, *J. Biol. Chem.* **1990**, *265*, 2588–2595.
130. J. A. Gustafsson, E. G. Hrycay, L. Ernster, *Arch. Biochem. Biophys.* **1976**, *174*, 440–453.
131. J. M. Pratt, T. I. Ridd, L. J. King, *J. Chem. Soc., Chem. Commun.* **1995**, *22*, 2297–2298.
132. L. M. Slaughter, J. P. Collman, T. A. Eberspacher, J. I. Brauman, *Inorg. Chem.* **2004**, *43*, 5198–5204.
133. J. P. Collman, H. Tanaka, R. T. Hembre, J. I. Brauman, *J. Am. Chem. Soc.* **1990**, *112*, 3689–3690.
134. M. M. Abu-Omar, *Dalton Trans.* **2011**, *40*, 3435–3444.
135. M. J. Zdilla, A. Q. Lee, M. M. Abu-Omar, *Angew. Chem. Int. Ed. Engl.* **2008**, *47*, 7697–7700.
136. M. J. Zdilla, A. Q. Lee, M. M. Abu-Omar, *Inorg. Chem.* **2009**, *48*, 2260–2268.
137. T. P. Umile, J. T. Groves, *Angew. Chem. Int. Ed. Engl.* **2011**, *50*, 695–698.
138. S. D. Hicks, J. L. Petersen, C. J. Bougher, M. M. Abu-Omar, *Angew. Chem. Int. Ed. Engl.* **2011**, *50*, 699–702.
139. W. D. Hewson, L. P. Hager, *J. Biol. Chem.* **1979**, *254*, 3175–3181.
140. S. Shahangian, L. P. Hager, *J. Biol. Chem.* **1982**, *257*, 1529–1533.
141. H. B. Dunford, R. A. Alberty, *Biochemistry* **1967**, *6*, 447.
142. M. A. Ator, S. K. David, P. R. O. De Montellano, *J. Biol. Chem.* **1987**, *262*, 14954–14960.
143. M. A. Ator, P. R. O. Demontellano, *J. Biol. Chem.* **1987**, *262*, 1542–1551.
144. P. R. O. Demontellano, S. K. David, M. A. Ator, D. Tew, *Biochemistry* **1988**, *27*, 5470–5476.
145. A. Wolterink, S. Kim, M. Muusse, I. S. Kim, P. J. M. Roholl, C. G. van Ginkel, A. J. M. Stams, S. W. M. Kengen, *Int. J. System. Evol. Microbiol.* **2005**, *55*, 2063–2068.
146. B. C. Okeke, W. T. Frankenberger, *Microbiol. Res.* **2003**, *158*, 337–344.

147. L. M. Steinberg, J. J. Trimble, B. E. Logan, *FEMS Microbiol. Lett.* **2005**, *247*, 153–159.
148. A. Wolterink, E. Schiltz, P. Hagedoorn, W. Hagen, S. Kengen, A. Stams, *J. Bacteriol.* **2003**, *185*, 3210–3213.
149. H. D. Thorell, N. H. Beyer, N. H. H. Heegaard, M. Ohman, T. Nilsson, *Eur. J. Biochem.* **2004**, *271*, 3539–3546.
150. K. Stenklo, H. D. Thorell, H. Bergius, R. Aasa, T. Nilsson, *J. Biol. Inorg. Chem.* **2001**, *6*, 601–607.
151. S. W. M. Kengen, G. B. Rikken, W. R. Hagen, C. G. van Ginkel, A. J. M. Stams, *J. Bact.* **1999**, *181*, 6706–6711.
152. F. Mehboob, A. F. M. Wolterink, A. J. Vermeulen, B. Jiang, P.-L. Hagedoorn, A. J. M. Stams, S. W. M. Kengen, *FEMS Microbiol. Lett.* **2009**, *293*, 115–121.
153. Å. Malmqvist, T. Welander, E. Moore, A. Ternström, G. Molin, I. Stenström, *Syst. Appl. Microbiol.* **1994**, *17*, 58–64.
154. H. Danielsson Thorell, K. Stenklo, J. Karlsson, T. Nilsson, *Appl. Environ. Microbiol.* **2003**, *69*, 5585–5592.
155. J. L. DuBois, C. J. Carrell, C. M. Wilmot, "Reactivity and Structure in the CDE Protein Superfamily: from O_2 Generation to Peroxidase Chemistry and Beyond", in *Handbook of Porphyrin Science*, Vol. 26, Ed G. Ferreira, World Scientific, Singapore, 2013, pp. 442–470.

Chapter 4
Respiratory Conservation of Energy with Dioxygen: Cytochrome *c* Oxidase

Shinya Yoshikawa, Atsuhiro Shimada, and Kyoko Shinzawa-Itoh

Contents

S. Yoshikawa (✉) • A. Shimada • K. Shinzawa-Itoh
Picobiology Institute, Graduate School of Life Science, University of Hyogo, Kamigohri Akoh Hyogo, 678-1297, Japan
e-mail: yoshi@sci.u-hyogo.ac.jp; ashima@sci.u-hyogo.ac.jp; shinzawa@sci.u-hyogo.jp

P.M.H. Kroneck, M.E. Sosa Torres (eds.), *Sustaining Life on Planet Earth: Metalloenzymes Mastering Dioxygen and Other Chewy Gases*, Metal Ions in Life Sciences 15, DOI 10.1007/978-3-319-12415-5_4

Abstract Cytochrome *c* oxidase (CcO) is the terminal oxidase of cell respiration which reduces molecular oxygen (O_2) to H_2O coupled with the proton pump. For elucidation of the mechanism of CcO, the three-dimensional location and chemical reactivity of each atom composing the functional sites have been extensively studied by various techniques, such as crystallography, vibrational and time-resolved electronic spectroscopy, since the X-ray structures (2.8 Å resolution) of bovine and bacterial CcO have been published in 1995.

X-ray structures of bovine CcO in different oxidation and ligand binding states showed that the O_2 reduction site, which is composed of Fe (heme a_3) and Cu (Cu_B), drives a non-sequential four-electron transfer for reduction of O_2 to water without releasing any reactive oxygen species. These data provide the crucial structural basis to solve a long-standing problem, the mechanism of the O_2 reduction.

Time-resolved resonance Raman and charge translocation analyses revealed the mechanism for coupling between O_2 reduction and the proton pump: O_2 is received by the O_2 reduction site where both metals are in the reduced state (R-intermediate), giving the O_2-bound form (A-intermediate). This is spontaneously converted to the P-intermediate, with the bound O_2 fully reduced to 2 O^{2-}. Hereafter the P-intermediate receives four electron equivalents from the second Fe site (heme *a*), one at a time, to form the three intermediates, F, O, and E to regenerate the R-intermediate. Each electron transfer step from heme *a* to the O_2 reduction site is coupled with the proton pump.

X-ray structural and mutational analyses of bovine CcO show three possible proton transfer pathways which can transfer pump protons (H) and chemical (water-forming) protons (K and D). The structure of the H-pathway of bovine CcO indicates that the driving force of the proton pump is the electrostatic repulsion between the protons on the H-pathway and positive charges of heme *a*, created upon oxidation to donate electrons to the O_2 reduction site. On the other hand, mutational and time-resolved electrometric findings for the bacterial CcO strongly suggest that the D-pathway transfers both pump and chemical protons. However, the structure for the proton-gating system in the D-pathway has not been experimentally identified. The structural and functional diversities in CcO from various species suggest a basic proton pumping mechanism in which heme *a* pumps protons while heme a_3 reduces O_2 as proposed in 1978.

Keywords cell respiration • cytochrome *c* oxidase • heme/copper terminal oxidase • membrane protein • O_2 reduction without forming ROS • proton pump

Please cite as: *Met. Ions Life Sci.* 15 (2015) 89–130

1 Introduction

Cytochrome *c* oxidase (CcO) is the terminal oxidase of aerobic cell respiration which reduces molecular dioxygen (O_2) to H_2O coupled to the process of proton pumping. Elucidation of the reaction mechanism of this enzyme at the atomic level is one of the most important subjects in Biological Science. The mechanism for O_2 reduction without releasing any reactive oxygen species (ROS) is also a long standing problem to be solved, in addition to the mechanism of coupling between the proton pump and O_2 reduction, and the mechanism of proton active transport [1, 2].

For the elucidation of the functional mechanism of a protein at the atomic level, the most basic information can be deduced from its high resolution X-ray structure. In the case of the CcO structure, a resolution at the hydrogen atom level will be necessary, since the CcO reaction is critically controlled by proton transfer. Usually, advances in our understanding of the reaction mechanism of a protein go along with the resolution of its X-ray structure [2]. X-ray structural analyses determine the three dimensional location of atoms composing the functional site of the protein. However, crystallography does not provide direct information on the chemical reactivity of the atoms in the protein. Thus, other physical techniques are needed, such as vibrational spectroscopy. Resonance Raman technique has given fundamental information for the mechanism of O_2 reduction to which two hemes in CcO critically contribute. Unfortunately, resonance Raman spectroscopy cannot be applied to examine proton transfer during the course of the catalytic cycle, since proton transfer is not directly driven by the Raman active chromophore. Thus, time-resolved infrared (IR) analysis is indispensable for the elucidation of the reaction mechanism. However, because of the strong IR absorption and of unavailability of a site-directed isotope labeling system producing sufficient amount of the sample, IR analysis has not been successfully applied for the CcO system except in a few cases. Nevertheless, our understanding of the reaction mechanism of CcO has improved remarkably [2] since the early reports on the X-ray structures of bovine and bacterial CcO in 1995 [3, 4]. Here, the recent structural understanding of CcO will be reviewed followed by a discussion on the O_2 reduction, the proton pump, and their coupling mechanisms based on the structural findings reported thus far.

2 The Structures of Bovine Heart Cytochrome *c* Oxidase

Recently X-ray structures of various bacterial CcO have been reported [5–10]. However, the structures of bovine CcO have been the most extensively studied [2] and therefore they are summarized here.

2.1 *Purification and Crystallization of Bovine Heart Cytochrome c Oxidase*

For X-ray structural analyses of any protein and the structural determination of cofactors and other constituents, such as lipids and hemes, purification and crystallization of the protein from the cell often represent the most difficult steps. Bovine heart CcO was solubilized with sodium cholate, a natural detergent, as early as 1941 [11], although the purified sample did not show any enzyme activity. However, 20 years later, cholate in the purified sample was replaced by non-ionic detergents to yield the enzyme activity [12]. The active sample was crystallized in 1961 [13]. Yet, another 34 years were required to establish conditions for obtaining crystals allowing the determination of the three-dimensional structure at 2.8 Å resolution [3]. Almost in parallel, a bacterial CcO was crystallized by an ingenious technique using a monoclonal antibody for promotion of the specific contacts between the CcO molecules in the crystal lattice. As a consequence, X-ray structures of both bovine and bacterial CcOs at 2.8 Å resolution were reported in the same week in 1995 [4].

The resolution of 2.8 Å is of course not at the hydrogen atom level. Therefore, extensive efforts for improving the resolution of the X-ray structure have been undertaken for bovine CcO. Bovine CcO contains 13 different subunits while bacterial CcOs have much less subunits, with a molecular mass of about 1/2 of the bovine enzyme [1, 3, 4]. Notably, bovine CcO shows one of the highest resolutions of X-ray structures of CcO reported thus far [2]. Simpler subunit composition might contribute to the stabilization of the isolated protein. However, a simple subunit composition does not necessarily assure the stability of the whole complex. In fact, the 10 nuclear coded subunits, which are not included in bacterial CcOs, surrounding the 3 core subunits which are homologous to the bacterial enzyme, are likely to stabilize the conformation of the core subunits to provide the high resolution X-ray structure.

2.2 *X-Ray Structure of the Protein Moiety*

The three dimensional arrangement of the 13 different subunits of bovine heart CcO is given in Figure 1. The biggest three core subunits I, II, and III are surrounded by 10 nuclear coded subunits [2]. The existence of 13 subunits in bovine heart CcO has been documented by careful SDS-PAGE analyses and later approved by the X-ray structure [14]. However, it should be noted that these two findings cannot exclude the possibility that these subunits, the physiological functions of which are unknown, are copurified and cocrystallized contaminant proteins. In fact, the physiological role of the 10 nuclear coded subunits is still unknown.

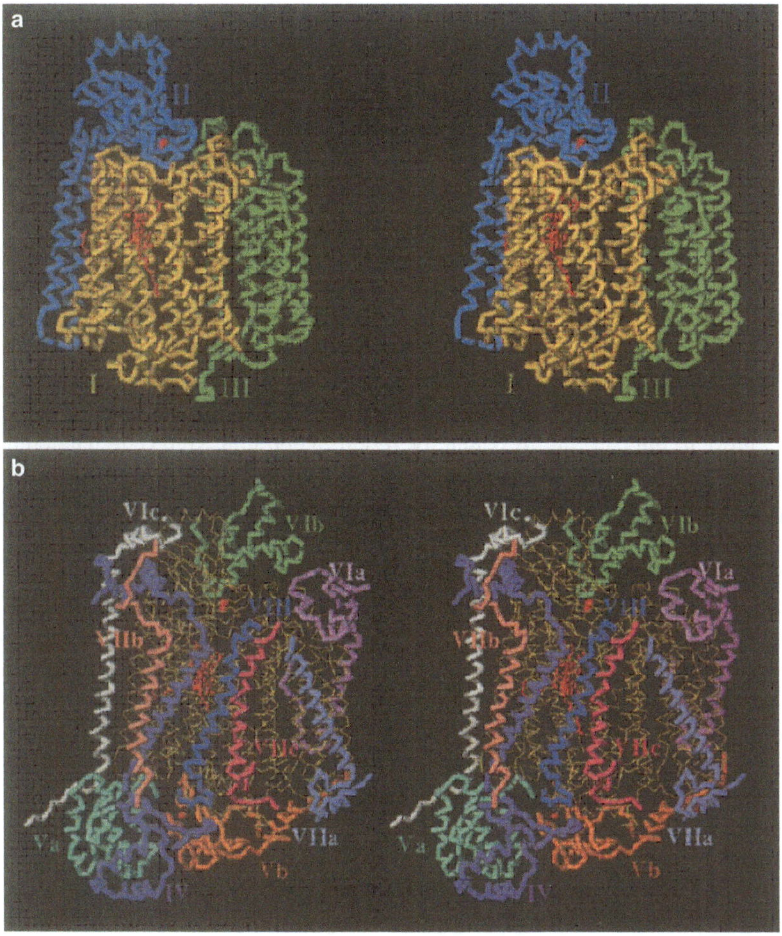

Figure 1 Stereo views of the Cα-backbone trace for the 13 subunits. (**a**) mitochondrially-encoded subunits I (yellow), II (blue), and III (green). (**b**) nuclear-encoded subunits IV (purple), Va (blue), Vb (dark yellow), VIa (pale reddish violet), VIb (blue green), VIc (gray), VIIa (lavender), VIIb (beige), VIIc (pink), VIII (indigo). A red ball in subunit Vb denotes the zinc atom. Red models and balls in subunits I and II, respectively, denote hemes and Cu atoms in Cu_A. Subunits I, II, and III are shown by yellow thin sticks in (**b**). Reproduced by permission from [85]; copyright 1996 American Association for the Advancement of Science.

2.3 Structure and Stoichiometry of the Metal Sites

Bovine heart CcO has four redox-active metal sites (Figure 2). The iron sites are occupied by heme A which is characterized by a formyl group at position 8 and a hydroxyfarnesylethyl group at position 2 (in the Fischer numbering of heme peripheral groups). Since heme A is quite unstable compared to heme B and has a significantly more complex structure, its molecular structure was determined as late as 1975, except for the absolute configuration of the asymmetric carbon of the hydroxyfarnesylethyl group [15]. This was determined to be an *S*-conformation in 2005 by X-ray structural analyses [16]. The two hemes A interact with the protein moiety (subunit I) in different ways to provide completely different functions. One heme A is in a low-spin state, with two axially coordinated histidine imidazole rings, and is designated heme *a* (or Fe_a) (Figure 2). The second heme A is located fairly close to the Cu_B site. It is in a five-coordinated high-spin state when reduced and ligand-free, and is designated heme a_3 (Fe_{a3}). The two heme planes are perpendicular to the membrane surface and fairly close to each other (Figure 2). The Cu_B site near Fe_{a3} has a trigonal planar geometry in the reduced state, with three histidines coordinated. Together with Fe_{a3} it forms the O_2 reduction site, where Fe_{a3} is the site for O_2 binding. The second copper site, Cu_A, consists of two copper ions bridged by two cysteine residues and is located in a cupredoxin configuration. This site reversibly accepts only one electron equivalent, the reduced and oxidized states of Cu_A are designated as ${Cu_A}^{1+}$ and ${Cu_A}^{2+}$, respectively, in this article. The cupredoxin configuration had been shown by EPR analysis before the determination of X-rays structures of bovine and bacterial CcOs [17]. The physiological requirement of the dinuclear center against a mononuclear center has not been elucidated. Cu_A is the initial electron acceptor from cytochrome *c*.

The X-ray structure of bovine heart CcO (Figure 3) indicates that Cu_A has three electron transfer pathways to heme a_3: (i) the His204-Arg438-D ring propionate of heme *a*-Fe_a-heme a_3 (the electron transfer between the two hemes is likely to be very fast because of the proximity of the two hemes), (ii) the His204-Arg438-C ring propionate of heme a_3, and (iii) the Glu198-Mg^{2+}-His368-D ring propionate of heme a_3 (numbering of amino acid sequences of bovine heart CcO is used unless otherwise noted). However, the redox-inactive Mg^{2+} site is unlikely to convey electrons, it rather blocks the electron transfer. No structural arguments against pathway (ii) have been found in the X-ray structure. However, no experimental results support the direct electron transfer from Cu_A to heme a_3 by-passing heme *a*. Thus, pathways (ii) and (iii) are inactive in CcO under physiological conditions.

Bovine CcO has two other redox-inactive metal sites, Zn^{2+} and Na^+, in addition to Mg^{2+}. It has been suggested that the Na^+ site contributes to facilitate proton transfer through a peptide bond in the hydrogen bond network of the H-pathway as described below. The physiological role of Zn^{2+} is still unknown.

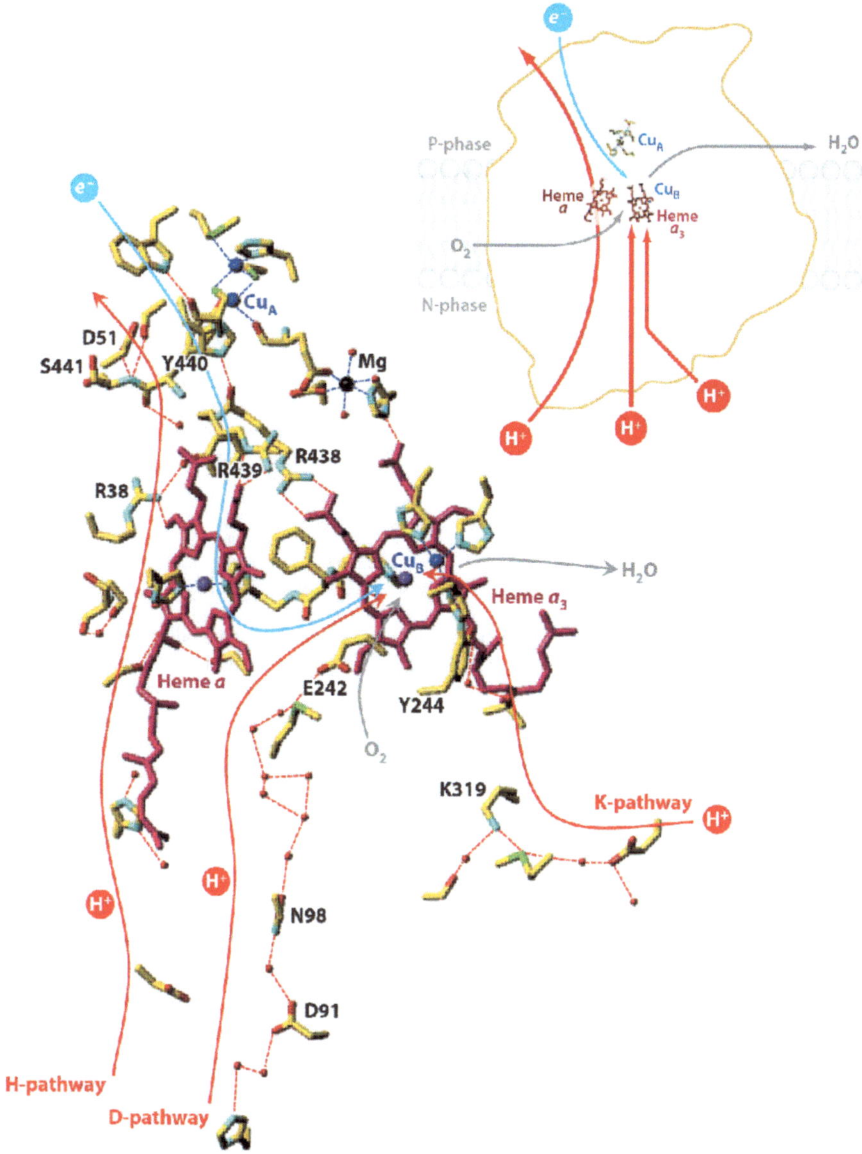

Figure 2 Structures and locations of redox active sites and possible pathways for protons, O_2, water, and electrons detectable in the X-ray structure of bovine heart CcO. The inset shows the location of the redox active metal sites in the overall structure. Reproduced by permission from [2]; copyright 2011 Annu. Rev. Biophys.

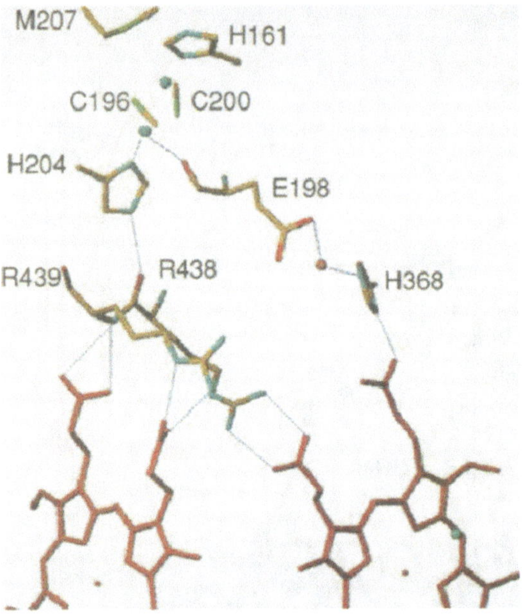

Figure 3 Hydrogen bond network between Cu_A and hemes *a* and a_3. Red models with and without a small blue ball (Cu_B) are hemes *a* and a_3, respectively. Two copper atoms in the Cu_A site are shown as two blue balls in the upper part. The Mg^{2+} site is denoted by a small yellow ball. The blue, red, and green portions of amino acids denote nitrogen, oxygen, and sulfur atoms, respectively. Dotted and broken lines denote hydrogen bonds and coordination bonds, respectively. Reproduced by permission from [85]; copyright 1996 American Association for the Advancement of Science.

2.4 *Lipid Structures and Contents*

Bovine CcO has lipids as intrinsic constituents which are likely to provide highly non-polar microenvironments in the interior of the protein. Thus, the structural determination is essential for the elucidation of its reaction mechanism. For lipid analysis of a membrane protein, discrimination of the intrinsic lipids from the weakly or accidentally bound lipids must be taken into account. Partial denaturation of the membrane protein would weaken the binding affinity of the intrinsic lipids. Thus, for lipid analysis, a significant amount of protein samples free of denatured molecules is absolutely necessary. One of the best methods for removing denatured molecules is crystallization. Using bovine CcO samples purified by repeated crystallizations, the structure determination of the phospholipids was achieved (Figure 4) [18]. The content and the three-dimensional structure of these phospholipids were determined by fitting these structures into the electron density of bovine heart CcO.

Subunit III, the second largest subunit, hosts the O_2 transfer pathway to the O_2 reduction site (Figure 5). Three phospholipids, phosphatidylethanolamine (PE2), phosphatidylglycerol (PG1 and PG2), form part of the wall of the O_2 transfer pathway (Figures 5a, 5b). These three phospholipids are tightly bound to the protein moiety so that the electron density maps of the unsaturated bonds of the two PGs identify the *cis*- against *trans*-configurations [18].

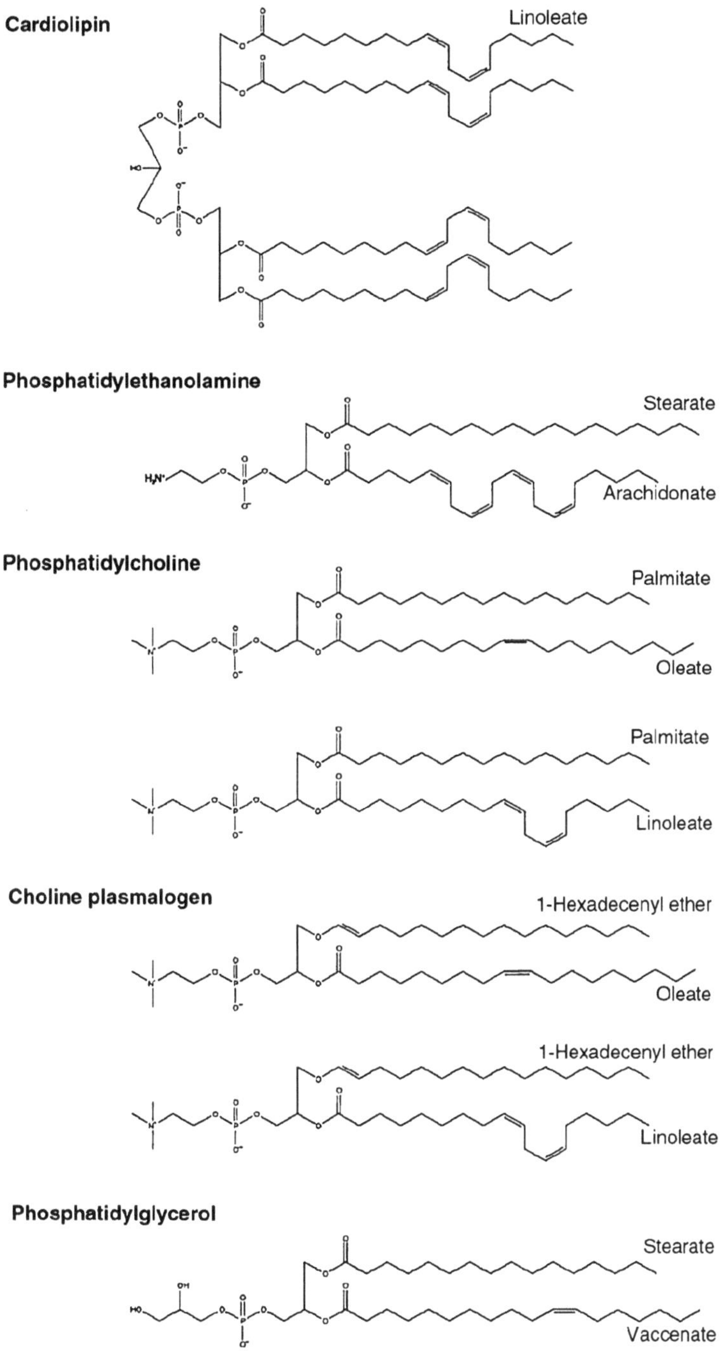

Figure 4 Chemical structures of phospholipids detected in crystalline bovine heart CcO. The major configuration (*cis*) is shown for vaccenate and oleate. The configuration of the other unsaturated fatty acids is provisionally assigned as the *cis* configuration. Reproduced by permission from [18]; copyright 2007 John Wiley and Sons.

Dicyclohexylcarbodiimide (DCCD) treatment, which gives the dicyclohexyl-*N*-acyl-urea (DCU) derivative of Glu90 in subunit III, influences the conformation of PG1 and PG2 to close the O_2 pathway (Figure 5c). The results suggest that phospholipids control the O_2 supply to the O_2 reduction site. X-ray structures of these fatty acid tails indicate that their flexibility is maximally utilized for a different space-filling function [18].

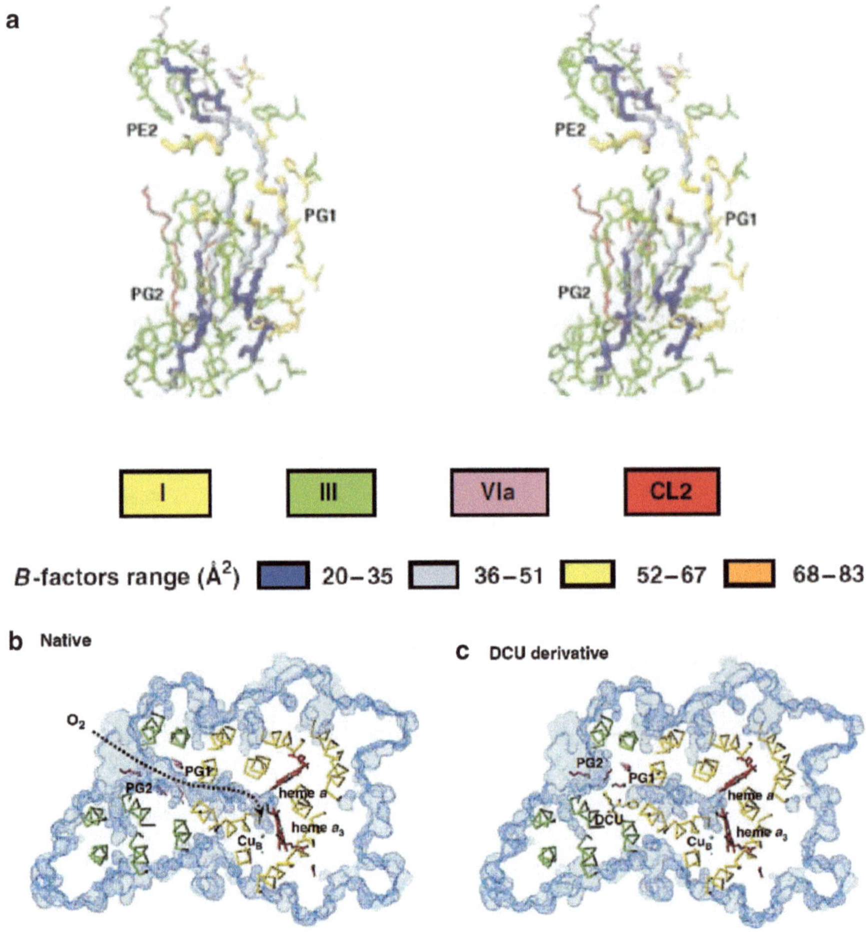

Figure 5 Phospholipids in subunit III. (**a**) Stereo view of atomic models of the three phospholipids within subunit III, including amino acid residues, hydrogen-bonded or under hydrophobic interactions to these phospholipids. The amino acids which belong to subunits I, III, and VIa are shown by yellow, green, and purple structures. A portion of cardiolipin (CL) as given in red interacts with PG2. (**b**) Cross section of CcO parallel to the membrane surface at the level of Fe_{a3} and Cu_B, denoted by cyan balls. Cα backbone structures of subunits I and III are shown by yellow and green thin sticks, respectively. Two PGs in subunit III, PG1 and PG2, are given in pink and magenta sticks, respectively. A possible O_2 channel is shown by a black dotted line. (**c**) Cross section of the DCU derivative CcO parallel to the membrane surface at the level of Fe_{a3} and Cu_B, denoted by cyan balls. The possible O_2 channel is blocked by the conformational changes of PG1 and PG2. Reproduced by permission from [18]; copyright 2007 John Wiley and Sons.

3 Mechanism of Dioxygen Reduction

As described above, CcO has four redox-active metal centers, each of which reversibly receives one electron equivalent. Thus, the fully reduced form of CcO, in which all four metal sites are in the reduced (one-electron received) state, will reduce O_2 completely to 2 H_2O and represents a suitable system to investigate the mechanism of the reduction of molecular oxygen to water. However, O_2 reduction by CcO is too fast to follow by conventional rapid mixing techniques which usually have a dead time in the range of 1 ms. Furthermore, CO does not block the O_2 access completely within the time range of the manual operation. Consequently, a flash-photolysis procedure has to be used, where CO-bound CcO is mixed with O_2-saturated buffer and CO is flashed off in the presence of O_2. CO is able to block the O_2 access of CcO within a few ms in the dark.

It is well-known that the one-electron reduction of O_2 to the superoxide anion, O_2^-, is energetically unfavorable while simultaneous (or non-sequential) two-electron reduction is energetically favorable [19]. This intrinsic property of O_2 contributes significantly to the stability of the oxygenated form of hemoglobin and myoglobin. However, IR data indicate that the O–O stretch band position is close to that of superoxide [20]. In other words, the oxygenated form of CcO can be best described by a resonance structure Fe_{a3}^{2+}-O_2 ↔ Fe_{a3}^{3+}-O_2^-; for the sake of simplicity, the structure of the oxygenated form is written as Fe_{a3}^{3+}-O_2^-. It should be noted that the Fe_{a3}^{3+}-O_2^- species cannot be compared to Fe^{3+}-Cl^- which will release Cl^- upon lowering the Cl^- concentration. On the other hand, oxygenated hemoglobin, or myoglobin, will release O_2 to form deoxygenated Fe(II)-globin. As described above, the O_2 reduction site of CcO is composed of a dinuclear site (Fe_{a3}^{2+}, Cu_B^{1+}). Thus, the second electron to the bound O_2 at Fe_{a3}^{2+} is readily available from Cu_B^{1+}, and the first intermediate of the O_2 reduction process is most likely not the O_2-bound form (Fe_{a3}^{2+}-O_2 ↔ Fe_{a3}^{3+}-O_2^-) as described above.

3.1 Resonance Raman Analysis

Since the historical report of the flow-flash analyses by Gibson and Greenwood [21], extensive work has been performed on the process of O_2 reduction by fully reduced CcO, following the UV-vis spectral changes. However, because of the limitation of absorption spectroscopy, the identification of the structure of the initial intermediate of the process remained unsuccessful until the resonance Raman technique was introduced by three research groups [22–24].

In contrast to absorption spectroscopy, the resonance Raman technique provides various structural findings, such as the chemical structure of the heme-bound ligand, which cannot be obtained from absorption spectral analyses. Another advantage is that the isotope shift effect is applicable for the assignment and identification of bands. A successful application of this technique to examine the process of the O_2 reduction by the fully reduced bovine heart CcO is given in Figure 6

[25]. Each spectrum represents a difference spectrum ($^{16}O_2$ against $^{18}O_2$) at each delay time after flash photolysis of the CO-bound fully reduced CcO in the presence of O_2. The initial intermediate detectable at the delay time of 0.1 ms shows the Raman band at 571 cm^{-1} for $^{16}O_2$ *versus* 544 cm^{-1} for $^{18}O_2$. As no other band is detected, this band must result from the initial intermediate in the reaction between O_2 and the fully reduced CcO. The band position is essentially identical to those observed for oxygenated hemoglobins and myoglobins. In other words, quite unexpectedly, the band position strongly suggests that the initial intermediate is Fe_{a3}^{3+}-O_2^-.

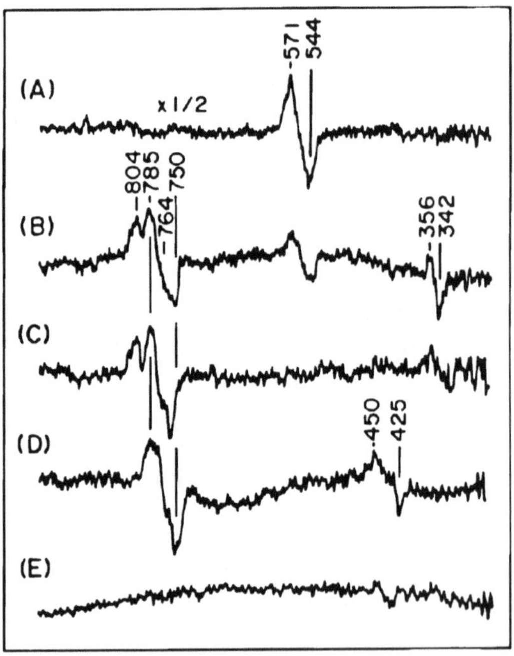

Figure 6 Time-resolved resonance Raman difference spectra of reaction intermediates of bovine heart CcO. The Raman difference spectra have been determined by subtracting the spectrum of the corresponding $^{18}O_2$ derivative from the spectrum of the $^{16}O_2$ derivative at each delay time, excited at the wave length of 423 nm at 3 °C. The delay time after the initiation of the reaction is 0.1 (A), 0.27 (B), 0.54 (C), 2.7 (D), and 5.4 (E) ms. Reproduced by permission from [25]; copyright 1996 American Chemical Society

In order to confirm this assignment, the isotope shifts of unevenly labeled O_2 ($^{18}O^{16}O$) have been examined. If this band is due to the O_2 bound at Fe_{a3}^{2+}, the $^{18}O^{16}O$ difference spectra *versus* spectra with $^{16}O_2$- and $^{18}O_2$-labeled dioxygen would show two bands located between the bands due to $^{16}O_2$ and $^{18}O_2$. However, if this band is due to $Fe^{4+}{=}O^{2-}$ (ferryloxide form) which is formed in the later stage of the O_2 reduction process, the two bands would appear at the same positions as the bands due to $^{16}O_2$ and $^{18}O_2$. The former results have been observed [26]. However, the results do not discriminate Fe^{3+}-O_2^- *versus* Fe^{3+}-O_2^{2-} (oxygenated *versus* peroxide-bound). However, it is well-known that the peroxide-bound form does not show a Fe–O stretch band near 570 cm^{-1}, while the oxygenated form does not show a O–O stretch band near 800 cm^{-1} [27]. In fact, the initial intermediate does not show the O–O band. Thus, it has been well accepted that the structure of the initial intermediate (A-intermediate) is Fe^{3+}-O_2^-.

The second intermediate shows the 804/764 cm^{-1} band which significantly overlaps with the 785/750 cm^{-1} band (Figure 6B). The former band disappears at a delay time of 2.7 ms (Figure 6D), when the 785/750 cm^{-1} band still exists. Thus, the 804/764 cm^{-1} and 785/750 cm^{-1} bands are due to the second and third intermediates following the A-intermediate. The fourth intermediate showing the 450/425 cm^{-1} band appears at a delay time of 2.7 ms (Figure 6D). It should be noted that the process at 3 °C is significantly slower than the results at room temperature reported thus far.

The resonance Raman spectral changes recorded during the course of the reaction between the mixed valence CcO and O_2 have been examined for the assignment of these bands. When CO is introduced to CcO in the two-electron reduced state, CO stabilizes both Fe_{a3} and Cu_B in the reduced state [Fe_{a3}^{2+}-CO and Cu_B^{1+}], which is designated as the mixed valence CO-bound form. The reaction between O_2 and the mixed valence CcO [Fe_{a3}^{2+}, Cu_B^{1+}, Fe_a^{3+}, Cu_A^{2+}] can be examined applying the flash photolysis technique to this form. In this case, the bound O_2 would not be completely reduced. The resonance Raman results have shown that the 571/544 cm^{-1} band appears as the initial intermediate, followed by the 804/764 cm^{-1} band without showing the 785/750 and 450/425 cm^{-1} bands detectable during the process of the fully reduced CcO with O_2 [28, 29]. These results indicate that the 804/764 cm^{-1} band is directly produced from the A-intermediate and that the intermediate giving the 804/764 cm^{-1} band is in one oxidation state higher than that of the 785/750 cm^{-1} band.

Taking into account the intrinsic chemical property of O_2, as described above [19], the bound O_2 in the A-intermediate [Fe_{a3}^{3+}-O_2^-, Cu_B^{1+}] is very likely to extract one electron from Cu_B^{1+} to form a peroxide-bound (or -bridged) form [Fe_{a3}^{3+}-O_2^{2-}-Cu_B^{2+}]. Therefore this intermediate is designated as P-intermediate. However, unexpectedly, the isotope shift effects detectable using unevenly labeled ($^{18}O^{16}O$) clearly have shown that Fe_{a3} is a mono-oxygen-bound species ($Fe^{4+}{=}O^{2-}$) [26]. In other words, the bound O_2 in the A-intermediate has been completely reduced to the O^{2-} level. Four electron equivalents are required for complete reduction of O_2. Two electron equivalents are available from Fe_{a3}^{2+} and one electron equivalent from Cu_B^{1+}. The origin of the fourth electron equivalent is still under debate although the OH group of Tyr244 which is covalently connected to one of the histidine ligands of Cu_B has been proposed [30, 31]. Similarly, the experiments with unevenly labeled ($^{18}O^{16}O$) indicate that both the 785/750 and 450/425 cm^{-1} bands favor a mono-oxygen-bound species. The observed band at 450/425 cm^{-1} strongly suggests that it is due to Fe_{a3}^{3+}-OH^-. Thus, the 785/750 cm^{-1} band is assigned to the ferryloxide form, $Fe^{4+}{=}O$. These intermediates are designated as O and F, respectively (Figure 7). In this figure, one of the oxidation equivalents in the P-intermediate is located on Fe_{a3} for the sake of simplicity.

The O_2 reduction process has been extensively examined by following the UV-vis spectral changes [32–36]. Because of the much wider bandwidths of the absorption spectra of these intermediates, compared with those of the Raman bands, straightforward interpretations for the absorption spectral changes remain difficult. However, these absorption spectral results essentially confirm the resonance Raman results.

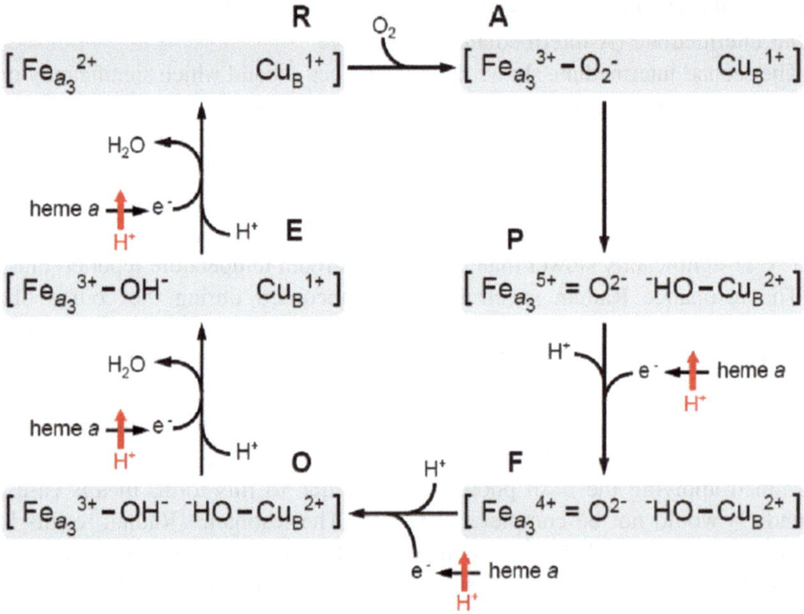

Figure 7 Schematic representation of the catalytic cycle of CcO. For the sake of simplicity, the low potential sites (Fe_a and Cu_A) and the possible electron donation site for the O_2 reduction process, Tyr244, are not included. The fourth oxidation equivalent created upon P formation is putatively located on Fe_{a3} in the P-intermediate.

3.2 X-Ray Structural Data

The resonance Raman results from above indicate that the P-intermediate, in which the bound oxygen has been fully reduced to the O^{2-} level, appears as the direct decay product of the A-intermediate. Thus, under normal turnover conditions, it is impossible to detect the elementary steps of the process $O_2 \rightarrow 2\ O^{2-}$. In other words, the decay of the intermediate species between O_2 and $2\ O^{2-}$ is much faster than their formations. Since the O_2 reduction is catalyzed by a protein, it is impossible to exchange the rate-limiting step in the physiological temperature with the other elementary step by a temperature change since the temperature range in which normal turnover proceeds is quite narrow. Thus, effects of O_2 analogues or respiratory inhibitors of the O_2 reduction site have been investigated extensively by X-ray crystallography.

NO, which is the best O_2 analogue, binds to $Fe_{a3}{}^{2+}$ in a bent end-on fashion (Figure 8a). As schematically shown, the distance between Cu_B and the O atom of NO is 2.5 Å, suggesting that the ligand-Cu_B interaction is very weak [37]. From the trigonal planar coordination geometry of the Cu_B site it can be deduced that Cu is both a poor ligand acceptor as well as a poor electron donor, which also will contribute to weaken the interaction between the ligand and Cu_B. Tyr244 is the possible fourth electron donor to produce the P-intermediate [30, 31]. However, as clearly shown in Figure 8b, the interaction between the bound NO and Tyr244-OH is effectively blocked by the imidazole ring of His240. Thus, a significant conformational change is required for the OH group to interact with the bound ligand to allow electron transfer. The X-ray data presented provide a strong structural basis for the stability of the O_2-bound form of CcO (A-intermediate, Figure 7).

The X-ray structure of the cyanide-bound fully reduced form (Figure 8c) shows fairly large rearrangements of the Cu_B site [37]. One of the three histidines, His290, coordinated to Cu_B in the fully reduced ligand-free state, dissociates from the Cu_B site and CN^- forms a new trigonal planar Cu_B center perpendicular to the heme plane (Figure 8d). The other end of the bound cyanide interacts with Tyr244 with a short hydrogen bond network including a fixed water (water510). A significant translational movement of the heme plane is necessary to form this hydrogen bond network. Water510 is likely to be transferred from a water storage site nearby, detectable in the X-ray structure of CcO (Supporting Information in [37]). This X-ray structure strongly suggests that the O_2 reduction site can form three possible electron transfer pathways to the bound O_2^-, that is, from Cu_B, Fe_{a3}, and Tyr244-OH via water510. The results provide a strong structural basis for the non-sequential three-electron donation to the bound O_2^-.

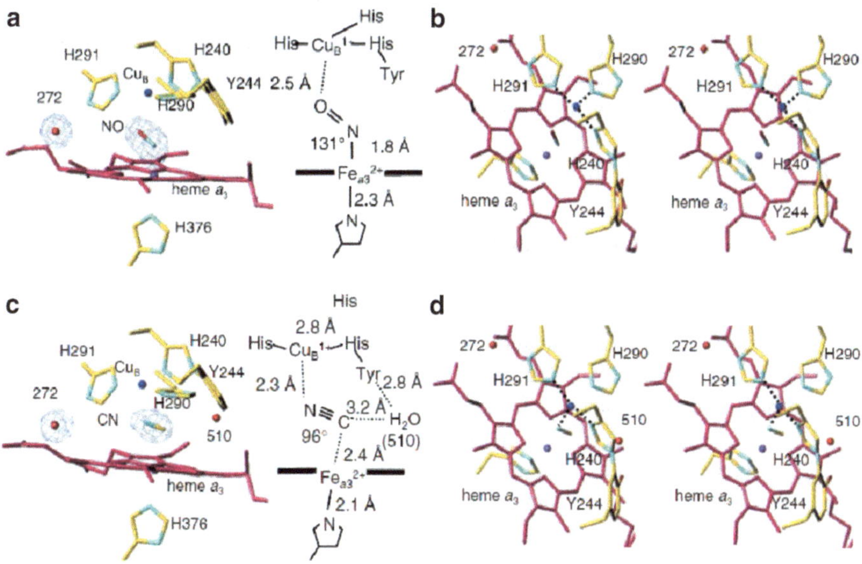

Figure 8 X-ray structures of the O_2 reduction site of the NO- and CN^--bound forms of fully reduced CcO. The NO- and CN^-- binding structures are given in *Fo-Fc* maps in panels (**a**) and (**c**), respectively. The digits without letters in the *Fo-Fc* maps indicate numbering of fixed water molecules. (**a**) The NO-bound form at 100 K and a schematic representation of its structural characteristics. (**b**) A stereo view of the NO-bound form. (**c**) The CN^--bound form at 100 K and a schematic representation of its structural characteristics. (**d**) A stereo view of the CN^--bound form. Reproduced by permission from [37]; copyright 2010 National Academy of Sciences USA.

These results indicate a possible O_2 reduction mechanism as described in Figure 9 [37]. When the O_2 reduction site is in the fully reduced state, O_2 is received by Cu_B (Figure 9b). The transient ligand binding to Cu_B has been shown by IR analyses [38, 39]. As discussed below, time-resolved IR data suggest that this transient binding is a prerequisite for effective collection of pump protons. At an appropriate timing sensed by Cu_B, O_2 is transferred to Fe_{a3}^{2+} to form Fe_{a3}^{3+}-O_2^- which induces the significant rearrangement of the Cu_B site, as CN^- does, to give the three possible electron transfer pathways to the bound O_2 (Figure 9d). The non-sequential three electron transfer steps to bound O_2^- will reduce O_2^- completely to provide the P-intermediate (Figure 9e) without releasing reactive oxygen species.

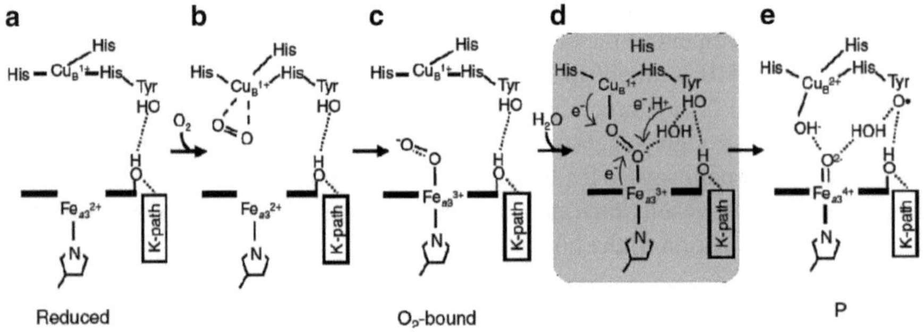

Figure 9 Schematic representation of the O_2 reduction mechanism in CcO. The proposed (experimentally undetected) intermediate is shown in the shadowed area. Reproduced by permission from [37]; copyright 2010 National Academy of Sciences USA.

3.3 *Biomimetic Studies*

Studies with synthetic model complexes have provided various crucial insights into the mechanism of O_2 reduction by CcO once the X-ray structures of CcOs had been published.

A model of the O_2 reduction site has been synthesized (Figure 10) [27, 40], which includes all functional groups interacting with both copper and iron. Hereby, Cu is trigonally planar coordinated whereas Fe is five-coordinated to a porphyrin, with an axial imidazole ligand. The iron and copper moieties are tethered together by three phenyl groups at the three methene bridges of the porphyrin. The fully reduced model complex binds O_2 to form a superoxide bound state [Fe^{3+}-O_2^-, Cu^{1+}], in agreement with the resonance Raman band at 570/544 cm^{-1} band for the ($^{16}O_2$/$^{18}O_2$) isotopes. The oxygenated compound is stable at room temperature under vacuum. However, when Cu is removed, low temperature is required to stabilize the O_2^--bound state. The intrinsic chemical property of O_2, i.e., thermodynamically unfavorable one-electron reduction and favorable two-electron reduction, does not predict these properties.

The model complex, with O_2 bound, oxidizes phenols producing two equivalents of the phenoxy radical species and the reduced complex [$Fe^{3+}OH^-$, Cu^{2+}-OH^-]. This stoichiometry indicates that O_2 has been reduced to two OH^- (or H_2O), without the release of ROS. Further examination of the reaction under

Figure 10 Structure of the biomimetic model complex of the O_2-reduction site (Fe/Cu[NMePr]). Reproduced by permission from [27]; copyright 2003 American Chemical Society.

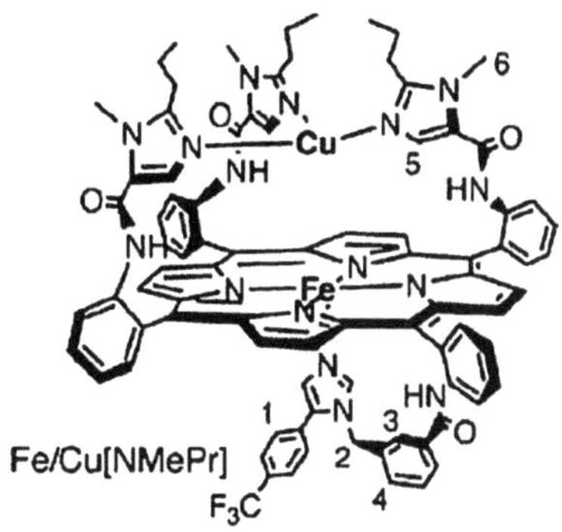

various experimental conditions suggest the formation of a peroxo intermediate [Fe^{3+}-OOH, Cu^{2+}] and a ferryloxo intermediate [$Fe^{4+}{=}O^{2-}$, Cu^{2+}-OH] [41]. As described above, the fourth electron donor for formation of the P-intermediate in the normal turnover of CcO is still debated. However, the biomimetic studies strongly suggest Tyr244 being the fourth electron donor.

Model compounds in which a phenol group is covalently attached to one of the three imidazoles coordinated to Cu have been synthesized [42]. The phenol group mimics Tyr244 of CcO directing toward the O_2 reduction site. The fully reduced model compound reduces O_2 to provide oxidized products containing a ferryloxo and a phenoxy radical. Esterification of the phenol OH group leads to a significant increase in ROS production. Note that the unmodified model complex produces small but still significantly higher amounts of ROS compared with CcO. It has been proposed that the specificity of O_2 binding to the reduced catalytic site, with both Fe and Cu reduced (Fe^{2+}, Cu^{+}), is critical for O_2 reduction and linked to this the release of potential ROS. In CcO, the catalytic site accepts O_2 only when both Fe and Cu are reduced (Fe^{2+}, Cu^{+}), whereas in the case of the model compounds O_2 also binds to the partially reduced (Fe^{2+}, Cu^{2+}) state [43].

The peroxide-bridged intermediate ($Fe_{a3}{}^{3+}$-O_2^{2-}-$Cu_B{}^{2+}$) has not been detected under normal CcO turnover conditions. However, it is likely for the peroxide-bridged species to appear as an intermediate in the reaction step from the A- to the P-intermediate during which O_2 receives four electrons. Thus, extensive model studies have been reported for peroxide compounds in order to examine the roles of each component of the O_2 reduction site. Cu complexes mimicking the Cu_B site, with three and four nitrogenous ligands (tri- and tetradentate, respectively, Figure 11a) have been synthesized [44]. These Cu complexes form high-spin peroxide-bridged [Fe^{3+}-O_2^{2-}-Cu^{2+}] species with Fe(II)-heme (Figure 11b) [44]. Both high-spin peroxide-bridged [Fe^{3+}-O_2^{2-}-Cu^{2+}] species, independent of whether Cu is coordinated to a tridentate or a tetradentate ligand, are more or less inactive towards phenols. In the case of the tridentate [Fe^{3+}-O_2^{2-}-Cu^{2+}] species, Fe is coordinated by the strong ligand 1,5-dicyclohexyl-imidazole (DCHIm) to produce a low-spin complex (Figure 11b, structure (2b)). In contrast, the tetradentate [Fe^{3+}-O_2^{2-}-Cu^{2+}] species, with DCHIm liganded to Fe, is unstable (Figure 11b).

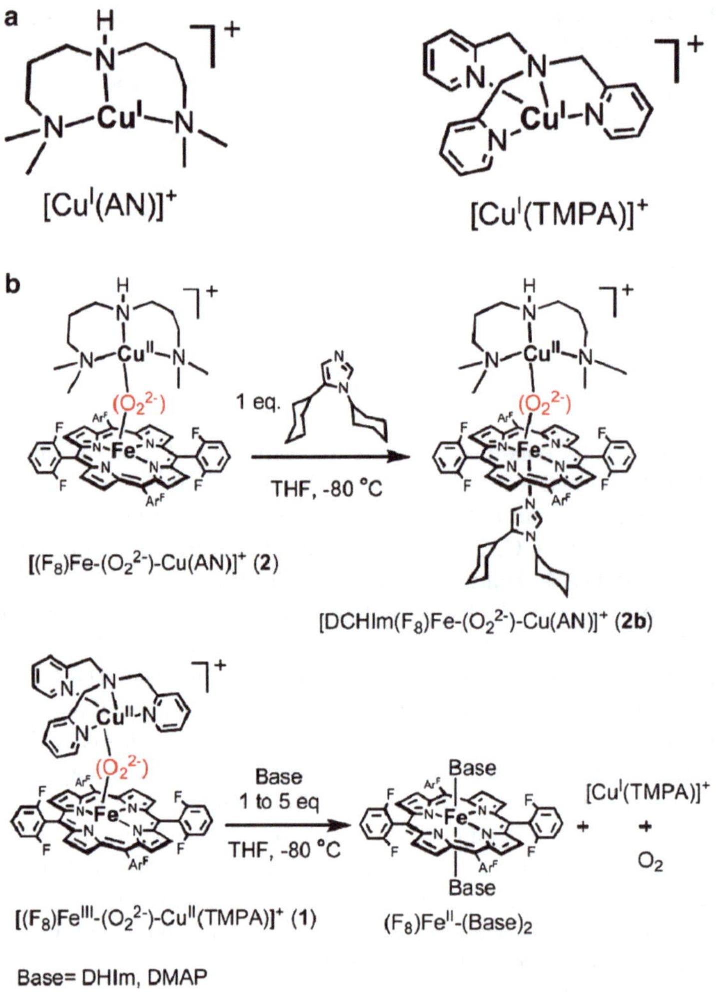

Figure 11 Generation of a low spin Fe-heme-peroxo-Cu complex. (**a**) The structures of tridentate and tetradentate Cu complexes used for untethered peroxide compounds. (**b**) **Top**: Generation of the low-spin heme-peroxo-Cu complex (2b) from the high-spin heme-peroxide-tridentate Cu complex by addition of 1,5-dicyclohexylimidazole (DCHIm). **Bottom**: Effect of the addition of base (1,5-dicyclohexylimidazole (DCHIm) or 4-(dimethylamino)pyridine (DMAP)) to the high-spin heme-peroxide-tetradentate Cu complex. Reproduced by permission from [44]; copyright 2010 American Chemical Society.

The results indicate that Cu complexes with tridentate ligands stabilize the low-spin peroxide species. In contrast to the high-spin species, the low-spin peroxide complex shows strong reactivity to phenols to receive one electron equivalent for reducing the bound peroxide completely to the 2 O^{2-} level giving $Fe^{4+}{=}O^{2-}$ (2c), Cu^{2+}-OH^{-} and a phenol radical (Figure 12) [44]. Thus, in order to reduce the bound peroxide in the compounds given in Figure 12, the tridentate Cu complex stabilizes

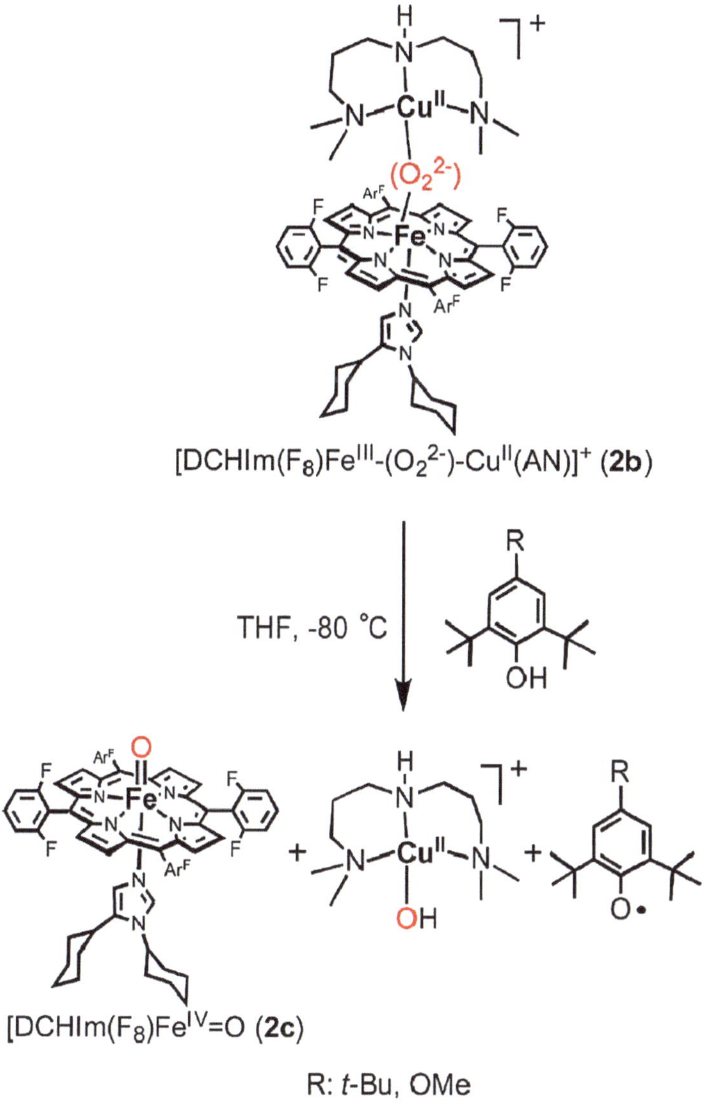

Figure 12 Formation of a phenoxy radical by addition of a phenol derivative to the heme-peroxy-tridentate copper complex (2b). Reproduced by permission from [44]; copyright 2010 American Chemical Society.

the low-spin state which has reactivity towards the phenolic OH group. However, Cu coordinated by a tetradentate ligand does not form the low-spin peroxide species. These experimental findings document the importance of the tridentate structure of Cu_B in CcO for stabilizing the low-spin peroxide species. The above accomplishments clearly show a typical successful example of the biomimetic model approach to metalloprotein chemistry giving unique and thoughtful insights.

A tetradentate tethered model complex (Figure 13) [45] has O_2 reduction activity in solution, using decamethylferrocene and trifluoroacetic acid as electron and proton donors, respectively. Remarkably, this system shows effective electron transfer, with turnover numbers up to 1000, without producing ROS. At room temperature, the fully reduced (Fe^{2+}, Cu^{1+}) state seems to be the steady state species while at −60 °C, a hydroperoxo state [$Fe^{3+}OOH$, Cu^{2+}] is the steady-state species. Thus, the rate-limiting step moves from O_2 binding to cleavage of hydroperoxo by the temperature decrease. A careful comparison of the structures of these model compounds with the X-ray structure of CcO at high resolution would provide various insights in the O_2 reduction mechanism.

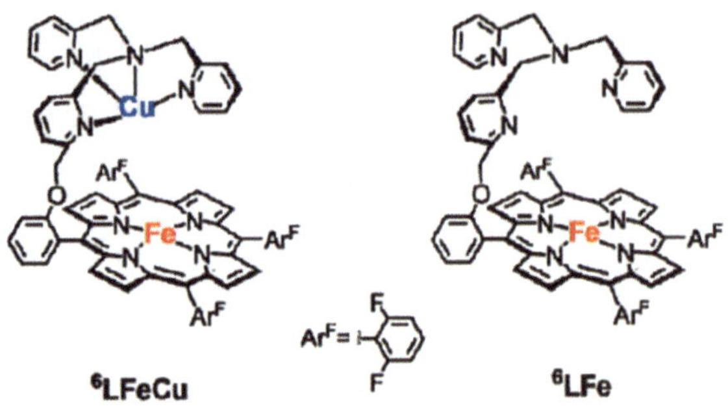

Figure 13 Heme/Cu synthetic models for CcO with copper (^{6}LFeCu) and without copper (^{6}LFe). Reproduced by permission from [45]; copyright 2011 National Academy of Sciences USA.

4 Proton Pump Mechanism

As schematically shown in Figure 7, the R-intermediate receives O_2 to form the A-intermediate which is spontaneously converted to the P-intermediate. The bound O_2 is completely reduced to O^{2-} and OH^- in the P-intermediate. The fourth electron equivalent and one proton equivalent are most likely from the Tyr244 OH group. The P-intermediate receives four electron equivalents, one at a time to generate the R-intermediate. Each electron transfer is coupled with the transfer of one chemical proton equivalent and one pump proton equivalent. Thus, in each catalytic cycle, one O_2 molecule is reduced to two water molecules by four electrons and four chemical protons, coupled with translocation of four protons across the membrane. The energy conversion (or coupling) efficiency is quantified by the ratio of pump proton/electron of unity in the case of bovine CcO. This coupling between O_2 reduction and the proton pump has been established by the Wikström group [46–48], as given below.

4.1 Coupling Between Dioxygen Reduction and Proton Pump

Here, the experimental results for the establishment of the coupling mechanism are introduced as one of the most ingenious accomplishments in Bioenergetics.

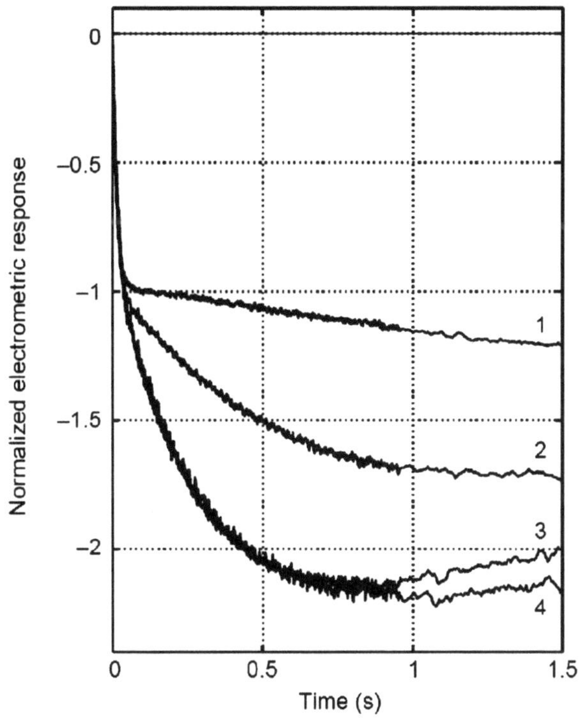

Figure 14 Charge translocation in CcO vesicles. The time-resolved development of the membrane potential of CcO at pH 7 and 25 °C in proteoliposomes at low O_2 concentration (10 nM) and high CO concentration (1 mM) was measured in the presence of various concentrations of Fe(II)-cytochrome *c* as follows: 20 (trace 1), 70 (trace 2), and 170 (trace 3) μM. In trace 4, 170 μM Fe(II)-cytochrome *c* was supplemented with 10 μM tetramethyl-*p*-phenylenediamine (TMPD) for ensuring full reduction of cytochrome *c* and CcO. Reproduced by permission from [46]; copyright 1999 Nature Publishing group.

In the catalytic cycle (Figure 7), the pathway from R to O is designated as the oxidative phase, while the second part is termed the reductive phase. The movement of electrical charge equivalents across the membrane during both phases has been determined by a system for time-resolved charge translocation during a single catalytic cycle of CcO starting from O_2 binding to the fully reduced CcO. Two phases are observed, a fast one (amplitude 44 %) and a slow one (amplitude 55 %) (Figure 14) [46]. The amplitude of the fast phase is independent of the Fe(II)-cytochrome *c* concentration while the maximum amplitude of the slow phase is obtained by an excess of Fe(II)-cytochrome *c*. Therefore, the fast and slow phases are assignable to the oxidative and reductive phases in the catalytic cycle. In a single catalytic cycle, 8 charges are transferred in total, i.e., 44 % and 55 % amplitudes correspond to 3.5 and 4.5 charges, respectively. Each of the charge translocation includes those due to electron transfer from Cu_A to heme *a*, chemical proton transfer coupled with each electron transfer from heme *a* to heme a_3 and the proton pump. The electron transfer pathway heme $a \rightarrow$ heme a_3 is parallel to the membrane surface, meaning no charge translocation is provided by this process. The charge translocation due to chemical proton transfer can be estimated by the charge translocation due to the electron transfer heme $a \rightarrow Cu_A$ upon photolysis of CO bound mixed valence CcO in the absence of O_2, since the amount of the electron transfer heme $a \rightarrow Cu_A$ can be quantified by UV-vis spectroscopy. Charge translocation induced by a single electron transfer $Cu_A \rightarrow$ heme *a* corresponds to 32 % of that induced by a single charge translocation across the membrane. This value is reasonable for the location of Cu_A and heme *a* [47]. Thus, the charge translocation for a process can be quantified from its electrometric response. The two hemes are located in parallel to the membrane surface so that transfer of a single chemical proton from the negative side of the membrane to heme a_3 induces charge translocation corresponding to 68 % (100 % – 32 %) of that induced by a single charge translocation across the membrane. In the oxidative phase, two chemical protons are transferred from the negative side (inside the proteoliposome) to the O_2 reduction site. The chemical protons provide 2 (100 % – 32 %) charge translocations induced by a single charge translocations across the membrane. Since the oxidative phase started from fully reduced CcO, only one electron is transferred from Cu_A to heme *a* during the oxidative phase. Consequently, 2 (100 % – 32 %) + 32 % charge translocations result from charge movement of the non-proton pump. This charge location (168 %) corresponds to the movement of 1.68 charges across the membrane. Thus, charge movement due to the proton pump can be estimated to be 1.82 charges (3.5 – 1.68). For the reductive phase, one electron equivalent must be transferred from Cu_A to heme *a* for regeneration of fully reduced CcO after a two-electron transfer for reducing the O_2 reduction site. Thus, (2 + 0.32) charge movements are due to non-proton pumping events. This value determines the charge translocation due to the proton pump as (4.5 – 2.32 = 2.18). These results indicate that both in the oxidative and in the reductive phases two equivalents of protons will be pumped across the membrane [46].

For a more quantitative evaluation of the efficiency of the proton pump, proton ejection from proteoliposomes of bovine CcO, determined with a sensitive pH meter, was titrated with electron equivalents added in the presence of O_2 (Figure 15) [48]. The results show that four electron equivalents reduced CcO ejects two proton equivalents, while three- and two-electron-reduced CcO eject one and zero proton equivalents, respectively. The results fit quite well to the calculated curve assuming that each of the $P \rightarrow F$ and $F \rightarrow O$ transitions is coupled to pumping of a single proton equivalent. The results from above obtained for the reductive phase are not sufficiently accurate to discriminate between the calculated curves, assuming that each of the $O \rightarrow E$ and $E \rightarrow R$ transitions is coupled with a single proton pump and assuming that $O \rightarrow E$ transition pumps two protons, while $E \rightarrow R$ transition is not coupled with the proton pump. However, a single electron injection experiment for the O-intermediate shows one equivalent of proton pumped in the $O \rightarrow E$ transition. Thus, another equivalent of proton in the reductive phase must be pumped in the $E \rightarrow R$ transition [48].

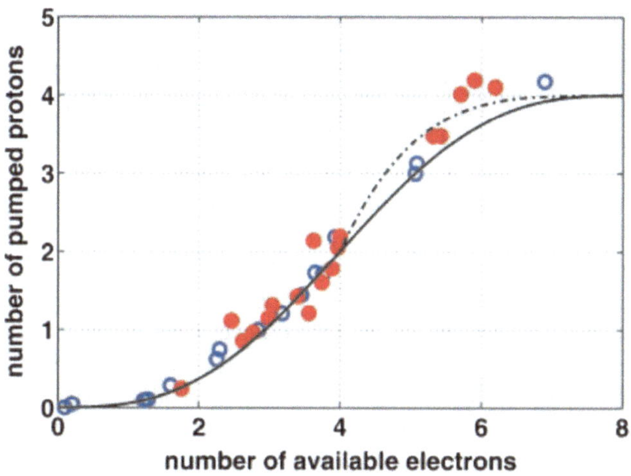

Figure 15 Proton ejection during oxidation and reduction of bovine heart CcO. Proton ejection was measured by a sensitive pH meter after addition of stoichiometric amounts of O_2 to CcO reduced anaerobically in various extents. The upper curve shows a calculated result assuming that one proton equivalent is ejected at each of the four transitions, P → F, F → O, O → E, and E → R, while the lower curve is the results of the calculation assuming one proton ejection at each of the transitions P → F and F → O, two proton ejections at O → E, and no ejection at E → R. Reproduced by permission from [48]; copyright 2004 National Academy of Sciences USA.

4.2 *Single Electron Injection Analyses of the Intermediates of the Catalytic Cycle*

Various methods for preparing or stabilizing the intermediate species (Figure 7) have been developed. Each of the individual electron transfer steps is coupled to the proton pump, and the role of these intermediates has been carefully evaluated by the single electron injection technique coupled to time-resolved electrometric and UV-vis spectroscopic analyses [49].

4.2.1 F → O Transition

The single injection experiments of CcO were made for the first time for the F → O transition of bovine CcO in proteoliposomes [50]. Treatment of resting oxidized bovine CcO with excess H_2O_2 provided a stable preparation with resonance Raman and UV-vis spectral characteristics of the F-intermediate. The F-intermediate, upon single photoreduction, showed three charge translocation phases with the time constants τ of 45 μs, 1.2 ms, and 4.5 ms. The initial phase proceeds with the same rate as that reported for the electron transfer Cu_A → heme *a* [50, 51]. The initial rapid phase does not include reduction of heme a_3. On the other hand, the X-ray structure shows that hemes *a* and a_3 are located very close to each other, suggesting extremely rapid electron transfer. This is confirmed by femtosecond

UV-vis spectroscopy for the electron transfer heme $a_3 \rightarrow$ heme a upon photolysis of the mixed valence CO complex [52]. Obviously, under single photoreduction conditions, the redox potential of heme a_3 is significantly lower than that of heme a, and protonation of heme a_3 in the next step is a prerequisite for the electron transfer heme $a \rightarrow$ heme a_3. The initial fast step is KCN-insensitive.

The following two charge translocation phases, with τ 1.2 and 4.5 ms, are KCN-sensitive. These phases must be due to proton translocations, since the electron transfer pathway heme $a \rightarrow$ heme a_3 runs parallel to the membrane plane. The amplitudes of the three phases are 20 %, 37 %, and 43 % of the total charge movement. The F → O transition is coupled, in total, with two electrical charge movements across the membrane as described above. Thus, the three phases are due to the movements of 0.40, 0.74 and 0.86 charges across the membrane [49, 53].

In order to examine which membrane potential generation is due to the transfer of pump protons, the properties of the Asn139Asp (*Rhodobacter sphaeroides* numbering) CcO variant, which shows no proton pumping without impairment of O_2 reduction activity, were examined electrometrically and spectrophotometrically. The charge movement consists of two phases, the KCN-insensitive (15 μs) one and the KCN-sensitive (0.6 ms) [54]. The 0.6 ms phase shows a strong H_2O/D_2O kinetic isotope effect (4-fold decrease in the rate in D_2O). In the wild-type enzyme the 4.5 ms phase reveals a strong H_2O/D_2O kinetic isotope effect, in contrast to the 1.2 ms phase which does not show any significant effect. The KCN-sensitive phase of the Asn139Asp variant must be due to chemical proton movement since it does not show any proton pump function. Thus, the H_2O/D_2O kinetic isotope effect of the Asn139Asp variant suggests that the slow (4.5 ms) phase is due to the chemical proton transfer.

4.2.2 The Other Transitions

All the other single electron reduction steps, P → F, O → E, and E → R, have been extensively examined by the single photoreduction technique coupled to electrometric and UV-vis measurements, providing essentially the same results as those for the F → O transition [49]. The results show three phases including a rapid KCN-insensitive phase followed by two slower (intermediate and slow) KCN-sensitive phases. The amplitude ratio is 2:4:4. The three phases are can be assigned to the vectorial electron transfer $Cu_A \rightarrow$ heme a, the pump proton transfer, and chemical proton transfer. Considering the nature of the chemical events happening in the O_2 reduction site between the four transitions, the similarity in the electrometric responses due to each single electron reduction strongly suggests a common proton pump system apart from the chemical proton transfer pathways such as the K- and D-pathways.

4.3 D-Pathway Mechanism

It has been believed that in the protein interior, protons are transferred through a hydrogen bond network or carried by mobile water molecules, in the hydronium ion H_3O^+ state. Thus, a potential proton transfer pathway can be identified in high resolution X-ray structures. Bovine heart CcO has three possible proton transfer pathways, K, D, and H (Figure 2) [2]. Figure 16 shows atomic models of the K- and D-pathways of bovine CcO in the fully oxidized state. These structures are well conserved in the CcO A family. An acidic amino acid, Glu242, is located near the upper end of the D-pathway, which does not have obvious structural features for proton transfer connections to the O_2 reduction Fe_{a3}/Cu_B site. The side chain of Glu242 has sufficient space for conformational changes. However, it is possible to induce a hydrogen bond network connecting to the O_2 reduction site by introducing two fixed water molecules as represented in the inset of Figure 16. This conformational change is assured by the flexibility of Gly239 which is well conserved. The two O^{2-} or OH^- anions in the network are likely to be provided by O_2 reduction at the Fe_{a3}/Cu_B site. Thus, the hydrogen bond network is likely to be set up when the P-intermediate is produced.

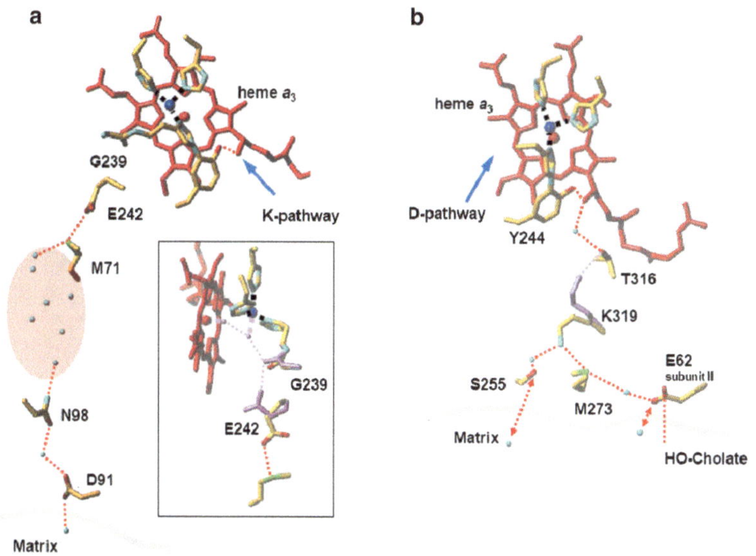

Figure 16 Atomic models of the D- and K-pathways. Red models denote heme a_3. The blue, red, and green portions of the amino acids denote nitrogen, oxygen, and sulfur atoms, respectively. Dotted and broken lines denote hydrogen bonds and coordination bonds, respectively. The K- and D-pathways are given in panels (**a**) and (**b**), respectively. The junction points for the K- and D-pathways are denoted by blue arrows in panels (**a**) and (**b**), respectively. The inset in panel a shows a possible hydrogen bond network from E242 to the O_2 reduction site.

A D-pathway variant, Glu242Gln, abolishes not only the proton pump function but also the chemical proton transfer in the oxidative phase. The K-pathway variant, Lys319Met, does not inhibit the chemical proton transfer in the oxidative phase, although it abolishes the turnover activity, reduction of O_2 to water (by blocking the chemical proton transfer in the reductive phase). When the mutation results were reported [55, 56], it was widely accepted that protons were pumped only in the oxidative phase. Therefore, it was considered that these mutation results confirmed the proposition that both pumping and chemical protons were transferred by D-pathway. Since then, most of the experimental results have been interpreted based on the above proposition.

4.3.1 Water-Gated Mechanism

If the D-pathway transfers both chemical and pumped protons, this pathway must have a branching point which sorts chemical protons from pump protons. Furthermore, it must have a proton-loading site for gating the proton transfer or facilitating unidirectional proton active transport. The glutamate Glu242 has been thought to be the candidate for such a branching point, while one of the propionate groups of heme a_3 is a candidate for the proton loading site. However, the X-ray structure suggests that this propionate forms a tight salt bridge with Arg438. The bridge seems too stable to have the gating function for the proton pump. Nevertheless, many proton pump mechanisms through the D-pathway have been proposed. The water-gated mechanism is one of

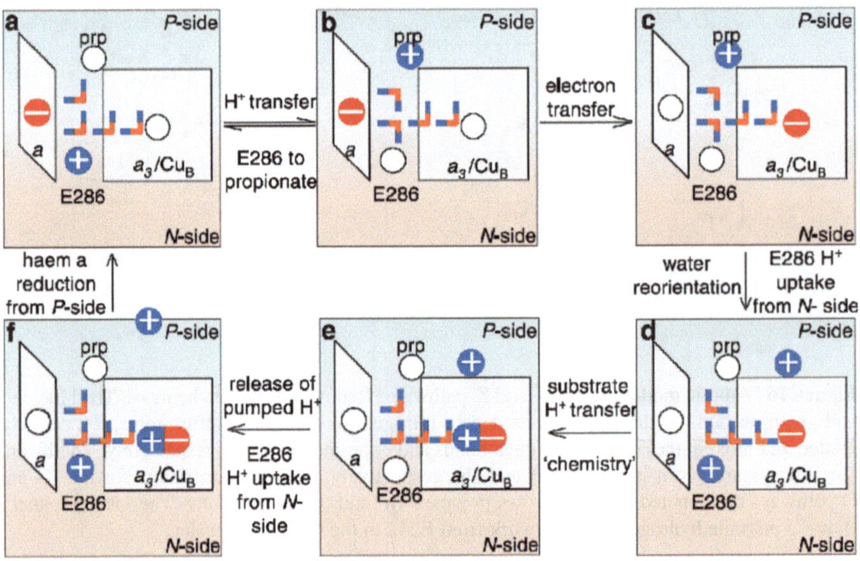

Figure 17 Schematic representation of the water-gated mechanism. In (**a**) heme *a* is reduced while the O_2 reduction site (heme a_3/Cu$_B$) is oxidized. The electric field between the two redox active sites determines the orientation of the water array to facilitate the proton transfer from Glu286 (*Rhodobacter sphaeroides* numbering) to the propionate (prp) of heme a_3 which is a putative proton-loading site. The proton transfer to the proton-loading site in (**b**) is coupled to the electron transfer to heme a_3/Cu$_B$ (**c**). The electric field change due to the electron transfer switches the orientation of the water array coupled with proton uptake from the negative side (*N*-side) (**d**). Then, protons on Glu286 are transferred to heme a_3/Cu$_B$ (**e**) to stimulate proton release to the positive side space (*P*-side) (**f**). Reproduced by permission from [57]; copyright 2003 Elsevier.

the most sophisticated ones which has been extensively examined theoretically [57]. A schematic representation of this mechanism is shown in Figure 17.

A water cluster is assumed in the hydrophobic cavity between the O_2 reduction site and Glu242 (Glu286 in the *Rhodobacter sphaeroides* numbering). The starting point of the proton pump cycle is depicted in Figure 17a, with heme *a* in the reduced state and Glu286 protonated. The water molecules are located regularly for proton transfer from Glu286 to prp, the propionate group of heme a_3. The proton transfer to prp influences the alignment of the water-cluster for preventing the reverse proton transfer (Figure 17b). The redox potential of the O_2 reduction site (abbreviated as a_3/Cu_B in the Figure 17) is increased by the protonation of prp to induce electron transfer to the Fea_3, Cu_B site (Figure 17c). Then, the reduction of a_3/Cu_B induces protonation of Glu286 (Figure 17c). Protonation of Glu286 stimulates proton release from prp to the hydrophilic domain of the positive side (P-side) of heme a_3 (Figure 17d) and also inhibits the back leak of the pumped protons. The chemical proton on Glu286 is transferred to a_3/Cu_B (Figure 17e). This positive charge on a_3/ Cu_B releases the pumped proton in the hydrophilic domain to the P-side (Figure 17f).

The flexibility of the Glu242 side chain is expected to contribute to blocking the spontaneous back leak of protons from prp. The possibility has been confirmed by computer simulation [58]. In addition to the function of Glu242, the water-gated mechanism as given above has been extensively examined by various simulation techniques. Recently, opposite conclusions have been reported [59, 60].

The possible proton-loading site, the heme propionate, forms a tight salt bridge to Arg439. At least, X-ray structures strongly suggest that this salt bridge is too strong to accept protons reversibly on a physiologically relevant time scale as stated above. However, extensive mutagenesis analyses for the site for bacterial aa_3-type and *E. coli* bo_3-type CcOs have shown that mutations of Arg439 (Lys, Gln, Asn, Leu) and Arg438 (Lys, Gln) do not significantly influence the proton pump activity, while the Arg438Gln/Arg439Gln double mutation and Arg439Pro and Arg438 (Asn, Leu) mutations impair the proton pump activity [61–63]. The results do not allow a straight forward interpretation. The modifications Arg $\rightarrow$ Gln, Asn, Leu, with serious impacts on the structure, do not lead to any significant influence on the enzyme activity, clearly a result speaking against the proposal for the proton-loading function of the heme propionate. On the other hand, the other mutations as described above, strongly support the proposal. Furthermore, a computer simulation analysis [64] suggests that the opening of the salt bridge, depending on the oxidation state of the metal sites, is possible in a physiologically relevant time scale.

It has been proposed that CcO must have a system for discharging the membrane potential to avoid deteriorative effects induced by an unusually high membrane potential [65–67]. As schematically shown, a space in which the water array (or cluster) is putatively located in the water-gated mechanism (Figure 17) connects the heme propionate and the O_2 reduction site. Thus, it is possible that the space connecting the propionate with the O_2 reduction site is the discharge system. Thus, one of the interpretations of the above mutation results is that the proton pump inhibition is caused by an increase in the reverse proton transfer through the discharge system by structural modification of the salt bridge.

4.3.2 Experimental Results Suggesting that Both Chemical and Pumped Protons Are Transferred Through the D-Pathway

It is impossible to identify the location of the proton pump system only by mutational analyses. For example, the Glu242Gln mutation abolishes the transfer of both pumped and chemical protons. The O_2 reduction is tightly coupled with chemical proton uptakes through the D- and K-pathways. Thus, blocking chemical proton transfer through the D-pathway by a Glu242 mutation impairs completely the O_2 reduction. Then, the proton pump cannot function because of lack of energy supply. Clearly, this mutation does not provide useful information on the location of the proton pump system. Nevertheless, the electrogenic analyses of various D-pathway mutations, together with the X-ray structure of the D-pathway as given below, have provided experimental results supporting that the D-pathway transfers both chemical and pumped protons.

The structural changes in the O_2 reduction site in the F → O and P → F transitions are $Fe_{a3}{}^{4+}{=}O^{2-}$ to $Fe_{a3}{}^{3+}$-OH^- and the Tyr244–O$^\bullet$ radical to Tyr244-OH, respectively. Each of these transitions requires transfer of one equivalent of chemical proton. Thus, abolishment of these transitions (the oxidative phase) by Glu242Gln mutation together with the location of Glu242 most likely results from a blockage of chemical proton transfer through the D-pathway [56]. It can be concluded that the D-pathway is used for transfer of the chemical protons.

The A → P transition process during reduction of O_2 by fully reduced CcO which is coupled to the oxidation of heme *a*, shows that the charge translocation corresponds to the one that would be induced by translocation of a single proton equivalent from Glu242 to the heme a_3 propionate [68]. The charge translocation from Glu242 has been established by mutation analysis [68]. This proton translocation is unlikely due to the proton donation to the O_2 reduction site which would induce the P → F transition. Most likely, the D-pathway facilitates proton transfer to a site other than the O_2 reduction site in addition to the proton transfer to the O_2 reduction site [69]. Furthermore, if Glu242 transfers protons to the propionate as discussed above, followed by the proton transfer to the O_2 reduction site giving the P → F transition, in a CcO with a mutation at the entrance of the D-pathway, Asp124Asn (in *Paracoccus denitrificans* numbering), the proton on Glu242 is used to protonate the heme propionate. However, this mutant shows a P → F transition, suggesting that the D-pathway has another proton-loading site [70]. The site has been identified with an Asp124Asn/Tyr35Phe double mutation which abolishes both the P → F and the F → O transition. Deprotonation of Tyr35 has been proven by IR analyses [70]. These results strongly suggest that the D-pathway transfers protons to two different sites. One of the sites is likely to be for loading pump protons. However, the gating function (or unidirectional proton transfer) of the proton loading site has not been shown.

Furthermore, the propionate group, if this is the proton-gating site, must have structures sorting pump protons from chemical protons as mentioned by Williams a long time ago [71]. Although a sophisticated water-gated mechanism has been proposed [57], the postulate of Williams has not been proven by concrete experimental evidence.

4.4 H-Pathway Mechanism

The redox-driven proton pump in CcO must be driven by redox-coupled conformational changes. Fairly large redox-coupled conformational changes around Asp51 near the positive side surface of bovine CcO were discovered at 2.3 Å resolution in 1998 [72]. The X-ray structures show that Asp51 is connected to the negative side surface of CcO by a hydrogen bond network and a water channel as schematically shown in Figure 18a. These structures strongly suggest that the system including the hydrogen bond network and the water channel represents the proton pump system of CcO. Later, this system was designated as the H-pathway. However, Asp51 is conserved only in the animal kingdom. Plant and bacterial CcO enzymes do not have this residue [2]. On the other hand, the D-pathway is conserved in animals, plants, and bacteria. Thus, the function of the H-pathway as the proton pump system has not been completely accepted. Note that, as described below, some bacterial CcOs do not have the D-pathway.

4.4.1 Structure and Function of the H-Pathway

The structure of the H-pathway is schematically shown in Figure 18a [73]. The upper half of the pathway is a hydrogen bond network connecting Asp51 at the positive side surface with Arg38. The Arg38 at the bottom end of the hydrogen bond network is connected with the positive side surface by a water channel (shadowed area) through which water molecules from the negative side space can access the Arg38 site. The shadowed ovals and circles denote the spaces in which at least one molecule of water can be stored. The space is designated as water cavity. Arg38 at the bottom end of the hydrogen bond network is hydrogen-bonded to the formyl group of heme *a* (Figure 18b). One of the fixed water molecules in the hydrogen bond network is also hydrogen-bonded to heme *a* via the propionate of the D ring.

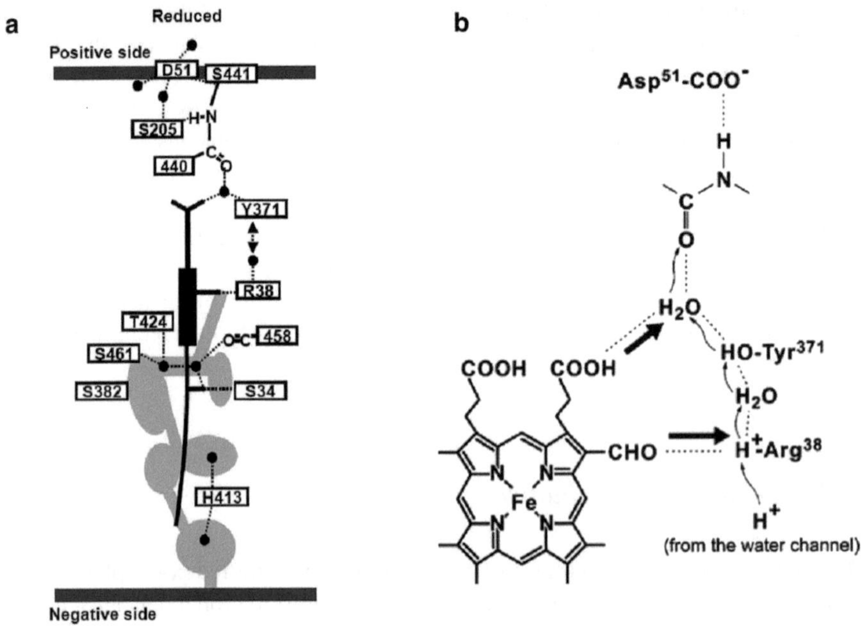

Figure 18 (**a**) Schematic representation of the H-channel of bovine heart CcO in the fully reduced state. The filled circles denote fixed water molecules detectable in the X-ray structure. The water channel is represented by the gray area. The side view of the heme *a* plane is shown by a rectangle with sticks which denote the peripheral groups. The dotted lines represent hydrogen bonds. (**b**) The interactions between the hydrogen bond network in the H-pathway and heme *a*. The thick arrows denote possible electrostatic repulsion to the proton transfer through the hydrogen bond network. Reproduced by permission from [86]; copyright 2012 Elsevier.

The redox-coupled conformational changes at Asp51 are given in Figure 19 [2]. Asp51 is buried inside the protein in the oxidized state (Figure 19b), the carboxyl group is hydrogen-bonded to two OH groups of serine residues and two peptide NH groups, suggesting that the carboxyl group is in a highly non-polar environment. However, upon reduction of CcO, Asp51 becomes exposed to the molecular surface (Figure 19b). In reduced CcO, Asp51 is hydrogen-bonded to 3 water molecules and one serine OH group. Thus, the microenvironment of the Asp51 carboxyl group is essentially the same as in water. This structural change, which indicates a large redox-coupled change in effective pK_a, strongly suggests that Asp51 is the proton-loading site of the proton pump system in CcO. The redox-coupled protonation state change has been confirmed by FTIR analyses (see Supporting Information in [74]). A redox titration of the protonation state change has shown that the oxidation state of one and only one of the low potential sites (heme *a* and Cu_A) controls the protonation state of Asp51.

In the oxidized state, the Asp51 COOH group is hydrogen-bonded to the NH group of the peptide bond between Tyr440 and Ser441 (Figures 18 and 19) [73]. Proton transfer through a peptide bond has been well established a long

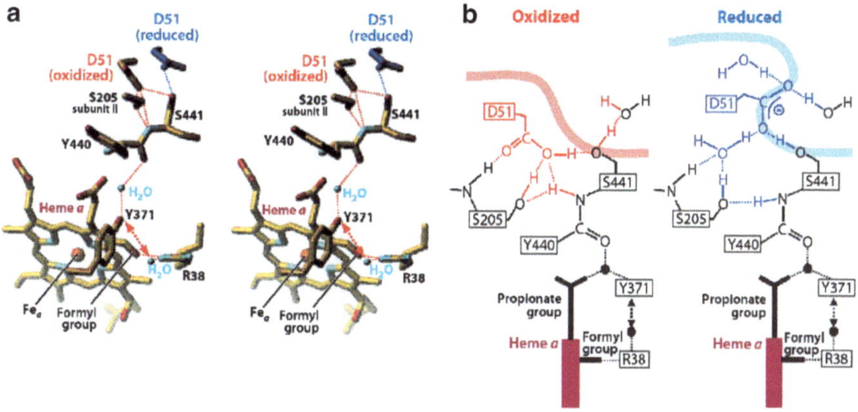

Figure 19 Redox-coupled conformational changes at Asp51. (**a**) stereo drawing of the hydrogen bond network of CcO in the fully oxidized and reduced states at 1.8 and 1.9 Å resolution, respectively, viewed from the positive side. (**b**) schematic representation of the structural characteristics of Asp51 in CcO in the oxidized and reduced states. The smooth thick curves denote the molecular surface to which the water molecules in the positive side are accessible. The conformational changes upon reduction of CcO are shown by the blue structure on the right. Reproduced by permission from [2]; copyright 2011 Annual Review of Biophysics.

time ago [75]. However, because the enol tautomer of the peptide bond (–C(OH)=N–) is less stable than its keto tautomer (–CO–NH–), the proton transfer is likely to be unidirectional. The function of this peptide bond as the proton transfer pathway has been examined by theoretical analyses, confirming that the unidirectional proton transfer through the peptide bond is possible on a physiologically relevant time scale. The results of the theoretical analyses also suggest a possible contribution of the Na^+ site for effective unidirectional proton transfer [76].

Conformations of the water channel and the heme *a* are significantly influenced by the oxidation state of bovine CcO (Figure 20a) [74]. The conformational changes are schematically shown in Figure 20b. In the reduced state, the water channel has six water cavities. One of the cavities near the upper end of the water channel is eliminated upon oxidation by formation of a bulge structure including Ser382 (Figure 20a). Another redox-coupled conformational change is detectable in the hydroxyfarnesylethyl group. The OH group is hydrogen-bonded to Ser382 in the oxidized state. The hydrogen bond is broken upon reduction, which could trigger the water cavity formation. In the oxidized state, the water channel has a long narrow pathway through which effective water migration seems impossible in the physiologically relevant time scale (Figure 20b). Thus, the conformations of the water channel in the oxidized and reduced states are designated as closed and open conformations, respectively. In the closed state, proton back leak from the hydrogen bond network is effectively blocked. X-ray structures of various respiratory inhibitor derivatives and the O_2 reduction intermediates show that P, F, and O intermediates as well as the strong ligand (CO and NO) bound derivatives of fully reduced

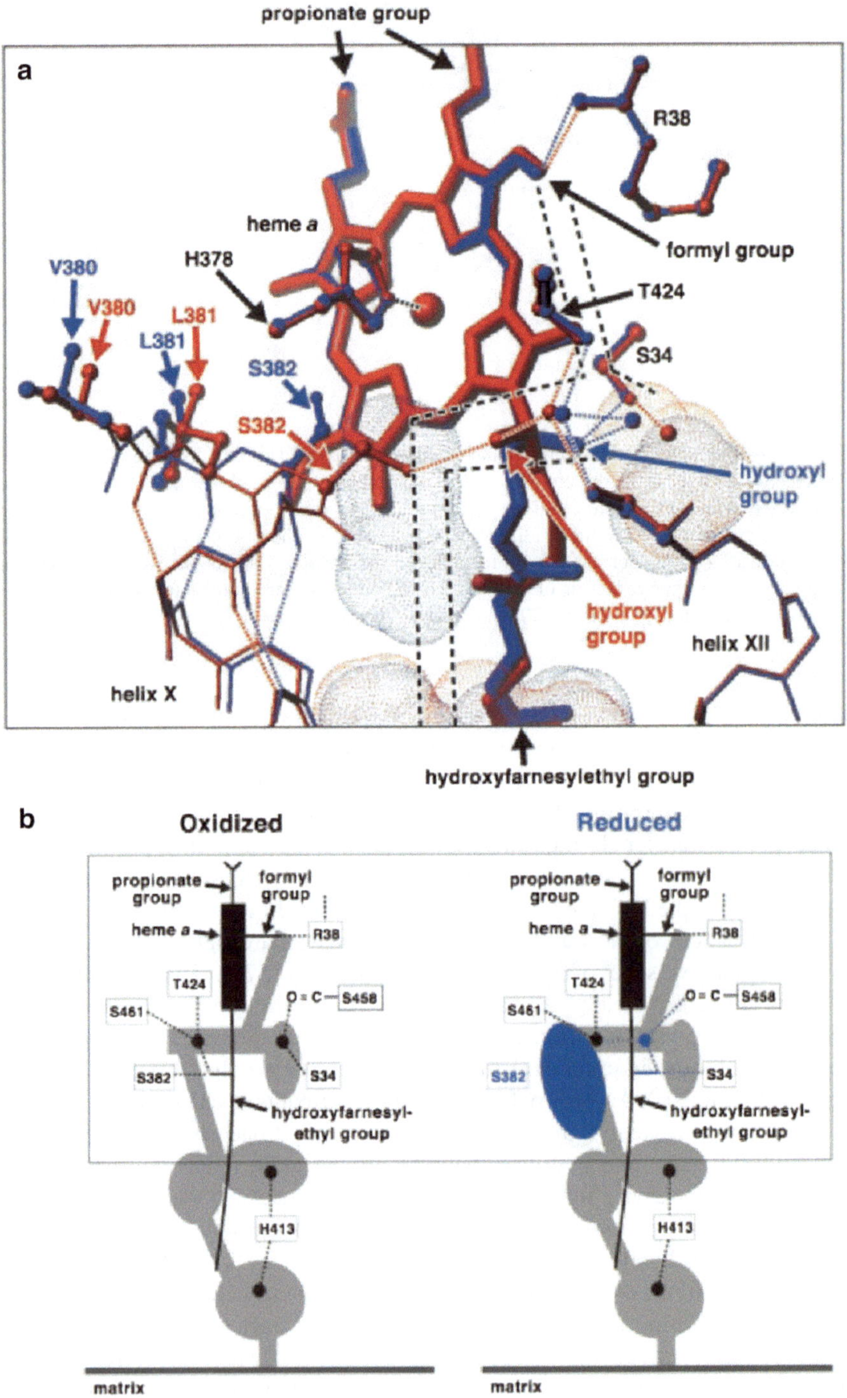

Figure 20 X-ray structure of the water channel in the H-pathway. (**a**) Redox-coupled conformational changes of the water channel detectable in the upper part of the channel. Red and blue denote oxidized and reduced states, respectively. Dotted surfaces indicate cavities detectable in the

CcO show the closed conformation of the water channel. In other words, the water channel is open only in the R-intermediate [37].

The proton pump is coupled to electron transfer heme $a \rightarrow$ heme a_3, one at a time after formation of the P-intermediate, giving the F, O, E, and R intermediates (Figure 7). Thus, all these proton pumping events occur when the water channel is closed. These results indicate that proton back leaks from the hydrogen bond network are blocked by both the upper and bottom ends of the H-pathway. The peptide bond has also a critical role for preventing collapse of the proton motive force by back leak from the positive side of the membrane.

4.4.2 The Structure for Proton Collection and Timely Closure of the Water Channel

The X-ray structural results given in the preceding section suggest that four equivalents of pumped protons must be collected when the O_2 reduction site is in the ligand-free fully reduced state [Fe_{a3}^{2+}, Cu_B^{1+}]. Also, the timely closure of the water channel after collection of the four equivalents of pumped protons is a prerequisite. The timely closure is most likely controlled by O_2 binding to the O_2 reduction site of the R-intermediate. In order to elucidate the mechanism of the timely closure of the water channel, the conformational change of the channel upon O_2 binding to the O_2 reduction site has been investigated by time-resolved IR techniques. The used instrument is sufficiently sensitive for analyses of the IR properties of a single peptide C=O group in the amide-I region of proteins in H_2O [77]. As a model system for O_2 binding to fully reduced CcO, CO-bound CcO has been investigated by time-resolved IR studies. Upon flash photolysis of CO-bound CcO, the photolyzed CO is stoichiometrically transferred to Cu_B within 50 ns. Then, the CO at Cu_B is released with the time constant of 2 μs from the protein. The most prominent IR spectral change detectable in the 2 μs phase can be assigned to the change in the bulge conformation including Ser382.

In the open state (the ligand-free fully reduced state) Ser382 is located very near the biggest cavity shown in Figure 20 but not involved in the wall of the cavity (Figure 21) [77]. Upon CO (or O_2) binding to the O_2 reduction site, the cavity is eliminated by the movement of Ser382 towards the cavity. The wall of the water cavity consists mostly of non-polar carbon atoms to provide a very high non-polar environment (Figure 21). Thus, Ser382OH interacts with the protons carried by H_2O inside the cavity to sense the overall protonation level inside the cavity. The

Figure 20 (continued) oxidized and reduced states, respectively. The broken lines indicate a possible water pathway connecting these cavities, estimated from the X-ray structure. The dotted lines show hydrogen bonds. The positions of the fixed water molecules are marked by small spheres. (**b**) A schematic representation of the redox-coupled conformational changes in the water channel. The area given in (**a**) is marked by a square. The water channel is represented by the gray and blue area. The spheres denote the fixed water molecules. Reproduced by permission from Ref. [74]; copyright 2003 National Academy of Sciences USA.

protonation level is likely to be equilibrated with the protonation state of the hydrogen bond network. As shown in Figure 21b, Ser382 is located in a helix turn which is an adjacent turn containing the fifth ligand of heme a_3, His376. The structure suggests a conformational relay system including Ser382, two helix turns, His376, Fe_{a3}, the ligand to Fe_{a3} and/or Cu_B and Cu_B. The relay system is expected to transfer information to the metal sites on the protonation state of the cavity sensed (or captured) by Ser382. Thus, the complete protonation of the hydrogen bond network is sensed by the protonation level of the water cavity by Ser382 which transfers the information to the metal site (Fe_{a3}^{2+}/Cu_B^{1+}). Upon receiving the

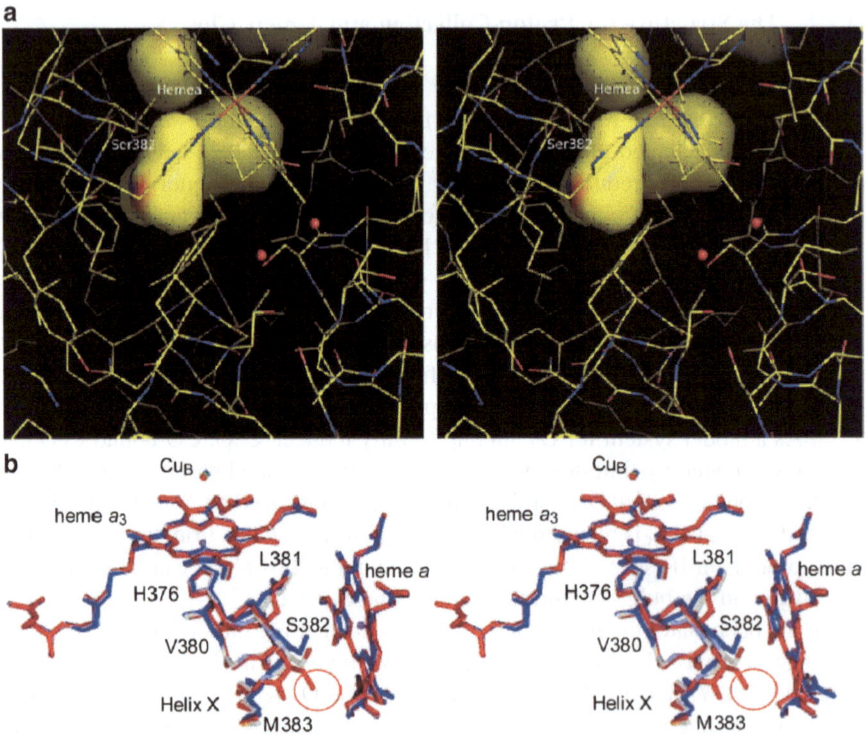

Figure 21 Conformation of Ser382 near the biggest water cavity in the water channel of bovine heart CcO. (**a**) X-ray structure of the water cavity near Ser382 in the fully reduced state; stereo view from the positive side perpendicular to the membrane surface. The water cavities are drawn on the surfaces calculated by the van der Waals radii of atoms exposed to the cavity surfaces. The cavity near Ser382 is closest to the positive side among the four cavities seen in the figure. The red and blue areas on the cavity surface are due to the peptide C=O of His378 and Ser382 and the peptide N–H of Met383. The Ser382 OH group is not part of the cavity surface. (**b**) Structural modeling of the possible conformational changes at Ser382. Top stereo view of the modeled structure of the intermediate forms (gray), superimposed with X-ray structures of the reduced (blue) and CO-bound (red) forms. The red circle indicates the location of the water cavity that is eliminated by Ser382 upon CO binding. Reproduced by permission from [77]; copyright 2013 American Society for Biochemistry and Molecular Biology.

information from Ser382 via the relay system, O_2 moves from Cu_B to Fe_{a3} to trigger the conformational change in Ser382 via the relay system to eliminate the water cavity.

When CO is bound to Cu_B, the IR data suggest that CO bound at Cu_B after the photolysis induces a migration of Ser382 to the location half-way between those of the closed and open conformations [77]. In other words, when CO is at Cu_B, the cavity still exists and Ser382 is closer to the cavity compared to the location of Ser382 in the open conformation. Thus, the negatively polarized Ser382 OH group is expected to contribute to proton collection from the negative side space of the membrane. It has been well known that O_2 is trapped first at Cu_B before binding to Fe_{a3}^{2+}. Therefore, in analogy to the CO binding results, O_2 bound at Cu_B is expected to induce the conformational change in Ser382 via the conformational relay system to the intermediate conformer to promote effective proton collection. At present, it is not clear whether the hydrogen bond network of the H-pathway has sufficient capacity for the storage of four equivalents of protons. A higher resolution in the X-ray structure of CcO is required.

4.4.3 Mutational Analyses of the H-Pathway

The X-ray structural and vibrational spectroscopic data clearly show that the H-pathway functions like the proton pump system in bovine CcO. However, it is indispensable to examine the function of the H-pathway to elucidate its role in proton pumping. For this purpose, various mutations of the H-pathway have been introduced by using the stable expression system of the subunit I gene of bovine heart CcO in HeLa cells (the H-pathway is included in subunit I). As mentioned above, the H-pathway sequence of bovine CcO is not completely conserved in bacterial CcO. Thus, mutations of bovine CcO are indispensable for examination of the active H-pathway function.

Three bovine CcO variants, Asp51Asn, Ser441Pro and Val386/Met390Trp, have been prepared to investigate the role of D51 and the peptide bond between Tyr440 and Ser441 in proton transfer, and the water transfer function of the water channel, respectively [74, 78]. It is well-known, that proline has a significant influence on the native protein conformation and the consequences of such a mutation are difficult to interpret. However, the dihedral angle of the peptide bond between Tyr440 and Ser441 in the X-ray structure indicates that a conformational change induced by the mutation is expected to be minimal. The exchange of Ser441 against Pro prevents the formation of the protonated imidic acid intermediate ($–C(OH)=NH^+–$). A molecular dynamics calculation [78] suggests that the double mutation of the amino acid residues in the water channel will lead to a complete closure of the channel. The phenotypes of these three mutations are identical, namely, no proton pump without affecting electron transfer activity. Three results clearly confirm the proposal for the function of the H-pathway as the proton pump system, based on X-ray and IR analyses.

4.5 *Diversity in Proton Transfer Pathways*

Recently CcOs from various origins have been purified and their crystal structures have been solved. These results reveal significant variances of the structures of the proton transfer pathways [79, 80]. The CcOs discovered thus far have been classified into 4 families, A_1, A_2, B, and C [79]. The A_1 CcO family, which includes bovine CcO, has active D- and K-pathways and Glu242 in the D-pathway. In the A_2 family, the position of the carboxyl group of Glu242 is occupied by an OH group of tyrosine residue located on the adjacent helix turn. Reversible deprotonation of the tyrosine has been confirmed by IR analysis [80]. CcOs belonging to the B and C families do not have an active D-pathway. The mutational analyses of the K-pathway of CcOs of the B family indicate that the K-pathway transfers chemical protons in both oxidative and reductive phases. Namely, the K-pathway is the only pathway for chemical proton transfer. CcOs of the C family do not have the tyrosine corresponding to Tyr244 in bovine CcO which is covalently linked to one of the three His ligands of Cu_B. However, the tyrosine is replaced by a tyrosine from an adjacent helix. Thus, Tyr244 is not conserved but the covalent link between His and Tyr is conserved [79].

This diversity is consistent with the difference in the two reactions driven by CcO, that is, the proton pump and the O_2 reduction. The former is a chemically simple process which various amino acid side chains can conduct. Thus, the proton transfer pathway system of CcOs shows significant diversity, although the function (proton transfer from the negative side to the O_2 reduction site) is conserved. However, O_2 reduction without releasing ROS is much more complex by comparison. No better system than the present Fe/Cu system has been found in the history of evolution. Thus, the structure of the O_2 reduction site is completely conserved in all CcOs discovered thus far.

The physiologically important amino acid residues are conserved in different species. Conservation of amino acid residues is a useful criterion to identify the functional sites of a protein. In the case of CcO, only the O_2 reduction site structure including the Tyr-His covalent linkage are conserved, and it has been proposed that O_2 reduction coupled to proton pumping is conducted only by the Fe/Cu dinuclear site including the His-Tyr covalent linkage, although no possible concrete mechanism has been given for the proposal [81]. However, note that the lack of the active D-pathway in CcOs of the B and C family does not prove that the D-pathway in CcOs of the A family is inactive. Similarly, lack of Asp51 in bacterial CcO does not automatically mean that the H-pathway is inactive in bovine CcO.

The proton pump activity of these pathways in bacterial CcO has been examined by various point mutations. In the case of the bovine CcO water channel, which is impaired by the double mutation [78], point mutation of the branched or bypassed proton transfer pathway would not influence the overall proton pump activity. It has been proposed that in the B and C family CcOs, both pump and chemical protons are transferred through the K-pathway. However, as described above, the location of the proton pump pathway cannot be identified just by mutational analysis. It should not be ignored that the low spin heme center is completely conserved in all CcOs identified thus far. The low spin heme center of bovine CcO drives the proton pump as described above. Thus, the low-spin heme center can also drive the proton pump in other CcOs as has been proposed a long time ago (Figure 22) [55]. That is, heme *a* and a_3 drive the proton pump and O_2 reduction coupled to chemical proton transfer, respectively [49, 55].

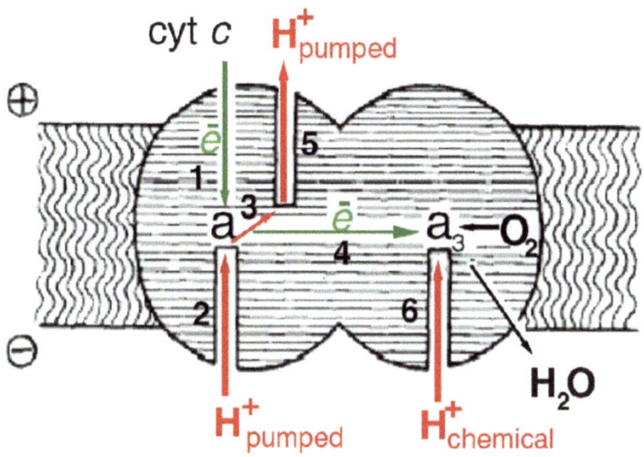

Figure 22 Model of the cytochrome *c* oxidase proton motive mechanism coupled to the transfer of a single electron through the enzyme. The postulated sequence of events is denoted by arrow numbers: (1) the initial electron transfer from cytochrome *c*; (2) proton uptake from the negative phase; (3) translocation of pump protons from the bottom of the input well to the bottom of the output well; (4) electron transfer from heme *a* to the O_2 reduction site; (5) release of the pump proton to the positive side; (6) uptake of the chemical proton from the negative side. Reproduced by permission from [49]; copyright 2012 Elsevier.

The recent resonance Raman results on the effect of H_2O/D_2O exchange in the propionate groups of hemes *a* and a_3 of bovine CcO strongly suggest that the propionate groups of hemes *a* and a_3 are well isolated from each other in terms of proton exchange so that direct proton transfer between the two heme propionates located side by side is completely blocked [82–84]. The results strongly suggest that proton active transport through the H-pathway including heme *a* proceeds without any interaction with the heme a_3 site. In other words, this property is expected to block the leaks of pumped protons to the O_2 reduction site to be used as chemical protons, which would dissipate the free energy for the proton pump. Thus, the resonance Raman results indicate that the postulate brought forward by Williams [71] is fulfilled in the proton pump driven by the H-pathway.

5 General Conclusions

The reaction mechanism of CcO is one of the biggest challenges to be tackled in modern Biological Science [2]. In order to understand the mechanism of any protein function on the atomic level, high resolution X-ray structures of the functional sites of the protein are a strict prerequisite. In the case of cytochrome *c* oxidase, the enzyme isolated from bovine heart has provided X-ray structures at 1.8 Å resolution, which provides electron density sufficiently accurate for the identification of the *cis*- (against *trans*-) configuration of the unsaturated bond of fatty acid tails of phospholipids tightly bound to the protein moiety [18].

The X-ray structures of various respiratory inhibitor (competitive inhibitors of O_2) derivatives suggest that the bound O_2 induces significantly large rearrangements in coordination structures of the O_2 reduction site forming three possible electron transfer pathways to the bound O_2^- resulting in a non-sequential three-electron reduction to O_2^- for complete reduction to the O^{2-} level without releasing any ROS [37]. Elegant studies using biomimetic complexes have allowed various unique insights on the mechanism of the O_2 reduction which cannot be obtained by the analyses of the natural protein system. For example, the tridentate (trigonal planar) structure of Cu_B is a prerequisite for stabilizing the Fe^{3+}-O_2^- state [27]. The copper structure is also critical for the facile electron transfer to the peroxide-bridged intermediates by stabilizing the low spin heme-peroxide-copper state [44]. Thus, the peroxide-bridged intermediates are undetectable in the normal enzyme catalytic cycle. Furthermore, various model compounds suggest the requirement of Tyr244 as electron donor to the bound O_2 [27, 44].

In the X-ray structure of bovine CcO the H-pathway has been detected which suggests a proton active transport driven by electrostatic repulsion between protons on the hydrogen bond network of the H-pathway and the positive charge of heme *a* created upon oxidation for electron transfer to heme a_3 for O_2 reduction [74]. Effective proton collection and proton gating systems have been identified in the X-ray structure. These X-ray structural findings have been confirmed by mutational analyses of bovine CcO [74, 78].

Diversity in the structure of the transfer systems for chemical and pump protons suggests a common mechanism of CcO, that is, the low-spin (*a* and *b*) and high-spin (a_3, b_3, and o_3) hemes drive proton pumping and O_2 reduction, respectively, consistent with the original proposal made a long time ago [49, 55].

Abbreviations and Definitions

AN	bis[3-(dimethylamino)propyl]amine
CcO	cytochrome *c* oxidase
CL	cardiolipin
DCCD	dicyclohexylcarbodiimide
DCHIm	1,5-dicyclohexylimidazole
DCU	dicyclohexyl-*N*-acyl-urea
DMAP	4-(dimethylamino)pyridine
D-pathway	a proton transfer pathway including E242, connecting the negative side with the O_2 reduction site

EPR	electron paramagnetic resonance
FTIR	Fourier transform infrared
K-pathway	a proton transfer pathway including K319, connecting the negative side with the O_2 reduction site via Y244
heme *a*	one of the heme A irons of the A family cytochrome *c* oxidase, which is in a low-spin state and 6-coordinated in both, the oxidized and reduced states
heme a_3	one of the heme A irons of the A family cytochrome *c* oxidase, which is in a high-spin state and 5-coordinated in the reduced state; this is the site for O_2 binding
heme *b*	a low-spin heme B iron of bacterial cytochrome *c* oxidase and quinol oxidase
heme b_3	a heme B iron of the C family cytochrome *c* oxidase, which is in a high-spin state and 5-coordinated in the reduced state; this is the site for O_2 binding
heme o_3	a heme O iron of bacterial quinol oxidase, which is in a high-spin state and 5-coordinated in the reduced state; this is the site for O_2 binding
H-pathway	a proton transfer pathway including R38 and D51, connecting the negative with the positive side
^{6}LFe	Cu-depleted form of complex ^{6}LFeCu, see [45]. NMePr=*N*-methyl-2-propylimidazole-4carboamide, the substituent on *o*-position of tetraphenylporphyrin (the whole model complex is abbreviated as Fe/Cu[NMePr])
^{6}LFeCu	a tethered porphyrin complexed with iron and copper. The "^{6}L" indicates that one end of the tethering group is located at the position 6 of one of the 3 pyridine groups coordinated to Cu
OMe	2,6-di-*tert*-butyl-4-methoxyphenol
PE	phosphatidylethanolamine
PG	phospatidylglycerol
prp	propionate
ROS	reactive oxygen species
SDS-PAGE	sodium dodecyl sulfate polyacrylamide gel electrophoresis
t-Bu	2,4,6-tri-*tert*-butylphenol
THF	tetrahydrofuran
TMPA	tris(2-pyridylmethylamine)
TMPD	tetramethyl-*p*-phenylenediamine

Acknowledgments This work is supported by a Grant-in-Aid for the Global Center of Excellence Program (to S. Yoshikawa) and for Scientific Research (A) 2247012 (to S. Yoshikawa), each provided by the Japanese Ministry of Education, Culture, Sports, Science and Technology, and supported by CREST. S. Yoshikawa is "Senior Visiting Scientist in the Riken Harima Institute".

References

1. S. Ferguson-Miller, G. T. Babcock, *Chem. Rev.* **1996**, *96*, 2889–2908.
2. S. Yoshikawa, K. Muramoto, K. Shinzawa-Itoh, *Annu. Rev. Biophys.* **2011**, *40*, 205–223.
3. T. Tsukihara, H. Aoyama, E. Yamashita, T. Tomizaki, H. Yamaguchi, K. Shinzawa-Itoh, R. Nakashima, R. Yaono, S. Yoshikawa, *Science* **1995**, *269*, 1069–1074.
4. S. Iwata, C. Ostermeier, B. Ludwig, H. Michel, *Nature* **1995**, *376*, 660–669.
5. C. Ostermeier, A. Harrenga, U. Ermler, H. Michel, *Proc. Natl. Acad. Sci. USA* **1997**, *94*, 10547–10553.
6. M. Svensson-Ek, J. Abramson, G. Larsson, S. Törnroth, P. Brzezinski, S. Iwata, *J. Mol. Biol.* **2002**, *321*, 329–339.
7. J. Abramson, S. Riistama, G. Larsson, A. Jasaitis, M. Svensson-Ek, L. Laakkonen, A. Puustinen, S. Iwata, M. Wikström, *Nat. Struct. Biol.* **2000**, *7*, 910–917.
8. J. A. Lyons, D. Aragão, O. Slattery, A. V Pisliakov, T. Soulimane, M. Caffrey, *Nature* **2012**, *487*, 514–518.
9. T. Tiefenbrunn, W. Liu, Y. Chen, V. Katritch, C. D. Stout, J. A. Fee, V. Cherezov, *PLoS One* **2011**, *6*, e22348.
10. S. Buschmann, E. Warkentin, H. Xie, J. D. Langer, U. Ermler, H. Michel, *Science* **2010**, *329*, 327–330.
11. E. Yakushiji, K. Okunuki, *Proc. Impact Acad. Japan* **1941**, *17*, 205–223.
12. T. Yonetani, S. Takemori, I. Sekuzu, K. Okunuki, *Nature* **1958**, *181*, 1339–1340.
13. T. Yonetani, *J. Biol. Chem.* **1961**, *236*, 1680–1688.
14. B. Kadenbach, M. Ungibauer, J. Jarausch, U. Buge, L. Kuhn-Nentwig, *Trends Biochem. Sci.* **1983**, *8*, 398–400.
15. W. S. Caughey, G. A. Smythe, D. H. O'Keeffe, J. E. Maskasky, M. I. Smith, *J. Biol. Chem.* **1975**, *250*, 7602–7622.
16. E. Yamashita, H. Aoyama, M. Yao, K. Muramoto, K. Shinzawa-Itoh, S. Yoshikawa, T. Tsukihara, *Acta Crystallogr. D. Biol. Crystallogr.* **2005**, *61*, 1373–1377.
17. P. M. Kroneck, W. A. Antholine, J. Riester, W. G. Zumft, *FEBS Lett.* **1988**, *242*, 70–74.
18. K. Shinzawa-Itoh, H. Aoyama, K. Muramoto, H. Terada, T. Kurauchi, Y. Tadehara, A. Yamasaki, T. Sugimura, S. Kurono, K. Tsujimoto, T. Mizushima, E. Yamashita, T. Tsukihara, S. Yoshikawa, *EMBO J.* **2007**, *26*, 1713–1725.
19. W. S. Caughey, W. J. Wallace, J. A. Volpe, S. Yoshikawa, in *Oxidation-Reduction, Part C*, Vol. 13 of *The Enzymes*, 3rd edn., Ed P. D. Boyer, Academic Press, New York, 1976, pp. 299–344.
20. W. T. Potter, M. P. Tucker, R. A. Houtchens, W. S. Caughey, *Biochemistry* **1987**, *26*, 4699–4707.
21. Q. H. Gibson, C. Greenwood, *Biochem. J.* **1963**, *86*, 541–554.
22. T. Ogura, S. Takahashi, K. Shinzawa-Itoh, S. Yoshikawa, T. Kitagawa, *J. Am. Chem. Soc.* **1990**, *112*, 5630–5631.
23. S. W. Han, Y. C. Ching, D. L. Rousseau, *Proc. Natl. Acad. Sci. USA* **1990**, *87*, 2491–2495.
24. C. Varotsis, W. H. Woodruff, G. T. Babcock, *J. Am. Chem. Soc.* **1989**, *111*, 6439–6440.
25. T. Ogura, S. Hirota, D. A. Proshlyakov, K. Shinzawa-Itoh, *J. Am. Chem. Soc.* **1996**, *118*, 5443–5449.
26. T. Ogura, S. Takahashi, S. Hirota, K. Shinzawa-Itoh, S. Yoshikawa, E. H. Appelman, T. Kitagawa, *J. Am. Chem. Soc.* **1993**, *115*, 8527–8536.
27. J. P. Collman, C. J. Sunderland, K. E. Berg, M. A. Vance, E. I. Solomon, *J. Am. Chem. Soc.* **2003**, *125*, 6648–9.
28. D. A. Proshlyakov, M. A. Pressler, G. T. Babcock, *Proc. Natl. Acad. Sci. USA* **1998**, *95*, 8020–8025.
29. T. Ogura, T. Kitagawa, *Biochim. Biophys. Acta* **2004**, *1655*, 290–297.
30. D. A. Proshlyakov, M. A. Pressler, C. DeMaso, J. F. Leykam, D. L. DeWitt, G. T. Babcock, *Science* **2000**, *290*, 1588–1591.
31. M. A. Yu, T. Egawa, K. Shinzawa-Itoh, S. Yoshikawa, V. Guallar, S.-R. Yeh, D. L. Rousseau, G. J. Gerfen, *J. Am. Chem. Soc.* **2012**, *134*, 4753–4761.

32. A. Sucheta, K. E. Georgiadis, O. Einarsdóttir, *Biochemistry* **1997**, *36*, 554–565.
33. A. Sucheta, I. Szundi, O. Einarsdóttir, *Biochemistry* **1998**, *37*, 17905–17914.
34. I. Szundi, G. L. Liao, O. Einarsdóttir, *Biochemistry* **2001**, *40*, 2332–2339.
35. O. Einarsdóttir, I. Szundi, N. Van Eps, A. Sucheta, *J. Inorg. Biochem.* **2002**, *91*, 87–93.
36. N. Van Eps, I. Szundi, O. Einarsdóttir, *Biochemistry* **2003**, *42*, 5065–5073.
37. K. Muramoto, K. Ohta, K. Shinzawa-Itoh, K. Kanda, M. Taniguchi, H. Nabekura, E. Yamashita, T. Tsukihara, S. Yoshikawa, *Proc. Natl. Acad. Sci. USA* **2010**, *107*, 7740–7745.
38. W. H. Woodruff, O. Einarsdóttir, R. B. Dyer, K. A. Bagley, G. Palmer, S. J. Atherton, R. A. Goldbeck, T. D. Dawes, D. S. Kliger, *Proc. Natl. Acad. Sci. USA* **1991**, *88*, 2588–2592.
39. J. O. Alben, P. P. Moh, F. G. Fiamingo, R. A. Altschuld, *Proc. Natl. Acad. Sci. USA* **1981**, *78*, 234–237.
40. J. P. Collman, C. J. Sunderland, R. Boulatov, *Inorg. Chem.* **2002**, *41*, 2282–2291.
41. J. P. Collman, R. A. Decréau, C. J. Sunderland, *Chem. Commun.* **2006**, 3894–3896.
42. J. P. Collman, N. K. Devaraj, R. A. Decréau, Y. Yang, Y.-L. Yan, W. Ebina, T. A. Eberspacher, C. E. D. Chidsey, *Science* **2007**, *315*, 1565–1568.
43. J. P. Collman, S. Ghosh, A. Dey, R. A. Decréau, Y. Yang, *J. Am. Chem. Soc.* **2009**, *131*, 5034–5035.
44. Z. Halime, M. T. Kieber-Emmons, M. F. Qayyum, B. Mondal, T. Gandhi, S. C. Puiu, E. E. Chufán, A. A. N. Sarjeant, K. O. Hodgson, B. Hedman, E. I. Solomon, K. D. Karlin, *Inorg. Chem.* **2010**, *49*, 3629–3645.
45. Z. Halime, H. Kotani, Y. Li, S. Fukuzumi, K. D. Karlin, *Proc. Natl. Acad. Sci. USA* **2011**, *108*, 13990–13994.
46. M. I. Verkhovsky, A. Jasaitis, M. L. Verkhovskaya, J. E. Morgan, M. Wikström, *Nature* **1999**, *400*, 480–483.
47. A. Jasaitis, M. I. Verkhovsky, J. E. Morgan, M. L. Verkhovskaya, M. Wikström, *Biochemistry* **1999**, *38*, 2697–2706.
48. D. Bloch, I. Belevich, A. Jasaitis, C. Ribacka, A. Puustinen, M. I. Verkhovsky, M. Wikström, *Proc. Natl. Acad. Sci. USA* **2004**, *101*, 529–533.
49. S. A. Siletsky, A. A. Konstantinov, *Biochim. Biophys. Acta* **2012**, *1817*, 476–488.
50. D. Zaslavsky, A. D. Kaulen, I. A. Smirnova, T. Vygodina, A. A. Konstantinov, *FEBS Lett.* **1993**, *336*, 389–393.
51. D. Zaslavsky, R. C. Sadoski, K. Wang, B. Durham, R. B. Gennis, F. Millett, *Biochemistry* **1998**, *37*, 14910–14916.
52. E. Pilet, A. Jasaitis, U. Liebl, M. H. Vos, *Proc. Natl. Acad. Sci. USA* **2004**, *101*, 16198–16203.
53. S. A. Siletsky, D. Han, S. Brand, J. E. Morgan, M. Fabian, L. Geren, F. Millett, B. Durham, A. A. Konstantinov, R. B. Gennis, *Biochim. Biophys. Acta* **2006**, *1757*, 1122–1132.
54. S. A. Siletsky, A. S. Pawate, K. Weiss, R. B. Gennis, A. A. Konstantinov, *J. Biol. Chem.* **2004**, *279*, 52558–52565.
55. V. Y. Artzatbanov, A. A. Konstantinov, V. P. Skulachev, *FEBS Lett.* **1978**, *87*, 180–185.
56. A. A. Konstantinov, S. Siletsky, D. Mitchell, A. Kaulen, R. B. Gennis, *Proc. Natl. Acad. Sci. USA* **1997**, *94*, 9085–9090.
57. M. Wikström, M. I. Verkhovsky, G. Hummer, *Biochim. Biophys. Acta* **2003**, *1604*, 61–65.
58. V. R. I. Kaila, M. I. Verkhovsky, G. Hummer, M. Wikström, *Proc. Natl. Acad. Sci. USA* **2008**, *105*, 6255–6259.
59. S. Yang, Q. Cui, *Biophys. J.* **2011**, *101*, 61–69.
60. T. Yamashita, G. A. Voth, *J. Am. Chem. Soc.* **2012**, *134*, 1147–1152.
61. G. Brändén, M. Brändén, B. Schmidt, D. A. Mills, S. Ferguson-Miller, P. Brzezinski, *Biochemistry* **2005**, *44*, 10466–10474.
62. A. Puustinen, M. Wikström, *Proc. Natl. Acad. Sci. USA* **1999**, *96*, 35–37.
63. J. Qian, D. A. Mills, L. Geren, K. Wang, C. W. Hoganson, B. Schmidt, C. Hiser, G. T. Babcock, B. Durham, F. Millett, S. Ferguson-Miller, *Biochemistry* **2004**, *43*, 5748–5756.
64. M. Wikström, C. Ribacka, M. Molin, L. Laakkonen, M. Verkhovsky, A. Puustinen, *Proc. Natl. Acad. Sci. USA* **2005**, *102*, 10478–10481.

65. J. Fetter, M. Sharpe, J. Qian, D. Mills, S. Ferguson-Miller, P. Nicholls, *FEBS Lett.* **1996**, *393*, 155–160.
66. J. R. Fetter, J. Qian, J. Shapleigh, J. W. Thomas, A. García-Horsman, E. Schmidt, J. Hosler, G. T. Babcock, R. B. Gennis, S. Ferguson-Miller, *Proc. Natl. Acad. Sci. USA* **1995**, *92*, 1604–1608.
67. D. A. Mills, B. Schmidt, C. Hiser, E. Westley, S. Ferguson-Miller, *J. Biol. Chem.* **2002**, *277*, 14894–14901.
68. I. Belevich, M. I. Verkhovsky, M. Wikström, *Nature* **2006**, *440*, 829–832.
69. P. R. Rich, A. Maréchal, *J. R. Soc. Interface* **2013**, *10*, 20130183.
70. I. Belevich, E. Gorbikova, N. P. Belevich, V. Rauhamäki, M. Wikström, M. I. Verkhovsky, *Proc. Natl. Acad. Sci. USA* **2010**, *107*, 18469–18474.
71. R. J. P. Williams, *Nature* **1995**, *376*, 643.
72. S. Yoshikawa, K. Shinzawa-Itoh, R. Nakashima, R. Yaono, E. Yamashita, N. Inoue, M. Yao, M. J. Fei, C. P. Libeu, T. Mizushima, H. Yamaguchi, T. Tomizaki, T. Tsukihara, *Science* **1998**, *280*, 1723–1729.
73. S. Yoshikawa, K. Muramoto, K. Shinzawa-Itoh, *Biochim. Biophys. Acta* **2011**, *1807*, 1279–1286.
74. T. Tsukihara, K. Shimokata, Y. Katayama, H. Shimada, K. Muramoto, H. Aoyama, M. Mochizuki, K. Shinzawa-Itoh, E. Yamashita, M. Yao, Y. Ishimura, S. Yoshikawa, *Proc. Natl. Acad. Sci. USA* **2003**, *100*, 15304–15309.
75. C. L. Perrin, *Acc. Chem. Res.* **1989**, *22*, 268–275.
76. K. Kamiya, M. Boero, M. Tateno, K. Shiraishi, A. Oshiyama, *J. Am. Chem. Soc.* **2007**, *129*, 9663–9673.
77. M. Kubo, S. Nakashima, S. Yamaguchi, T. Ogura, M. Mochizuki, J. Kang, M. Tateno, K. Shinzawa-Itoh, K. Kato, S. Yoshikawa, *J. Biol. Chem.* **2013**, *288*, 30259–30269.
78. K. Shimokata, Y. Katayama, H. Murayama, M. Suematsu, T. Tsukihara, K. Muramoto, H. Aoyama, S. Yoshikawa, H. Shimada, *Proc. Natl. Acad. Sci. USA* **2007**, *104*, 4200–4205.
79. M. M. Pereira, F. L. Sousa, A. F. Veríssimo, M. Teixeira, *Biochim. Biophys. Acta* **2008**, *1777*, 929–934.
80. M. M. Pereira, F. L. Sousa, M. Teixeira, R. M. Nyquist, J. Heberle, *FEBS Lett.* **2006**, *580*, 1350–1354.
81. H.-Y. Chang, J. Hemp, Y. Chen, J. A. Fee, R. B. Gennis, *Proc. Natl. Acad. Sci. USA* **2009**, *106*, 16169–16173.
82. P. V Argade, Y. C. Ching, M. Sassaroli, D. L. Rousseau, *J. Biol. Chem.* **1986**, *261*, 5969–5973.
83. T. Egawa, H. J. Lee, H. Ji, R. B. Gennis, S.-R. Yeh, D. L. Rousseau, *Anal. Biochem.* **2009**, *394*, 141–143.
84. T. Egawa, S.-R. Yeh, D. L. Rousseau, *PLoS One* **2013**, *8*, e63669.
85. T. Tsukihara, H. Aoyama, E. Yamashita, T. Tomizaki, H. Yamaguchi, K. Shinzawa-Itoh, R. Nakashima, R. Yaono, S. Yoshikawa, *Science* **1996**, *272*, 1136–1144.
86. S. Yoshikawa, K. Muramoto, K. Shinzawa-Itoh, M. Mochizuki, *Biochim. Biophys. Acta* **2012**, *1817*, 579–589.

Chapter 5
Transition Metal Complexes and the Activation of Dioxygen

Gereon M. Yee and William B. Tolman

Contents

G.M. Yee • W.B. Tolman (✉)
Department of Chemistry, University of Minnesota, 207 Pleasant St. SE, Minneapolis, MN 55455, USA
e-mail: wtolman@umn.edu

P.M.H. Kroneck, M.E. Sosa Torres (eds.), *Sustaining Life on Planet Earth: Metalloenzymes Mastering Dioxygen and Other Chewy Gases*, Metal Ions in Life Sciences 15, DOI 10.1007/978-3-319-12415-5_5

Abstract In order to address how diverse metalloprotein active sites, in particular those containing iron and copper, guide O_2 binding and activation processes to perform diverse functions, studies of synthetic models of the active sites have been performed. These studies have led to deep, fundamental chemical insights into how O_2 coordinates to mono- and multinuclear Fe and Cu centers and is reduced to superoxo, peroxo, hydroperoxo, and, after O-O bond scission, oxo species relevant to proposed intermediates in catalysis. Recent advances in understanding the various factors that influence the course of O_2 activation by Fe and Cu complexes are surveyed, with an emphasis on evaluating the structure, bonding, and reactivity of intermediates involved. The discussion is guided by an overarching mechanistic paradigm, with differences in detail due to the involvement of disparate metal ions, nuclearities, geometries, and supporting ligands providing a rich tapestry of reaction pathways by which O_2 is activated at Fe and Cu sites.

Keywords copper • iron • oxo complexes • peroxo • superoxo

Please cite as: *Met. Ions Life Sci.* 15 (2015) 131–204

1 Introduction

1.1 Overview

The use of dioxygen as a source of energy and as a reagent for functionalization of organic molecules is critical to life on Earth. Since the photosynthetic cyanobacteria began producing $O_2 \sim 10^9$ years ago, life forms have evolved by means to capture O_2 and use it for diverse functions. Dioxygen serves as a sink for electrons in respiration (to yield H_2O) and the source of O-atoms in critical biosyntheses of essential organic compounds. The thermodynamics for such uses are quite favorable, but the paramagnetic triplet state of O_2 imposes a kinetic barrier that enables it to coexist with combustible diamagnetic substrates. Nature overcomes this kinetic barrier by using reactive centers that can undergo single-electron transfer reactions or that contain unpaired electrons. Such centers include organic radicals, photochemically produced triplet excited states, and transition metal centers. Of these, transition metal centers (Cu and Fe, in particular) in proteins play especially important roles in the binding and reduction of O_2 during metabolic processes (Table 1) [1]. Such metalloprotein active sites exquisitely control the dioxygen-utilizing reaction pathways in order to promote the desired biological function and prevent deleterious formation of damaging byproducts (reactive oxygen species, ROS, such as H_2O_2 or $HO^{\bullet}$).

Table 1 Illustrative metalloproteins that use transition metal sites to activate dioxygen.

Entry	Enzyme	Active Site Metal	Function
1	Hemoglobin (Hb)	Fe	O_2 binding and transport
2	Myoglobin (Mb)	Fe	
3	Hemerythrin	Fe_2	
4	Hemocyanin	Cu_2	
5	Catechol 1,2-dioxygenase	Fe	Oxidative ring cleavage of catechol
6	Tyrosinase	Cu_2	Oxidation of catechol to quinone
7	Cytochrome P450	Fe	Hydroxylation of organic compounds
8	Taurine/α-ketoglutarate dioxygenase (TauD)	Fe	
9	Naphthalene-1,2-dioxygenase	Fe	
10	Phenylalanine hydroxylase	Fe	
11	Soluble methane monooxygenase (sMMO)	Fe_2	
12	Peptidylglycine-α-hydroxylating monooxygenase (PHM)	Cu	
13	Particulate methane monooxygenase (pMMO)	Cu_2/Cu_3?	
14	Isopenicillin N synthase	Fe	Reduction of O_2 to water
15	Ascorbate oxidase	Cu_3	
16	Cytochrome *c* oxidase (C*c*O)	Cu, Fe	

What is the basis of this control? How do structurally diverse metalloprotein active sites guide the O_2 binding and activation processes so as to perform diverse functions with the requisite rates and selectivities? These and related questions have driven extensive research over many decades that has led to deep insights into metalloprotein structure/function relationships. Among the myriad research approaches followed, studies of synthetic models of the active sites have been notably useful for understanding the fundamental chemistry involved in O_2 binding and activation [2, 3]. In addition, such studies have informed and stimulated the development of new reagents and catalysts for oxidation reactions.

In this chapter, we summarize the current understanding of dioxygen activation by transition metal centers gleaned from studies of synthetic complexes relevant to metalloprotein active sites, with particular emphasis on recent advances. We focus on the chemistry of Fe and Cu because of their prevalence in biology. For context, we first outline a general mechanistic paradigm that underlies research in the field. The subsequent discussion is organized by the type of metalloprotein active site, and in each case begins with a brief summary of older work into which readers can delve by consulting various cited reviews. In more expanded discussion, we then focus on recent work that has provided notable new understanding of dioxygen activation mechanisms. While aiming to be as comprehensive as feasible, we cannot hope to fully cover the breadth of the field. Thus, we have chosen to focus on specific mechanistic issues that we view subjectively as being of particular significance. We apologize to those whose work we neglect, and hope that the cited reviews will make up for our deficiencies and point readers to a more complete understanding of the field.

1.2 Paradigm for Dioxygen Activation by Metalloprotein Active Sites

The diverse and sometimes quite complicated mechanisms by which specific metalloprotein active sites bind and activate dioxygen can best be evaluated and understood by placing them within the context of an overarching, general mechanistic paradigm (Figure 1). According to this picture, a site with available electrons, either in the form of one or more metal ions (M) in a reduced oxidation state and/or with ligands (not shown) that are "non-innocent" and thus capable of providing reducing equivalents, binds to the O_2 molecule and reduces it to superoxide ($O_2^{-\bullet}$) to yield adduct **A**. This metal-superoxide adduct may be reduced further to the peroxide level, with concomitant delivery of a proton or involvement of an additional metal ion (species **B**). Scission of the O-O bond may then occur to yield metal-oxo or -oxyl species (**C**), again with possible involvement of additional protons or metal ions. At each stage of the process, electrons flow to the coordinated O_2 moiety, either from the ligand/metal unit or from an external source. We emphasize that this picture is highly schematic and generalized; a variety of structures, bonding patterns, and oxidation levels are possible for **A–C**.

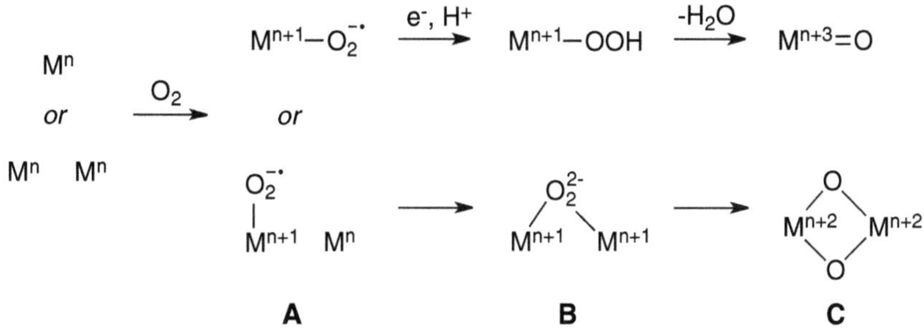

Figure 1 Generalized mechanistic scheme for O_2 activation by transition metal sites in proteins, illustrating mononuclear (**top**) and dinuclear (**bottom**) pathways.

In considering any generalized mechanistic picture, it is of particular interest to evaluate to what extent it is followed as a function of the specific attributes of the systems under study. Researchers thus have focused on how the nature of the metal ion(s) and its coordinating ligands, the geometric and electronic structures of the sites, the role of additional cofactors, and the protein environment influence the structures, stabilities, interconversions, and reactivity of **A–C**. Studies of synthetic complexes designed to replicate aspects of the metalloprotein active sites provide important insights into these influences, including protein environment effects as reflected by secondary-sphere hydrogen bonding involving suitable designed ligands, for example.

As discussed in more detail below, studies of the structures, spectroscopic properties, reactivity, and reaction mechanisms traversed by synthetic compounds reveal to what extent the general dioxygen activation paradigm is followed, whether diversions from this pathway occur, and, perhaps most importantly, the fundamental chemical basis for the observed properties and reactivity. In the end, one sees a rich diversity in the dioxygen activation mechanisms, with differences in the properties of the intermediates involved and the energy landscapes traversed. Understanding the fascinating chemistry underlying these differences continues to be a goal of research, with current progress reflected by the discoveries described in the following sections.

2 Dioxygen Activation by Iron Porphyrins

Heme containing enzymes are ubiquitous in nature, many of which bind and activate O_2 or which utilize O_2 intermediates in order to perform a variety of functions (Table 1, entries 1,2,7,16). These functions include oxygen transport, the oxidation of organic substrates via hydroxylation or O-atom transfer reactions, and the reduction of O_2 to water. It is no wonder that such enzymes have gained so much attention over the last three to four decades. Modeling of heme enzyme active intermediates generally involves the treatment of Fe(II)- and Fe(III)-porphyrin precursors with various O_2 and O-atom transfer reagents such as O_2, O_2^-, ROOH (R=H, alkyl, acyl), and PhIO.

The reactive nature of many of these compounds often demands low-temperature conditions and inert solvents; consequently, characterization often requires stopped-flow techniques, and in many cases structural details can only be inferred from spectroscopic methods. Despite these difficulties, synthetic modeling has led to key insights regarding the heme enzymatic systems as well as the potential of these synthetic systems to perform as catalysts in synthetic chemistry. In the following sections, recent advances in synthetic oxy-heme systems will be the focus, being prefaced with a brief discussion about O_2 activation by heme enzymes. For a more in depth discussion of previous work in oxy-heme chemistry, both biological and synthetic, the reader can look to numerous reviews on this topic [4–11].

2.1 Dioxygen Activation at Heme-Iron Centers: Lessons from Cytochrome P450$_{cam}$

Much of what is known about O_2 activation by heme centers has been elucidated through the study of cytochrome P450$_{cam}$, which carries out the stereo- and regio-selective hydroxylation of camphor to yield 5-*exo*-hydroxy-camphor [12]. Extensive structural and spectroscopic data [12–16] on this enzyme has provided a clear picture of the mechanism of O_2 activation, which serves as the paradigm for O_2 activation by heme enzymes and models. As shown in Figure 2, the resting state of the enzyme is a six-coordinate, low-spin Fe(III)-OH_2 adduct (**1**). Binding of the substrate (RH) triggers the loss of the axial water molecule and results in a change in spin state to give a five-coordinate high-spin Fe(III) complex (**2**). Reduction by one electron of the Fe(III) center generates the Fe(II) adduct **3**, which is now poised to activate O_2. Binding of O_2 generates an Fe(III)-superoxo intermediate (**4**), which can be further reduced by one electron to give the Fe(II)-superoxo complex **5**. Intramolecular electron transfer from the Fe(II) center to the O_2^- ligand followed by protonation results in the formation of Fe(III)-hydroperoxo intermediate **6** (also known as Cpd 0), and heterolysis of the O-O bond leads to the formation of a highly reactive Fe(IV)-oxo π* cation-radical (**7**). Species **7** (known as Cpd I) is the proposed active species that performs the initial C-H bond activation of camphor, which leads to the alcohol product bound to the Fe(III) center **8**; displacement by a water molecule regenerates the resting state [12].

The nature of the axial ligand as well as the distal residues are crucial in facilitating the O-O bond breaking process. In the case of cytochrome P450$_{cam}$, axial ligation of an anionic thiolate ligand provides a strong electronic "push" that inductively strengthens the Fe-O bond of **6** [17]. At the same time, it is proposed that a well ordered water molecule, which is hydrogen-bonded to

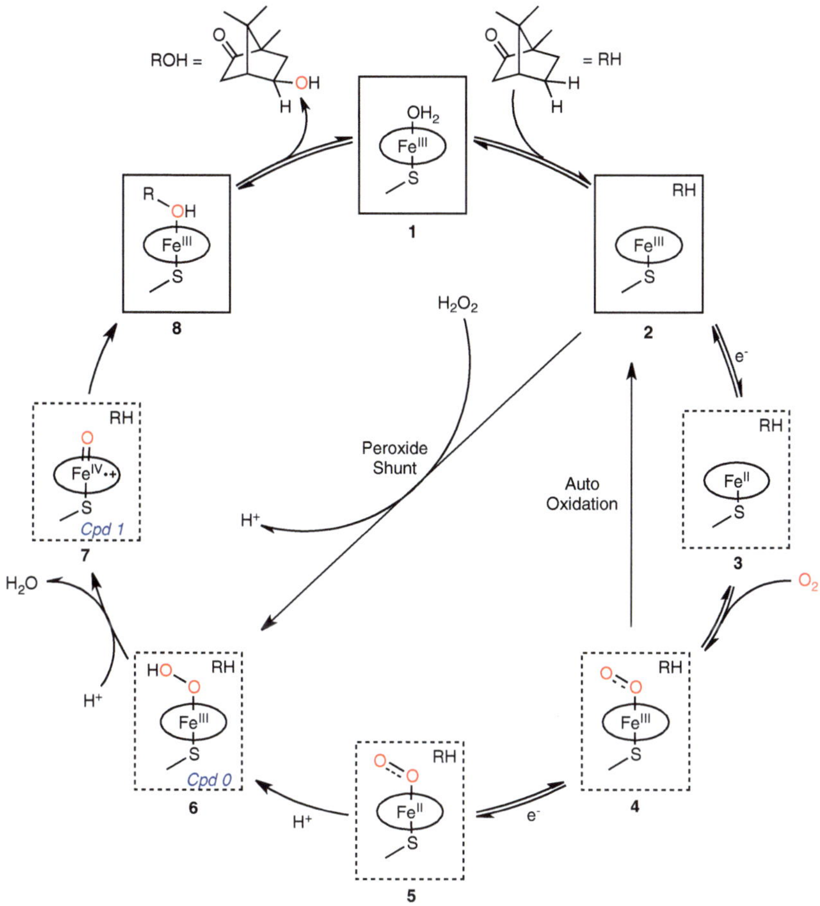

Figure 2 Mechanism of O_2 activation by cytochrome $P450_{cam}$ (adapted from [12]).

conserved Asp251 and Thr252 residues, serves to protonate the distal oxygen atom of the hydroperoxide ligand, pulling away the resulting water molecule [12]. The strengthening of the Fe-O bond coupled with the simultaneous formation/loss of water has been termed the "push-pull" mechanism [18] and results in the cleavage of the O-O bond and the formation of Cpd I (**7**).

Similar structural features are found in catalase and peroxidase enzymes, which form Cpd I by cleavage of the O-O bond in H_2O_2 (Figure 3). In both of these enzymes deprotonation of the proximal O-atom of the H_2O_2 ligand is performed by a distal histidine residue upon coordination. This proton is transferred to the distal O-atom, and the resulting water molecule is pulled away by either an asparagine or an arginine residue in the case of catalase (Figure 3a) or the peroxidases (Figure 3b), respectively [19]. In contrast, the active sites of heme-O_2 carriers lack the push-pull machinery for O-O bond cleavage. For example, in the case of oxy-myoglobin (oxyMb) (Figure 3c) a non-hydrogen bonded neutral histidine residue serves as a much weaker axial ligand, and the distal pocket lacks the polar residues necessary to pull the O-O bond apart [19].

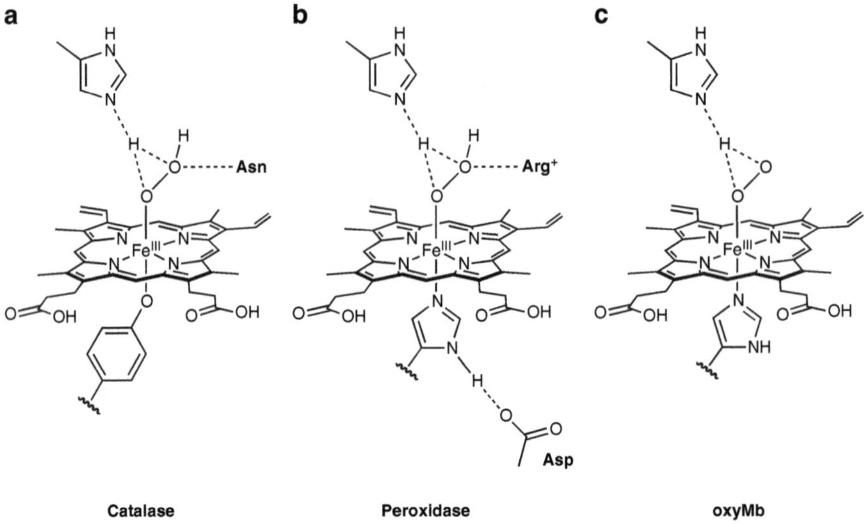

Figure 3 Active site interactions in catalase (**a**), the peroxidases (**b**), and oxyMb (**c**).

Over the past four decades significant progress has been made in characterizing the various intermediates shown in Figure 2 for a number of enzymatic systems, yet many questions still remain. In addition to being of fundamental interest, a better understanding of how nature is able to harness the oxidizing power of O_2 and H_2O_2 will enable the development of greener, more efficient catalysts for industrial purposes. With these goals in mind, chemists have turned to synthetic models.

2.2 *Iron-Porphyrin Complexes as Heme Models*

Generally, iron-porphyrin models have been developed with the dual purpose of (1) providing a better understanding of the spectroscopic signatures of the enzymatic systems and consequently the electronic nature of these intermediates, and (2) reproducing the reactivity of these systems, both as a means of understanding the structure-function relationship for these systems and for the purpose of catalyst development. The following sections focus first on the structural and spectroscopic features of these systems, beginning with models that bind O_2 and proceeding in order of the O_2 reduction process described above in Figure 2. Discussion of the reactivity of these models with specific regard to O-O bond cleavage and C-H bond activation will follow.

2.2.1 Fe(III)-Superoxo Intermediates

The first step in the activation of O_2 by iron-porphyrin complexes involves binding to the Fe(II) ion. Initial attempts to prepare such 1:1 Fe-O_2 adducts were met with difficulty due to the propensity of these early models to undergo auto-oxidation via the intermediacy of (μ-η^1:η^1-peroxo)diiron intermediates [10]. Success in model studies was achieved by inhibiting dimerization, often through the use of sterically hindered porphyrin rings, and using aprotic organic solvents as well as low temperatures in order to mitigate auto-oxidation pathways. Generally, oxygenation of a high-spin (S = 2) Fe(II) precursor yields a low-spin (S = 0) six coordinate oxy-heme complex [10, 20].

Building upon the early, now classic work in this area [10], recent studies have provided new insights. For example, in work aimed at providing benchmark information for studies of models of cytochrome *c* oxidase (Section 2.3.1), reversible O_2 binding to the Fe(II) precursor of **9** was observed at −80 °C (Figure 4) [21]. Manometry confirmed the formation of a 1:1 O_2/Fe adduct in tetrahydrofuran; however, in non-coordinating solvents the reversible formation of a μ-peroxo dimer was observed, as supported by a 1:2 O_2/Fe stoichiometry [21]. More recently, compounds **10–12** have also been synthesized in which the axial solvent molecule has been replaced by a pendant imidazole or pyridyl ligand [22, 23]. Compounds **9**, **10**, and **12** have been assigned as low-spin Fe(III)-superoxide species, primarily on the basis of their ^{1}H NMR and UV-vis spectra, as well as resonance Raman (rR) data in the case of **9** [22]. The assignment of **11** was inferred based on its conversion to an Fe(III)-peroxo intermediate upon one electron reduction [23].

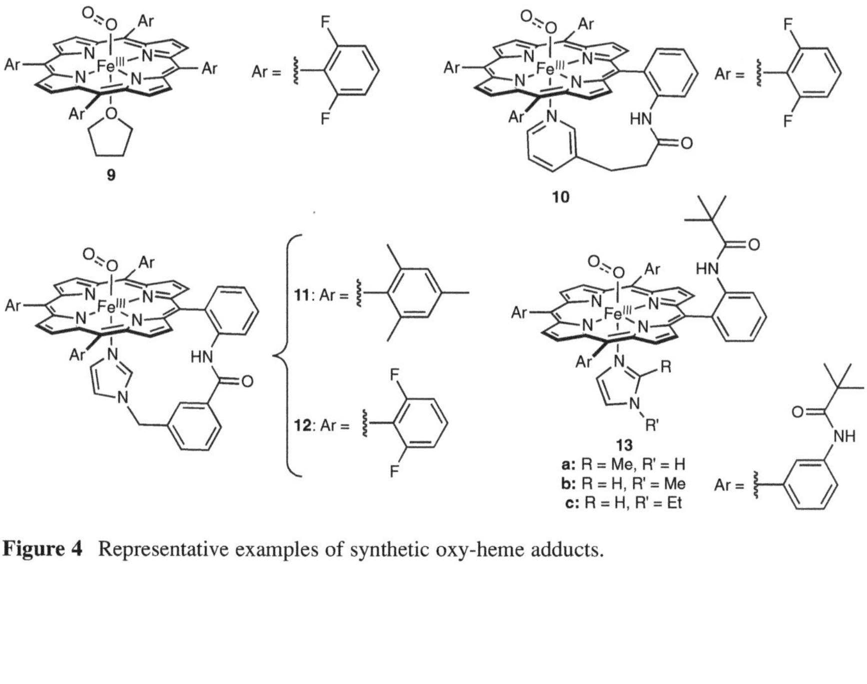

Figure 4 Representative examples of synthetic oxy-heme adducts.

Crystallographic characterization of synthetic oxyheme compounds has remained elusive, and to date, compounds **13a–c** (termed the "picket-fence porphyrin") are the only crystallographically characterized examples. The structure of **13b** was first solved in 1978, but was plagued by severe disorder of the O_2 ligand as well as high thermal motion [24, 25]. Very recently, data for this structure, along with those of **13a,c**, was recollected at low temperature (100–300 K) [26]. Important features of these structures include a bent η^1-coordination of the O_2 ligand to the iron center and disorder of the O_2 ligand over 4 positions, the occupancies of which are temperature-dependent [24–26]. At low temperatures (80 K), a 6.2° tilt in the Fe-O bond relative to the normal of the porphyrin plane is resolved for **13c**, which results in an O-O bond length of approximately 1.28 Å. This distance is consistent with values of ν(O-O) and with bond lengths proposed by DFT calculations [27–29]. Taken together, the crystallographic data are in satisfying agreement with that of oxyMb [30].

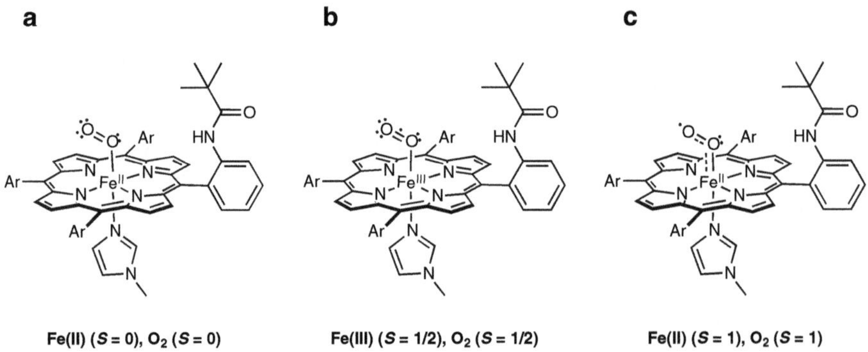

Figure 5 Proposed Fe-O_2 bonding descriptions for oxy-heme species, shown for compounds **13a–c** (adapted from [35]).

Despite the availability of crystal structures and evidence from rR experiments, the nature of the Fe-O_2 bonding in oxy-heme species is still not fully understood. Three main descriptions have persisted since the 1960's and these are summarized in Figure 5 [31–33]. In models **b** and **c**, anti-ferromagnetic coupling is proposed to account for the observed diamagnetism of these species. Recent theoretical studies support model **b** for oxyMb [34]. The results of these studies indicate a π-interaction for the Fe-O_2 moiety resulting from single electron transfer from the iron center to the O_2 ligand. Even more recently, iron L-edge X-ray absorption studies on **13b** have also revealed a significant π-interaction which the authors interpret as involving donation from the π^* orbital of an O_2^- ligand into the empty d_π orbital of the iron center [35]. However, comparison of the data to a set of Fe(II) and Fe(III) standards show the oxidation state of the iron center to be most consistent with an $S = 0$ Fe(II) configuration. Mössbauer spectra, which are characterized by large quadrupole splittings and small isomer shifts, also support an Fe(II) center with a strong π-interaction to the O_2 ligand [36]. Taken collectively, these results point towards an oxidation state of the iron center that is intermediate with respect to low-spin Fe(II) and Fe(III), which results from significant charge-transfer facilitated by a strong π-interaction for the Fe-O_2 moiety.

2.2.2 Fe(III)-Peroxo Intermediates

The second step in O_2 activation by iron-porphyrins is the one-electron reduction of the initial Fe-O_2 adduct formed upon O_2 binding, which yields species best formulated as Fe(III)-O_2^{2-} [12, 37, 38]. Alternatively, treatment of the Fe(II) precursor with O_2^- can also generate Fe(III)-O_2^{2-} adducts [39]. Some representative examples are shown in Figure 6.

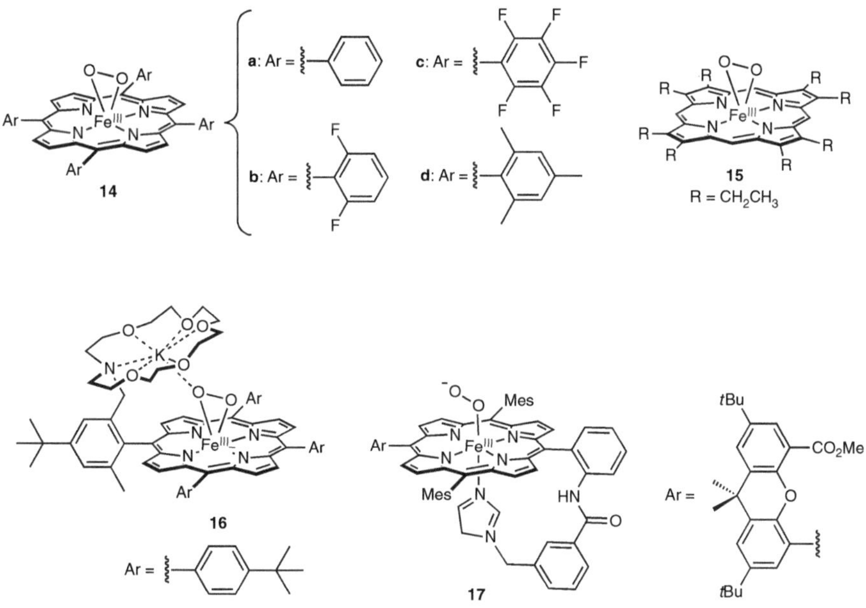

Figure 6 Representative examples of Fe(III)-O_2^{2-} adducts.

Early studies performed on compounds **14a** and **15** led researchers to formulate these species as side-on bound (η^2-peroxo)Fe(III) species on the basis of spectroscopic evidence from EPR, Mössbauer, and magnetic measurements, which confirmed the high-spin Fe(III) state of the iron center [39, 40]. Further characterization of the O-O and Fe-O bonding by IR, rR, and EXAFS spectroscopy supported a symmetrically bound peroxo ligand [39, 41]. These data have served as benchmarks for the characterization of other model compounds (**14b–d, 16**) (Figure 6), which all show similar spectroscopic features [42–44]. In a more recent report, compound **16** was found to be in equilibrium with an Fe(II)-O_2^- form. Preparation of this species was achieved by the treatment of the Fe(II) precursor with two equivalents of O_2^- [42]. Kinetic and equilibrium studies showed the reaction between the Fe(II) precursor and O_2^- to be fast, with the equilibrium strongly in favor of the bound species. New experimental and theoretical evidence suggests that **16** actually exists in equilibrium with its (η^1-superoxo)Fe(II)(DMSO) *isomer* rather than a resonance form of **16** [45].

The high-spin state and side-on binding of these previous examples are in contrast to what has been observed in heme enzymes [38]. Recently, the first example of a low-spin (η^1-peroxo)Fe(III) intermediate (**17**) (Figure 6) was prepared which utilized an imidazole-tailed hang-man porphyrin motif to sterically force the end-on binding of the O_2^{2-} ligand [38]. The synthesis of this species was achieved by oxygenation of the Fe(II) precursor followed by one electron reduction with cobaltocene at −70 °C. Spectroscopic comparison of the end-on and side-on imidazole-tailed Fe(III)-O_2^{2-} species revealed some significant differences for the end-on adduct. These differences include a blue shifted Soret band, as well as a low-spin Fe(III) EPR spectrum and higher-energy Fe-O stretching frequencies for the end-on complex. Interestingly, both species were suitable precursors for the formation of the corresponding Fe(III)-OOH species, suggesting the possibility that protonation of the η^2 adduct involves the initial transition to the η^1 conformation [23, 38].

2.2.3 Fe(III)-Hydroperoxo (Alkylperoxo) Intermediates

In the P450 and peroxidase enzymes, as well as in catalase, an Fe(III)-hydroperoxo intermediate (Cpd 0, **6** in Figure 2) is the precursor to Cpd I (**7** in Figure 2), forming the reactive Fe(IV)-oxo π* cation-radical upon O-O bond heterolysis. Additionally, Fe(III)-OOR intermediates have been proposed as being competent oxidants themselves [46–48]. However, literature reports of well-characterized Fe(III)-hydroperoxo (alkylperoxo) models are uncommon, owing to their general instability in solution (even at low temperatures) and the rapid formation of Fe(IV)-oxo/Fe(IV)-oxo π* cation-radical species. Synthetic methods targeting Fe(III)-hydroperoxo (alkylperoxo) complexes generally involve the addition of HOOR (R = alkyl, acyl, or H) to an Fe(III) precursor at low temperatures (< −60 °C) in organic solvents (Figure 7). These reactions are often performed under basic conditions to generate ROO^- to yield either a five- or six-coordinate product. It has also been demonstrated that reduction of an Fe(II)-superoxide followed by protonation can yield a hydroperoxide intermediate, although these examples are rare [23, 54].

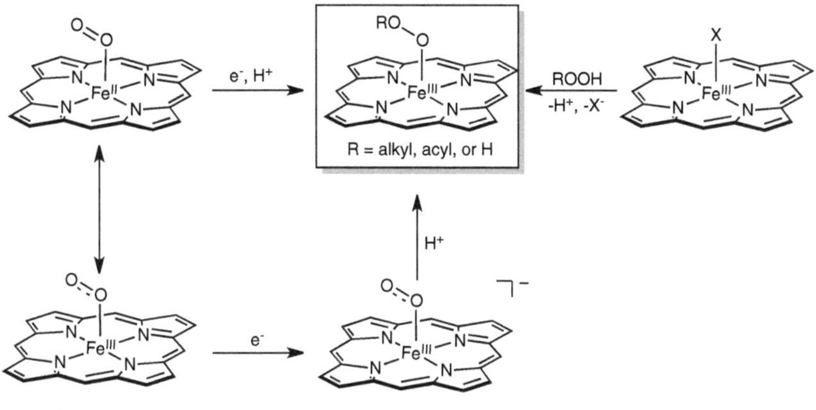

Figure 7 Synthetic routes to porphyrin Fe-OOR adducts.

Early model compound studies observed putative acyl and alkylperoxo intermediates in reactions leading to Cpd I type species [49]. However, conclusive characterization of these intermediates was lacking. These studies were able to show that the observed species were five-coordinate and high-spin on the basis of their electronic absorption, EPR, and ^{1}H NMR spectra [50–52]. Compounds **18a–19** (Figure 8) were the first six-coordinate examples [53–55] and differed from their five-coordinate counterparts insofar as they displayed EPR spectra characterized by a low g dispersion, consistent with a low-spin Fe(III) center and similar to spectra reported for hydroperoxo (alkylperoxo) adducts of hemoglobin, horseradish peroxidase, and bleomycin [56–58].

Axial ligand exchange was also demonstrated in the case of complex **20** (Figure 8), which was prepared by the addition of imidazole to a solution of the hydroxide precursor [59].

Notwithstanding the above data, until fairly recently conclusive characterization of the O-O and Fe-O bonding in Fe(III)-hydroperoxo (alkylperoxo) species was lacking. The characterization of complexes **21a** and **21b** marked the first reports of reliable rR parameters for an Fe(III)-hydroperoxide intermediate [23, 38]. Bands assigned to ν(Fe-O) and ν(O-O) were observed for both complexes near 570 and 810 cm^{-1}, respectively, consistent with an η^1 binding mode for the HOO$^-$ ligand. Additionally, the EPR spectra of these complexes exhibited the low g-dispersion characteristic of previously reported six-coordinate Fe(III)-hydroperoxide intermediates.

Interestingly, formation of **21a** (Figure 8) required the presence of the pendant imidazole, underscoring the importance of axial ligation in these systems. A role of the axial ligand has been proposed to be to enhance the nucleophilicity of the O_2^{2-} moiety. This has been recently demonstrated for a non-heme Mn(III)-peroxo complex, for which the authors postulate the axial ligand to facilitate the conversion between the η^2 and η^1 conformations [60]. It was also shown for the

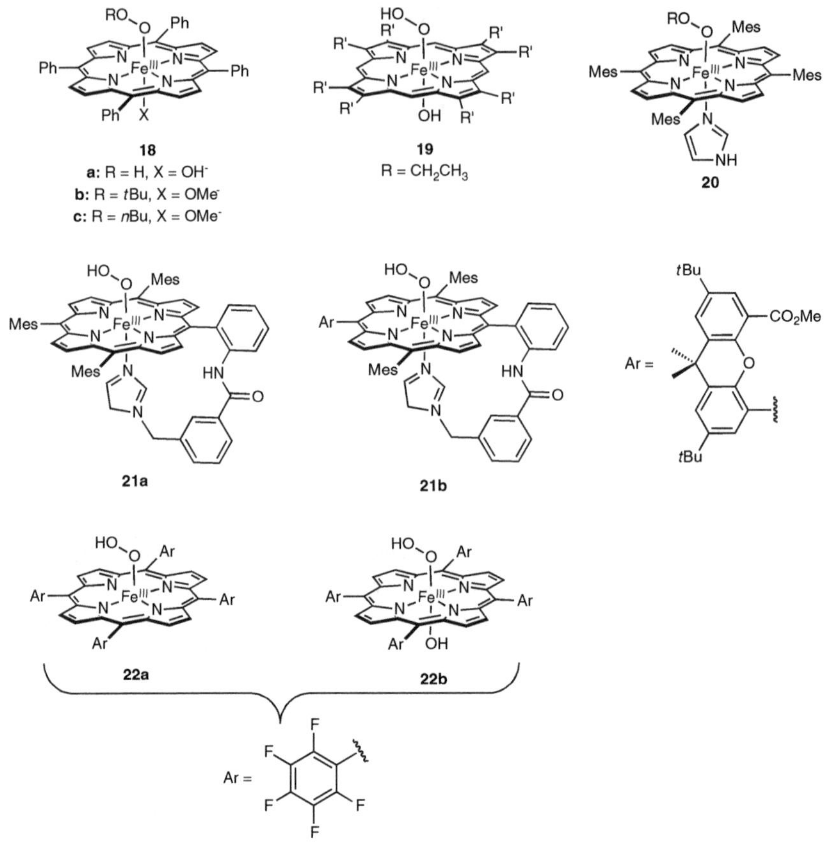

Figure 8 Well characterized examples of Fe-OOR complexes.

electron-deficient heme complex [(tpp)Fe^{III}(O_2^{2-})] (tpp = 5,10,15,20-tetrakis-(pentafluorophenyl)porphyrin) that epoxidation of the olefin 2-methyl-1,4-naphthoquinone only proceeds in the presence of 20 % dimethyl sulfoxide, which the authors attributed to increased axial coordination at the Fe(III) center [61].

EPR studies have also demonstrated that changes in spin-state tend to accompany coordination of an additional axial ligand for these systems, with five- and six-coordinate species often being in high- and low-spin states, respectively [55]. These differences can have significant mechanistic implications for O-O bond cleavage. Product analysis for the decomposition of **22a** and **22b** (Figure 8) by O-O bond cleavage revealed different mechanisms for each complex. Invoking a thermodynamic argument, the authors reasoned that **22a**, being in the high-spin state, favors the homolytic pathway which proceeds via a high-spin transition state to yield a high-spin Fe(IV)-oxo product, minimizing the energetic cost associated with changes in spin. Similarly, **22b** is low-spin and favors the heterolytic pathway yielding an Fe(IV)-oxo π*-cation radical species as the final product [62].

2.2.4 Fe(IV)-Oxo Intermediates

Synthetic (porphyrin)Fe(IV)-oxo complexes have long been sought in efforts to understand and harness the oxidation chemistry of enzymes such as the cytochrome P450's. Over the past three to four decades, thorough spectroscopic characterization has provided a good understanding of the electronic nature and structure of these species. Representative examples of these complexes are shown below in Figure 9. A brief overview of this early work will be given here as a preface to the discussion of the reactivity of these species which has largely been the focus of more recent efforts.

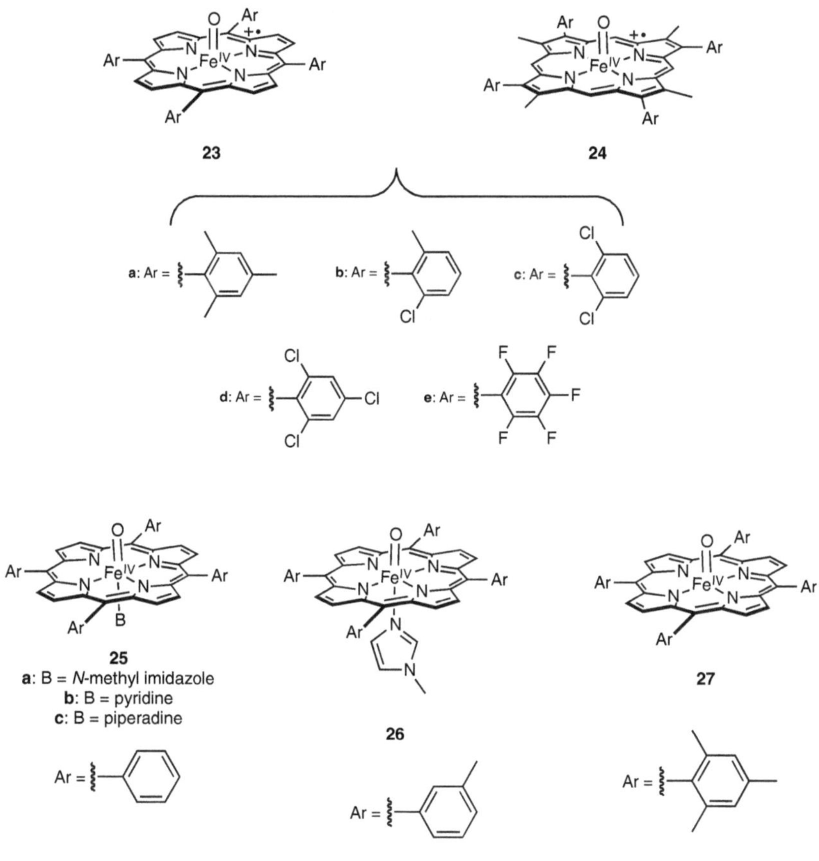

Figure 9 Some examples of the most fully characterized Fe(IV)-oxo π* cation-radical and Fe(IV)-oxo intermediates.

Compound **23a** (Figure 9) is the earliest and most studied examples of a Cpd I model, which was shown by Mössbauer and EPR studies to be an S = 3/2 system, in which a triplet Fe(IV) center is ferromagnetically coupled ($J = 43\ \mathrm{cm}^{-1}$)

to an a_{2u} porphyrin radical [63, 64]. Electron-withdrawing aryl rings (**23b–e**) have been shown to weaken ferromagnetic interaction and even result in a change in symmetry of the porphyrin radical (**23e**) from a_{2u} to a_{1u} as confirmed by ^{1}H NMR and EPR studies [65, 66]. Electron-withdrawing substituents have also been shown to prolong the lifetimes of these intermediates with respect to oxidative degradation, allowing for the study of these compounds at temperatures well above −50 °C [62]. In contrast to their *meso*-substituted counterparts, β-substituted Fe(IV)-oxo π* cation-radicals (**24a–e**) (Figure 9) have been characterized as a_{1u} porphyrin radicals weakly coupled with the Fe(IV) triplet state [65–67]. Interestingly, the electronic differences of the porphyrin systems in **23** and **24** do not significantly affect the Fe-O bond strength as reflected by their similar ν(Fe-O) values [68]. The Fe-O bonding has been characterized by EXAFS and rR, which have shown the Fe-O bond distance to be approximately 1.6 Å with values of ν(Fe-O) in the range of 800–850 cm^{-1} [69, 70]. These latter values are also sensitive to axial coordination as well as electron-withdrawing aryl substituents, which tend to result in negative shifts of ν(Fe-O) [63, 71–74].

Fe(IV)-oxo intermediates containing neutral porphyrin rings have also been prepared (**25a–c**, **26,** and **27**) (Figure 9), and were first observed in the decomposition of (η^1:η^2-peroxo)diiron intermediates [75, 76]. More recently, electrochemical preparation of **27** has been achieved by the one-electron oxidation of the Fe(III)-OH precursor [77]. Many structural similarities exist between these species and their Cpd I cousins, particularly with regard to their Fe-O stretching frequencies. Thus, the oxidation state of the porphyrin ring does not significantly affect the Fe-O bond strength [6].

2.2.5 Reactivity of Iron-Porphyrin Intermediates

Factors Affecting O-O Bond Cleavage: Heterolysis *versus* Homolysis

In Nature and in a number of synthetic systems, Fe(IV)-oxo intermediates are the O-O bond cleavage products of Fe(III)-OOR species. As such, significant efforts have been made to find out what factors determine the mode of O-O bond scission. Product analysis studies carried out in the late 1990's and early 2000's demonstrated that electron-deficient porphyrin systems and protic/acidic reaction conditions favored O-O bond heterolysis, while the opposite was true for O-O bond homolysis [78–80]. The nature of the substituent of the hydroperoxide oxidant also was shown to influence the mechanism of cleavage, with electron-withdrawing and electron-donating groups favoring heterolytic and homolytic cleavage, respectively [81]. A number of subsequent studies on the reactivity of Fe(IV)-oxo species have indirectly confirmed these results [82–84].

Two more recent studies [85, 86] have sought to further characterize the nature of the observed pH dependence of these systems. While it had been postulated that a mechanistic changeover was the result of acid-base effects or changes in speciation of the reacting Fe(III) complexes, it was found that the observed reactivity correlates well with the pH dependence of $E^{\circ\prime}(Fe^{IV/III})$ and $E^{\circ\prime}(P^{+\bullet}/P)$ (P = porphyrin). These results were interpreted to be consistent with the notion that pH-dependent redox equilibria may in fact mask the true identity of the oxidation products, especially if product analysis is the method of identification [85]. Interestingly, further investigation using rapid-scan UV-vis experiments performed at low temperature and under conditions of excess oxidant *only* showed formation of the Fe(IV)-oxo π* cation-radical in the pH ranges studied (pH = 6.3–11.4), indicating that, for the complexes studied, O-O bond cleavage only proceeds via heterolysis when *meta*-chlorobenzoic acid (*m*-CPBA) is the oxidant [86].

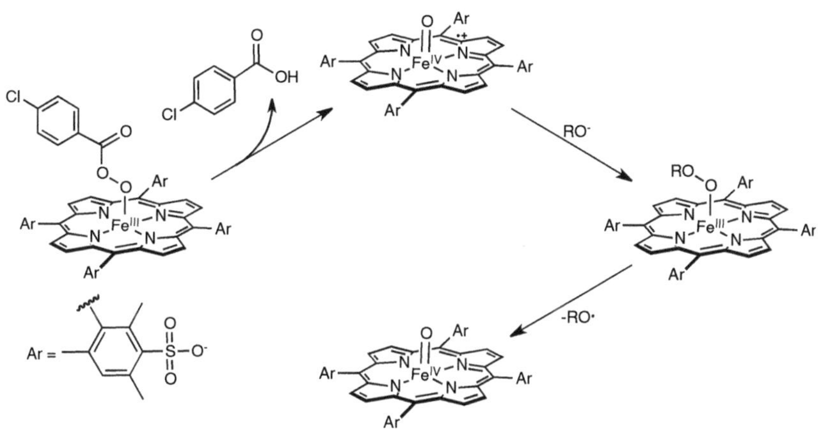

Figure 10 Proposed mechanism of O-O bond cleavage and formation of Fe(IV)-oxo intermediate under basic conditions.

Under basic conditions (pH >8, Figure 10) further reaction of the Fe(IV)-oxo π* cation-radical was observed to generate the one electron reduced Fe(IV)-oxo species as well as regenerate some of the starting Fe(III) precursor. These findings are consistent with the fact that under basic conditions (pH >7) the Fe(IV)-oxo form is electrochemically more stable than the Fe(IV)-oxo π* cation-radical [85]. The authors propose that under basic conditions, reaction of the Fe(IV)-oxo π* cation-radical with RO^- (R=H, CH_3) likely generates a reduced Fe(III)-OOR species which can then undergo homolysis to form the observed Fe(IV)-oxo intermediate or lose ROO^- to reform the starting Fe(III) precursor [86]. These findings highlight the importance of the nature of the oxidant as well as the reaction conditions in dictating the mechanism of O-O bond cleavage.

Alkane Hydroxylation by Fe(IV)-Oxo Intermediates

Alkane hydroxylation by Fe(IV)-oxo π* cation-radical species has been well studied, although few models have come close to reproducing the reactivity towards C-H bonds observed in the cytochrome P450 enzymes. Recently, compound **29**

(Figure 11) has been shown to be a highly reactive hydroxylating agent for xanthene, fluorene, and benzoic acid substrates, with rate constants similar to those observed for Cpd I of the cytochrome P450 enzyme CYP119 [87].

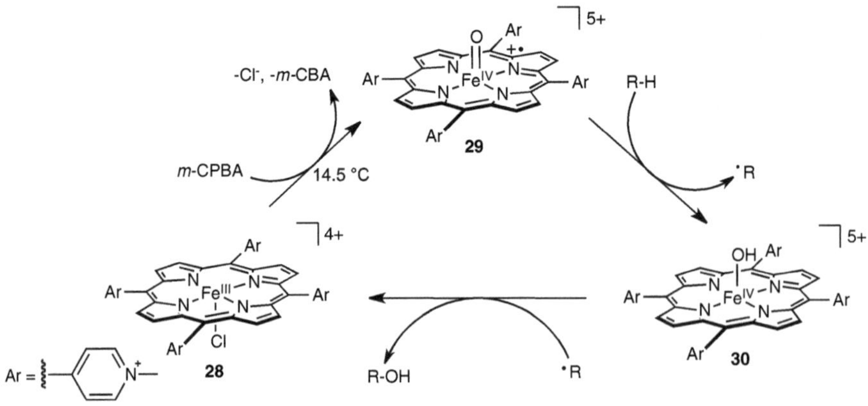

Figure 11 Formation of **29** from **28** and subsequent oxidation of organic substrates.

Kinetic analyses were performed using single-mixing stopped-flow UV-vis measurements to observe the reaction between mixtures of **28** and *m*-CPBA with substrate at 14.5 °C. A linear correlation was observed between log(k) for alkane hydroxylation and the bond dissociation enthalpy (BDE) of the C-H bond, indicative of rate-limiting C-H bond breaking [87]. While the origin of high kinetic reactivity is unknown, the authors postulate that both a low-lying a_{2u} porphyrin HOMO and higher spin states (Figure 12), made accessible from the electron deficiency of the porphyrin system, may result in diminished reaction barriers via spin-crossing along the reaction coordinate, similar to that observed for Mn(V)-oxo complexes supported by analogous porphyrin systems [87, 88]. Theoretical calculations performed for a variety of P450 models support the possibility of such a scenario provided that significant lowering of transition state energies can be achieved [89], as is proposed for **29**.

Carbon-hydrogen bond activation by Fe(IV)-oxo species (Figure 13) has also recently been demonstrated, albeit with more activated substrates and at much lower rates than their Cpd I analogues [90]. Compounds **31–33** were generated at 15 °C and subsequently treated with substrate. Large kinetic isotope effects (KIEs) were observed (KIE = 20 and 21 for DHA and xanthene, respectively) and linear correlations between log(k) and BDE_{CH} were measured, indicative of an H-atom abstraction mechanism [90]. In comparing **31–33**, compound **31** with the more electron-deficient porphyrin system was found to be the stronger oxidant, while by the same token, **33** was shown to be the weakest. Interestingly, in a previous report the order of reactivity for **31–33** was reversed for the oxidation of styrene and benzyl alcohol. The authors also observed a suppression of reactivity when an excess of the starting Fe(III) precursors was added. It was proposed that these results together seemed to indicate a disproportionation of the formed Fe(IV)-oxo

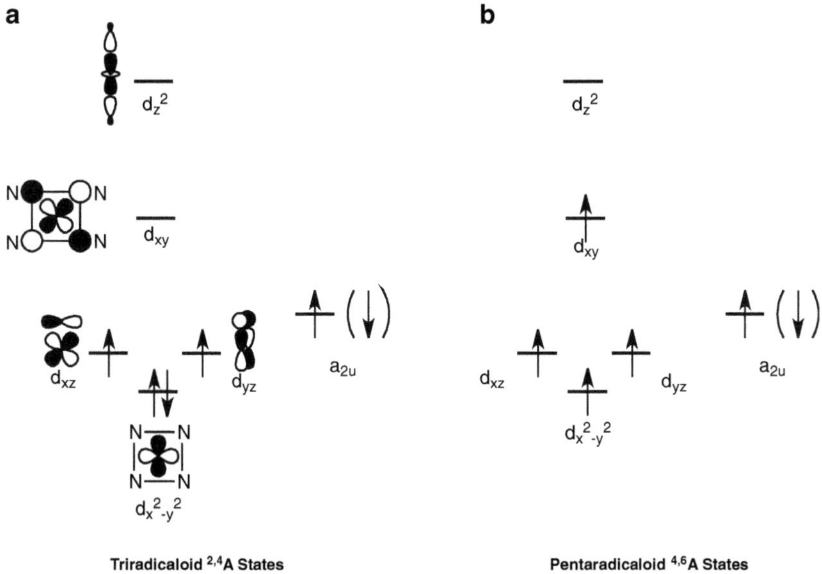

Figure 12 Orbital diagram depicting potential spin states of **29** (adapted from [89]).

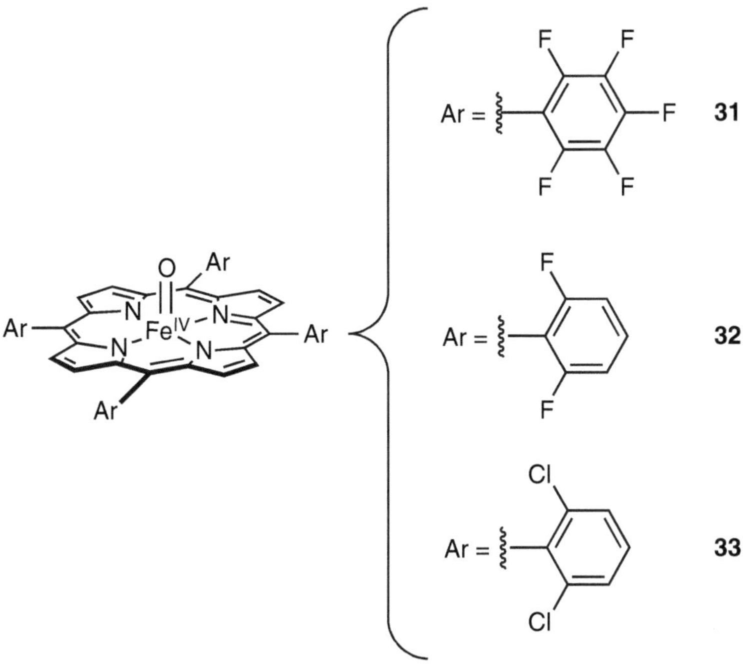

Figure 13 Fe(IV)-oxo species for which C-H bond activation has been demonstrated.

species to give the corresponding Fe(IV)-oxo π* cation-radical intermediate and the Fe(III) precursor, and that in fact the Fe(IV)-oxo π* cation-radical species was the active oxidizing agent. However, this hypothesis is speculative, as this species was not directly observed [91].

The generally accepted mechanism for alkane hydroxylation by Fe(IV)-oxo π* cation-radical intermediates is known as the "oxygen rebound" pathway [92, 93] and is outlined in Figure 14. The mechanism proceeds by initial, rate-limiting H-atom abstraction from the substrate to generate an intermediate Fe(IV)-OH (Cpd II) and alkyl-radical species. Kinetic studies have revealed large KIEs indicative of a significant tunneling contribution and consistent with the involvement of rate limiting C-H bond cleavage [92, 94–96]. Subsequent recombination of the alkyl and hydroxyl radicals (the latter originating from the Fe(IV)-OH intermediate) is proposed to occur via a "rebound" process, yielding the hydroxylated product bound to the Fe(III) center as has been isolated for $P450_{cam}$ (Section 2.1) [94]. The involvement of a short-lived radical intermediate is supported on the basis of stereochemical and allylic scrambling observed during the oxidation of a number of substrates by both enzymatic [92] and synthetic systems [97].

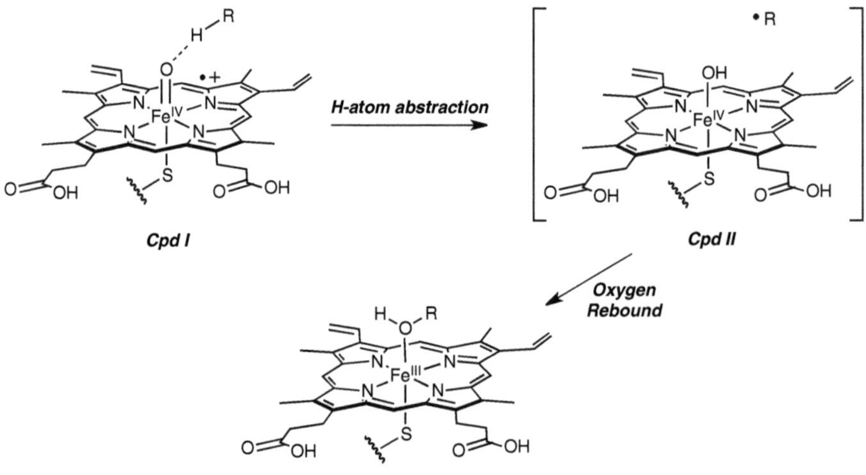

Figure 14 Proposed oxygen rebound mechanism.

There has been a long-standing controversy as to whether or not hydroxylation occurs exclusively via a radical rebound process. Some have proposed the involvement of multiple oxidants on the basis of ultra-fast radical clock experiments, which have provided a disparate range of rate constants for the rebound process [93]. In addition, these studies have provided evidence for the involvement of cationic intermediates in addition to radical species. Another theory which has been argued on the basis of computational evidence invokes the involvement of multiple spin-states for which the activation barriers to rebound differ significantly, and as a result yield significantly different rates [98]. A multi-state reactivity scenario has also been argued for non-heme Fe(IV)-oxo systems [99, 100].

O-Atom Transfer Reactivity

Fe(III) heme complexes have long been known to be efficient O-atom transfer catalysts, particularly for olefin epoxidation [101]. Examples of sulfur [82, 102], nitrogen [103], and phosphorus [76] oxidation have also been observed. It was originally proposed that epoxidation was carried out by an Fe(IV)-oxo π* cation-radical species [101]. Since then, this chemistry has been more thoroughly explored and demonstrated for a variety of catalysts and oxidants including PhIO [101, 104], H_2O_2 [105], and HOOR [106, 107] in which the involvement of Fe(IV)-oxo and Fe(III)-OOH intermediates have also been proposed [103, 108, 109].

Recent studies of Fe(III)X/H_2O_2 systems in alcoholic solvents have provided insight regarding the epoxidation reactivity of Fe(IV)-oxo π* cation-radical intermediates [83] (Figure 15). Binding of H_2O_2 was shown to be preceded by loss of X^- and coordination by the alcohol solvent. O-O bond cleavage follows, which has been proposed to occur both hetero- and homolytically to generate Fe(IV)-oxo π* cation-radical and putative Fe(IV)-hydroxo/Fe(IV)-oxo intermediates, respectively [80]. In these systems, the homolysis products are not believed to be involved in the epoxidation chemistry. As for other peroxide oxidants, the mechanism of O-O bond cleavage was found to be dependent on the acidity of the solvent [110] as well as on the electron-withdrawing/donating natures of the oxidant [81] and porphyrin system [108]. Additionally, both the rate of H_2O_2 consumption and O-O bond heterolysis were shown to increase with increasing alcohol concentration [110], supporting the intermediacy of a Fe(III) alcohol adduct. A mechanism for the O-atom transfer and epoxide formation has been proposed from both experimental and DFT studies, which involves initial rate determining electrophilic attack at the olefin by the oxo moiety to form an Fe(IV)-substrate radical intermediate [111, 112]. Subsequent ring closure yields the epoxide product coordinated to the Fe(III) center of the resting state. Similar to the hydroxylation reactivity for these models, DFT studies also suggest a two-state reactivity scenario for epoxidation, with major differences between the two pathways being the barrier to ring closure, which is almost negligible for the doublet state and slightly higher for the quartet state [112–114].

O-atom transfer by Fe(IV)-oxo and Fe(III)-OOR intermediates has also been demonstrated [109]. Recent investigation into the reactivity of Fe(III)(*m*-CPBA) (compound **34**, Figure 16) has shown this species to be capable of O-atom transfer to *cis*-stilbene and dimethyl sulfide at −15 °C in acetonitrile [82]. Both reactions were found to be highly dependent on the reaction conditions (aprotic solvents, temperature, equivalents of oxidant) and were shown to be accelerated under conditions of excess substrate. O-atom transfer occurs in competition with O-O bond heterolysis, which predominates under conditions of excess oxidant to generate the corresponding Fe(IV)-oxo π* cation-radical species.

On the basis of the observed rate constants, the authors argue that competitive substrate oxidation by **34** in the presence of **35** (Figure 16) is only possible at extremely high equivalents of substrate (>5000 eq). Additional DFT studies provided evidence that argues against the co-participation of an Fe(III)-OOH intermediate alongside Fe(IV)-oxo π* cation-radical species [115], so the involvement of this species is only relevant in aprotic solvents and under conditions that inhibit O-O bond heterolysis.

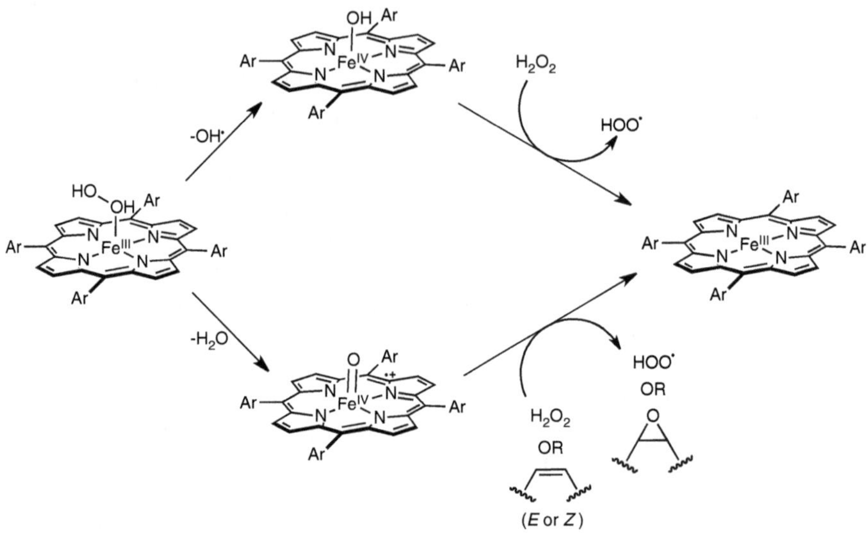

Figure 15 Proposed mechanism for epoxidation of olefins by catalyzed by Fe(III) heme chloride complexes (adapted from [83])

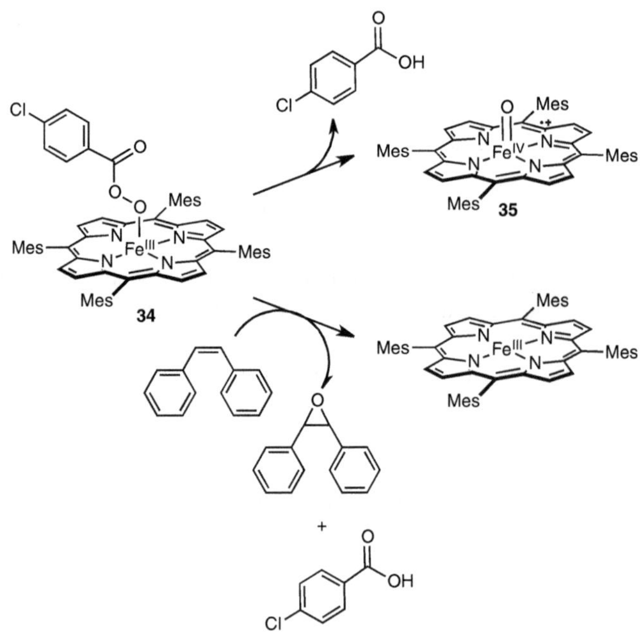

Figure 16 Competitive oxidation/O-O bond cleavage reactivity of **34**.

Role of the Axial Ligand in Hydroxylation and O-Atom Transfer Reactions

For a number of model systems, coordination of an anionic axial ligand has been shown to have a sometimes drastic effect on epoxidation and hydroxylation reactivity. Consistently these findings seem to demonstrate that increased electron donation to the iron center results in markedly faster rates [116, 118]. However, the origin of this increased reactivity is still somewhat unclear. Electrochemical studies [112] have shown that axial ligands which are known to be stronger donors actually increase the value of $E_{1/2}(P^{+\bullet}/P)$ for these intermediates, which is both counterintuitive and in disagreement with calculated values of $E_{1/2}(P^{+\bullet}/P)$ [117] that display an opposite trend.

Other DFT studies have found that the major effect of the axial ligand is to increase the basicity of the oxo moiety, thereby strengthening the FeO-H bond that results from H-atom abstraction, and weakening the Fe-O bonding, which encourages O-atom transfer [118]. However, the lack of a linear correlation between values of ν(Fe-O) and rates of epoxidation still remains to be explained. Recent calculations have confirmed a linear correlation between the O-H BDE of the corresponding Fe(IV)-hydroxo species and the barrier heights of a variety of reactions, including aromatic hydroxylation [119], olefin epoxidation [120], and sulfoxidation [121], all of which in part implicate the basicity of the oxo moiety as being a key factor. Still other recent experimental and theoretical evidence suggests that strong axial ligation serves to stabilize the ground state of the Fe(III) product resulting from these reactions, which in turn leads to a stabilization of the transition state and an enhancement of reactivity. This hypothesis is supported by an observed linear correlation between values of $E_{1/2}(Fe^{III}/Fe^{II})$ and $\ln(k_2)$ for the epoxidation of cyclooctene [112].

2.3 *Iron-Porphyrin/Copper Complexes as Cytochrome* c *Oxidase Models*

In biology, cytochrome *c* oxidase (C*c*O) plays a crucial role as the terminal oxidant in aerobic respiration. Specifically, the enzyme utilizes the electrons from the electron transport chain to carry out the four-electron reduction of dioxygen to water, which is performed at a dinuclear active site consisting of an iron heme center and a copper center. Remarkably, the native enzyme performs the $4e^-$ reduction of O_2 to H_2O with >99 % selectivity relative to the generation of partially reduced oxygen species (PROS). Much as in the case of heme-O_2 chemistry, synthetic modeling has focused on two primary areas: the characterization of (μ-peroxo)iron-copper intermediates and studies of catalytic O_2 reduction and reactivity. As will be discussed, both approaches use very different methods and offer complementary perspectives.

2.3.1 (μ-Peroxo)Iron-Copper Intermediates

Initial efforts to model C*c*O were focused on preparing analogs of the "resting state" of the enzyme, which features Fe(III) and Cu(II) centers [122]. These early studies were more directed towards electronic characterization of the resting state, however, and it was not until the *reduced* form of the enzyme was modeled that reactivity with O_2 was observed. One of the first examples of O_2 reduction for such model compounds is described in Figure 17, in which a mixture of **36** and **37** in dichloromethane was treated with a stoichiometric amount of O_2 at −80 °C [123]. Compound **38** was isolated upon warming to 0 °C followed by treatment of the reaction solution with heptane. Other μ-oxo complexes with iron and copper have been prepared in addition to **38** [124, 125], which may suggest that such species have roles as intermediates in the catalytic cycle of C*c*O. Importantly, this example suggested the efficacy of these complexes in promoting complete O_2 reduction.

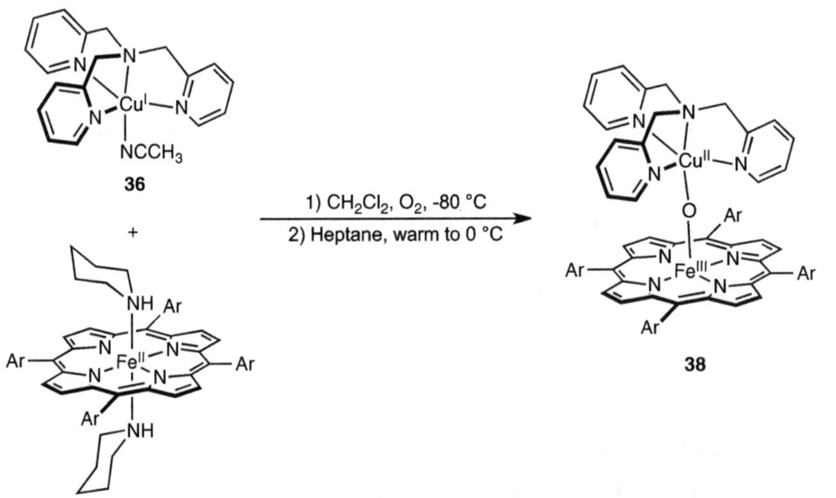

Figure 17 Formation of the μ-oxo iron-copper complex (**38**) from the reaction of **36** and **37** with O_2.

Since this early work, a number of examples of (μ-peroxo)iron-copper intermediates have been generated and studied spectroscopically for a variety of ligand systems, some of which are featured in Figure 18. The first and only crystallographic report of a (μ-peroxo)iron-copper complex (**44**) appeared in 2003, and revealed an η^2:η^1 coordination for the iron and copper centers, respectively [126]. Resonance Raman and EXAFS spectroscopy have also confirmed such coordination in solution for complexes **39–41** [127]. Reducing the denticity of the ligands on the Cu site from four to three resulted in η^2:η^2 coordination of the peroxo bridge, as shown for complexes **42** and **43**. This change in coordination is evidenced by a decrease in ν(O-O) by approximately 30–50 cm^{-1}, indicative of O-O bond weakening [128, 129]. Further study of complexes **40** and **42** by DFT calculations showed this decrease in ν(O-O) to be the result of efficient backbonding from the copper center into the σ* orbital of the peroxo ligand in the case of the η^2:η^2 binding configuration. Given the conformational restraints of the copper supporting ligand for compound **40**, such backbonding does not occur [130].

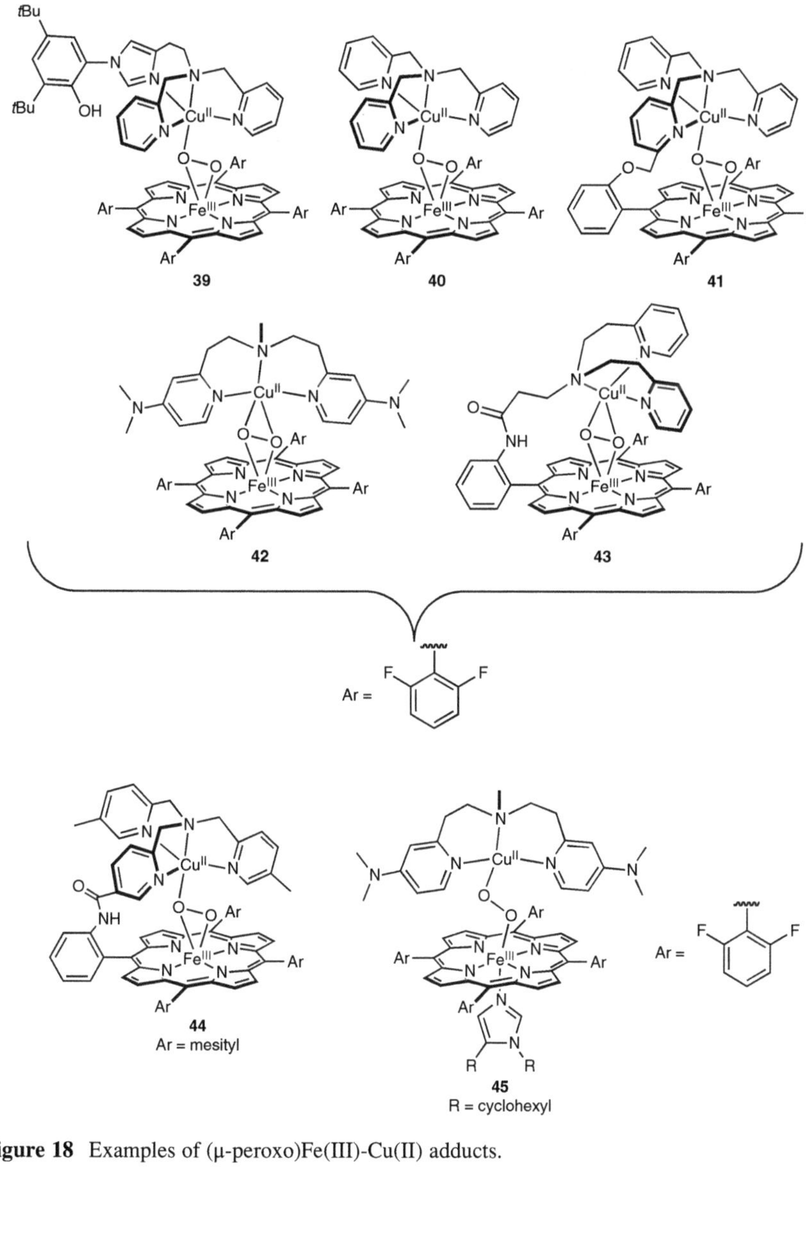

Figure 18 Examples of (μ-peroxo)Fe(III)-Cu(II) adducts.

Axial ligation at the heme center has also been demonstrated to have an effect on the binding mode of the peroxo bridge. In a recent example, addition of 1,5-dicyclohexylimidazole (DCHim) to compound **42** (Figure 18) was shown to result in the fully end-on $\eta^1:\eta^1$ coordination of the O_2^{2-} moiety to both metal centers (**45**), which was supported on the basis of rR, EXAFS, and UV-vis spectroscopy [131]. Values of ν(O-O) and ν(Fe-O) for **45** are similar to those observed for the $\eta^2:\eta^1$ bound species. Interestingly, the addition of dicyclohexylimidazole (DCHim) and dimethylaminopyridine (DMAP) to compound **40** only resulted in the dissociation of the complex and the loss of O_2 [130].

In each of these examples, the Fe(III) and Cu(II) centers have been shown to be anti-ferromagnetically coupled via the peroxo bridge, which serves as the conduit for superexchange between iron and copper [130]. In the case of compound **45** (Figure 18), axial ligation of the heme iron center leads to the unique example of a low-spin (μ-$\eta^1:\eta^1$-peroxo)Fe(III)-Cu(II) species, analogous to that observed in heme-hydroperoxo chemistry (see Section 2.2.3). Modeling of this species by DFT has demonstrated the groundstate for this complex to be dependent on the Fe-O-O-Cu dihedral angle (Θ), which effectively dictates the nature of the Cu-peroxo interaction, and thus the nature of the coupling between the two metal centers (Figure 19) [132]. For $\Theta < 150°$, the spins located on the iron and the copper interact with the same π^* orbital of the peroxo bridge (π^*_ν); however, for $150° < \Theta < 180°$, the Cu $3d_{x^2-y^2}$ orbital interacts with the π^*_σ orbital of the peroxo moiety, effectively closing the superexchange pathway and resulting in a ferromagnetically coupled adduct [132].

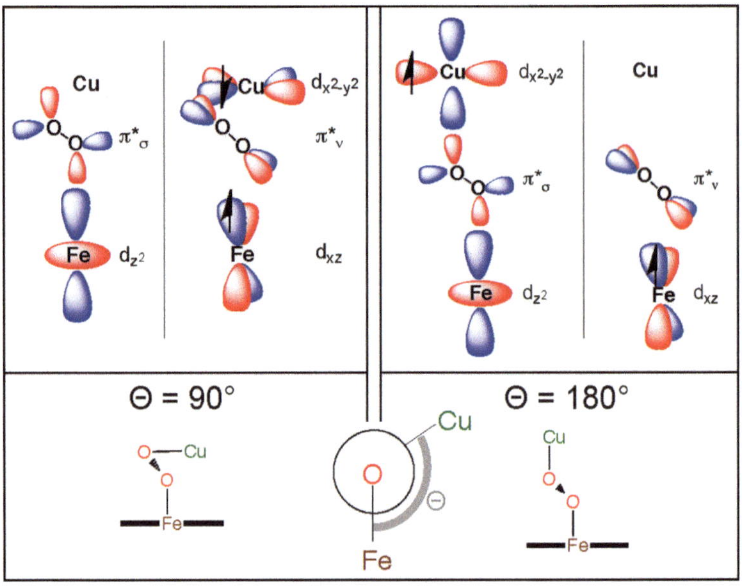

Figure 19 Origin of groundstate dependence on the Fe-O-O-Cu dihedral angle. Reproduced with permission from [132]; copyright 2011, American Chemical Society.

Recent kinetic studies have sought to better characterize the mechanism of O_2 activation to yield (μ-peroxo)iron-copper adducts, a key question being whether O_2 binds first to Fe or to Cu. Stopped-flow UV-vis measurements below −90 °C revealed the formation of a heme-superoxo adduct upon addition of O_2 to the non-oxygenated precursors of **40** and **43** (Figure 18), which were confirmed on the basis of rR and Mössbauer spectroscopy [133, 134]. Such intermediates have also been observed in other model systems [140]. In contrast, some studies of the native enzyme have proposed that initial O_2 binding occurs instead at the copper center, and that the subsequent formation of a myoglobin-like heme-O_2 adduct results from O_2 transfer from the copper site to the Fe(II) heme center [135].

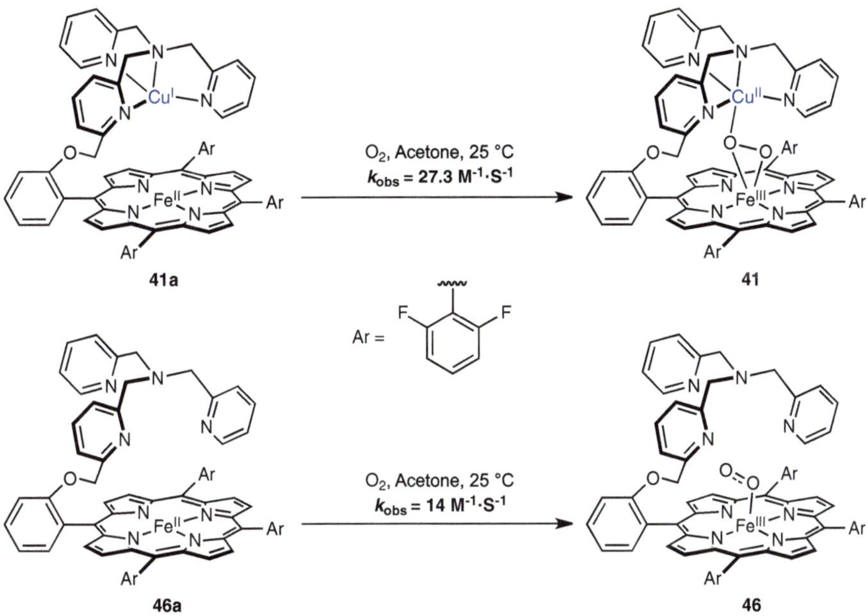

Figure 20 Comparison of reactivity between **41a** and **46a** with O_2.

Results from a more recent study of **41a** (Figure 20) suggest that a primary role of the copper center is to assist in the binding of O_2 to the iron center. This conclusion is largely supported on the basis of comparison to **46a**, which reacts slower than its dinuclear counterpart [136]. In aqueous systems, the presence of copper has also been suggested to play a role in protecting the iron center against the coordination of water, which has been shown to significantly inhibit the binding of O_2 by effecting a high- to low-spin crossover [137].

2.3.2 Reactivity of Iron-Copper Dioxygen Intermediates

For a number of iron-copper systems, catalytic and electrocatalytic O_2 reduction have been demonstrated, with some primary examples shown in Figure 21. Complexes based on the structural motif illustrated by **47a–c** represent some of the earliest and most studied functional C*c*O models, boasting perhaps the closest structural and functional likeness of C*c*O for any model system. In a seminal study, the electrocatalytic reduction of O_2 was demonstrated for all three compounds under physiologically relevant conditions (i.e., at potentials between 0 and 300 mV *versus* NHE and pH = 7) [138].

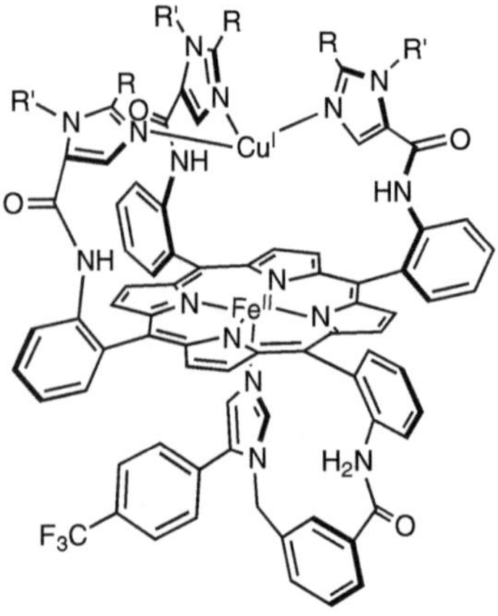

47
a: R = H, R' = CH_3
b: R = $(CH_2)_2CH_3$, R' = H
c: R = $(CH_2)_2CH_3$, R' = CH_3

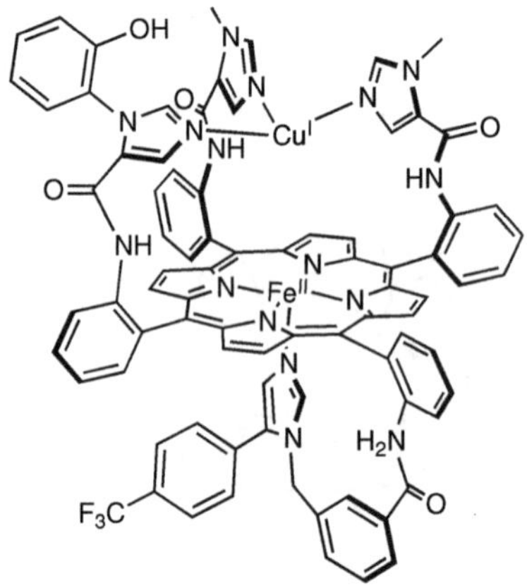

48

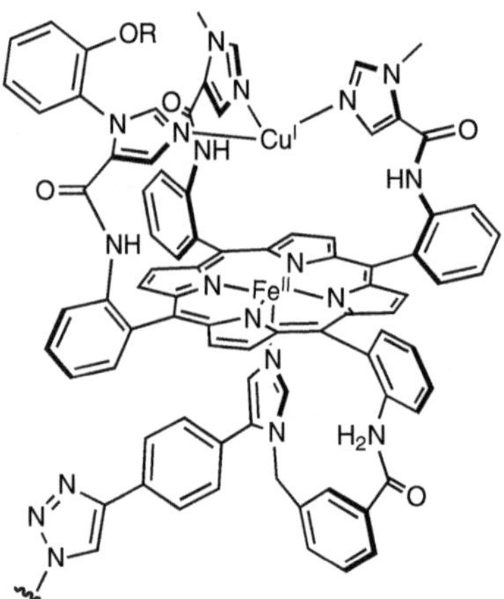

49
a: R = H
b: R = Me

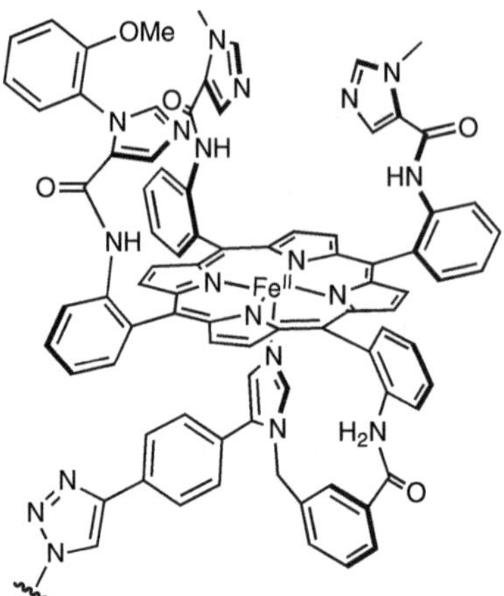

49c

Figure 21 Some well-studied functional cytochrome *c* oxidase (C*c*O) models.

Two key features of these catalysts are their high selectivity for the 4e$^-$ reduction of O_2 (and thus their limited generation of PROS) as well as their robustness as evidenced by their high turnover numbers ($1.2(2) \times 10^4 \leq$ TON at 200 mV, pH <8). From these studies, a catalytic mechanism was proposed (Figure 22) [139]. Importantly, O_2 reduction occurs as proposed for mononuclear heme complexes, with the copper center functioning mainly to enhance O_2 binding (*vide supra*) and transfer electrons (*vide infra*). Also of note is that protonation of the peroxo species to generate the hydroperoxo intermediate is slow relative to O-O bond cleavage. This disparity has the effect of maintaining a low steady state concentration of the hydroperoxo species, mitigating the loss of H_2O_2 which might occur via hydrolysis of this species [122, 138].

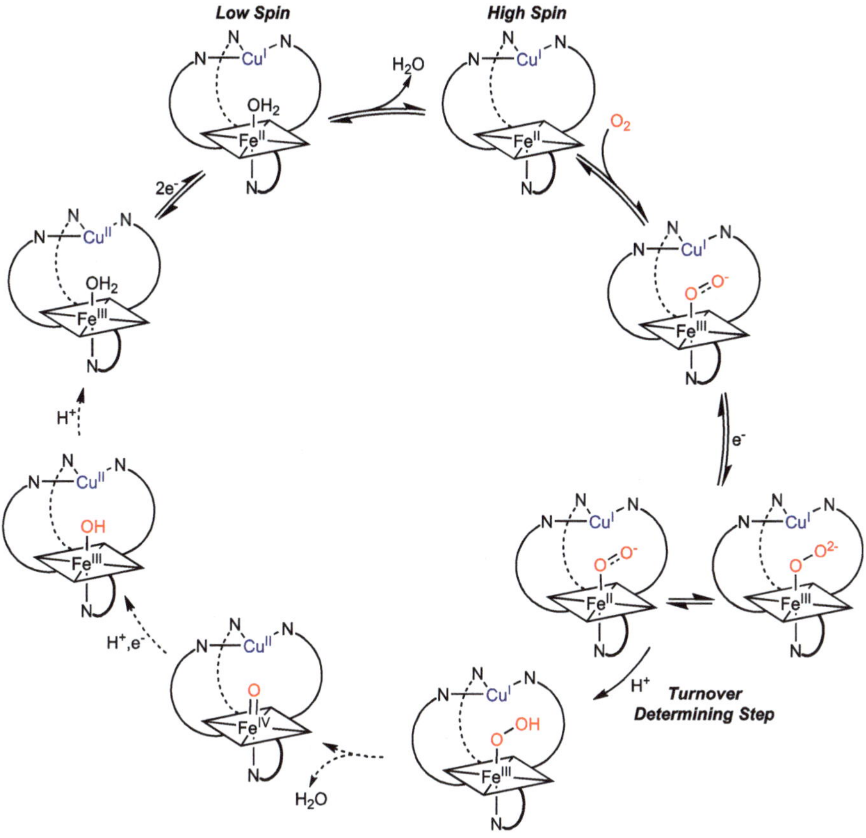

Figure 22 Proposed mechanism for the electrocatalytic 4e$^-$ reduction of O_2 carried out by **47a-c** under steady-state conditions (adapted from [139]). The dashed arrows represent steps that are kinetically invisible under the studied conditions.

More recent efforts have been focused on studying the role of the covalently linked tyrosine (Tyr244) in the C*c*O catalytic cycle by single turnover experiments performed in the presence of exogenous and appended phenols. In one of the first of these studies, compound **47c** (Figure 21) was shown by EPR spectroscopy to generate phenoxyl radicals when treated with exogenous 2,6-di-*tert*-butylphenol derivatives in dichloromethane at room temperature [140]. A KIE of 2 was measured, implicating proton transfer in the rate determining step, which the authors argued as evidence of an intermediate hydroperoxo species. In a subsequent study, phenol appended **48** (Figure 21) was also able to achieve full reduction of O_2 in a single-turnover process, and it was argued that the pendant phenol participates as an H-atom donor in a manner similar to that proposed for exogenously added phenols [141]. In both cases, the authors invoke the formation of an Fe(IV)-oxo species, resulting from heterolysis of the proposed hydroperoxo intermediate that is facilitated by electron transfer from the nearby Cu(I) center. The presence of an Fe(IV)-oxo intermediate was supported on the basis of its O-atom transfer reactivity with triphenylphosphine, which gave triphenylphosphine oxide in high yields, in addition to evidence from mass spectrometry [140, 141]. Taken together, the results of these studies demonstrate the efficacy of tyrosine mimics to participate as redox centers in the $4e^-$ reduction of O_2.

Furthering this implication of Tyr244 as a key player in catalysis by C*c*O, catalytic O_2 reduction was investigated for a series of complexes (**49a–c**) (Figure 21) appended onto mixed self-assembled monolayer (SAM) coated gold electrodes under conditions of rate limiting electron-transfer in conjunction with rotating ring-disk voltammetry. It was found that under conditions of rate-limiting electron transfer, both the pendant phenol and copper center were necessary for minimizing the generation of PROS, suggesting that a key role of these redox sites is to serve as electron reservoirs in order to mitigate their buildup. Such a scenario is similar to that observed in the native enzyme, in which electron transfer to the active site is believed to be relatively slow on the timescale of O_2 reduction. In contrast, when electron transfer was not rate-limiting, the selectivities among the different catalysts were much less disparate, further supporting the electron transfer role of the distal Tyr244 and the copper center [142].

Catalytic O_2 reduction has also been demonstrated in the case of compounds **48** (Figure 21) and **41a** (Figure 20) using chemical reductants, namely cytochrome *c* and decamethylferrocene (Fc*). Mechanistically, the $4e^-$ reduction of O_2 proceeds in essentially the same manner regardless of the electron source (chemical *versus* electrolytic; Figure 22). Catalysis with **48** performed at a 2 % catalyst loading in a 1:1 water:acetonitrile mixture (pH = 7) at 25 °C was shown to be stoichiometric with respect to the reductant [143]. Interestingly, O_2 binding was shown to be rate limiting as opposed to electron transfer in this case, in contrast to what is proposed in the native enzyme [142]. Under the experimental conditions, the poor solubility of cytochrome *c* limited the number of turnovers to 25. The catalytic mechanism for **41a** (Figure 20) is thought to proceed in much the same way (Figure 23), with O_2 binding being rate limiting at 25 °C [136]. In contrast to **48** (Figure 21) however, catalysis was carried out in acetone, which allowed for the reaction to be studied at −60 °C.

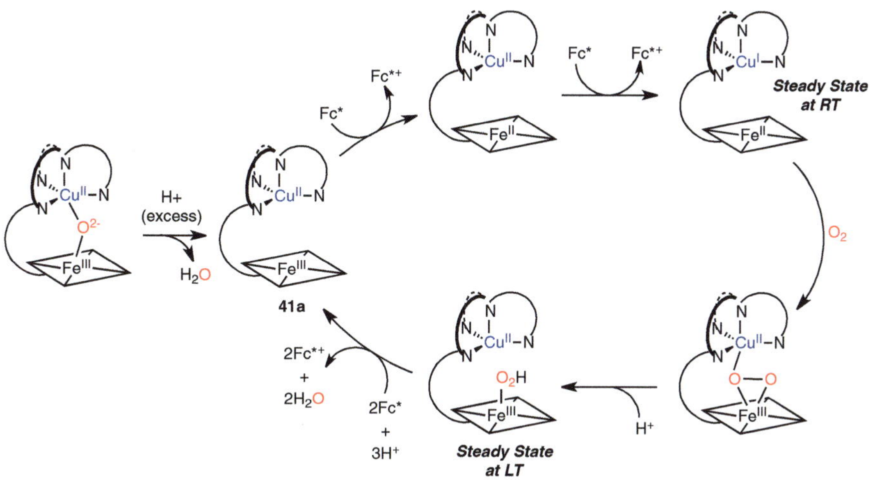

Figure 23 Catalytic mechanism of **41a** (adapted from [136]).

At low temperature, the rate-determining step switches from O_2 binding to O-O bond cleavage, which allowed for the monitoring and characterization of the hydroperoxo intermediate by UV-vis spectroscopy. Importantly, comparison of the rates of O-O bond cleavage in the case of **41a** (Figure 20) and its copper-free analog revealed negligible differences, further evidence that the distal copper center is not involved in O-O bond heterolysis [136].

3 Dioxygen Activation by Non-heme Iron Complexes

Non-heme iron active sites are found in numerous enzymes that play critical roles in life processes. The literature describing structural, spectroscopic, and mechanistic studies of these enzymes is expansive, with much of it accessible through a number of comprehensive reviews [144–149] and more focused accounts [150–165]. The non-heme active sites may be divided into two broad categories defined by whether they contain either one or two iron atoms. Typically, non-heme iron centers are bound to the protein by a combination of histidine and carboxylate ligands, with additional ligation by tyrosines, water molecules, hydroxide/oxo groups, and/or a cofactor such as α-ketoglutarate. The structural diversity and varied functions of these active sites are impressive, providing a fertile area for research aimed at understanding structure/function relationships. We summarize a few highlights here, with the specific aim of providing context for the most current synthetic modeling work targeting the dioxygen activation chemistry of these sites.

In general, the O_2 binding and activation pathways followed by non-heme monoiron enzymes parallel that of their heme counterparts (Section 2), but there are notable differences that arise from the presence of *cis* coordination sites and/or bound cofactors in the non-heme systems. For example, differences in active site structures correlate with divergent regiospecificity and mechanisms in the intradiol and extradiol catechol oxygenases (Figure 24) [145, 146, 155]. In the former, the iron remains as Fe(III) throughout the catalytic cycle, and upon binding as a dianion the substrate catecholate is activated by the Fe(III) ion for direct attack by O_2 (e.g., via the form semiquinone-Fe(II)). The resulting alkylperoxo intermediate then undergoes a Criegee-type rearrangement to afford an anhydride that opens to the final diacid product. In the extradiol oxygenases, the substrate is proposed to bind as a monoanion to an Fe(II) center, which then binds O_2 to yield an adduct identified as an Fe(III)-superoxide on the basis of X-ray crystallographic [166], spectroscopic [167], and theoretical [168] work.

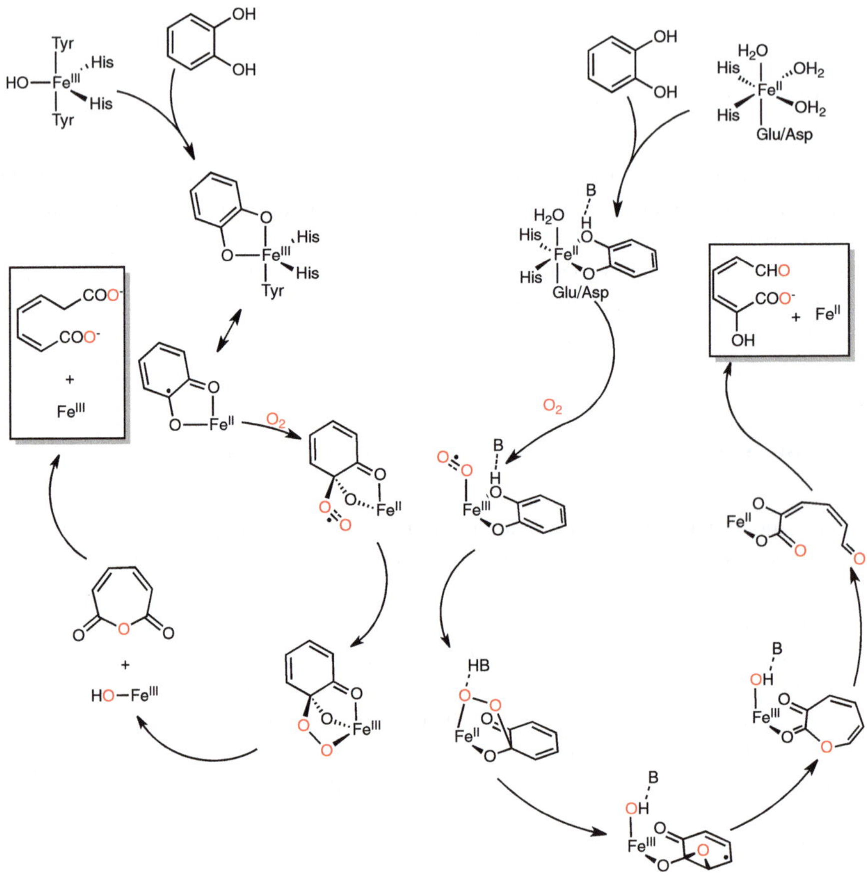

Figure 24 Proposed mechanisms for the intradiol (**left**) and extradiol (**right**) dioxygenases (adapted from [145] and [168], respectively).

Similar adducts also are proposed for other members of the family of enzymes that feature the supporting 2-His-1-carboxylate facial triad as a common structural motif. The subsequent fate of the Fe(III)-superoxo moiety varies (Figure 25), however, and includes rearrangement and reduction to an η^2-(hydro)peroxo species (Rieske dioxygenases, "Intraprotein ET") [169], hydrogen atom abstraction from substrate (isopenicillin N-synthase, "HAT") [170], and intramolecular attack at bound α-ketoglutarate (α-KG) cofactor to induce its oxidative decarboxylation (α-KG-dependent monooxygenases, "Nucleophilic Attack") [145, 146, 171]. In the Rieske dioxygenases and the α-KG-dependent monooxygenases, O-O bond scission processes yield the active oxidants, proposed for the former to be an iron-oxo-hydroxo unit and for the latter as an Fe(IV)-oxo.

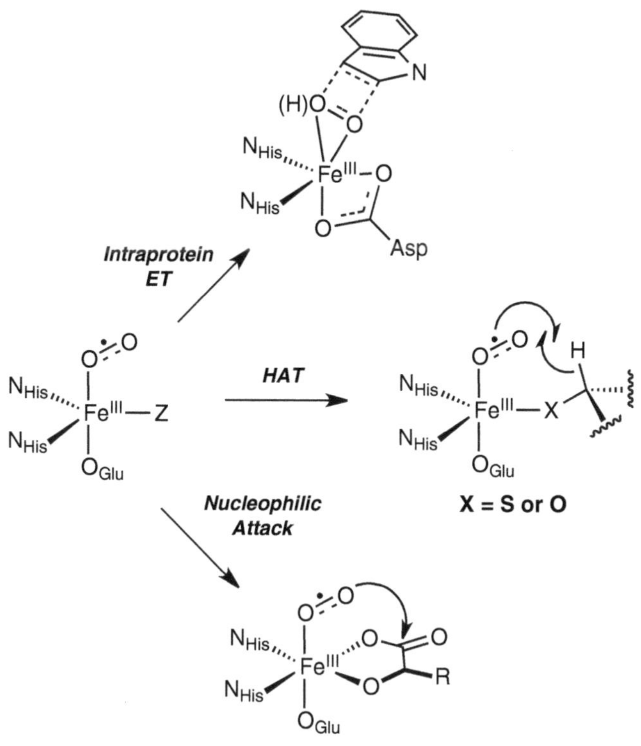

Figure 25 Reactivity of monoiron-superoxo adducts in non-heme enzymes (adapted from [163]).

A milestone discovery was the trapping and unambiguous identification of the reactive Fe(IV)-oxo unit in the taurine/α-ketoglutarate dioxygenase (TauD) [172, 173]. Related pathways are implicated for non-heme iron enzymes that are supported by three histidine ligands instead of the 2-His-1-carboxylate triad [174],

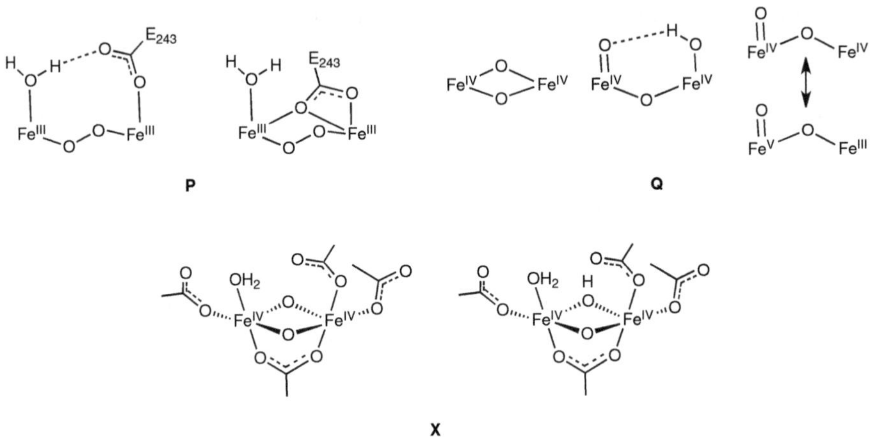

Figure 26 Proposed reactive intermediates in non-heme diiron enzymes.

such as cysteine dioxygenase [175–177] and β-diketone dioxygenase [178]. Also notable is the chemistry of a low-spin Fe(III)-OOH species identified in the anticancer drug bleomycin, which may perform hydrogen atom abstraction from deoxyribonucleic acid directly or via prior O-O bond scission pathways [179, 180]. The diverse mechanisms by which O_2 is activated by the non-heme monoiron active sites is striking, and raises many fundamental questions about how ligand environment influences O_2 binding, reactivity, and reduction that model studies have aimed to address [181].

Proposed structures for key intermediates involved in dioxygen activation by non-heme diiron enzymes are shown in Figure 26. Intermediate P is the result of dioxygen binding to the diiron(II) active sites of several non-heme diiron enzymes [158], compound Q is generated from P and is proposed to be the active oxidant in soluble methane monooxygenase (sMMO) [158, 182], and intermediate X is the species proposed to generate a tyrosyl radical involved in catalysis by the diiron-containing ribonucleotide reductase [149, 183]. Identification of each of these intermediates, the detailed structures of which continue to be debated, has relied upon interpretation of spectroscopic data that has been greatly informed by detailed studies of synthetic model complexes.

3.1 Monoiron Models of Mononuclear Non-heme Iron Active Sites

Efforts to synthesize and characterize models of non-heme monoiron active sites involved in dioxygen activation have been extensive and wide-ranging, with numerous strategies having been taken to isolate reactive intermediates relevant to the biological systems. A number of reviews are available that describe the advances made to date in quite comprehensive fashion [145, 155, 161, 184–187]. We focus here on selected recent contributions that describe the characterization of putative superoxo, hydroperoxo, and oxo intermediates involved in O_2 activation pathways, and on efforts to understand the O_2 reactivity of monoiron-cofactor complexes.

3.1.1 Monoiron Superoxo and Hydroperoxo Complexes

Stimulated by evidence from studies of various biological systems that non-heme monoiron-superoxo complexes may be capable of attacking substrate C-H bonds [150], attempts have been made to identify such complexes in synthetic systems and characterize their reactivity. In early work, an iron-superoxo moiety was characterized as a transient intermediate in the oxygenation of a bis(μ-hydroxo) diiron(II,II) complex to ultimately yield a (1,2-peroxo)diiron species [188]. This intermediate exhibited a notably high ν(O-O) band at 1310 cm^{-1}, and end-on binding of the superoxo ligand was proposed on the basis of the results of ^{18}O-isotope labeling experiments. A related iron-superoxo complex was proposed upon oxygenation of Fe(II) species bound to dendrimer-appended carboxylates, although no direct evidence of a superoxo ligand was provided [189].

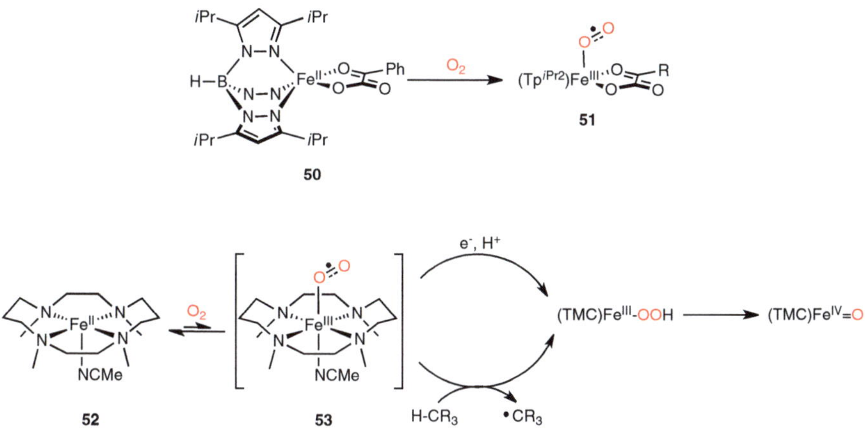

Figure 27 Proposed non-heme iron(II)-superoxo complexes (adapted from [163] and [192]).

In a more recent study, a variety of observations were interpreted to indicate the intermediacy of an iron(II)-superoxo species from the oxygenation of an iron(II)-benzoylformate complex **51** (Figure 27) that models α-KG-dependent monooxygenase active sites [163]. A similar intermediate was proposed in studies of an Fe(II)-acetylacetone model of β-diketone dioxygenase [190]. Indirect evidence for the intermediacy of an iron(II)-superoxo complex was also cited in studies of **52** (Figure 27), which reacted with O_2 only in the presence of reductant and proton donors [191] or substrates with weak C-H bonds [192] to yield iron(III)-hydroperoxo or iron(IV)-oxo species. Mechanistic studies were interpreted to support a pathway involving pre-equilibrium formation of **53**, which then reacts with reductants/proton sources or attacks the allylic positions of olefin substrates. Insights into this reactivity have been provided by theory [193]. In sum, while some data on non-heme iron(II)-superoxo species have been obtained, full characterization remains an unachieved goal [194].

Many examples of monoiron(III)-hydroperoxo species have been reported, with typical syntheses involving reactions of Fe(III) precursors with H_2O_2 [145, 195, 196]. The reactivity of such intermediates prepared in this way has come under particular scrutiny of late [197] due to the discovery of selective catalytic oxidations by iron complexes with H_2O_2 [198–200]. More rare is the identification of monoiron(III)-hydroperoxo species from reactions of Fe(II) complexes with O_2 via processes that directly model dioxygen activation by metalloprotein active sites. In one example, the aforementioned monoiron-superoxo complex formed from oxygenation of a bis(μ-hydroxo)diiron(II,II) complex decays to a transient species hypothesized on the basis of resonance Raman spectroscopy to be a highly reactive complex with a Fe(III)-OOH unit; it decays rapidly to a (1,2-peroxo)diiron(III,III) product [188]. In another notable case, reaction of a multi-iron(II) polyoxometalate with O_2 yielded a stable complex that was identified as having end-on coordination of two hydroperoxo groups in unprecedented linear geometries at the two terminal positions of the polyoxometalate "belt" (Figure 28) [201]. A rather long O-O distance in the reported X-ray crystal structure (1.612(9) Å) was rationalized by the presence of multiple hydrogen bonding interactions with water molecules.

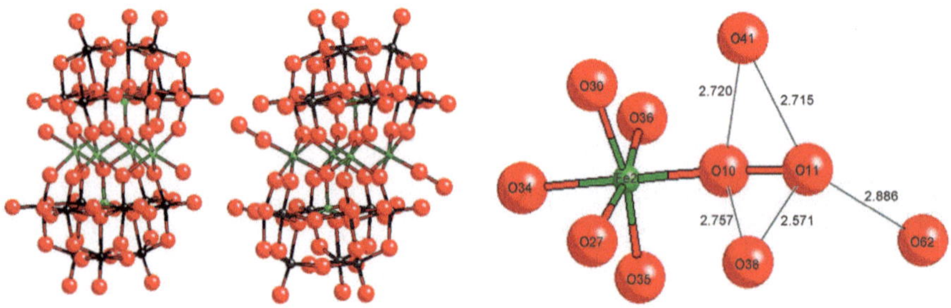

Figure 28 Structure of a polyoxometalate with Fe(III)-OOH units derived from reaction of Fe(II) precursors with O_2. Reproduced with permission from [201]; copyright 2008, Wiley-VCH Verlag GmbH & Co. KGaA, Weinheim.

While alkyl- or acylperoxo complexes are clearly related to hydroperoxo species, we choose to forego discussion of them here because they are not derived directly from O_2. However, in recent work that is relevant to the reactivity of FeOOH species, the decay of a high-spin non-heme Fe(III) alkylperoxo complex to a Fe(IV)=O species was studied and shown to involve O-O bond homolysis induced by binding of an axial anionic ligand [202].

3.1.2 Monoiron(IV)-Oxo Complexes

The definition of the structure of a non-heme monoiron(IV)-oxo complex by X-ray crystallography in 2003 [203] has heralded more than a decade of intense research into the properties and reactivity of such species. Results from these studies have been summarized most thoroughly in a recent review [187], but also in others [186, 200, 204, 205]; we will not reproduce those discussions here, and instead will briefly mention selected, more recently reported advances.

Notable attention has been drawn to studies of the reactivity of discrete monoiron(IV)-oxo complexes. Both low-spin ($S = 1$) and high-spin ($S = 2$) complexes have been characterized, and the latter has been found to be more reactive than the former with respect to reactions with C-H bonds. Interestingly, it has recently been reported that addition of Sc^{3+} or $HClO_4$ significantly perturbs the reactivity of the originally reported $S = 1$ complexes **54** and **55** (Figure 29). For example, a 120-fold acceleration in the rate of oxidation of a benzyl alcohol was observed upon addition of $Sc(OTf)_3$ to **54** in CH_3CN, with an accompanying significant decrease in the H/D KIE [206]. These and other results were interpreted to support a change in mechanism from concerted HAT in the absence of Sc^{3+} to stepwise Sc^{3+}-coupled electron-transfer followed by proton-transfer. A similar mechanistic shift was proposed for the oxidations of toluene derivatives by **54** in the presence of $HClO_4$ [207]. In addition, **55** performed as a 1-electron oxidant of ferrocene in the absence of added Sc^{3+}, but as a 2-electron oxidant in its presence [208]. All of these effects were attributed to the coordination of Sc^{3+} or H^+ to the Fe(IV)=O unit, with support for this notion coming from an X-ray crystal structure proposed to show $[(OTf)_4ScOH]^{2-}$ bound to the Fe(IV)=O moiety in **56** (Figure 29) [208]. This interpretation has been called into question on the basis of theoretical calculations that indicate the complex is more likely an Fe(III)-oxo bound to $[(OTf)_4Sc(OH_2)]^{1-}$ (**57**) [209].

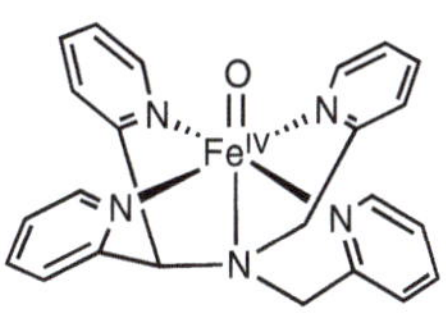

54

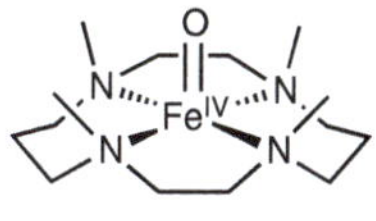

55

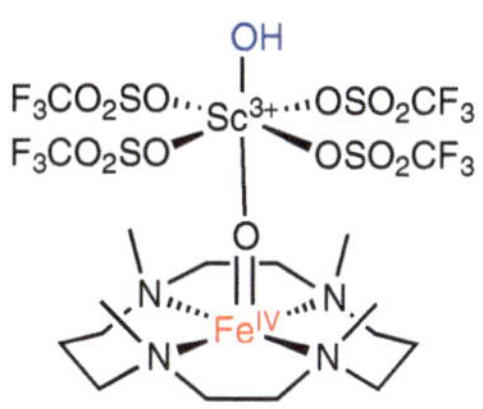

56

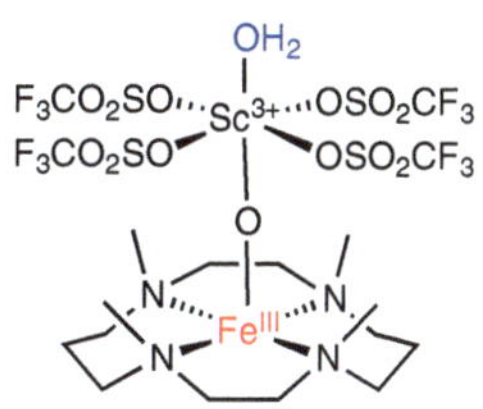

57

Figure 29 Fe(IV)=O complexes **54** and **55** (**left**) and structural hypotheses for Sc^{3+} adducts of **55** (**right**).

In another recent mechanistic study, the reactivity of Fe(IV)=O and Fe(IV)=NTs (Ts = tosyl) units at parity of supporting ligand (N_4Py = N,N-bis(2-pyridyl-methyl)-N-bis(2-pyridyl)methylamine) was compared [210]. The oxo complex was found to perform HAT faster than the tosylimido compound, but to be less efficient at sulfoxidation of thioanisoles. These differences were attributed to steric effects associated with approach of substrate to the reactive Fe(IV)=O and Fe(IV)=NTs units.

Interest also has been focused on even more oxidized species, such as complexes with Fe(V)=O or (L$^{\bullet}$)Fe(IV)=O (akin to the heme relative Cpd I; Figure 2), particularly in view of the postulates that such species may be involved in dihydroxylations by the Rieske dioxygenases and related model compounds [169, 200]. In a notable recent example, the complexes **58–60** (Figure 30) were found to catalyze the oxidation of light alkanes by potassium peroxymonosulfate (oxone), and spectroscopic and theoretical evidence was presented in support of the involvement of **61** (Figure 30) as the active species [211]. In another example, evidence from variable temperature mass spectrometry led to the identification of a metastable Fe(V)(O)(OH) complex generated from reaction of an Fe(II) precursor with H_2O_2 [212].

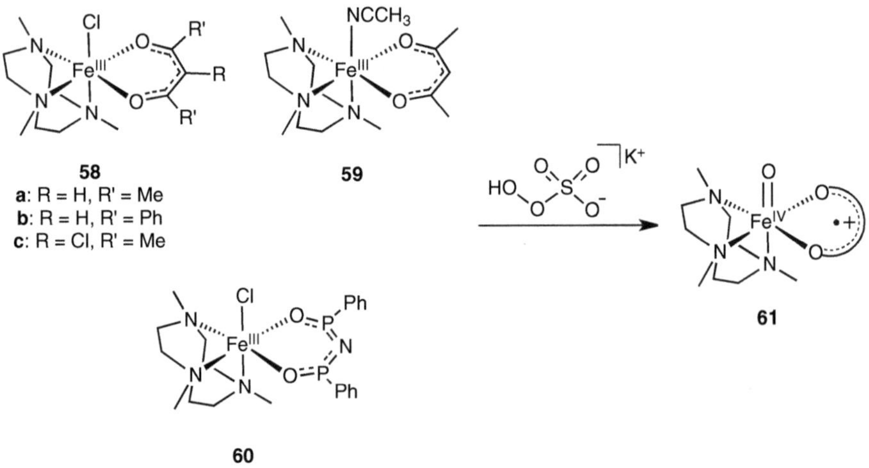

Figure 30 Catalysts for the oxidation of light alkanes and proposed active species.

3.1.3 Monoiron-Cofactor Complexes: Catecholates and α-Ketoglutarates

Many examples of mononuclear iron complexes of catecholates have been prepared as models of the intra- and extradiol oxygenase active sites, with a key aim being to evaluate their ability to undergo diol cleavage upon reaction with O_2. The effects of varying the donor properties, denticity, and steric profiles of supporting ligands and the solvent environment on the rates and intradiol/extradiol cleavage selectivities in O_2 reactions of Fe(III) complexes of bound catecholates (usually 2,4-di-tertbutyl-catecholate, DBC^{2-}) have been explored extensively in the past few years [213–225] (Figure 31). Notable findings include a positive correlation of the reaction rates with the energy of catecholate → Fe(III) LMCT transitions [213, 222], an inverse correlation with the redox potentials of the DBSQ/DBC^{2-} couple (DBSQ = 2,4-di-tertbutyl-semiquinone) in the complexes [213, 225], and significant changes in extradiol *versus* intradiol selectivity as a function of solvent medium [216, 217]. Notwithstanding this work, intermediates in these reactions have not been identified and the fundamental mechanistic basis for intradiol *versus* extradiol cleavage selectivity remains unclear.

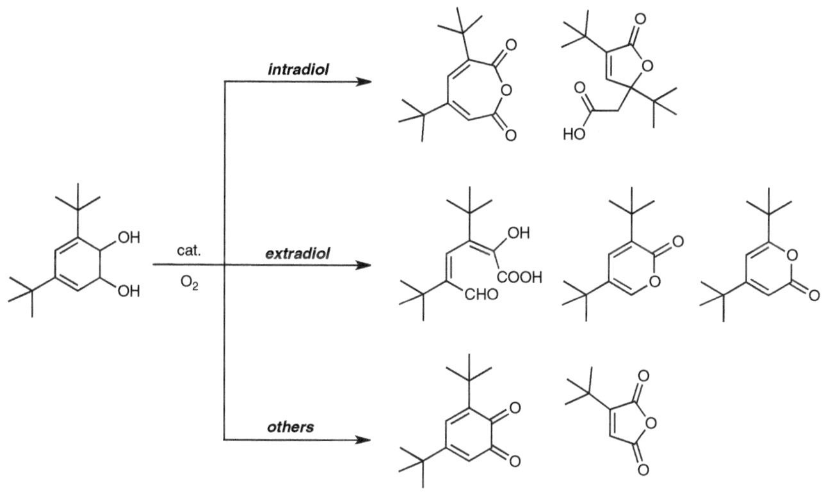

Figure 31 Typical products of reactions of Fe(III)-catecholate complexes with O_2 (adapted from [218]).

In efforts to model the α-KG-dependent enzymes, numerous complexes of α-ketocarboxylates, with benzoylformate (BF) being the most commonly used, have been prepared and their reactions with O_2 studied [145]. Yet only in a few cases has decarboxylation of the bound α-ketocarboxylate been accompanied by oxidation of a substrate in a reaction that would model enzyme function (Figure 32). In an early example, reaction of complex **62** with O_2 in the presence of cyclohexene or *cis*-stilbene resulted in decarboxylation of BF and formation of epoxides [226]. Complex **63a** reacted with O_2 to yield a product in which the ortho position of one of the phenyl substituents was hydroxylated [227]; a similar intramolecular hydroxylation was observed for a phenylpyruvate analog (**63b**) [228].

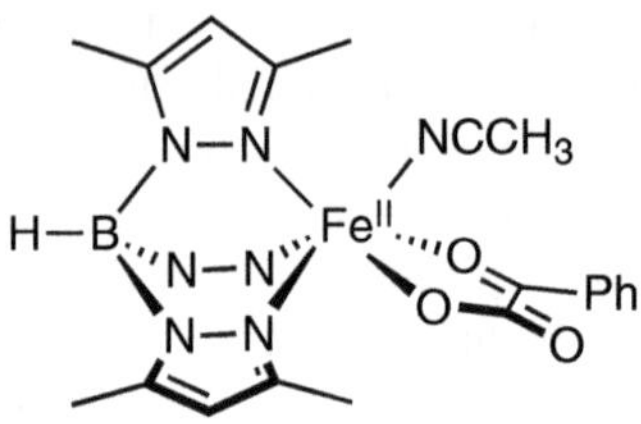

62

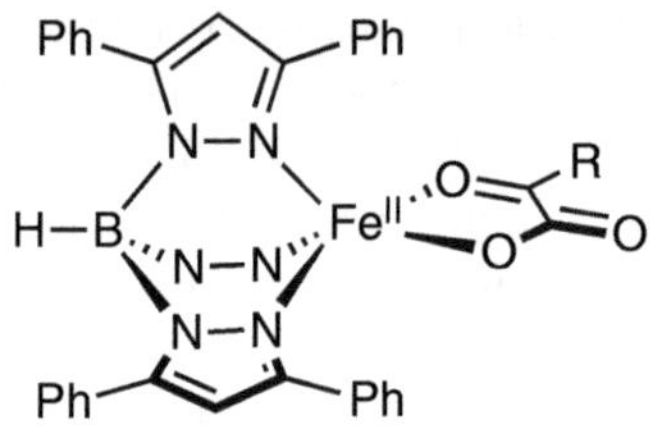

63

a: R = Me
b: R = CH_2Ph

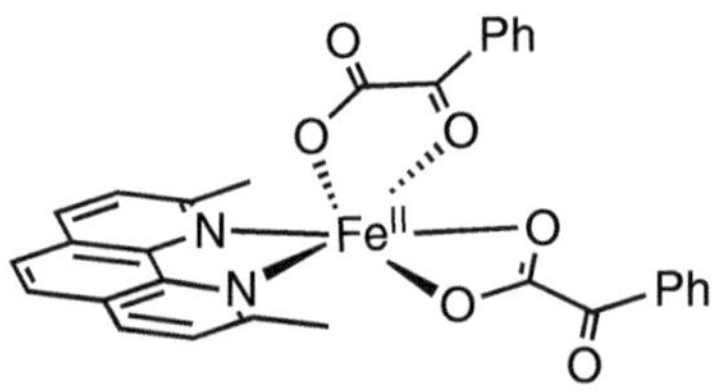

64

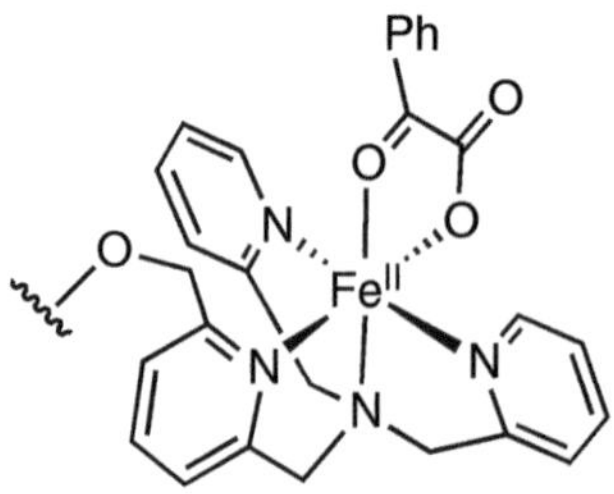

65

Figure 32 Complexes with coordinated α-ketocarboxylates that undergo decarboxylation and oxidize substrates upon reaction with O_2.

In subsequent work, the oxidant responsible for this intramolecular process (presumably an Fe(IV)-oxo species) was trapped by exogenous substrates, and the extent of trapping was found to be determined by the molecular shape of the substrate and the strength of the C-H bond being attacked [229]. More recently, decarboxylation of BF and sulfoxidation reactions were observed upon oxygenation of **64** (Figure 32) in the presence of dimethyl sulfide or dimethyl sulfoxide [230]. In a potentially useful extension of these studies, **65** immobilized on gold nanoparticles reacted with O_2 in the presence of various substrates to yield oxidized products, and initial demonstration of catalytic activity in the presence of excess BF was demonstrated [231].

3.2 Diiron Models of Dinuclear Non-heme Diiron Active Sites

With the aim of shedding light on the properties of the intermediates proposed during O_2 activation by the non-heme diiron enzymes, numerous strategies have been used to target peroxo- and oxo-diiron complexes for synthesis and characterization. We point the reader to comprehensive reviews for summaries of the approaches taken and successes achieved [232–237]. In the sections below, recent advances toward the characterization of (peroxo)diiron(III,III) complexes that model intermediate P (or H_{peroxo}) in the enzymes, diiron(III,IV) complexes that model intermediate X in ribonucleotide reductase, and diiron(IV,IV) complexes relevant to the active oxidant Q in sMMO are summarized. Particular focus is placed on notable new insights into the interconversions among such species that are directly relevant to dioxygen activation pathways in the enzymes.

3.2.1 (Peroxo)diiron Complexes

Building upon earlier reports of the structures and properties of (μ-1,2-peroxo)diiron complexes such as **66–68** (Figure 33), efforts more recently have focused on identifying such species as intermediates in catalytic oxidations, probing rearrangements involving other bridging ligands, and examining their O-O bond breaking reactions that lead to higher valent diiron compounds. For example, a transient species formulated as a (μ-oxo)(μ-1,2-peroxo)diiron(III,III) complex was identified via spectroscopy in studies of oxidations using H_2O_2 with a tridentate, meridional N-donor ligand [238]. However, the authors surmised that an unidentified species derived from the observed intermediate was likely the actual oxidant in the catalytic reactions. In related work, a (μ-1,2-peroxo)diiron(III,III) species proposed to lack the additional oxo-bridge was identified in studies of H_2O_2 disproportionation [239].

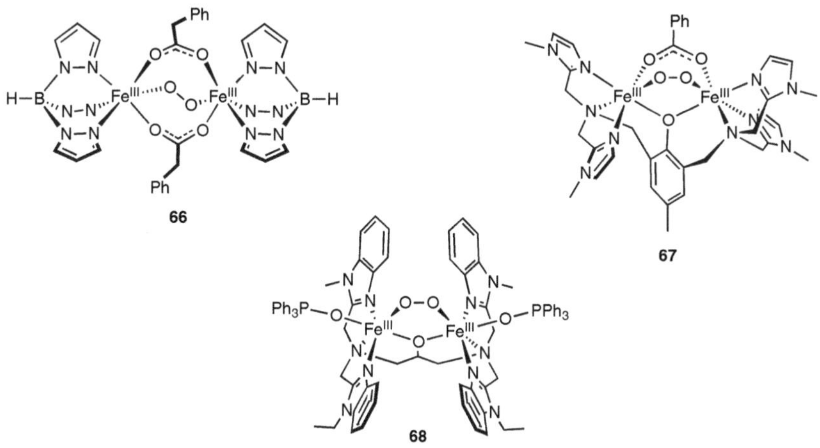

Figure 33 Examples of structurally characterized (μ-1,2-peroxo)diiron(III,III) complexes.

The effects of varying the bridging ligand on the properties of (μ-1,2-peroxo)-diiron(III,III) complexes **69** were examined (Figure 34), and in early work it was proposed that rearrangement occurred such that the bridging ligand shifted from bidentate bridging to monodentate on one metal ion [240]. The propensity for these rearrangements, which resemble "carboxylate shifts" proposed to be important in diiron enzyme reactivity [241], was linked to the bite angle preferred by the

XO_2^- ligand, with the larger angle for $X = Me_2As$ better stabilizing the triply bridged form of **69a**. Subsequent spectroscopic and kinetic studies explored these and other factors in greater detail [242]. In addition, a similar rearrangement involving conversion of a carboxylate from bridging to terminal monodentate in (μ-1,2-peroxo)diiron(III,III) species **69b** was reported to be induced by protonation [243].

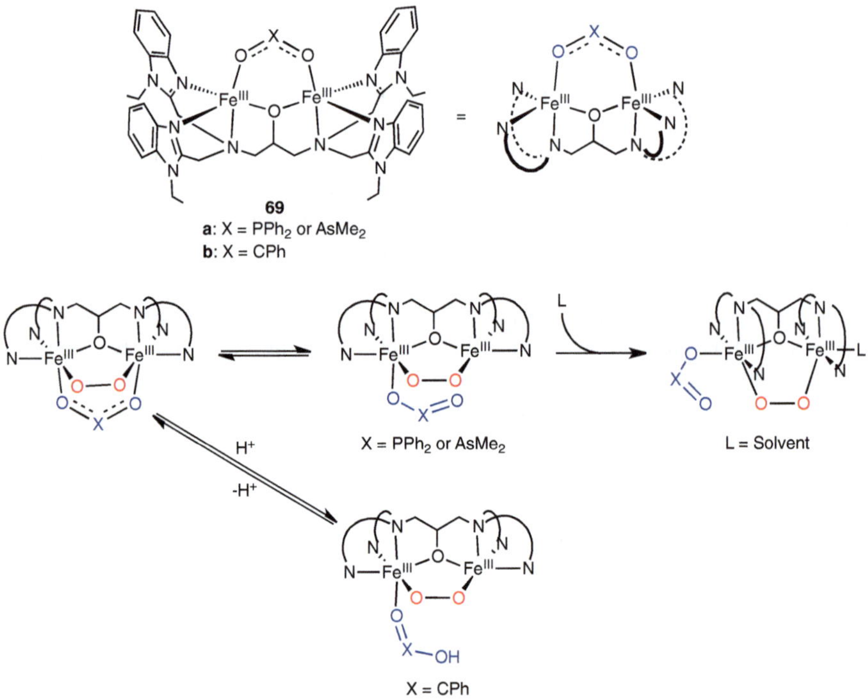

Figure 34 Rearrangements of (μ-1,2-peroxo)diiron(III,III) complexes (adapted from [240] and [243]).

Protonation is thought to facilitate scission of the O-O bond in (peroxo)diiron intermediates to generate active oxidants in enzymes like soluble methane monooxygenase (sMMO) [244]. In work aimed at understanding how this might occur, reactions of (μ-1,2-peroxo)diiron(III,III) complexes with protic acids have been explored. Treatment of **70** (Figure 35) with $HClO_4$ or HNO_3 was shown to result in initial protonation of the oxo bridge (**71**), followed by decay to the (μ-oxo)-diiron(III,IV) complex **72** that models intermediate X in ribonucleotide reductase [245]. Mechanistic details such as the source of the reducing equivalent required or how proton transfer from the hydroxo group induces breaking of the peroxo O-O bond in this system have yet to be elucidated.

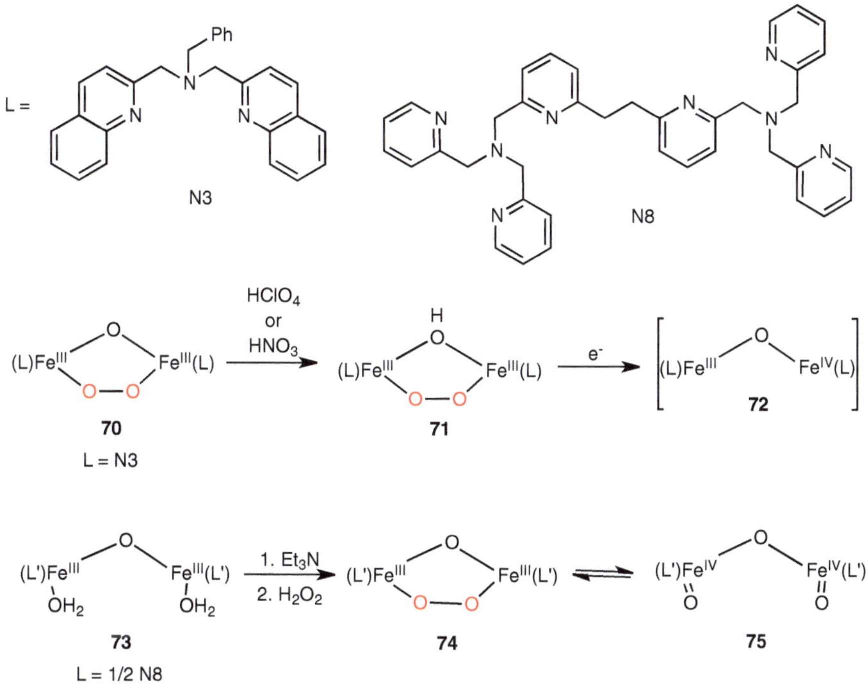

Figure 35 Proton-induced O-O bond breaking reactions of (μ-1,2-peroxo)diiron(III,III) complex **70** (top) and proposed O-O bond cleavage of **74** to give diiron(IV,IV) complex **75** (bottom) (adapted from [245] and [246], respectively).

In another particularly provocative report, addition of H_2O_2/NEt_3 to diiron(III,III) complex **73** (Figure 35) results in formation of (μ-1,2-peroxo)-diiron(III,III) species **74**, which was proposed to undergo *reversible* O-O bond scission to yield a high spin $S = 2$ diiron(IV,IV) complex **75** [246] on the basis of solution and solid-state spectroscopic data. The conclusion that both **74** and **75** co-exist in solution has been questioned, however, although separate more detailed studies of an analog of **75** (Section 3.2.2) are in line with its formulation as denoted in Figure 35 [247].

3.2.2 High Valent Diiron-Oxo Complexes

Investigations of iron complexes supported by ligand **76** (Figure 36) have led to the identification of an intriguing set of diiron(III,IV) and diiron(IV,IV) complexes and elucidation of pathways by which they may be interconverted [187]. These results have provided key precedence for postulated high valent diiron enzyme intermediates and insights into possible ways they may react. The high valent compounds are accessed through the reaction of diiron(III,III) species with H_2O_2, originally proposed to involve conversion of hydroxo-aqua species **A'** to the hydroperoxo intermediate **C'** [248], but more recently shown to involve initial dehydration to **B'** which then generated **C'** [249]. Complex **C'** rapidly undergoes O-O bond scission to yield diiron(IV,IV) complex **D'**, which on the basis of Mössbauer spectroscopy and DFT calculations was shown to contain inequivalent low-spin $S=1$ Fe(IV) sites coupled ferromagnetically to yield an overall $S=2$ state [250]. Complex **D'** yields bis(μ-oxo)diiron species **E'** or **F'** upon reaction with hydrogen atom donors or strong acids, respectively, and **E'** and **F'** may be interconverted electrochemically. These two "diamond core" species serve as important structural models for enzyme intermediates X and Q, but their rates of reactions with weak C-H bonds (cf., dihydroanthracene, DHA) are relatively slow, which for **F'** raises a key question about its viability as the active oxidant of methane by sMMO.

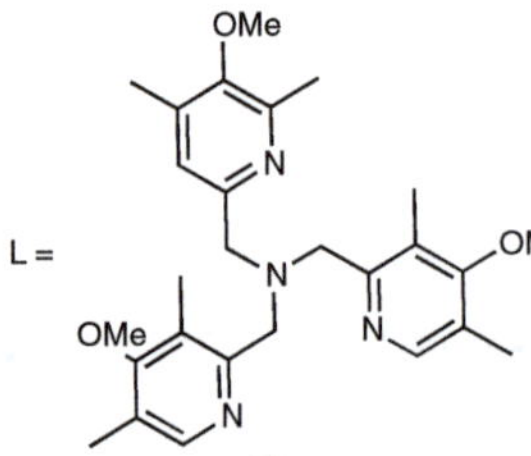

76

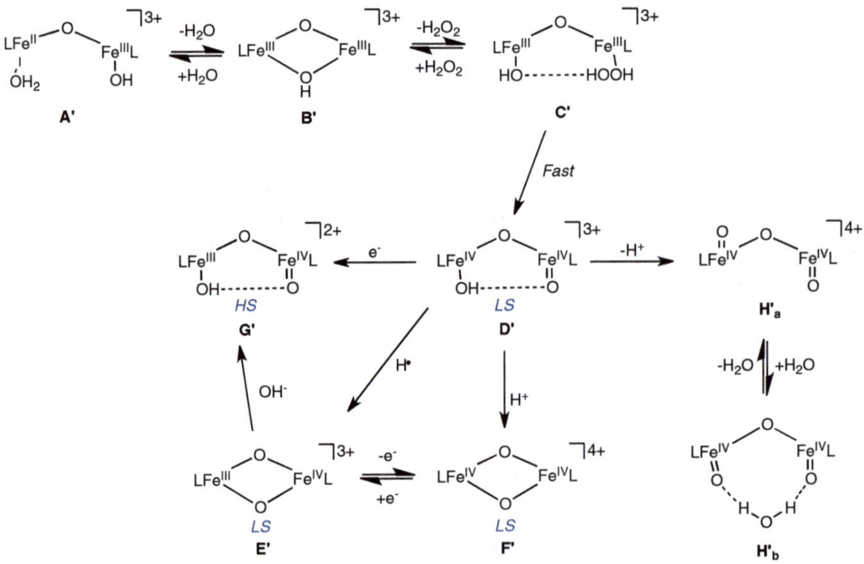

Figure 36 High-valent diiron(III,IV) and -(IV,IV) complexes and their interconversions.

Interestingly, either through 1-electron reduction of **D'** or treatment of **E'** with hydroxide, a high spin diiron(III,IV) species **G'** forms that exhibits greatly enhanced reactivity with dihydroanthracene (10^6-fold increase in rate relative to low-spin relative **E'**) [251]. Finally, deprotonation of **D'** has been shown to yield a rare example of a diiron(IV,IV) complex (**H'**) with two terminal high-spin Fe(IV)=O units [247]. Spectroscopy and theory data were interpreted to indicate that **H'** exists in two forms in frozen solution, differing with respect to the spatial relationships of the Fe(IV)=O units and the presence or absence of a 'bridging' water molecule. Intriguing similarities between Mössbauer parameters for **H'**$_{\mathbf{a}}$ and compound Q in sMMO were noted, as was the need for synthetic diiron(IV,IV) complexes that feature O-donor supporting ligands more akin to those found in the enzyme.

4 Dioxygen Activation by Copper Complexes

Many enzymes that activate dioxygen contain copper ions in their active sites, which feature a variety of nuclearities, structures, and functional attributes (Table 1, entries 4,6,12,13,15,16) [252–260]. Efforts to understand the dioxygen activation pathways and the chemistry of reactive intermediates for these enzymes have included extensive studies of the reactions of O_2 with Cu(I) complexes with a variety of supporting ligands. Such studies are often performed at low temperature (<−50 °C) in organic solvents in order to best enable identification of highly reactive species relevant to postulated catalytic intermediates. Using this approach, and complementary ones such as the reactions of ROOH with Cu(II) precursors, complexes with the core structures shown in Figure 37 have been characterized in depth and key aspects of their reactivity explored. These studies have provided important knowledge relevant to possible enzyme mechanisms and pathways of Cu-promoted catalytic oxidations.

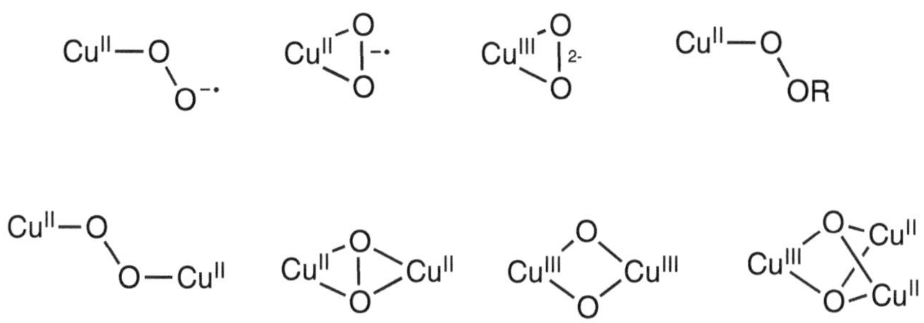

Figure 37 Copper-oxygen cores characterized in synthetic complexes.

Much of the work on these complexes is presented in comprehensive reviews that summarize work completed prior to approximately 2005 [261–265]. In the following sections, we focus on more recent research results reported since that date, organized by the nuclearity of the copper active site targeted. Some aspects are described in a recent review [266].

4.1 Mononuclear Models of Monocopper Active Sites in Enzymes

Mechanistic hypotheses for dioxygen activation by monocopper enzyme active sites typically invoke binding of O_2 to a Cu(I) center to yield a 1:1 adduct (Figure 38) [253, 267–270]. Various fates for such an adduct have been proposed, including (a) reduction by one electron to yield a Cu(I)-O_2^- species [271], or (b) reaction with substrate via HAT to yield a Cu(II)-OOH unit [268, 272, 273]. The latter might also be formed by protonation of the Cu(I)-O_2^- species [274]. The Cu(II)-OOH intermediate may react with substrate via HAT or be reduced and protonated with loss of H_2O, both of which would yield a $[CuO]^+$ moiety that theoretical calculations suggest is best formulated as an S = 1 Cu(II)-O$^•$ [275, 276]. This unit may itself be responsible for attacking the C-H bond of the substrate via HAT [277], or might be trapped by a substrate-derived radical to yield an alcoholate product (i.e., “rebound”).

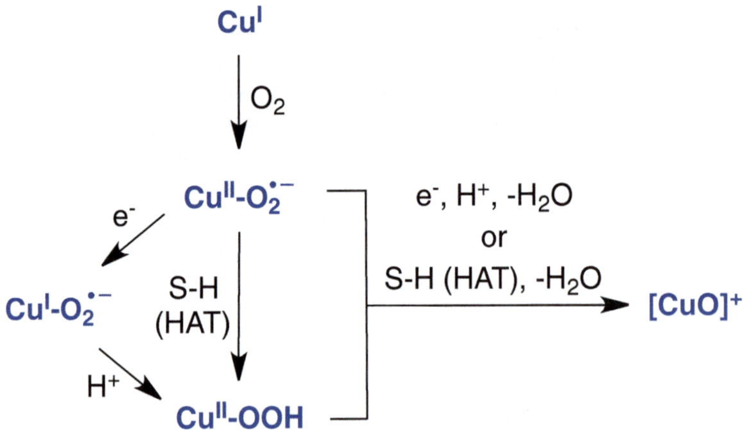

Figure 38 Overview of possible interconversions of copper-oxygen intermediates (S-H = substrate; HAT = hydrogen atom abstraction)

The only species that have been identified in the enzymes are a peroxide bound to Cu in an amine oxidase [278], and a peroxide complex [274] and a weakly bound 1:1 Cu/O_2 adduct in peptidylglycine α-amidating monooxygenase [279]. The $[CuO]^+$ unit has been identified in the gas phase and shown to oxidize CH_4 [276, 280], but has not been identified in biology or in a synthetic complex. Importantly, a general lack of understanding of the structures, spectroscopic properties, and reactivity of the various aforementioned monocopper-oxygen species provides the impetus for synthetic modeling studies.

4.1.1 1:1 Cu/O_2 Adducts

The most fully characterized examples of 1:1 Cu/O_2 complexes are shown in Figure 39. Complexes **77–80** have been structurally defined by X-ray crystallography, and their geometries and electronic structures fully investigated [281–283]. In brief, while complexes **77–79** feature side-on binding of the O_2^{n-} fragment, they differ in their bonding description. DFT calculations and spectroscopy results indicate that while **77–79** are singlets (S = 0 ground state), **77** is best described as a Cu(II)-superoxide while **78** and **79** are Cu(III)-peroxides. Key data include Cu K- and L-edges consistent with the indicated copper oxidation state assignments [282] and ν(O-O) values indicative of superoxide (1112 cm^{-1} for **77a**) or peroxide (968 cm^{-1} for **78a**) character, respectively [281]. Complex **80** exhibits end-on coordination with structural and rR parameters indicative of a Cu(II)-superoxide (O-O = 1.280(3) Å, ν(O-O) = 1120 cm^{-1}) [283, 284], but unlike **77–79**, it has an S = 1 ground state, as shown by variable temperature, variable field magnetic circular dichroism (VTVH-MCD) spectroscopy and DFT studies [283]. While not yet characterized by X-ray crystallography, adducts **81–85** have been studied by spectroscopy and theory, which also support end-on Cu(II)-superoxo formulations [285–291]. Like **80**, all are best described as having triplet ground states, with an initial report of **81** being a singlet [285] having been subsequently refuted through VTVH-MCD and NMR spectroscopy [291]. Intramolecular hydrogen bonding between the superoxide ligand and the amide NH of the ligand side-arm is a unique attribute of **84** [289], with interaction with either the α or β oxygen of the superoxide being indicated by DFT calculations. It is worth mentioning that an analog of **84** featuring two pivalamido groups was identified [292], and that an earlier report of an X-ray structure of an analog with three such substituents [293] was subsequently found to be in error [294].

Reactivity studies of end-on adducts **80–85** have provided fundamental insights into ligand effects on 1:1 Cu/O_2 adduct properties and have shed light on the question of whether the initially formed Cu(II)-O_2^- species proposed for the monocopper enzymes is capable of attacking the C-H bonds of substrates. In support of this capability, complex **82** (R = phenethyl group) decays to yield **86** (Figure 40) via a pathway shown by kinetic experiments and DFT calculations to proceed by rate-limiting intramolecular hydrogen atom abstraction by the coordinated superoxide from the benzylic C-H bond of the ligand phenethyl arm [286, 287].

A similar oxidation product with a hydroxylated ligand substituent forms upon reaction of complex **80** (Figure 39) with phenols [295]. In this case, it is postulated that H-atom abstraction from the phenol by the Cu(II)-O_2^- unit yields a Cu(II)-OOH species, which then attacks the ligand C-H bond (see below) [295, 296]. Complex **83** (Figure 39) also abstracts H-atoms from phenols [288]. Interestingly, while **84** (Figure 41) is unreactive with weak C-H bonds of substrates like DHA or xanthene, it oxidizes the NADH analog 1-benzyl-1,4-dihydronicotinamide (BNAH, Figure 41) [289]. It was suggested on the basis of kinetic studies that this reaction proceeds not via hydride transfer, but by rate-limiting hydrogen atom abstraction from substrate by the Cu(II)-O_2^- moiety.

The protonation/reduction of **80** and **84** (Figure 39) also has been studied [297, 298]. In the reaction of **80** to yield H_2O_2 with CF_3CO_2H, initial formation

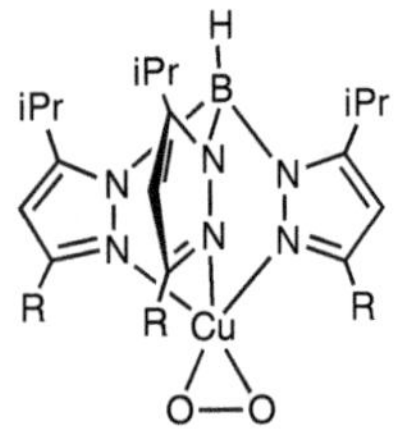

77
a, R = tBu
b, R = Ad

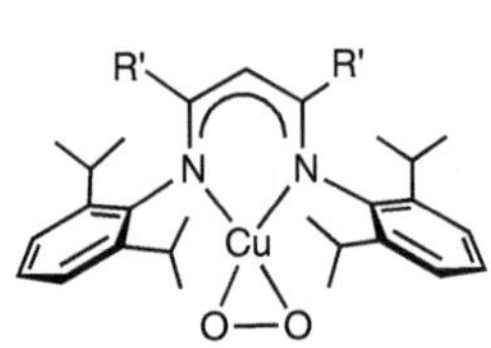

78
a, R' = Me
b, R' = tBu

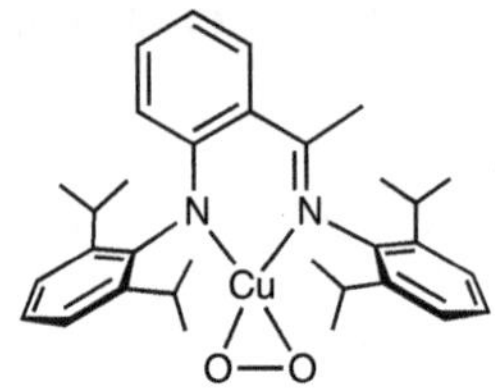

79

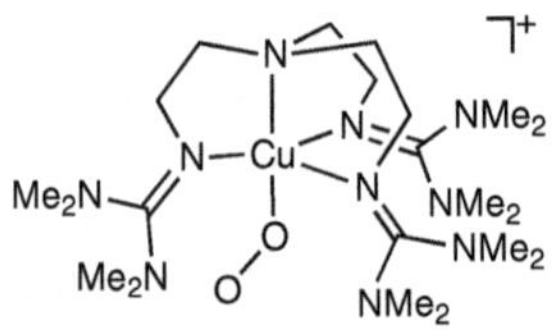

80

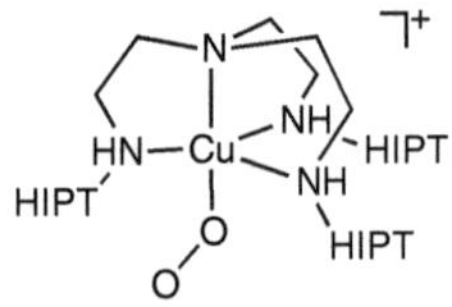

81

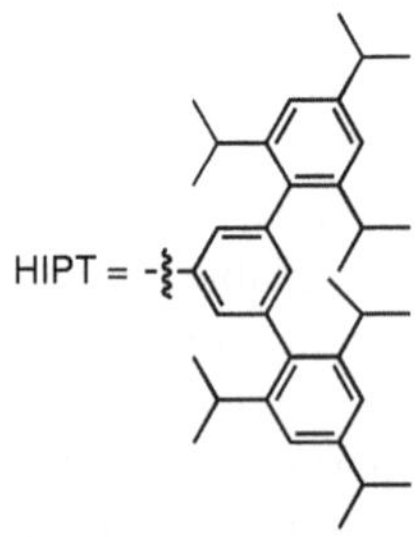

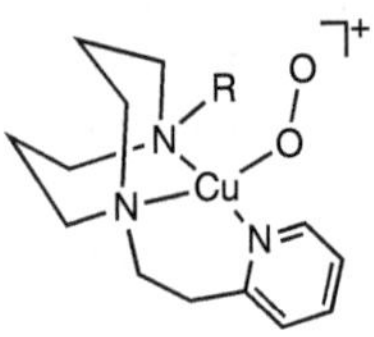

82

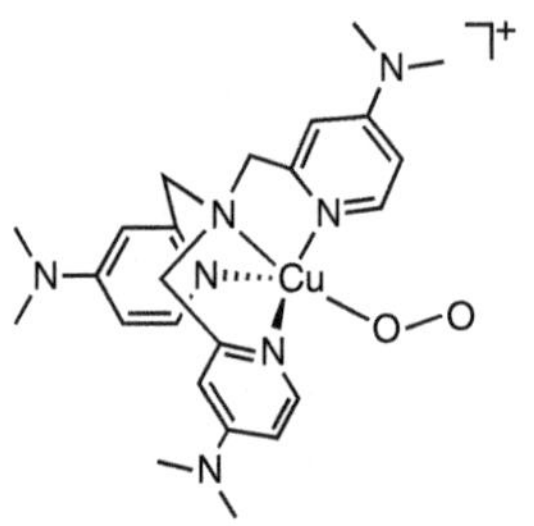

83

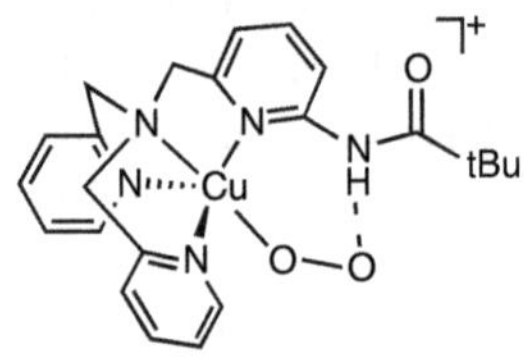

84

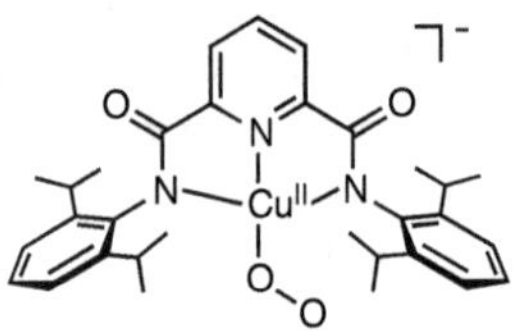

85

Figure 39 1:1 Cu/O_2 complexes.

of a hydrogen-bonded Cu(II)-$O_2^-$$\cdots$$HO_2CCF_3$ adduct was proposed on the basis of spectroscopy and theory results [297]. Subsequent reduction by alkyl-substituted ferrocenes yields H_2O_2 and a Cu(II) product. The 4-electron reduction of O_2 by Fc* was catalyzed by **84**, and the reaction was shown to proceed faster than with the analog lacking the pivalamido substituent [298]. Mechanistic experiments led to the proposed mechanism shown in Figure 42.

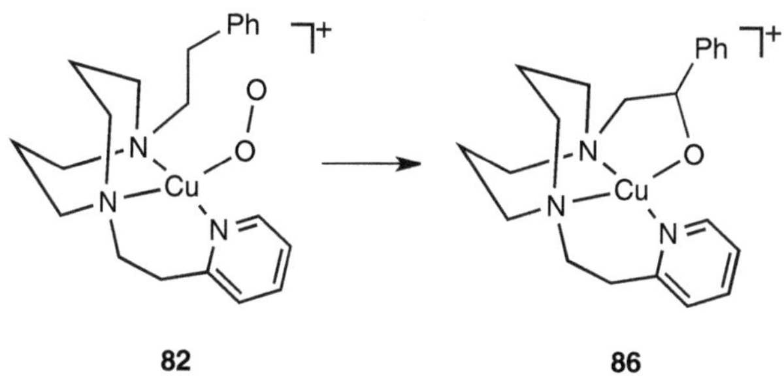

Figure 40 Intramolecular oxidation reaction of a Cu(II)-O_2^- complex.

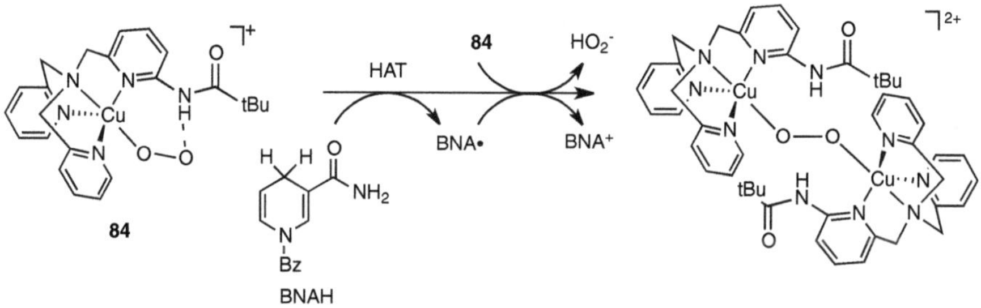

Figure 41 Reaction of a Cu(II)-O_2^- complex with 1-benzyl-1,4-dihydronicotinamide (BNAH). Adapted from [289].

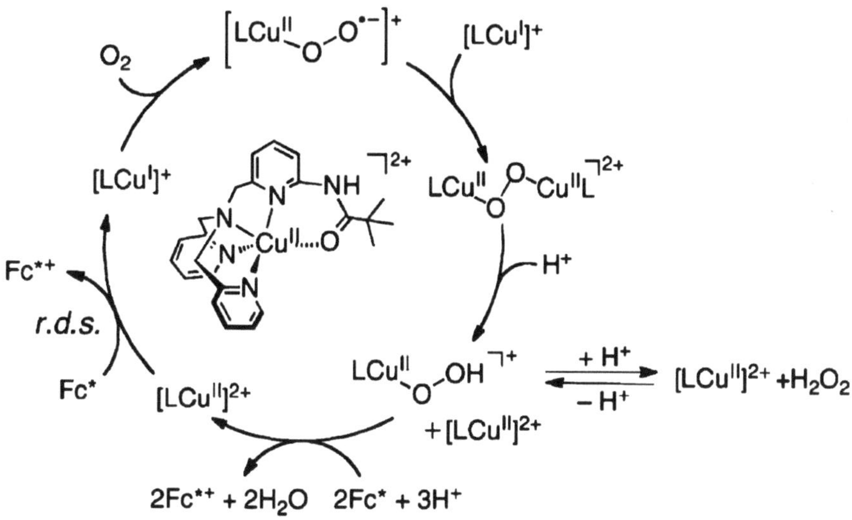

Figure 42 Proposed mechanism for the catalytic 4-electron reduction of O_2 catalyzed by **84**. Reproduced with permission from [295]; copyright 2009, American Chemical Society.

In contrast to the generally electrophilic reactivity of **80–84**, complex **85** (Figure 39) acts as a base and was surmised to have nucleophilic character, presumably because of the overall negative charge of the complex and the strong electron donating properties of the pyridine(dicarboxamide) supporting ligand [290]. Thus, **85** is unreactive with alkyl-substituted phenols, appears to be protonated by *p*-nitrophenol, does not oxidize PPh_3, and may be trapped by [(TMPA)-Cu(I)]OTf to yield a *trans*-η^1:η^1-peroxo-dicopper(II,II) complex [290].

4.1.2 CuOOR Complexes

In addition to being proposed as a product of reactions of 1:1 Cu/O_2 adducts with substrates, monocopper-hydroperoxo species are themselves considered to be viable oxidants in enzymes. Building upon previous work focused primarily on the characterization of monocopper-hydroperoxo and -alkylperoxo complexes [299–304], more recent research has emphasized reactivity studies. For example, the complexes **87a, b** (Figure 43) decompose to yield mixtures of products resulting from oxidation of substituents R (e.g., N-alkylation, formation of aldehydes or ketones) [299, 305]. Mechanistic studies suggest that these oxidations involve intramolecular HAT from the position α to the N-atom of the ligand, either directly by the Cu(II)-OOH unit or by a $[CuO]^+$ moiety formed first via O-O bond homolysis. In indirect support of the latter, similar oxidized products formed in reactions of PhIO with Cu(I) complexes of the supporting ligand in **87a**, with electrospray ionization mass spectrometry evidence supporting the involvement of $[CuO]^+$ or adduct $[CuOIPh]^+$ cores [299].

In related chemistry, hydroxylation at the ortho position of the pendant aryl ring in **87c** (Figure 43) occurred upon warming [306]. The involvement of bis(oxo) dicopper species in this reaction was ruled out by evaluation of the reactivity of such a species generated independently. The oxidative reactivity of complexes **87a–c** stands in contrast to the Cu(II)-OOH complex supported by N_4Py, which while competent to oxidize PEt_3, does not attack substrates such as thioanisole or

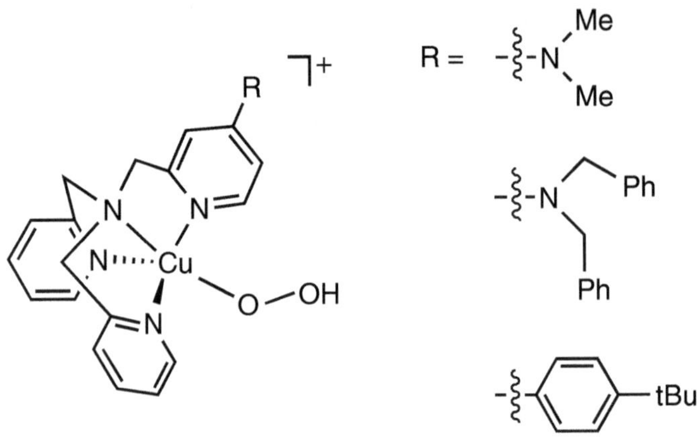

87a

87b

87c

Figure 43 Copper-hydroperoxo complexes.

cyclohexadiene [307]. Evaluation by theory indicated that O-O bond homolysis in this system is energetically unfavorable.

Homolytic scission of the O-O bond in a Cu(II)-OOR species (R = alkyl or aryl group) to yield a $[CuO]^+$ intermediate has been invoked in several instances. For example, the Cu(II)-cumylperoxo complex **88** decayed to yield a bis(hydroxo) dicopper complex **89** and acetophenone, implicating an O-O bond homolysis pathway (Figure 44) [308]. Further evidence in favor of this proposal came from observation of hydrogen atom transfer reactivity in the presence of external substrates and the formation of **90** upon addition of the radical trap 5,5-dimethyl-1-pyrroline N-oxide. In later work, however, kinetic studies and DFT calculations supported a pathway for reactions of **88** and an analog supported by tris [2-(2-pyridyl)methyl]amine that involved concerted HAT and O-O bond homolysis, rather than via prior formation of a discrete $[CuO]^+$ intermediate [309].

In another example, a $[CuO]^+$ species was evaluated as a possible intermediate in the ligand hydroxylation outlined in Figure 45 [310]. Mimicking the O_2 reactivity of Fe(II)-α-ketocarboxylate units in proteins and model complexes (Section 3), Cu(I)-α-ketocarboxylates **91** reacted with O_2 to yield products derived from ortho

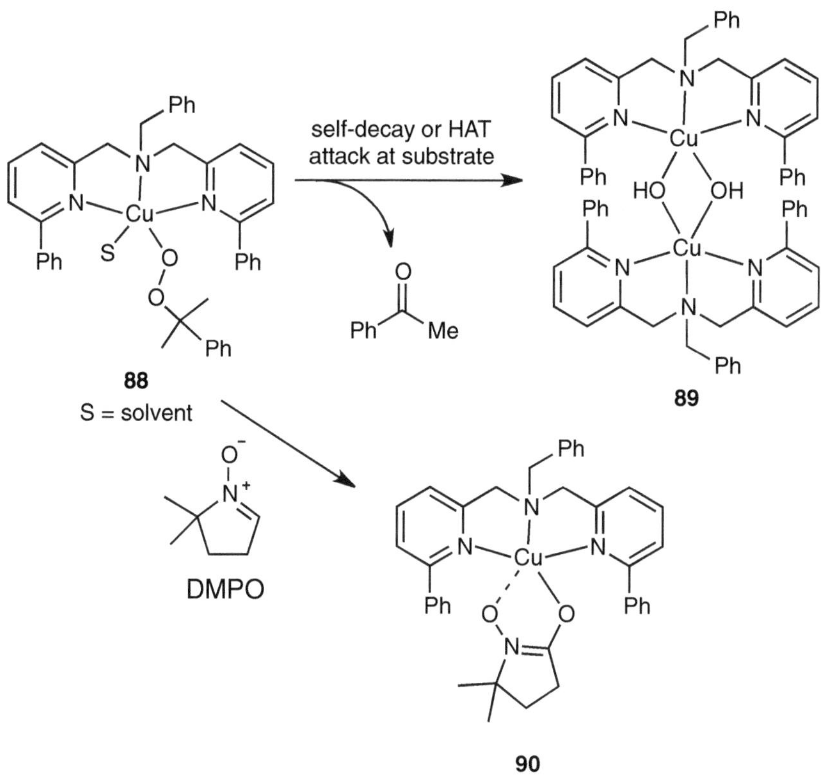

Figure 44 Reactions of a copper-cumylperoxide complex.

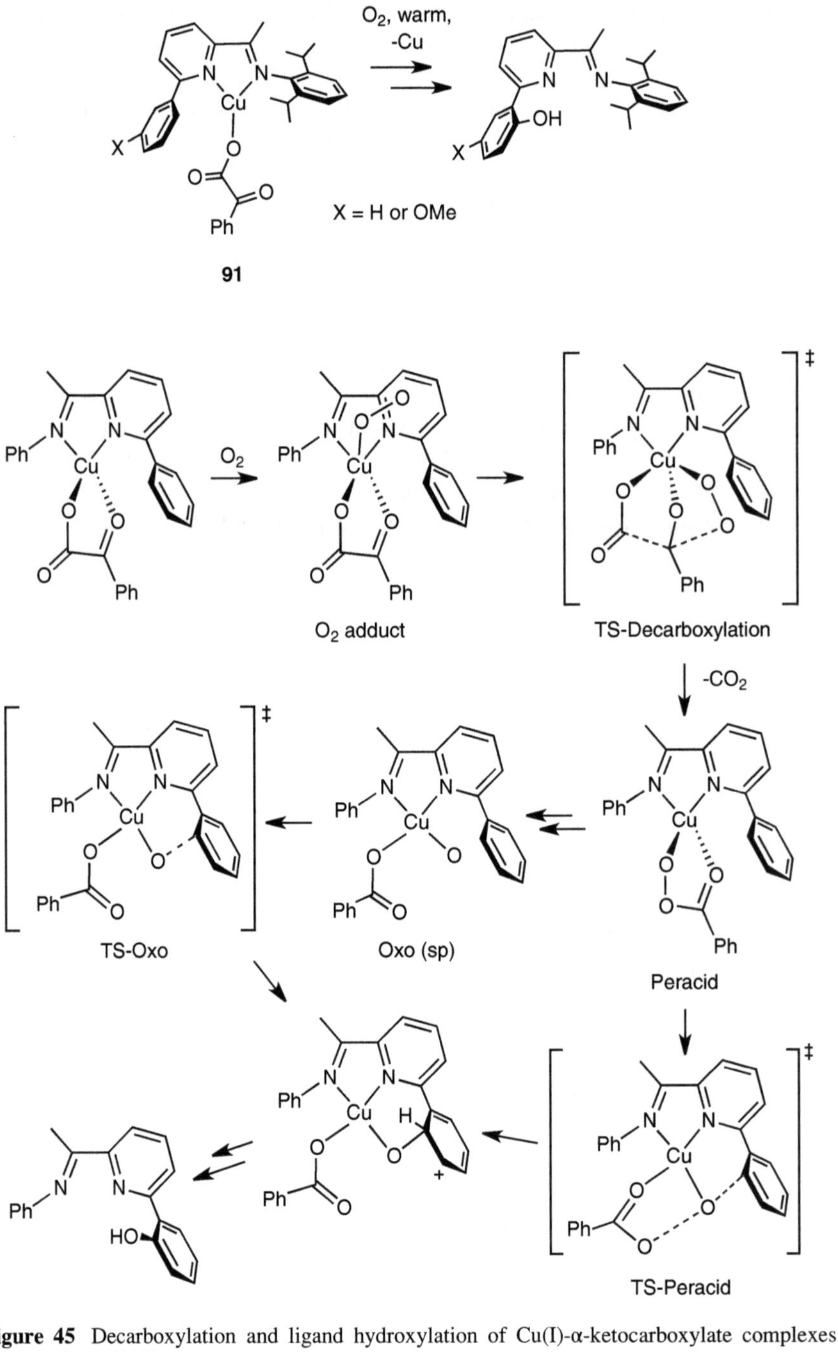

Figure 45 Decarboxylation and ligand hydroxylation of Cu(I)-α-ketocarboxylate complexes. Adapted from [310].

hydroxylation of the aryl ring of the supporting ligand. DFT calculations [310, 311] support a mechanism involving initial decarboxylation to yield a "peracid" intermediate, which can attack the aromatic ring directly or through initial conversion to a reactive $[CuO]^+$ species (Figure 45).

Finally, monocopper-hydroperoxo species are also implicated in catalytic reductions of dioxygen, as noted above in Figure 42 for a 4-electron process that generates H_2O [295]. Another example is in the 2-electron reduction of O_2 to H_2O_2 by ferrocene or Fc* catalyzed by a copper complex of tris[2-(2-pyridyl)ethyl]amine in the presence of $HClO_4$ [312]. A detailed mechanistic study of this system revealed pre-equilibrium binding of O_2 to a Cu(I) complex prior to formation of a $[CuOOH]^+$ intermediate that is not reduced, but is instead protonated to yield H_2O_2.

4.1.3 Targeting $[CuO]^+$ and Related Monocopper Species

As noted above, $[CuO]^+$ species have been invoked as intermediates in reactions of $[CuOOR]^+$ cores. They have also been proposed to be generated as transient intermediates from reactions of PhIO with various copper complexes [295, 299, 313, 314], as well as to be the active species responsible for aromatic arene hydroxylations performed by Cu/Me_3NO systems [315–319]. A mechanism for the latter was proposed on the basis of theory that involves homolysis of the O-N bond in Cu(II)-$ONMe_3$ complexes to yield the $[CuO]^+$ core and $Me_3N^{+\bullet}$ [320, 321]. Here we describe some other recent attempts to prepare complexes comprising the $[CuO]^+$ unit and a protonated variant.

Reasoning that O-N bond heterolysis in Cu(I)-ONR_3 complexes might yield $[CuO]^+$ species, reactions of pyridine- and trialkylamine-*N*-oxides with illustrative Cu(I) complexes **92–94** supported by various bidentate N-donor ligands were explored (Figure 46) [322]. For some cases, stable adducts that were not susceptible to O-N bond scission formed (e.g., **95** or **96**). An exception was the generation of bis (hydroxo)dicopper(II) (**97**) or bis(oxo)dicopper(III) complexes (**98**) in the reaction of **94** with Me_3NO at room temperature or −80 °C, respectively. It is unclear whether these products form via O-N heterolysis followed by $[CuO]^+$ dimerization, or by dimerization of an initially formed Cu(I)-$ONMe_3$ adduct followed by loss of Me_3N.

In a different approach, monocopper-hydroxide complexes **99** (R = *i*Pr or Me) were shown to undergo 1-electron oxidation at low potential (for R = *i*Pr, −0.076 V *versus* ferrocenium/ferrocene (Fc^+/Fc) in acetone) to yield highly colored reactive intermediates (Figure 47) [323, 324]. These intermediates were identified as Cu(III)-hydroxide species **100** on the basis of DFT calculations and X-ray absorption spectroscopy for the system with R = *i*Pr [323].

A significant shift in Cu K-edge features in the X-ray absorption spectrum to higher energy compared to its copper(II)-hydroxide precursor were diagnostic for assignment of a Cu(III) oxidation state in **100** (R = *i*Pr), which may be viewed as comprising a protonated version of the $[CuO]^+$ core. In order to draw comparisons to other metal-oxo/hydroxo complexes, the reactivity of **100** with DHA was studied. Kinetics studies revealed clean conversion to anthracene and a Cu(II)-aqua complex that proceeded with a second-order rate law, a large KIE, and high

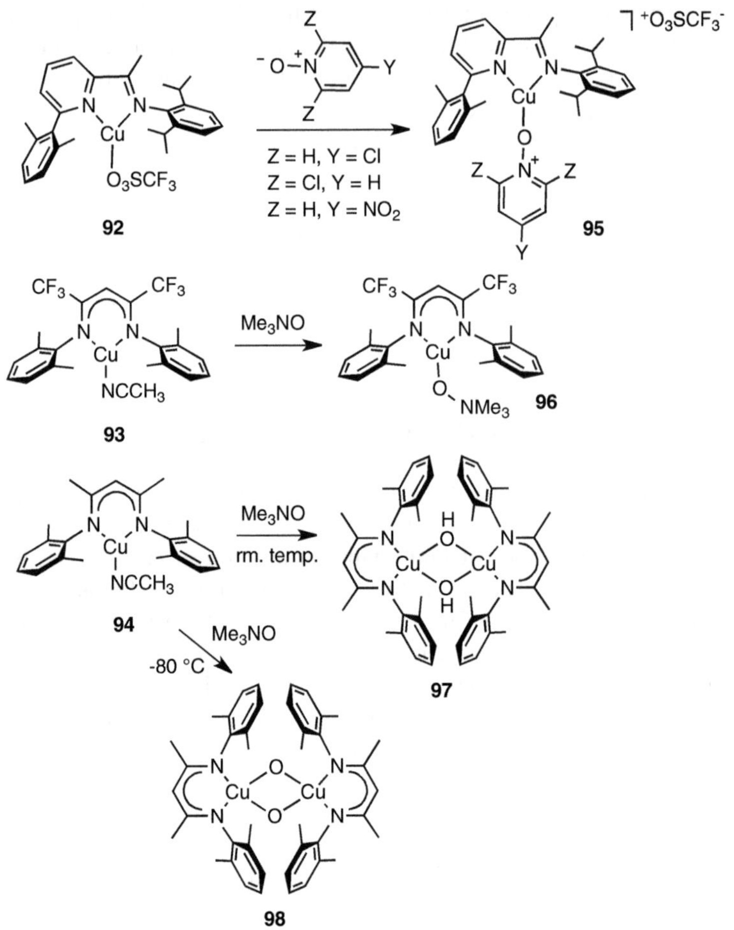

Figure 46 Reactions of Cu(I) complexes with pyridine- and trialkylamine-*N*-oxides.

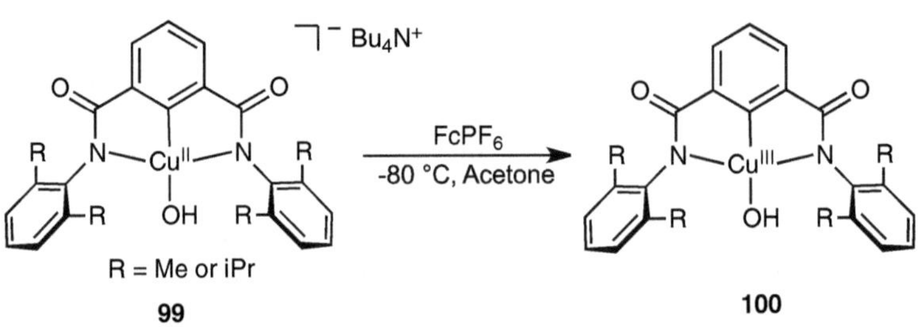

Figure 47 Oxidation of a copper(II)-hydroxide complex.

rates. Indeed, a comparison among rate constants at −80 °C for reactions of **100** (R = *i*Pr) with DHA and those of other copper, iron, and manganese oxo/hydroxo complexes revealed the reaction rate of **100** to be higher than most, and exceeded only by a particularly reactive diiron-oxo complex. Evaluation of the underlying reasons for this high reactivity are ongoing.

4.2 Dicopper Models of Dicopper Active Sites in Enzymes

Two copper ions in close proximity (within 5 Å) act to bind and/or activate dioxygen in the O_2 carrier protein hemocyanin and the enzymes tyrosinase and catechol oxidase (Table I, entries 4, 6, and 13). The oxygenated forms of these systems have been characterized by spectroscopy and X-ray crystallography [325–329], and have been formulated conclusively as (μ-η^2:η^2-peroxo)dicopper(II) cores (Figure 48). As discussed below, a key question that has inspired much synthetic modeling research concerns whether this stable, identifiable core or an intermediate derived therefrom is responsible for attacking substrate in tyrosinase or catechol oxidase.

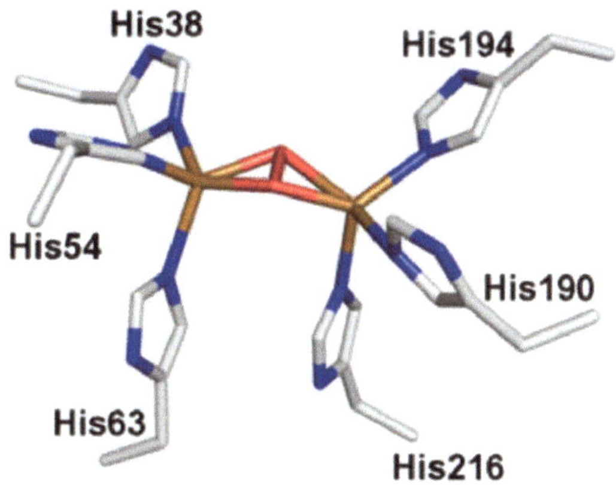

Figure 48 Active site structure of oxytyrosinase from *Streptomyces castaneoglobisporus* (PDB ID 1WX2). Reproduced with permission from [329]; copyright 2013, Springer.

Dicopper-oxygen species have also been suggested as oxidizing intermediates in particulate methane monooxygenase (pMMO) [330], although the involvement of tricopper species has also been argued [331]. While some spectroscopic results point to the formation of a (μ-η^2:η^2-peroxo)dicopper(II) moiety in pMMO [332], other species such as those shown in Figure 49 have been postulated on the basis of theory [260, 333, 334] or merely speculated [335] to be more reactive oxidants that are more likely to be capable of attacking the strong C-H bond of substrate methane. In view of their novel structures and proposed reactivity, these species are prime targets for synthetic modeling studies. Another noteworthy target is the (μ-oxo)-dicopper(II) unit, which has been proposed recently on the basis of spectroscopy and theory to be the active oxidant in Cu-doped zeolite catalysts that perform the same reaction as pMMO [336, 337].

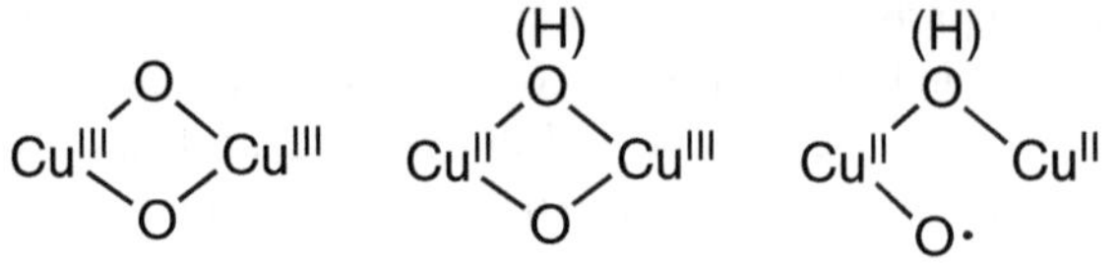

Figure 49 Dicopper-oxygen cores proposed as active oxidants in particulate methane monooxygenase (pMMO).

Many studies of synthetic dicopper-oxygen intermediates have aimed to address the aforementioned issues (as well as others), and these have been summarized in extensive reviews, to which readers interested in work appearing prior to early 2012 are pointed [261–266, 338–341]. Most recently, Rolff et al. [340] describe work showing (a) that both the (μ-η^2:η^2-peroxo)- and bis(μ-oxo)dicopper cores are capable of hydroxylating phenolates like in tyrosinase in both stoichiometric and catalytic reactions, (b) that reactions of phenols with these cores often yield radical coupling products, pointing to the importance of proton 'management' by the enzyme, and (c) how the orbital interactions involved in arene hydroxylation reactions of both cores underly their reactivity. Halvagar et al. [266] describe work with focus on revealing the factors that favor the formation of either or both of these cores in oxygenation reactions, while Fukuzumi and Karlin [341] summarize efforts to reveal the kinetics and thermodynamics of the formation of dicopper-oxygen species and their mechanisms of reduction by hydride and electron-transfer reagents. Herein, rather than duplicating the discussions presented in these (and earlier) review articles, we focus on selected pertinent recent work published in 2012 and 2013.

4.2.1 Peroxo- and Bis(μ-oxo)dicopper Complexes

New examples of peroxo- and bis(μ-oxo)dicopper complexes were recently described using the supporting ligands shown in Figure 50. Low temperature oxygenation of the Cu(I) complex of **101** yielded a bis(μ-oxo)dicopper species, identified as such on the basis of diagnostic UV-vis and rR spectroscopic features [342]. Self-assembly of the (μ-η^2:η^2-peroxo)dicopper core was observed upon oxygenation of the Cu(I) complex supported by three imidazole ligands **102** in 2-methyltetrahydrofuran at −125 °C, demonstrating that the oxytyrosinase core may be prepared bereft of the protein matrix using simple biomimics of histidine residues [343]. Functional modeling of tyrosinase was also achieved with this system, as revealed through mechanistic studies of hydroxylations of added phenolates. The (μ-η^2:η^2-peroxo)dicopper core was also observed upon low temperature oxygenation of the Cu(I) complex of **103** (R = *i*Pr) [344], in contrast to the previous reports of bis(μ-oxo)dicopper complex formation for the cases of R = Et, Bn, *i*Bu [345]. These results are consistent with previous findings that increased steric bulk of supporting ligands disfavors formation of the more compact bis(μ-oxo)dicopper core [261, 262, 266]. A (μ-η^2:η^2-peroxo)dicopper complex was also formed upon oxygenation of the Cu(I) complex of **104** (Figure 50) and this

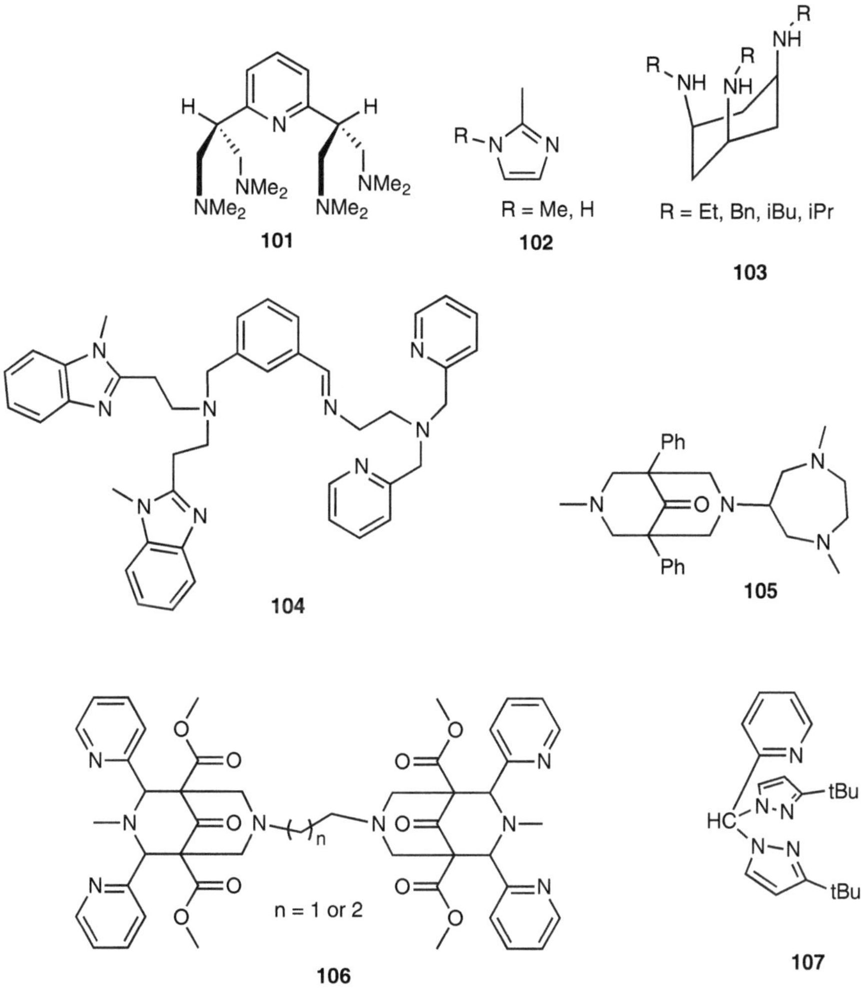

Figure 50 Ligands used in recent syntheses of dicopper-oxygen species.

species was sufficiently stable to be used to catalytically oxidize phenols to quinones in the presence of added base [346].

The "asymmetric" ligand **105** (Figure 50) favors formation of a (μ-η^1:η^1-peroxo)-dicopper core, which was shown to be competent to hydroxylate added phenolates via a pathway that involves detectable phenolate-bound intermediates [347]. These electrophilic hydroxylations are unusual for such end-on peroxo cores, which typically act as nucleophiles instead [339]. In addition, (μ-η^1:η^1-peroxo)dicopper cores were implicated in oxygenation reactions of Cu(I) complexes of bispidine ligands **106** [348] and **107** [349], and the pathway for catecholase activity of such oxygenated complexes was evaluated by theory [349].

Two recent contributions exclusively use theoretical approaches to tackle issues associated with the properties of (μ-η^2:η^2-peroxo)- and bis(μ-oxo)dicopper complexes.

Following up on previous work that underlined the difficulties associated with using theory to understand the electronic structural properties and relative stabilities of these cores [350, 351], a further evaluation of the reliability of various functionals was reported [352]. The results confirmed that no single functional is optimal for describing systems with different supporting ligands. Another theoretical study focused on evaluating the mechanism for phenolate hydroxylation by the (μ-η^2:η^2-peroxo)dicopper complex supported by *N,N'*-di-*tert*-butylethylenediamine [353]. In the original work on this system [354, 355], evidence had been provided to support O-O bond scission to yield a bis(μ-oxo)dicopper intermediate with bound phenolate which subsequently underwent electrophilic attack, thus providing precedent for the notion that the bis(μ-oxo)dicopper core might be the active oxidant in tyrosinase. In the more recent theoretical study, O-O bond scission was found to precede phenolate coordination, and details of the hydroxylation reaction were described [353].

Intramolecular hydroxylation of bridging arene groups in dinucleating ligands is a well-known process in copper chemistry [261–264], but recent work has provided some new insights into the nature of the species that might be responsible for this type of reaction. In one study, bis(μ-oxo)dicopper complex **108** (Figure 51) was identified as an intermediate and shown to both attack the bridging *m*-xylyl group and to react with exogenous phenols (predominantly via a radical type pathway) [356]. In contrast, DFT calculations calibrated by experimental Cu L-edge spectroscopy data that shed light on metal-ligand covalency showed that the peroxo unit in complex **109** attacks the *m*-xylyl group without the intermediacy of a bis(μ-oxo)dicopper complex [357].

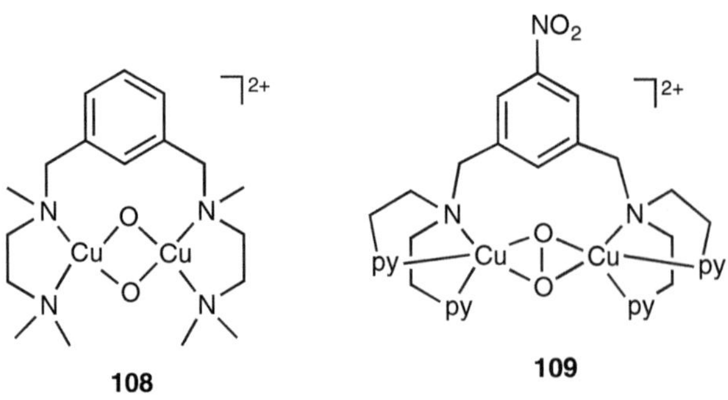

Figure 51 Complexes that undergo intramolecular hydroxylation of the bridging *m*-xylyl group.

4.2.2 (μ-Oxo)dicopper Complexes

The recent postulate of a (μ-oxo)dicopper(II) core as the active oxidant in heterogeneous Cu-doped zeolite catalysts that convert methane to methanol has stimulated consideration of this unit as an intermediate in pMMO [260]. As described in a recent review [358], synthetic complexes with this core are relatively rare, in part because of their tendency to react readily with water to yield bis(μ-hydroxo)dicopper(II) complexes [359–362]. In a recent resurgence of interest in these species, ligands **110** and **111** (Figure 52) have been used to prepare and characterize species proposed to contain (μ-oxo)dicopper(II) cores [363, 364]. In both cases, these species formed upon treatment of dicopper(I) complexes with O_2 or PhIO. For the system comprising **110**, the (μ-oxo)dicopper(II) product was identified on the basis of UV-vis absorption spectroscopy and resonance Raman spectroscopy, but more complete characterization was hindered by its thermal instability [363]. Ligand **111** led to a more stable complex **112**, which was more fully characterized, including by EXAFS (Cu-Cu = 2.91 Å), NMR spectroscopy, high-resolution electrospray ionization mass spectrometry, and DFT calculations [364]. In initial reactivity studies, **112** induced coupling of 2,4-di-*tert*-butylphenol, a process reminiscent of the H-atom abstraction reactivity known for bis(μ-oxo)-dicopper(III) complexes [262].

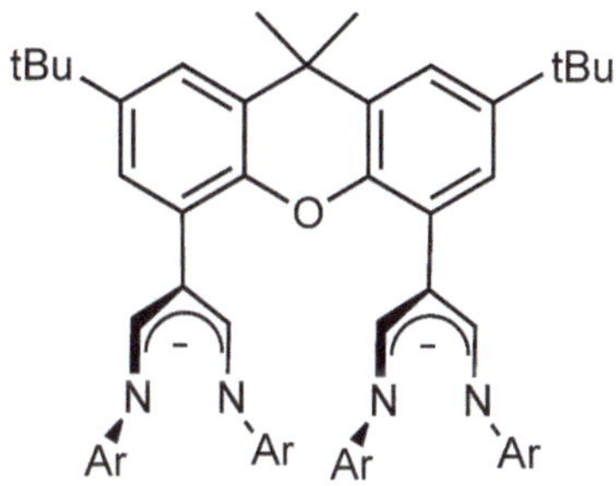

110

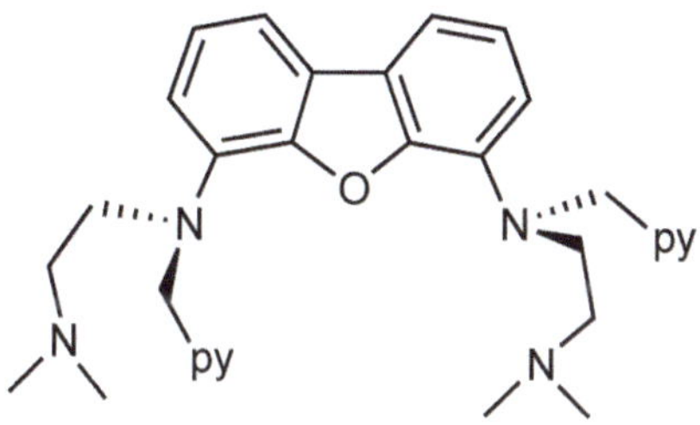

111

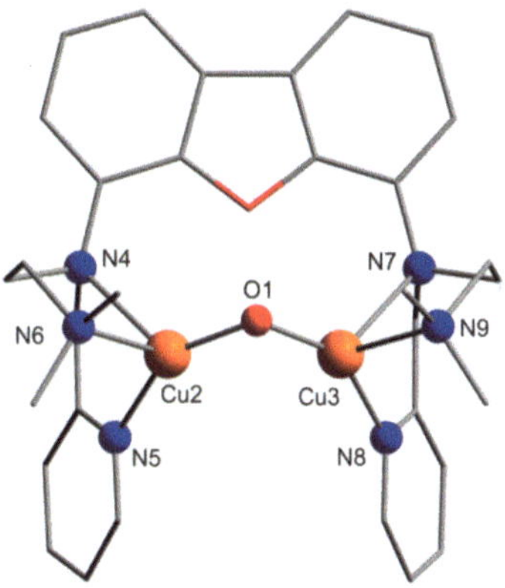

112

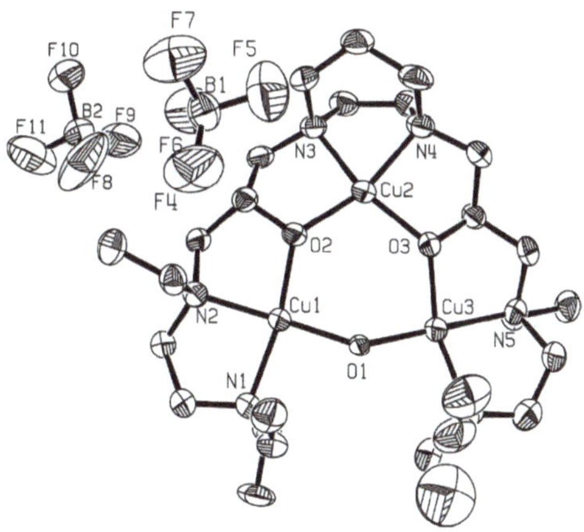

113

Figure 52 Ligands **110** and **111** used to prepare (μ-oxo)dicopper(II) species, the calculated structure of the one derived from **111** (reproduced with permission from [364]; copyright 2013, American Chemical Society), and the X-ray crystal structure of a tricopper cluster (**113**) featuring the (μ-oxo)dicopper(II) unit (reproduced with permission from [365]; copyright 2007, National Academy of Sciences of the USA).

The (μ-oxo)dicopper(II) unit has been characterized by X-ray crystallography in only one instance, to our knowledge, within the tricopper cluster **113** (Figure 52) [365]. Touted as a model of the oxidized form of pMMO, reduction of this cluster by α-diketones followed by oxygenation results in the formation of the corresponding carboxylic acid. A mechanistic rationale for this process involving O_2 activation by a tricopper(I) species was invoked.

4.3 Tricopper Models of Multicopper Active Sites in Enzymes

Interest in tricopper-oxygen species is inspired by (a) the involvement of tricopper active sites in the multicopper oxidases [252, 254], ubiquitous enzymes that play a central role in oxidative metabolism and other key biological processes, and (b) the proposal that the active site of pMMO is a tricopper center [331]. A variety of multicopper catalysts for hydrocarbon oxidations have been described [366, 367], but while stable Cu(I) or Cu(II) precursors or 'resting states' are often identified, the nature of the multicopper-oxygen intermediates responsible for attacking the C-H bonds are usually only speculated on the basis of indirect and/or ambiguous spectroscopic or theoretical evidence [368–370]. In studies of dioxygen activation by more than two Cu(I) sites that target possible intermediates, tricopper species often comprise the bis(μ-oxo)tricopper(II,II,III) core (**114**, Figure 53). Initially prepared and fully characterized upon low temperature oxygenation of Cu(I) complexes of ligands **115** [371–373], such cores have since been reported using several other N-donors (**115–118**) [374–376]. Complex **119** was prepared by low temperature oxygenation of a tricopper(I) complex of ligand **118**. Interestingly, dipicolylamine-copper(I) complexes in which the dipicolyl ligand is not 'pre-organized' like in **118** do not afford $[Cu_3(\mu\text{-}O)_2]^{3+}$ cores, showing that the preassembly of the Cu(I) centers by the ligand scaffold of **118** is key to the formation of this core in this case.

Interesting reactivity differences have been identified for the $[Cu_3(\mu\text{-}O)_2]^{3+}$ core supported by different N-donor ligands. The complexes of **115** and **116** react with 2,4-di-*tert*-butylphenol to yield the radical coupling product 3,3',5,5'-tetra-*tert*-butyl-2,2'-biphenol, consistent with electrophilic character and H-atom abstraction reactivity similar to related bis(μ-oxo)dicopper(III) complexes [262]. In contrast, the complex supported by **117** was unreactive with phenols or DHA, but oxidized PPh_3 to $O{=}PPh_3$ and was quenched by CO_2 [375]. Apparently, the charge and strong electron donating properties of **117** perturb the core so as to induce 'nucleophilic' reactivity similar to that identified for (*trans*-1,2-peroxo)dicopper complexes [264].

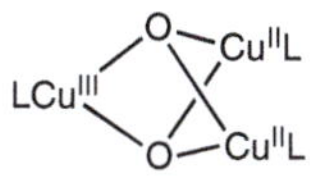

114

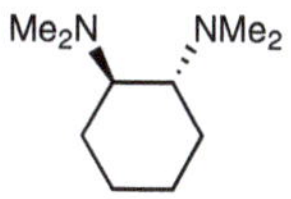

115

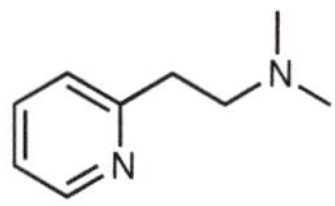

116

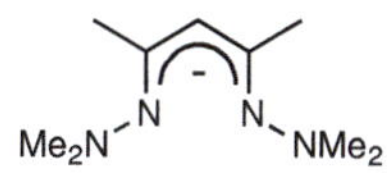

117

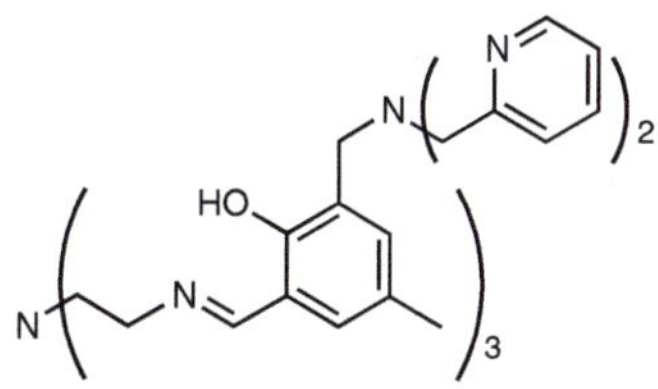

118

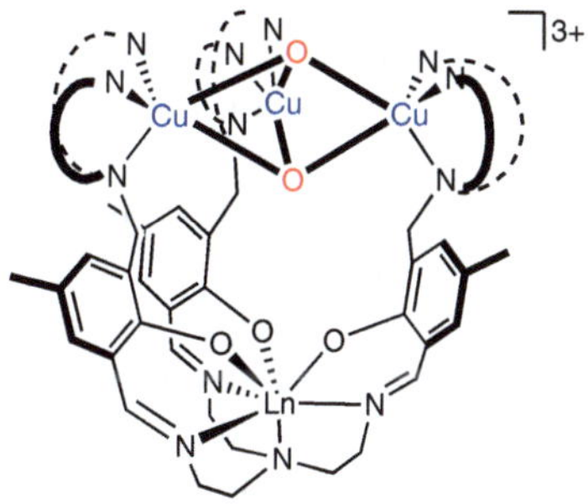

119

Figure 53 The structure of bis(μ-oxo)tricopper(II,II,III) complexes (**114**), ligands used to prepare such complexes (**115–118**), and the complex **119** prepared using **118**.

5 Concluding Remarks

Recent advances in the study of synthetic Fe and Cu complexes has led to greater understanding of the detailed chemistry involved in the binding and activation of O_2 by Fe and Cu sites in biological and other catalysts. Adherence to the overarching mechanistic paradigm shown in Figure 1 appears general, notwithstanding fascinating differences in detail and the complicating involvement of redox noninnocent supporting ligands and/or cofactors.

Conclusive identification of many key intermediates has been achieved, opening the door to reactivity studies that are beginning to provide mechanistic insights useful for understanding metalloprotein function and for developing new and useful catalysts. Yet many putative intermediates suggested to be involved in some of the most challenging oxidation reactions, such as the conversion of methane to methanol, have not been prepared and characterized.

Many questions remain about how O-O bond breaking occurs and is controlled (cf., by secondary-sphere hydrogen bonding and other environmental influences), and about the structures and properties of high valent Fe(IV), Fe(V), and Cu(III) species thought to be the active oxidants in oxidation reactions. These and other questions continue to inspire efforts to isolate, characterize, and evaluate the reactivity of synthetic complexes that bind O_2 and/or incorporate O_2-derived moieties, high valent metal centers, and redox active ligands. Perhaps most evident from the survey of the field is the fact that creative ligand design and application of advanced spectroscopic, isolation, and mechanistic tools is key to enabling the discovery of new compounds and mechanistic ideas.

Abbreviations and Definitions

Ar	aryl
BDE	bond dissociation enthalpy
BF	benzoylformate
Bn	benzyl
BNAH	1-benzyl-1,4-dihydronicotinamide
cat	catalyst
C*c*O	cytochrome *c* oxidase
m-CPBA	*meta*-chloroperbenzoic acid
Cpd 0	compound 0
Cpd I	compound I
Cpd II	compound II
CPh	benzyl
DBC	2,4-di-*tert*-butyl-catecholate
DBSQ	2,4-di-*tert*-butyl-semiquinone
DFT	density functional theory
DHA	9,10-dihydroanthracene
DMSO	dimethyl sulfoxide
EPR	electron paramagnetic resonance
Et	ethyl
EXAFS	extended X-ray absorption fine structure
Fc^+/Fc	ferrocenium/ferrocene
Fc*	decamethylferrocene
HAT	hydrogen atom transfer
Hb	hemoglobin
HOMO	highest occupied molecular orbital
*i*Pr	*iso*-propyl
IR	infrared
KIE	kinetic isotope effect
α-KG	α-ketoglutarate
LMCT	ligand-to-metal charge transfer
Mb	myoglobin
Me	methyl
Mes	mesityl

NADH	reduced nicotinamide adenine dicnucleotide
*n*Bu	*n*-butyl
NHE	normal hydrogen electrode
NMR	nuclear magnetic resonance
N_4Py	*N*,*N*-bis(2-pyridyl-methyl)-N-bis(2-pyridyl)methylamine
NTs	*N*-toluene sulfonyl
OMe	methoxy
OTf	trifluoromethanesulfonate
oxyMb	oxy-myoglobin
PhIO	iodosylbenzene
pMMO	particulate methane monooxygenase
PROS	partially reduced oxygen species
py	pyridine
r.d.s.	rate-determining step
rR	resonance Raman
RT	room temperature
sMMO	soluble methane monooxygenase
*t*Bu	*tert*-butyl
TMC	tetramethylcyclam
TMPA	tris[(2-pyridyl)methyl]amine
TON	turnover number
tpp	5,10,15,20-tetrakis(pentafluorophenyl)porphyrin
Ts	tosyl
UV-vis	ultraviolet-visible
VTVH-MCD	variable temperature variable field magnetic circular dichroism
XAS	X-ray absorption spectroscopy

Acknowledgment We thank the NIH (GM47365) for financial support of our work in the area of O_2 activation described herein.

References

1. R. H. Holm, P. Kennepohl, E. I. Solomon, *Chem. Rev.* **1996**, *96*, 2239–2314.
2. J. A. Ibers, R. H. Holm, *Science* **1980**, *209*, 223–235.
3. K. D. Karlin, *Science* **1993**, *261*, 701–708.
4. S. Shaik, D. Kumar, S. P. de Visser, A. Altun, W. Thiel, *Chem. Rev.* **2005**, *105*, 2279–2328.
5. I. G. Denisov, T. M. Makris, S. G. Sligar, I. Schlichting, *Chem. Rev.* **2005**, *105*, 2253–2277.
6. H. Fujii, *Coord. Chem. Rev.* **2002**, *226*, 51–60.
7. B. Meunier, S. P. de Visser, S. Shaik, *Chem. Rev.* **2004**, *104*, 3947–3980.
8. E. Rose, M. Quelquejeu, A. Kossanyi, B. Boitrel, *Coord. Chem. Rev.* **1998**, *178–180*, 1407–1431.
9. M. Rivera, Y. Zeng, *J. Inorg. Biochem.* **2005**, *99*, 337–354.
10. M. Momenteau, C. A. Reed, *Chem. Rev.* **1994**, *94*, 659–698.

11. M. Sono, M. P. Roach, E. D. Coulter, J. H. Dawson, *Chem. Rev.* **1996**, *96*, 2841–2887.
12. I. Schlichting, *Science* **2000**, *287*, 1615–1622.
13. T. L. Poulos, B. C. Finzel, I. C. Gunsalus, G. C. Wagner, J. Kraut, *J. Biol. Chem.* **1985**, *260*, 16122–16130.
14. T. L. Poulos, B. C. Finzel, A. J. Howard, *Biochemistry* **1986**, *25*, 5314–5322.
15. H. Li, S. Narasimhulu, L. M. Havran, J. D. Winkler, T. L. Poulos, *J. Am. Chem. Soc.* **1995**, *117*, 6297–6299.
16. T. L. Poulos, R. Raag, *FASEB J.* **1992**, *6*, 674–679.
17. T. L. Poulos, *Adv. Inorg. Biochem.* **1988**, *7*, 1–36.
18. J. H. Dawson, *Science* **1988**, *240*, 433–439.
19. L. Que, Jr., *Dioxygen Activating Enzymes*, in *Biological Inorganic Chemistry: Structure and Reactivity*, Eds I. Bertini, H. B. Gray, E. I. Stiefel, J. S. Valentine, University Science Books, Sausalito, CA, USA, 2007.
20. L. Pauling, C. D. Coryell, *Proc. Natl. Acad. Sci. USA* **1936**, *22*, 210–216.
21. R. A. Ghiladi, R. M. Kretzer, I. Guzei, A. L. Rheingold, Y. M. Neuhold, K. R. Hatwell, A. D. Zuberbüehler, K. D. Karlin, *Inorg. Chem.* **2001**, *40*, 5754–5767.
22. Y. Li, S. K. Sharma, K. D. Karlin, *Polyhedron* **2013**, *58*, 190–196.
23. J. G. Liu, T. Ohta, S. Yamaguchi, T. Ogura, S. Sakamoto, Y. Maeda, Y. Naruta, *Angew. Chem. Int. Ed.* **2009**, *48*, 9262–9267.
24. G. B. Jameson, G. A. Rodley, W. T. Robinson, R. R. Gagne, C. A. Reed, J. P. Collman, *Inorg. Chem.* **1978**, *17*, 850–857.
25. G. B. Jameson, F. S. Molinaro, J. A. Ibers, J. P. Collman, J. I. Brauman, E. Rose, K. S. Suslick, *J. Am. Chem. Soc.* **1978**, *100*, 6769–6770.
26. J. Li, B. C. Noll, A. G. Oliver, C. E. Schulz, W. R. Scheidt, *J. Am. Chem. Soc.* **2013**, *135*, 15627–15641.
27. H. Nasri, Y. Wang, B. H. Huynh, W. R. Scheidt, *J. Am. Chem. Soc.* **1991**, *113*, 717–719.
28. C. Rovira, M. Parrinello, *Biophys. J.* **2000**, *78*, 93–100.
29. J. Li, Q. Peng, A. Barabanschikov, J. W. Pavlik, E. E. Alp, W. Sturhahn, J. Zhao, C. E. Schulz, J. T. Sage, W. R. Scheidt, *Chem. Eur. J.* **2011**, *17*, 11178–11185.
30. S. E. V. Phillips, *J. Mol. Biol.* **1980**, *142*, 531–554.
31. L. Pauling, *Nature* **1964**, *203*, 182–183.
32. J. J. Weiss, *Nature* **1964**, *202*, 83–84.
33. R. D. Harcourt, *J. Biol. Inorg. Chem.* **2014**, *19*, 113–123.
34. H. Chen, M. Ikeda-Saito, S. Shaik, *J. Am. Chem. Soc.* **2008**, *130*, 14778–14790.
35. S. A. Wilson, T. Kroll, R. A. Decreau, R. K. Hocking, M. Lundberg, B. Hedman, K. O. Hodgson, E. I. Solomon, *J. Am. Chem. Soc.* **2013**, *135*, 1124–1136.
36. H. Nasri, Y. Wang, B. H. Huynh, W. R. Scheidt, *J. Am. Chem. Soc.* **1991**, *113*, 717–719.
37. C. H. Welborn, D. Dolphin, B. R. James, *J. Am. Chem. Soc.* **1981**, *103*, 2869–2871.
38. J. G. Liu, Y. Shimizu, T. Ohta, Y. Naruta, *J. Am. Chem. Soc.* **2010**, *132*, 3672–3673.
39. E. McCandlish, A. R. Miksztal, M. Nappa, A. Q. Sprenger, J. S. Valentine, J. D. Stong, T. G. Spiro, *J. Am. Chem. Soc.* **1980**, *102*, 4268–4271.
40. J. N. Burstyn, J. A. Roe, A. R. Miksztal, B. A. Shaevitz, G. Lang, J. S. Valentine, *J. Am. Chem. Soc.* **1988**, *110*, 1382–1388.
41. P. Friant, J. Goulon, J. Fischer, L. Ricard, M. Schappacher, R. Weiss, M. Momenteau, *Nouv. J. Chim.* **1985**, *9*, 33–40.
42. K. Durr, B. P. Macpherson, R. Warratz, F. Hampel, F. Tuczek, M. Helmreich, N. Jux, I. Ivanovic-Burmazovic, *J. Am. Chem. Soc.* **2007**, *129*, 4217–4228.
43. E. E. Chufan, K. D. Karlin, *J. Am. Chem. Soc.* **2003**, *125*, 16160–16161.
44. T. Ohta, J. G. Liu, Y. Naruta, *Coord. Chem. Rev.* **2013**, *257*, 407–413.
45. K. Duerr, O. Troeppner, J. Olah, J. Li, A. Zahl, T. Drewello, N. Jux, J. N. Harvey, I. Ivanovic-Burmazovic, *Dalton Trans.* **2012**, *41*, 546–557.
46. K. Machii, Y. Watanabe, I. Morishima, *J. Am. Chem. Soc.* **1995**, *117*, 6691–6697.
47. W. Nam, Y. O. Ryu, W. J. Song, *J. Biol. Inorg. Chem.* **2004**, *9*, 654–660.

48. W. Nam, M. H. Lim, S. K. Moon, C. Kim, *J. Am. Chem. Soc.* **2000**, *122*, 10805–10809.
49. J. T. Groves, Y. Watanabe, *Inorg. Chem.* **1987**, *26*, 785–786.
50. R. D. Arasasingham, A. L. Balch, L. Latos-Grazynski, *J. Am. Chem. Soc.* **1987**, *109*, 5846–5847.
51. J. T. Groves, Y. Watanabe, *J. Am. Chem. Soc.* **1988**, *110*, 8443–8452.
52. J. T. Groves, Y. Watanabe, *Inorg. Chem.* **1987**, *26*, 785–786.
53. K. Tajima, *Inorg. Chim. Act.* **1989**, *163*, 115–122.
54. K. Tajima, M. Shigematsu, J. Jinno, K. Ishizu, H. Ohya-Nishiguchi, *J. Chem. Soc., Chem. Commun.* **1990**, 144.
55. K. Tajima, J. Jinno, K. Ishizu, H. Sakurai, H. Ohyanishiguchi, *Inorg. Chem.* **1989**, *28*, 709–715.
56. J. Stubbe, J. W. Kozarich, *Chem. Rev.* **1987**, *87*, 1107–1136.
57. M. C. R. Symons, R. L. Petersen, *Biochim. Biophys. Acta, Protein Struct.* **1978**, *535*, 241–246.
58. Z. Gasyna, *FEBS Lett.* **1979**, *106*, 213–218.
59. K. Tajima, S. Oka, T. Edo, S. Miyake, H. Mano, K. Mukai, H. Sakurai, K. Ishizu, *J. Chem. Soc., Chem. Commun.* **1995**, 1507–1508.
60. J. Annaraj, J. Cho, Y. M. Lee, S. Y. Kim, R. Latifi, S. P. de Visser, W. Nam, *Angew. Chem. Int. Ed.* **2009**, *48*, 4150–4153.
61. M. Selke, J. S. Valentine, *J. Am. Chem. Soc.* **1998**, *120*, 2652–2653.
62. A. Franke, C. Fertinger, R. van Eldik, *Chem. Eur. J.* **2012**, *18*, 6935–6949.
63. D. Mandon, R. Weiss, K. Jayaraj, A. Gold, J. Terner, E. Bill, A. X. Trautwein, *Inorg. Chem.* **1992**, *31*, 4404–4409.
64. B. Boso, G. Lang, T. J. McMurry, J. T. Groves, *J. Chem. Phys.* **1983**, *79*, 1122–1126.
65. H. Fujii, T. Yoshimura, H. Kamada, *Inorg. Chem.* **1996**, *35*, 2373–2377.
66. H. Fujii, *J. Am. Chem. Soc.* **1993**, *115*, 4641–4648.
67. H. Fujii, K. Ichikawa, *Inorg. Chem.* **1992**, *31*, 1110–1112.
68. K. Czarnecki, L. M. Proniewicz, H. Fujii, J. R. Kincaid, *J. Am. Chem. Soc.* **1996**, *118*, 4680– 4685
69. J. E. Penner-Hahn, E. K. Smith, T. J. McMurry, M. Renner, A. L. Balch, J. T. Groves, J. H. Dawson, K. O. Hodgson, *J. Am. Chem. Soc.* **1986**, *108*, 7819–7825.
70. S. Hashimoto, Y. Tatsuno, T. Kitagawa, *J. Am. Chem. Soc.* **1987**, *109*, 8096–8097.
71. S. Hashimoto, Y. Mizutani, Y. Tatsuno, T. Kitagawa, *J. Am. Chem. Soc.* **1991**, *113*, 6542– 6549.
72. S. E. J. Bell, P. R. Cooke, P. Inchley, D. R. Leanord, S. J. R. Lindsay, A. Robbins, *J. Chem. Soc., Perkin Trans. 2* **1991**, 549–559.
73. J. R. Kincaid, A. J. Schneider, K. J. Paeng, *J. Am. Chem. Soc.* **1989**, *111*, 735–737.
74. K. Czarnecki, S. Nimri, Z. Gross, L. M. Proniewicz, J. R. Kincaid, *J. Am. Chem. Soc.* **1996**, *118*, 2929–2935.
75. D.H. Chin, A. L. Balch, G. N. La Mar, *J. Am. Chem. Soc.* **1980**, *102*, 1446–1448.
76. D.H. Chin, G. N. La Mar, A. L. Balch, *J. Am. Chem. Soc.* **1980**, *102*, 5945–5947.
77. J. T. Groves, Z. Gross, M. K. Stern, *Inorg. Chem.* **1994**, *33*, 5065–5072.
78. W. Nam, H. J. Choi, H. J. Han, S. H. Cho, H. J. Lee, S.-Y. Han, *Chem. Commun.* **1999**, 387–388.
79. Y. M. Goh, W. Nam, *Inorg. Chem.* **1999**, *38*, 914–920.
80. N. A. Stephenson, A. T. Bell, *J. Am. Chem. Soc.* **2005**, *127*, 8635–8643.
81. W. Nam, H. J. Han, S.-Y. Oh, Y. J. Lee, M.-H. Choi, S.-Y. Han, C. Kim, S. K. Woo, W. Shin, *J. Am. Chem. Soc.* **2000**, *122*, 8677–8684.
82. A. Franke, C. Fertinger, R. van Eldik, *Angew. Chem. Int. Ed.* **2008**, *47*, 5238–5242.
83. N. A. Stephenson, A. T. Bell, *J. Mol. Catal. A: Chem.* **2007**, *275*, 54–62.
84. W. Nam, *Acc. Chem. Res.* **2007**, *40*, 522–531.
85. M. Wolak, R. van Eldik, *Chem. Eur. J.* **2007**, *13*, 4873–4883.
86. A. Franke, M. Wolak, R. van Eldik, *Chem. Eur. J.* **2009**, *15*, 10182–10198.

87. S. R. Bell, J. T. Groves, *J. Am. Chem. Soc.* **2009**, *131*, 9640–9641.
88. N. Jin, J. T. Groves, *J. Am. Chem. Soc.* **1999**, *121*, 2923–2924.
89. A. Altun, S. Shaik, W. Thiel, *J. Am. Chem. Soc.* **2007**, *129*, 8978–8987.
90. Y. J. Jeong, Y. Kang, A.-R. Han, Y.-M. Lee, H. Kotani, S. Fukuzumi, W. Nam, *Angew. Chem. Int. Ed.* **2008**, *47*, 7321–7324.
91. Z. Pan, M. Newcomb, *Inorg. Chem.* **2007**, *46*, 6767–6774.
92. J. T. Groves, G. A. McClusky, *J. Am. Chem. Soc.* **1976**, *98*, 859–861.
93. M. Costas, *Coord. Chem. Rev.* **2011**, *255*, 2912–2932.
94. J. T. Groves, *Proc. Natl. Acad. Sci. USA* **2003**, *100*, 3569–3574.
95. L. M. Hjelmeland, L. Aronow, J. R. Trudell, *Biochem. Biophys. Res. Commun.* **1977**, *76*, 541–549.
96. Z. Pan, J. H. Horner, M. Newcomb, *J. Am. Chem. Soc.* **2008**, *130*, 7776–7777.
97. T. G. Traylor, F. Xu, *J. Am. Chem. Soc.* **1988**, *110*, 1953–1958.
98. H. Hirao, D. Kumar, W. Thiel, S. Shaik, *J. Am. Chem. Soc.* **2005**, *127*, 13007–13018.
99. H. Hirao, D. Kumar, L. Que, Jr., S. Shaik, *J. Am. Chem. Soc.* **2006**, *128*, 8590–8606.
100. H. Hirao, L. Que, Jr., W. Nam, S. Shaik, *Chem. Eur. J.* **2008**, *14*, 1740–1756.
101. J. T. Groves, T. E. Nemo, *J. Am. Chem. Soc.* **1983**, *105*, 5786–5791.
102. T. Ueda, H. Kitagishi, K. Kano, *Inorg. Chem.* **2014**, *53*, 543–551.
103. Y. Watanabe, *J. Biol. Inorg. Chem.* **2001**, *6*, 846–856.
104. A. A. Guedes, A. C. M. A. Santos, M. D. Assis, *Kinet. Catal.* **2006**, *47*, 555–563.
105. W. Nam, H. J. Lee, S. Y. Oh, C. Kim, H. G. Jang, *J. Inorg. Biochem.* **2000**, *80*, 219–225.
106. K. Machii, Y. Watanabe, I. Morishima, *J. Am. Chem. Soc.* **1995**, *117*, 6691–6697.
107. Y. Watanabe, K. Yamaguchi, I. Morishima, K. Takehira, M. Shimizu, T. Hayakawa, H. Orita, *Inorg. Chem.* **1991**, *30*, 2581–2582.
108. W. Nam, M. H. Lim, H. J. Lee, C. Kim, *J. Am. Chem. Soc.* **2000**, *122*, 6641–6647.
109. W. Nam, S.-E. Park, I. K. Lim, M. H. Lim, J. Hong, J. Kim, *J. Am. Chem. Soc.* **2003**, *125*, 14674–14675.
110. N. A. Stephenson, A. T. Bell, *Inorg. Chem.* **2006**, *45*, 2758–2766.
111. D. Kumar, R. Latifi, S. Kumar, E. V. Rybak-Akimova, M. A. Sainna, S. P. de Visser, *Inorg. Chem.* **2013**, *52*, 7968–7979.
112. A. Takahashi, T. Kurahashi, H. Fujii, *Inorg. Chem.* **2011**, *50*, 6922–6928.
113. D. Kumar, B. Karamzadeh, G. N. Sastry, S. P. de Visser, *J. Am. Chem. Soc.* **2010**, *132*, 7656–7667.
114. D. Kumar, S. P. de Visser, S. Shaik, *Chem. Eur. J.* **2005**, *11*, 2825–2835.
115. T. Kamachi, Y. Shiota, T. Ohta, K. Yoshizawa, *Bull. Chem. Soc. Jpn.* **2003**, *76*, 721–732.
116. A. Takahashi, T. Kurahashi, H. Fujii, *Inorg. Chem.* **2009**, *48*, 2614–2625.
117. L. Castro, M. Buhl, *J. Chem. Theory Comput.* **2014**, *10*, 243–251.
118. Y. Kang, H. Chen, Y. J. Jeong, W. Lai, E. H. Bae, S. Shaik, W. Nam, *Chem. Eur. J.* **2009**, *15*, 10039–10046.
119. D. Kumar, G. N. Sastry, S. P. de Visser, *J. Phys. Chem. B.* **2012**, *116*, 718–730.
120. D. Kumar, R. Latifi, S. Kumar, E. V. Rybak-Akimova, M. A. Sainna, S. P. de Visser, *Inorg. Chem.* **2013**, *52*, 7968–7979.
121. D. Kumar, G. N. Sastry, S. P. de Visser, *Chem. Eur. J.* **2011**, *17*, 6196–6205.
122. E. Kim, E. E. Chufan, K. Kamaraj, K. D. Karlin, *Chem. Rev.* **2004**, *104*, 1077–1133.
123. K. D. Karlin, A. Nanthakumar, S. Fox, N. N. Murthy, N. Ravi, B. H. Huynh, R. D. Orosz, E. P. Day, *J. Am. Chem. Soc.* **1994**, *116*, 4753–4763.
124. H. V. Obias, G. P. F. van Strijdonck, D.-H. Lee, M. Ralle, N. J. Blackburn, K. D. Karlin, *J. Am. Chem. Soc.* **1998**, *120*, 9696–9697.
125. S. C. Lee, R. H. Holm, *J. Am. Chem. Soc.* **1993**, *115*, 11789–11798.
126. T. Chishiro, Y. Shimazaki, F. Tani, Y. Tachi, Y. Naruta, S. Karasawa, S. Hayami, Y. Maeda, *Angew. Chem. Int. Ed.* **2003**, *42*, 2788–2791.
127. E. E. Chufan, S. C. Puiu, K. D. Karlin, *Acc. Chem. Res.* **2007**, *40*, 563–572.

128. E. Kim, J. Shearer, S. Lu, P. Moenne-Loccoz, M. E. Helton, S. Kaderli, A. D. Zuberbühler, K. D. Karlin, *J. Am. Chem. Soc.* **2004**, *126*, 12716–12717.
129. E. Kim, M. E. Helton, S. Lu, P. Moënne-Loccoz, C. D. Incarvito, A. L. Rheingold, S. Kaderli, A. D. Zuberbühler, K. D. Karlin, *Inorg. Chem.* **2005**, *44*, 7014–7029.
130. Z. Halime, M. T. Kieber-Emmons, M. F. Qayyum, B. Mondal, T. Gandhi, S. C. Puiu, E. E. Chufan, A. A. N. Sarjeant, K. O. Hodgson, B. Hedman, E. I. Solomon, K. D. Karlin, *Inorg. Chem.* **2010**, *49*, 3629–3645.
131. M. T. Kieber-Emmons, M. F. Qayyum, Y. Li, Z. Halime, K. O. Hodgson, B. Hedman, K. D. Karlin, E. I. Solomon, *Angew. Chem. Int. Ed.* **2012**, *51*, 168–172.
132. M. T. Kieber-Emmons, Y. Li, Z. Halime, K. D. Karlin, E. I. Solomon, *Inorg. Chem.* **2011**, *50*, 11777–11786.
133. R. A. Ghiladi, E. E. Chufan, D. del Rio, E. I. Solomon, C. Krebs, B. H. Huynh, H. W. Huang, P. Moenne-Loccoz, S. Kaderli, M. Honecker, A. D. Zuberbühler, L. Marzilli, R. J. Cotter, K. D. Karlin, *Inorg. Chem.* **2007**, *46*, 3889–3902.
134. E. Kim, J. Shearer, S. Lu, P. Moenne-Loccoz, M. E. Helton, S. Kaderli, A. D. Zuberbühler, K. D. Karlin, *J. Am. Chem. Soc.* **2004**, *126*, 12716–12717.
135. K. D. Karlin, E. Kim, *Chem. Lett.* **2004**, *33*, 1226–1231.
136. Z. Halime, H. Kotani, Y. Li, S. Fukuzumi, K. D. Karlin, *Proc. Natl. Acad. Sci. USA* **2011**, *108*, 13990–13994.
137. J. P. Collman, R. A. Decreau, A. Dey, Y. Yang, *Proc. Natl. Acad. Sci. USA* **2009**, *106*, 4101–4105.
138. R. Boulatov, J. P. Collman, I. M. Shiryaeva, C. J. Sunderland, *J. Am. Chem. Soc.* **2002**, *124*, 11923–11935.
139. J. P. Collman, R. Boulatov, C. J. Sunderland, L. Fu, *Chem. Rev.* **2004**, *104*, 561–588.
140. J. P. Collman, R. A. Decreau, C. J. Sunderland, *Chem. Commun.* **2006**, 3894–3896.
141. J. P. Collman, R. A. Decreau, Y. Yan, J. Yoon, E. I. Solomon, *J. Am. Chem. Soc.* **2007**, *129*, 5794–5795.
142. J. P. Collman, N. K. Devaraj, R. A. Decreau, Y. Yang, Y.-L. Yan, W. Ebina, T. A. Eberspacher, C. E. D. Chidsey, *Science* **2007**, *315*, 1565–1568.
143. J. P. Collman, S. Ghosh, A. Dey, R. A. Decreau, Y. Yang, *J. Am. Chem. Soc.* **2009**, *131*, 5034–5035
144. E. I. Solomon, T. C. Brunold, M. I. Davis, J. N. Kemsley, S.-K. Lee, N. Lehnert, F. Neese, A. J. Skulan, Y.-S. Yang, J. Zhou, *Chem. Rev.* **2000**, *100*, 235–349.
145. M. Costas, M. P. Mehn, M. P. Jensen, L. Que, Jr. *Chem. Rev.* **2004**, *104*, 939–986.
146. M. M. Abu-Omar, A. Loaiza, N. Hontzeas, *Chem. Rev.* **2005**, *105*, 2227–2252.
147. B. J. Wallar, J. D. Lipscomb, *Chem. Rev.* **1996**, *96*, 2625–2658.
148. M.-H. Baik, M. Newcomb, R. A. Friesner, S. J. Lippard, *Chem. Rev.* **2003**, *103*, 2385–2419.
149. A. B. Tomter, G. Zoppellaro, N. H. Andersen, H.-P. Hersleth, M. Hammerstad, Å. K. Røhr, G. K. Sandvik, K. R. Strand, G. E. Nilsson, C. B. Bell III, A.-L. Barra, E. Blasco, L. Le Pape, E. I. Solomon, K. K. Andersson, *Coord. Chem. Rev.* **2013**, *257*, 3–26.
150. J. J. M. Bollinger, C. Krebs, *Curr. Opin. Chem. Biol.* **2007**, *11*, 151–158.
151. J. J. M. Bollinger, C. Krebs, *J. Inorg. Biochem.* **2006**, *100*, 586–605.
152. C. Krebs, D. F. Galonı́c, D. Fujimori, C. Walsh, J. Bollinger, *Acc. Chem. Res.* **2007**, *40*, 484–492.
153. E. Kovaleva, M. Neibergall, S. Chakrabarty, J. Lipscomb, *Acc. Chem. Res.* **2007**, *40*, 475–483.
154. M. Y. M. Pau, J. D. Lipscomb, E. I. Solomon, *Proc. Natl. Acad. Sci. USA* **2007**, *104*, 18355–18362.
155. P. C. A. Bruijnincx, G. v. Koten, R. J. M. Klein-Gebbink, *Chem. Soc. Rev.* **2008**, *37*, 2716–2744.
156. J. D. Lipscomb, *Curr. Opin. Struct. Biol.* **2008**, *18*, 644–649.
157. M. H. Sazinsky, S. J. Lippard, *Acc. Chem. Res.* **2006**, *39*, 558–566.
158. C. E. Tinberg, S. J. Lippard, *Acc. Chem. Res.* **2011**, *44*, 280–288.

159. L. Murray, S. Lippard, *Acc. Chem. Res.* **2007**, *40*, 466–474.
160. V. Guallar, B. F. Gherman, S. J. Lippard, R. A. Friesner, *Curr. Opin. Chem. Biol.* **2002**, *6*, 236–242.
161. L. Que, Jr. *Acc. Chem. Res.* **2007**, *40*, 493–500.
162. K. D. Koehntop, J. P. Emerson, L. Que, Jr. *J. Biol. Inorg. Chem.* **2005**, *10*, 87–93.
163. A. Mukherjee, M. A. Cranswick, M. Chakrabarti, T. K. Paine, K. Fujisawa, E. Münck, L. Que, Jr., *Inorg. Chem.* **2010**, *49*, 3618–3628.
164. T. D. H. Bugg, S. Ramaswamy, *Curr. Opin. Chem. Biol.* **2008**, *12*, 134–140.
165. A. Bassan, T. Borowski, P. E. M. Siegbahn, *Dalton Trans.* **2004**, 3153–3162.
166. E. G. Kovaleva, J. D. Lipscomb, *Science* **2007**, *316*, 453–457.
167. M. M. Mbughuni, M. Chakrabarti, J. A. Hayden, E. L. Bominaar, M. P. Hendrich, E. Münck, J. D. Lipscomb, *Proc. Natl. Acad. Sci. USA* **2010**, *107*, 16788–16793.
168. G. J. Christian, S. Ye, F. Neese, *Chem. Sci.* **2012**, *3*, 1600–1611.
169. M. R. A. Blomberg, P. E. M. Siegbahn, A. Bassan, *J. Biol. Inorg. Chem.* **2004**, *9*, 439–452.
170. W. A. Schenk, *Angew. Chem. Int. Ed.* **2000**, *39*, 3409–3411.
171. C. Loenarz, C. J. Schofield, *Nat. Chem. Biol.* **2008**, *4*, 152–156.
172. J. C. Price, E. W. Barr, B. Tirupati, J. M. Bollinger, C. Krebs, *Biochemistry* **2003**, *42*, 7497–7508.
173. P. J. Riggs-Gelasco, J. C. Price, R. B. Guyer, J. H. Brehm, E. W. Barr, J. J. M. Bollinger, C. Krebs, *J. Am. Chem. Soc.* **2004**, *126*, 8108–8109.
174. G. D. Straganz, B. Nidetzky, *Chem. Bio. Chem.* **2006**, *7*, 1536–1548.
175. J. D. Gardner, B. S. Pierce, B. G. Fox, T. C. Brunold, *Biochemistry* **2010**, *49*, 6033–6041.
176. J. A. Crawford, W. Li, B. S. Pierce, *Biochemistry* **2011**, *50*, 10241–10253.
177. D. Kumar, W. Thiel, S. P. de Visser, *J. Am. Chem. Soc.* **2011**, *133*, 3869–3882.
178. G. D. Straganz, A. Glieder, L. Brecker, D. W. Ribbons, W. Steiner, *Biochem. J.* **2003**, *369*, 573–581.
179. A. Decker, E. I. Solomon, *Curr. Opin. Chem. Biol.* **2005**, *9*, 152–163.
180. D. Kumar, H. Hirao, S. Shaik, P. M. Kozlowski, *J. Am. Chem. Soc.* **2006**, *128*, 16148 – 16158.
181. J. Hohenberger, K. Ray, K. Meyer, *Nature Comm.* **2012**, *3*, 720–713.
182. L. Shu, J. C. Nesheim, K. Kauffmann, E. Münck, J. D. Lipscomb, L. Que, Jr., *Science* **1997**, *275*, 515–518.
183. L. M. K. Dassama, A. Silakov, C. M. Krest, J. C. Calixto, C. Krebs, J. M. Bollinger, Jr, M. T. Green, *J. Am. Chem. Soc.* **2013**, *135*, 16758–16761.
184. S. V. Kryatov, E. V. Rybak-Akimova, S. Schindler, *Chem. Rev.* **2005**, *105*, 2175–2226.
185. A. Borovik, *Acc. Chem. Res.* **2005**, *38*, 54–61.
186. S. P. de Visser, J.-U. Rohde, Y.-M. Lee, J. Cho, W. Nam, *Coord. Chem. Rev.* **2013**, *257*, 381–393.
187. A. R. McDonald, L. Que, Jr., *Coord. Chem. Rev.* **2013**, *257*, 414–428.
188. X. Shan, L. Que, Jr., *Proc. Natl. Acad. Sci. USA* **2005**, *102*, 5340–5345.
189. M. Zhao, B. Helms, E. Slonkina, S. Friedle, D. Lee, J. DuBois, B. Hedman, K. O. Hodgson, J. M. J. Frechet, S. J. Lippard, *J. Am. Chem. Soc.* **2008**, *130*, 4352–4363.
190. H. Park, M. M. Bittner, J. S. Baus, S. V. Lindeman, A. T. Fiedler, *Inorg. Chem.* **2012**, *51*, 10279–10289.
191. S. Hong, Y.-M. Lee, W. Shin, S. Fukuzumi, W. Nam, *J. Am. Chem. Soc.* **2009**, *131*, 13910–13911.
192. Y.-M. Lee, S. Hong, Y. Morimoto, W. Shin, S. Fukuzumi, W. Nam, *J. Am. Chem. Soc.* **2010**, *132*, 10668–10670.
193. H. Chen, K.-B. Cho, W. Lai, W. Nam, S. Shaik, *J. Chem. Theor. Comp.* **2012**, *8*, 915–926.
194. D. Mandon, H. Jaafar, A. Thibon, *New J. Chem.* **2011**, *35*, 1986–2000.
195. J.-J. Girerd, F. Banse, A. J. Simaan, *Struct. Bond.* **2000**, *97*, 145–177.
196. M. Martinho, P. Dorlet, E. Rivière, A. Thibon, C. Ribal, F. Banse, J.-J. Girerd, *Chem. Eur. J.* **2008**, *14*, 3182–3188.
197. W. N. Oloo, K. K. Meier, Y. Wang, S. Shaik, E. Münck, L. Que, *Nature Commun.* **2014**, *5*, 1–9.
198. C. L. Sun, B. J. Li, Z. J. Shi, *Chem. Rev.* **2011**, *111*, 1293–1314.

199. M. C. White, *Science* **2012**, *335*, 807–809.
200. O. Y. Lyakin, R. V. Ottenbacher, K. P. Bryliakov, E. P. Talsi, *Top. Catal.* **2013**, *56*, 939– 949.
201. D. Barats, G. Leitus, R. Popovitz-Biro, L. J. W. Shimon, R. Neumann, *Angew. Chem. Int. Ed.* **2008**, *47*, 9908–9912.
202. S. Hong, Y.-M. Lee, K.-B. Cho, M. S. Seo, D. Song, J. Yoon, R. Garcia-Serres, M. Clémancey, T. Ogura, W. Shin, J.-M. Latour, W. Nam, *Chem. Sci.* **2013**, *5*, 156–162.
203. J.-U. Rohde, H.-H. In, M. H. Lim, W. W. Brennessel, M. R. Bukowski, A. Stubna, E. Münck, N. W., L. Que, Jr., *Science* **2003**, *299*, 1037–1039.
204. J. Hohenberger, K. Ray, K. Meyer, *Nature Comm.* **2012**, *3*, 720–713.
205. K. Ray, F. Heims, F. F. Pfaff, *Eur. J. Inorg. Chem.* **2013**, *2013*, 3784–3807.
206. Y. Morimoto, J. Park, T. Suenobu, Y.-M. Lee, W. Nam, S. Fukuzumi, *Inorg. Chem.* **2012**, *51*, 10025–10036.
207. J. Park, Y.-M. Lee, W. Nam, S. Fukuzumi, *J. Am. Chem. Soc.* **2013**, *135*, 5052–5061.
208. S. Fukuzumi, Y. Morimoto, H. Kotani, P. Naumov, Y.-M. Lee, W. Nam, *Nature Chem.* **2010**, *2*, 756–759.
209. M. Swart, *Chem. Commun.* **2013**, *49*, 6650–6652.
210. A. K. Vardhaman, P. Barman, S. Kumar, C. V. Sastri, D. Kumar, S. P. de Visser, *Angew. Chem. Int. Ed.* **2013**, *52*, 12288–12292.
211. C.-W. Tse, T. W.-S. Chow, Z. Guo, H. K. Lee, J.-S. Huang, C.-M. Che, *Angew. Chem. Int. Ed.* **2014**, *53*, 798–803.
212. I. Prat, J. S. Mathieson, M. Güell, X. Ribas, J. M. Luis, L. Cronin, M. Costas, *Nature Chem.* **2011**, *3*, 788–793.
213. R. Mayilmurugan, K. Visvaganesan, E. Suresh, M. Palaniandavar, *Inorg. Chem.* **2009**, *48*, 8771–8783.
214. S. Paria, P. Halder, T. K. Paine, *Inorg. Chem.* **2010**, *49*, 4518–4523.
215. R. Mayilmurugan, M. Sankaralingam, E. Suresh, M. Palaniandavar, *Dalton Trans.* **2010**, *39*, 9611–9625.
216. N. Anitha, M. Palaniandavar, *Dalton Trans.* **2010**, *39*, 1195–1197.
217. K. Sundaravel, E. Suresh, M. Palaniandavar, *Inorg. Chim. Acta* **2010**, *363*, 2768–2777.
218. K. Sundaravel, M. Sankaralingam, E. Suresh, M. Palaniandavar, *Dalton Trans.* **2011**, *40*, 8444–8458.
219. K. Sundaravel, E. Suresh, K. Saminathan, M. Palaniandavar, *Dalton Trans.* **2011**, *40*, 8092–8107.
220. N. Anitha, M. Palaniandavar, *Dalton Trans.* **2011**, *40*, 1888–1901.
221. M. Palaniandavar, K. Visvaganesan, *J. Chem. Sci.* **2011**, *123*, 145–162.
222. K. Visvaganesan, S. Ramachitra, M. Palaniandavar, *Inorg. Chim. Acta* **2011**, *378*, 87–94.
223. P. Halder, S. Paria, T. K. Paine, *Chem. Eur. J.* **2012**, *18*, 11778–11787.
224. S. Chatterjee, D. Sheet, T. K. Paine, *Chem. Commun.* **2013**, *49*, 10251–10253.
225. T. Váradi, J. S. Pap, M. Giorgi, L. Párkányi, T. Csay, G. Speier, J. Kaizer, *Inorg. Chem.* **2013**, *52*, 1559–1569.
226. E. H. Ha, R. Y. N. Ho, J. F. Kisiel, J. S. Valentine, *Inorg. Chem.* **1995**, *34*, 2265–2266.
227. M. P. Mehn, K. Fujisawa, E. L. Hegg, L. Que, Jr., *J. Am. Chem. Soc.* **2003**, *125*, 7828–7842.
228. T. K. Paine, H. Zheng, L. Que, Jr., *Inorg. Chem.* **2005**, *44*, 474–476.
229. A. Mukherjee, M. Martinho, E. L. Bominaar, E. Münck, L. Que, Jr., *Angew. Chem. Int. Ed.* **2009**, *48*, 1780–1783.
230. O. Das, S. Chatterjee, T. Paine, *J. Biol. Inorg. Chem.* **2013**, *18*, 401–410.
231. D. Sheet, P. Halder, T. K. Paine, *Angew. Chem. Int. Ed.* **2013**, *52*, 13314–13318.
232. L. Que, Jr., Y. Dong, *Acc. Chem. Res.* **1996**, *29*, 190–196.
233. E. Y. Tshuva, S. J. Lippard, *Chem. Rev.* **2004**, *104*, 987–1012.
234. S. Friedle, E. Reisner, S. J. Lippard, *Chem. Soc. Rev.* **2010**, *39*, 2768.
235. I. Siewert, C. Limberg, *Chem. Eur. J.* **2009**, *15*, 10316–10328.
236. L. H. Do, S. J. Lippard, *J. Inorg. Biochem.* **2011**, *105*, 1774–1785.
237. A. A. Shteinman, *Russ. Chem. Bull.* **2011**, *60*, 1290–1300.

238. J. S. Pap, M. A. Cranswick, É. Balogh-Hergovich, G. Baráth, M. Giorgi, G. T. Rohde, J. Kaizer, G. Speier, L. Que, Jr. *Eur. J. Inorg. Chem.* **2013**, *2013*, 3858–3866.
239. J. S. Pap, A. Draksharapu, M. Giorgi, W. R. Browne, J. Kaizer, G. Speier, *Chem. Commun.* **2014**, *50*, 1326–1329.
240. J. R. Frisch, V. V. Vu, M. Martinho, E. Münck, L. Que, Jr., *Inorg. Chem.* **2009**, *48*, 8325–8336.
241. R. L. Rardin, W. B. Tolman, S. J. Lippard, *New J. Chem.* **1991**, *15*, 417–430.
242. J. R. Frisch, R. McDonnell, E. V. Rybak-Akimova, L. Que, Jr., *Inorg. Chem.* **2013**, *52*, 2627–2636.
243. L. H. Do, T. Hayashi, P. Moënne-Loccoz, S. J. Lippard, *J. Am. Chem. Soc.* **2010**, *132*, 1273–1275.
244. C. E. Tinberg, S. J. Lippard, *Biochemistry* **2009**, *48*, 12145–12158.
245. M. A. Cranswick, K. K. Meier, X. Shan, A. Stubna, J. Kaizer, M. P. Mehn, E. Münck, L. Que, *Inorg. Chem.* **2012**, *51*, 10417–10426.
246. M. Kodera, Y. Kawahara, Y. Hitomi, T. Nomura, T. Ogura, Y. Kobayashi, *J. Am. Chem. Soc.* **2012**, *134*, 13236–13239.
247. S. A. Stoian, G. Xue, E. L. Bominaar, L. Que, Jr., E. Münck, *J. Am. Chem. Soc.* **2014**, *136*, 1545–1558.
248. G. Xue, A. T. Fiedler, M. Martinho, E. Münck, L. Que, Jr., *Proc. Natl. Acad. Sci. USA* **2008**, *105*, 20615–20620.
249. L. H. Do, G. Xue, L. Que, Jr., S. J. Lippard, *Inorg. Chem.* **2012**, *51*, 2393–2402
250. M. Martinho, G. Xue, A. T. Fiedler, L. Que, Jr., E. L. Bominaar, E. Münck, *J. Am. Chem. Soc.* **2009**, *131*, 5823–5830.
251. G. Xue, R. De Hont, E. Münck, L. Que, Jr., *Nat. Chem.* **2010**, *2*, 400–405.
252. E. I. Solomon, U. M. Sundaram, T. E. Machonkin, *Chem. Rev.* **1996**, *96*, 2563–2605.
253. J. P. Klinman, *Chem. Rev.* **1996**, *96*, 2541–2561.
254. *Multi-copper Oxidases*, Ed A. Messerschmidt, World Scientific, **1997**.
255. S. Itoh, in *Comprehensive Coordination Chemistry II, Vol. 8*, Eds J. A. McCleverty, T. J. Meyer, Elsevier, Amsterdam, The Netherlands, **2004**, pp. 369–393.
256. I. Bento, M. Carrondo, P. Lindley, *J. Biol. Inorg. Chem.* **2006**, *11*, 539–547.
257. M. A. Culpepper, A. C. Rosenzweig, *Crit. Rev. Biochem. Mol. Biol.* **2012**, *47*, 483–492.
258. A. C. Rosenzweig, M. H. Sazinsky, *Curr. Opin. Struct. Biol.* **2006**, *16*, 729–735.
259. *Handbook of Metalloproteins: 2 Volume Set*, Eds A. Messerschmidt, R. Huber, T. Poulos, K. Wieghardt, Wiley, **2001**.
260. E. I. Solomon, J. W. Ginsbach, D. E. Heppner, M. T. Kieber-Emmons, C. H. Kjaergaard, P. J. Smeets, L. Tian, J. S. Woertink, *Faraday Disc.* **2010**, *148*, 11.
261. L. M. Mirica, X. Ottenwaelder, T. D. P. Stack, *Chem. Rev.* **2004**, *104*, 1013–1045.
262. E. A. Lewis, W. B. Tolman, *Chem. Rev.* **2004**, *104*, 1047–1076.
263. L. Hatcher, K. D. Karlin, *J. Biol. Inorg. Chem.* **2004**, *9*, 669–683.
264. L. Q. Hatcher, K. D. Karlin, *Adv. Inorg. Chem.* **2006**, *58*, 131–184.
265. S. Itoh, S. Fukuzumi, *Bull. Chem. Soc. Jpn.* **2002**, *75*, 2081–2095.
266. M. R. Halvagar, D. J. Salmon, W. B. Tolman, in *Comprehensive Inorganic Chemistry II, Vol. 3*, Eds J. Reedijk, K. Poeppelmeier, Elsevier, Oxford, UK, **2013**, pp. 455–486.
267. T. Kamachi, N. Kihara, Y. Shiota, K. Yoshizawa, *Inorg. Chem.* **2005**, *44*, 4226–4236.
268. J. W. Whittaker, *Arch. Biochem. Biophys.* **2005**, *433*, 227–239.
269. J. W. Whittaker, *Chem. Rev.* **2003**, *103*, 2347–2363.
270. A. Mukherjee, V. V. Smirnov, M. P. Lanci, D. E. Brown, E. M. Shepard, D. M. Dooley, J. P. Roth, *J. Am. Chem. Soc.* **2008**, *130*, 9459–9473.
271. A. Crespo, M. A. Marti, A. E. Roitberg, L. M. Amzel, D. A. Estrin, *J. Am. Chem. Soc.* **2006**, *128*, 12817–12828.
272. J. P. Klinman, *J. Biol. Chem.* **2006**, *281*, 3013–3016.
273. P. Chen, J. Bell, B. A. Eipper, E. I. Solomon, *Biochemistry* **2004**, *43*, 5735–5747.

274. K. Rudzka, D. M. Moreno, B. Eipper, R. Mains, D. A. Estrin, L. M. Amzel, *J. Biol. Inorg. Chem.* **2012**, *18*, 223–232.
275. D. Schroder, M. C. Holthausen, H. Schwarz, *J. Phys. Chem. B* **2004**, *108*, 14407–14416.
276. N. Dietl, C. van der Linde, M. Schlangen, M. K. Beyer, H. Schwarz, *Angew. Chem. Int. Ed.* **2011**, *50*, 4966–4969.
277. K. Yoshizawa, N. Kihara, T. Kamachi, Y. Shiota, *Inorg. Chem.* **2006**, *45*, 3034–3041.
278. C. M. Wilmot, J. Hajdu, M. J. Mcpherson, P. F. Knowles, S. E. V. Phillips, *Science* **1999**, *286*, 1724–1728.
279. S. T. Prigge, B. A. Eipper, R. E. Mains, L. M. Amzel, *Science* **2004**, *304*, 864–867.
280. N. Dietl, M. Schlangen, H. Schwarz, *Angew. Chem. Int. Ed.* **2012**, *51*, 5544–5555.
281. C. J. Cramer, W. B. Tolman, *Acc. Chem. Res.* **2007**, *40*, 601–608.
282. R. Sarangi, N. Aboelella, K. Fujisawa, W. B. Tolman, B. Hedman, K. O. Hodgson, E. I. Solomon, *J. Am. Chem. Soc.* **2006**, *128*, 8286–8296.
283. J. S. Woertink, L. Tian, D. Maiti, H. R. Lucas, R. A. Himes, K. D. Karlin, F. Neese, C. Würtele, M. C. Holthausen, E. Bill, J. Sundermeyer, S. Schindler, E. I. Solomon, *Inorg. Chem.* **2010**, *49*, 9450–9459.
284. C. Würtele, E. Gaoutchenova, K. Harms, M. C. Holthausen, J. Sundermeyer, S. Schindler, *Angew. Chem. Int. Ed.* **2006**, *45*, 3867–3869.
285. Y. Kobayashi, K. Ohkubo, T. Nomura, M. Kubo, N. Fujieda, H. Sugimoto, S. Fukuzumi, K. Goto, T. Ogura, S. Itoh, *Eur. J. Inorg. Chem.* **2012**, *2012*, 4574–4578.
286. A. Kunishita, M. Kubo, H. Sugimoto, T. Ogura, K. Sato, T. Takui, S. Itoh, *J. Am. Chem. Soc.* **2009**, *131*, 2788–2789.
287. A. Kunishita, M. Z. Ertem, Y. Okubo, T. Tano, H. Sugimoto, K. Ohkubo, N. Fujieda, S. Fukuzumi, C. J. Cramer, S. Itoh, *Inorg. Chem.* **2012**, *51*, 9465–9480.
288. D. Maiti, H. C. Fry, J. S. Woertink, M. A. Vance, E. I. Solomon, K. D. Karlin, *J. Am. Chem. Soc.* **2007**, *129*, 264–265.
289. R. L. Peterson, R. A. Himes, H. Kotani, T. Suenobu, L. Tian, M. A. Siegler, E. I. Solomon, S. Fukuzumi, K. D. Karlin, *J. Am. Chem. Soc.* **2011**, *133*, 1702–1705.
290. P. J. Donoghue, A. K. Gupta, D. W. Boyce, C. J. Cramer, W. B. Tolman, *J. Am. Chem. Soc.* **2010**, *132*, 15869–15871.
291. J. W. Ginsbach, R. L. Peterson, R. E. Cowley, K. D. Karlin, E. I. Solomon, *Inorg. Chem.* **2013**, *52*, 12872–12874.
292. T. Fujii, S. Yamaguchi, Y. Funahashi, T. Ozawa, T. Tosha, T. Kitagawa, H. Masuda, *Chem. Commun.* **2006**, 4428–4430.
293. M. Harata, K. Jitsukawa, H. Masuda, H. Einaga, *J. Am. Chem. Soc.* **1994**, *116*, 10817–10818.
294. L. M. Berreau, S. Mahapatra, J. A. Halfen, V. G. Young, Jr., W. B. Tolman, *Inorg. Chem.* **1996**, *35*, 6339–6342.
295. D. Maiti, D.-H. Lee, K. Gaoutchenova, C. Würtele, M. C. Holthausen, A. A. N. Sarjeant, J. Sundermeyer, S. Schindler, K. D. Karlin, *Angew. Chem. Int. Ed.* **2008**, *47*, 82–85.
296. A. Poater, L. Cavallo, *Inorg. Chem.* **2009**, *48*, 4062–4066.
297. R. L. Peterson, J. W. Ginsbach, R. E. Cowley, M. F. Qayyum, R. A. Himes, M. A. Siegler, C. D. Moore, B. Hedman, K. O. Hodgson, S. Fukuzumi, E. I. Solomon, K. D. Karlin, *J. Am. Chem. Soc.* **2013**, *135*, 16454–16467.
298. S. Kakuda, R. L. Peterson, K. Ohkubo, K. D. Karlin, S. Fukuzumi, *J. Am. Chem. Soc.* **2013**, *135*, 6513–6522.
299. D. Maiti, A. A. Narducci Sarjeant, K. D. Karlin, *Inorg. Chem.* **2008**, *47*, 8736–8747, and references cited therein.
300. T. Ohta, T. Tachiyama, K. Yoshizawa, T. Yamabe, T. Uchida, T. Kitagawa, *Inorg. Chem.* **2000**, *39*, 4358–4369.
301. A. Wada, M. Harata, K. Hasegawa, K. Jitsukawa, H. Masuda, M. Mukai, T. Kitagawa, H. Einaga, *Angew. Chem. Int. Ed.* **1998**, *37*, 798–799.
302. S. Yamaguchi, A. Wada, S. Nagatomo, T. Kitagawa, K. Jitsukawa, H. Masuda, *Chem. Lett.* **2004**, *33*, 1556–1557.

303. S. Yamaguchi, S. Nagatomo, T. Kitagawa, Y. Funahashi, T. Ozawa, K. Jitsukawa, H. Masuda, *Inorg. Chem.* **2003**, *42*, 6968–6970.
304. T. Fujii, A. Naito, S. Yamaguchi, A. Wada, Y. Funahashi, K. Jitsukawa, S. Nagatomo, T. Kitagawa, H. Masuda, *Chem. Commun.* **2003**, 2700–2701.
305. D. Maiti, A. A. Narducci Sarjeant, K. D. Karlin, *J. Am. Chem. Soc.* **2007**, *129*, 6720–6721.
306. D. Maiti, H. R. Lucas, A. A. N. Sarjeant, K. D. Karlin, *J. Am. Chem. Soc.* **2007**, *129*, 6998–6999.
307. T. Kamachi, Y.-M. Lee, T. Nishimi, J. Cho, K. Yoshizawa, W. Nam, *J. Phys. Chem. A* **2008**, *112*, 13102–13108.
308. A. Kunishita, H. Ishimaru, S. Nakashima, T. Ogura, S. Itoh, *J. Am. Chem. Soc.* **2008**, *130*, 4244–4245.
309. T. Tano, M. Z. Ertem, S. Yamaguchi, A. Kunishita, H. Sugimoto, N. Fujieda, T. Ogura, C. J. Cramer, S. Itoh, *Dalton Trans.* **2011**, *40*, 10326.
310. S. Hong, S. M. Huber, L. Gagliardi, C. J. Cramer, W. B. Tolman, *J. Am. Chem. Soc.* **2007**, *129*, 14190–14192.
311. S. M. Huber, M. Z. Ertem, F. Aquilante, L. Gagliardi, W. B. Tolman, C. J. Cramer, *Chem. Eur. J.* **2009**, *15*, 4886–4895.
312. D. Das, Y.-M. Lee, K. Ohkubo, W. Nam, K. D. Karlin, S. Fukuzumi, *J. Am. Chem. Soc.* **2013**, *135*, 2825–2834.
313. M. Cvetkovic, S. R. Batten, B. Moubaraki, K. S. Murray, L. Spiccia, *Inorg. Chim. Acta* **2001**, *324*, 131–140.
314. M. Reglier, E. Amadei, R. Tadayoni, B. Waegell, *J. Chem. Soc., Chem. Commun.* **1989**, 447–450.
315. T. Kametani, M. Ihara, *J. Chem. Soc., Perkin Trans. I* **1980**, 629–632.
316. P. Capdevielle, D. Sparfel, J. Baranne-Lafont, N. K. Cuong, M. Maumy, *J. Chem. Soc., Chem. Commun.* **1990**, 565–566.
317. O. Reinaud, P. Capdevielle, M. Maumy, *J. Chem. Soc., Chem. Commun.* **1990**, 566–568.
318. O. Reinaud, P. Capdevielle, M. Maumy, *J. Mol. Cat.* **1991**, *68*, L13–L15.
319. G. Rousselet, P. Capdevielle, M. Maumy, *Tetrahedron Lett.* **1995**, *36*, 4999–5002.
320. W. Buijs, P. Comba, D. Corneli, H. Pritzkow, *J. Organomet. Chem.* **2002**, *641*, 71–80.
321. P. Comba, S. Knoppe, B. Martin, G. Rajaraman, C. Rolli, B. Shapiro, T. Stork, *Chem. Eur. J.* **2008**, *14*, 344–357.
322. S. Hong, A. K. Gupta, W. B. Tolman, *Inorg. Chem.* **2009**, *48*, 6323–6325.
323. P. J. Donoghue, J. Tehranchi, C. J. Cramer, R. Sarangi, E. I. Solomon, W. B. Tolman, *J. Am. Chem. Soc.* **2011**, *133*, 17602–17605.
324. J. Tehranchi, P. J. Donoghue, C. J. Cramer, W. B. Tolman, *Eur. J. Inorg. Chem.* **2013**, 4077–4084.
325. K. A. Magnus, H. Ton-That, J. E. Carpenter, *Chem. Rev.* **1994**, *94*, 727–735.
326. M. E. Cuff, K. I. miller, K. E. v. Holde, W. A. Hendrickson, *J. Mol. Biol.* **1998**, *278*, 855–870.
327. Y. Matoba, T. Kumagai, A. Yamamoto, H. Yoshitsu, M. Sugiyama, *J. Biol. Chem.* **2006**, *281*, 8981–8990.
328. A. Rompel, H. Fischer, D. Meiwes, K. B. Karentzopoulos, R. Dillinger, F. Tuczek, H. Witzel, B. Krebs, *J. Biol. Inorg. Chem.* **1999**, *4*, 56– 63.
329. N. Hakulinen, C. Gasparetti, H. Kaljunen, K. Kruus, J. Rouvinen, *J. Biol. Inorg. Chem.* **2013**, *18*, 917–929.
330. M. A. Culpepper, A. C. Rosenzweig, *Crit. Rev. Biochem. Mol. Biol.* **2012**, *47*, 483–492.
331. S. I. Chan, S. S. F. Yu, *Acc. Chem. Res.* **2008**, *41*, 969–979.
332. M. A. Culpepper, G. E. Cutsail III, B. M. Hoffman, A. C. Rosenzweig, *J. Am. Chem. Soc.* **2012**, *134*, 7640–7643.
333. K. Yoshizawa, Y. Shiota, *J. Am. Chem. Soc.* **2006**, *128*, 9873–9881.
334. Y. Shiota, K. Yoshizawa, *Inorg. Chem.* **2009**, *48*, 838–845.
335. R. A. Himes, K. D. Karlin, *Curr. Opin. Chem. Biol.* **2009**, *13*, 119–131.
336. P. Vanelderen, R. G. Hadt, P. J. Smeets, E. I. Solomon, R. A. Schoonheydt, B. F. Sels, *J. Catal.* **2011**, *284*, 157–164.

337. P. Vanelderen, J. Vancauwenbergh, B. F. Sels, R. A. Schoonheydt, *Coord. Chem. Rev.* **2013**, *257*, 483–494.
338. L. Que, Jr., W. B. Tolman, *Angew. Chem. Int. Ed.* **2002**, *41*, 1114–1137.
339. L. Hatcher, K. D. Karlin, *J. Biol. Inorg. Chem.* **2004**, *9*, 669–683.
340. M. Rolff, J. Schottenheim, H. Decker, F. Tuczek, *Chem. Soc. Rev.* **2011**, *40*, 4077–4098.
341. S. Fukuzumi, K. D. Karlin, *Coord. Chem. Rev.* **2013**, *257*, 187–195.
342. A. Jozwiuk, E. A. Ünal, S. Leopold, J. P. Boyd, M. Haryono, N. Kurowski, F. V. Escobar, P. Hildebrandt, J. Lach, F. W. Heinemann, D. Wiedemann, E. Irran, A. Grohmann, *Eur. J. Inorg. Chem.* **2012**, 3000–3013.
343. C. Citek, C. T. Lyons, E. C. Wasinger, T. D. P. Stack, *Nature Chem.* **2012**, *4*, 317–322.
344. J. Matsumoto, Y. Kajita, H. Masuda, *Eur. J. Inorg. Chem.* **2012**, 4149–4158.
345. Y. Kajita, H. Arii, T. Saito, Y. Saito, S. Nagatomo, T. Kitagawa, Y. Funahashi, T. Ozawa, H. Masuda, *Inorg. Chem.* **2007**, *46*, 3322–3335.
346. A. Hoffmann, C. Citek, S. Binder, A. Goos, M. Rübhausen, O. Troeppner, I. Ivanovic-Burmazovic, E. C. Wasinger, T. D. P. Stack, S. Herres-Pawlis, *Angew. Chem. Int. Ed.* **2013**, *52*, 5398–5401.
347. I. Garcia-Bosch, X. Ribas, M. Costas, *Chem. Eur. J.* **2012**, *18*, 2113–2122.
348. P. Comba, C. Haaf, S. Helmle, K. D. Karlin, S. Pandian, A. Waleska, *Inorg. Chem.* **2012**, *51*, 2841–2851.
349. P. Comba, B. Martin, A. Muruganantham, J. Straub, *Inorg. Chem.* **2012**, *51*, 9214–9225.
350. B. F. Gherman, C. J. Cramer, *Coord. Chem. Rev.* **2009**, *253*, 723–753.
351. D. G. Liakos, F. Neese, *J. Chem. Theor. Comp.* **2011**, *7*, 1511–1523.
352. M. Rohrmüller, S. Herres-Pawlis, M. Witte, W. G. Schmidt, *J. Comp. Chem.* **2013**, *34*, 1035–1045.
353. Y.-F. Liu, J.-G. Yu, P. E. M. Siegbahn, M. R. A. Blomberg, *Chem. Eur. J.* **2013**, *19*, 1942–1954.
354. L. M. Mirica, M. Vance, D. J. Rudd, B. Hedman, K. O. Hodgson, E. I. Solomon, T. D. P. Stack, *Science* **2005**, *308*, 1890–1892.
355. B. T. Op't Holt, M. A. Vance, L. M. Mirica, D. E. Heppner, T. D. P. Stack, E. I. Solomon, *J. Am. Chem. Soc.* **2009**, *131*, 6421–6438.
356. S. Mandal, J. Mukherjee, F. Lloret, R. Mukherjee, *Inorg. Chem.* **2012**, *51*, 13148–13161.
357. M. F. Qayyum, R. Sarangi, K. Fujisawa, T. D. P. Stack, K. D. Karlin, K. O. Hodgson, B. Hedman, E. I. Solomon, *J. Am. Chem. Soc.* **2013**, *135*, 17417–17431.
358. P. Haack, C. Limberg, *Angew. Chem. Int. Ed.*, **2014**, *53*, 4282–4293.
359. K. D. Karlin, Y. Gultneh, J. C. Hayes, J. Zubieta, *Inorg. Chem.* **1984**, *23*, 519–521.
360. I. Sanyal, M. Mahroof-Tahir, M. S. Nasir, P. Ghosh, B. I. Cohen, Y. Gultneh, R. W. Cruse, A. Farooq, K. D. Karlin, S. Liu, J. Zubieta, *Inorg. Chem.* **1992**, *31*, 4322–4332.
361. H. V. Obias, Y. Lin, N. N. Murthy, E. Pidcock, E. I. Solomon, M. Ralle, N. J. Blackburn, Y.-M. Neuhold, A. D. Zuberbühler, K. D. Karlin, *J. Am. Chem. Soc.* **1998**, *120*, 12960–12961.
362. N. Kitajima, T. Koda, S. Hashimoto, T. Kitagawa, Y. Moro-oka, *J. Am. Chem. Soc.* **1991**, *113*, 5664–5671.
363. P. Haack, C. Limberg, K. Ray, B. Braun, U. Kuhlmann, P. Hildebrandt, C. Herwig, *Inorg. Chem.* **2011**, *50*, 2133–2142.
364. P. Haack, A. Kärgel, C. Greco, J. Dokic, B. Braun, F. F. Pfaff, S. Mebs, K. Ray, C. Limberg, *J. Am. Chem. Soc.* **2013**, *135*, 16148–16160.
365. P. P. Y. Chen, R. B. G. Yang, J. C. M. Lee, S. I. Chan, *Proc. Natl. Acad. Sci. USA* **2007**, *104*, 14570–14575.
366. A. M. Kirillov, M. V. Kirillova, A. J. L. Pombeiro, *Coord. Chem. Rev.* **2012**, *256*, 2741–2759.
367. A. M. Kirillov, M. V. Kirillova, A. J. L. Pombeiro, *Adv. Inorg. Chem.* **2013**, *65*, 1–31.
368. S. Maji, J. C. M. Lee, Y.-J. Lu, C.-L. Chen, M.-C. Hung, P. P. Y. Chen, S. S. F. Yu, S. I. Chan, *Chem. Eur. J.* **2012**, *18*, 3955–3968.

369. P. Nagababu, S. Maji, M. P. Kumar, P. P. Y. Chen, S. S. F. Yu, S. I. Chan, *Adv. Synth. Catal.* **2012**, *354*, 3275–3282.
370. S. I. Chan, Y.-J. Lu, P. Nagababu, S. Maji, M.-C. Hung, M. M. Lee, I. J. Hsu, P. D. Minh, J. C. H. Lai, K. Y. Ng, S. Ramalingam, S. S. F. Yu, M. K. Chan, *Angew. Chem. Int. Ed.* **2013**, *52*, 3731–3735.
371. A. P. Cole, D. E. Root, P. Mukherjee, E. I. Solomon, T. D. P. Stack, *Science* **1996**, *273*, 1848–1850.
372. D. E. Root, M. J. Henson, T. Machonkin, P. Mukherjee, T. D. P. Stack, E. I. Solomon, *J. Am. Chem. Soc.* **1998**, *120*, 4982–4990.
373. T. Machonkin, P. Mukherjee, M. Henson, T. D. P. Stack, E. I. Solomon, *Inorg. Chim. Acta* **2002**, *341*, 39–44.
374. M. Taki, S. Teramae, S. Nagatomo, Y. Tachiu, T. Kitagawa, S. Itoh, S. Fukuzumi, *J. Am. Chem. Soc.* **2002**, *124*, 6367–6377.
375. A. K. Gupta, W. B. Tolman, *Inorg. Chem.* **2012**, *51*, 1881–1888.
376. D. Lionetti, M. W. Day, T. Agapie, *Chem. Sci.* **2013**, *4*, 785–790.

Chapter 6
Methane Monooxygenase: Functionalizing Methane at Iron and Copper

Matthew H. Sazinsky and Stephen J. Lippard

Contents

M.H. Sazinsky
Department of Chemistry, Pomona College, Claremont, CA 91711, USA
e-mail: matthew.sazinsky@pomona.edu

S.J. Lippard (✉)
Department of Chemistry, Massachusetts Institute of Technology,
Cambridge, MA 02139, USA
e-mail: lippard@mit.edu

P.M.H. Kroneck, M.E. Sosa Torres (eds.), *Sustaining Life on Planet Earth: Metalloenzymes Mastering Dioxygen and Other Chewy Gases*, Metal Ions in Life Sciences 15, DOI 10.1007/978-3-319-12415-5_6

Abstract Methane monooxygenases (MMOs) catalyze the conversion of methane to methanol as the first committed step in the assimilation of this hydrocarbon into biomass and energy by methanotrophs, thus playing a significant role in the biogeochemistry of this potent greenhouse gas. Two distinct enzymes, a copper-dependent membrane protein, particulate methane monooxygenase (pMMO), and an iron-dependent cytosolic protein, soluble methane monooxygenase (sMMO), carry out this transformation using large protein scaffolds that help to facilitate the timely transport of hydrocarbon, O_2, proton, and electron substrates to buried dimetallic active sites. For both enzymes, reaction of the reduced metal centers with O_2 leads to intermediates that activate the relatively inert C–H bonds of hydrocarbons to yield oxidized products. Among synthetic and biological catalysts, MMOs are unique because they are the only ones known to hydroxylate methane at ambient temperatures. As a need for new industrial catalysts and green chemical transformations increases, understanding how the different MMO metal centers efficiently accomplish this challenging chemistry has become the focus of intense study. This chapter examines current understanding of the sMMO and pMMO protein structures, their methods for substrate channeling, and mechanisms for the dimetallic activation of O_2 and C–H bonds.

Keywords bacterial multicomponent monooxygenase • dicopper • diiron • mechanism • particulate methane monooxygenase • protein complexes • soluble methane monooxygenase

Please cite as: *Met. Ions Life Sci.* 15 (2015) 205–256

1 Introduction

Methanogens living in anaerobic environments produce greater than 200–250 million metric tons of methane per year as a byproduct of catabolizing biomass [1]. Taking advantage of this unique waste product, methanotrophs evolved to survive at the interface between anaerobic and aerobic worlds where they require both methane and O_2 for the metabolic assimilation into macromolecules and stored energy [2]. By using methane as a primary carbon and energy source, methanotrophs limit the escape of this potent greenhouse gas into the atmosphere and thereby lessen its global environmental impact [2, 3]. Responsible for catalyzing the conversion of methane to methanol and initiating biomass assimilation in methanotrophs are methane monooxygenases (MMOs), of which two distinct forms exist. The most prevalent, and least understood, form is the copper-utilizing membrane-bound or particulate MMO (pMMO), which can account for up to 20 % of the total protein in methanotrophs [4–7]. The iron-containing cytosolic or soluble MMO (sMMO), found only in some methanotrophic species, is better characterized [8–11]. Both enzymes have received significant attention because they routinely carry out one of the most difficult chemistries in nature, the breaking of a relatively inert and high energy C–H bond (104 kcal/mol) at ambient temperatures. Other iron, copper, and cytochrome P450 monooxygenases that readily hydroxylate larger, more reactive hydrocarbons cannot oxidize methane, making MMOs unique catalysts.

Isolated from a variety of different environments, most acidophilic, alkaliphilic, psychrophilic, thermophilic, and mesophilic methanotrophs except for the *Methyocella* genus contain genes for pMMO, suggesting that it is the predominate MMO in the environment [2, 12]. Some species, which include extensively studied *Methylococcus capsulatus* (Bath) and *Methylosinus trichosporium* OB3b, differentially express both pMMO and sMMO based on copper availability [13–16]. High environmental copper suppresses sMMO translation in these organisms and induces the expression of pMMO as well as the synthesis of intracytoplasmic membranes into which pMMO is incorporated. Conversely, low copper concentrations lead to sMMO expression and synthesis of methanobactins, copper scavenging compounds or chalkophores, with properties and functions like those of the widely studied siderophores [17–20].

Structurally and mechanistically, there are few similarities between pMMO and sMMO. However, both of these enzymes share one important task, the efficient generation of oxidized hydrocarbon products by coordinating the reactivity of four substrates (hydrocarbon, oxygen, electrons, and protons) at a highly reactive dimetallic center. Methane is the primary MMO substrate that can most readily be funneled into lower metabolic pathways for the purposes of carbon assimilation and energy production via the serine and ribulose monophosphate pathways (Figure 1) [2].

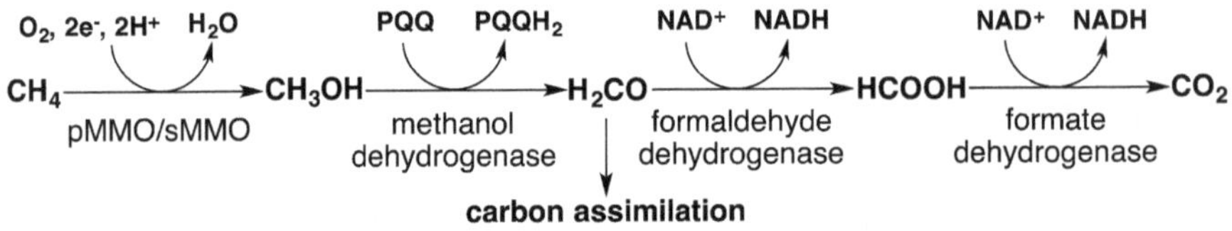

Figure 1 Metabolic pathway for hydrocarbon assimilation in methanotrophs.

pMMO and sMMO, however, are capable of regio- and stereospecifically hydroxylating a wide variety of hydrocarbon substrates. Whereas sMMO can act on linear and branched alkanes and alkenes of up to eight carbons in length, in addition to aromatic, heterocyclic, and halogenated compounds [21–26], pMMO is more selective toward alternative substrates with preferences for alkanes and alkenes of up to five and four carbons, respectively [3, 27–29]. Because they are such powerful catalysts, the iron and copper centers of sMMO and pMMO, respectively, have been investigated intensely for their potential wide-ranging applications to synthesis and environmental remediation [3, 22]. This chapter focuses on the current knowledge of the pMMO and sMMO systems and how they carry out their respective enzymatic functions.

2 Particulate Methane Monooxygenase

pMMO comprises three polypeptide chains, a 45-kDa α-subunit (pmoB), a 26-kDa β-subunit (pmoA), and a 23-kDa γ-subunit (pmoC), that assemble into a homotrimer with an $(\alpha\beta\gamma)_3$ configuration (Figure 2 and Figure 3 below) [30]. Because pMMO is vital to the survival of most methanotrophs, several species contain duplicate copies of the *pmoCAB* genes [31]. The closely related pMMO homolog, ammonia monooxygenase (AMO), is the only other known enzyme that can oxidize methane, and its composition is similar to that of pMMO [32]. Over the years, different approaches by several laboratories toward the purification and biochemical and spectroscopic characterization of pMMO have generated much debate and confusion in the literature about the metal content, metal type, subunit stoichiometry, and active site location. For example, purifications of pMMO from *M. capsulatus* (Bath) and *M. trichosporium* OB3b have generated results suggesting that it contains 2, 2–3, 8–10, and 15–20 coppers per αβγ protomer as well as significant quantities of iron, ~0.75–2.5 ions/protomer [14, 33–39].

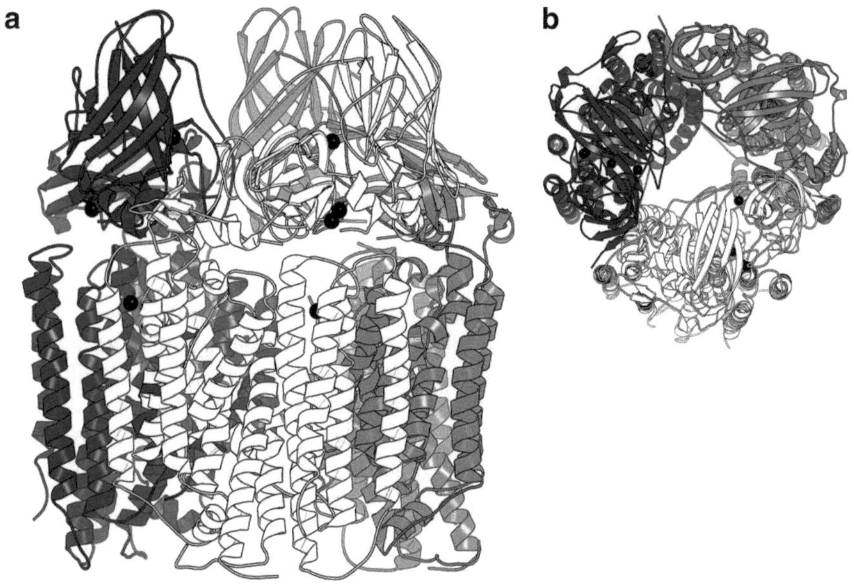

Figure 2 (**a**) Global structure of pMMO (PDB entry 3RGB) depicting each of the three αβγ protomers. (**b**) Top down view of the cytosolic pMMO pmoB cupredoxin-like domains depicting the channel through the protein. The metal centers are shown as black spheres.

From spectroscopic investigations, mononuclear, dinuclear, and trinuclear copper sites, as well as a diiron center, have all been proposed for the pMMO active site [5, 12, 40]. The inherent complexities with working on a membrane protein coupled with the difficulty of identifying the nature of the pMMO active site explain why our understanding of pMMO has lagged behind that of sMMO. Fortunately, significant strides have been made recently that bring clarity to the problem and a better understanding of pMMO. The following sections are not intended to be an exhaustive treatment of the pMMO literature, but instead aim to present the most reasonable and clear picture to date as to how this protein operates.

2.1 Architecture

The first crystal structure of pMMO from *Methylococcus capsulatus* (Bath) was solved to 2.8 Å resolution [30, 40]. The structure confirmed that pMMO is an $(\alpha\beta\gamma)_3$ homotrimer in which each αβγ protomer contains a single copy of the pmoB, pmoA, and pmoC subunits arranged as a 3-fold cylinder with a channel through the middle of the barrel (Figure 2). The pmoC and pmoA subunits comprise the bulk of the transmembrane region, which is ~45 Å in length. The N- and C-terminal cytosolic domains of the pmoB subunit have a cupredoxin-like β-barrel fold and are linked by two transmembrane helices. The global structure is highly consistent with earlier cryo-electron microscopy images at 23 Å of *M. capsulatus* (Bath) pMMO [41]. The electrostatic surfaces of the pmoB cupredoxin domains are strongly negative and may facilitate docking to methanol dehydrogenase (MDH) [42], the second enzyme in the methane catabolic pathway.

2.2 *Metal Centers*

The initial *M. capsulatus* (Bath) pMMO structure identified three metal binding sites and a fourth hypothetical site derived from a clustering of potential metal ligands. At each of the occupied sites, the type of bound metal was identified by anomalous difference electron density maps [30]. The cytosolic domains of pmoB harbor two of these sites. On the N-terminal domain, His33, His137, and His139 and the N-terminal amine of His33 coordinate a dinuclear copper site located close to the membrane surface (Figure 3).

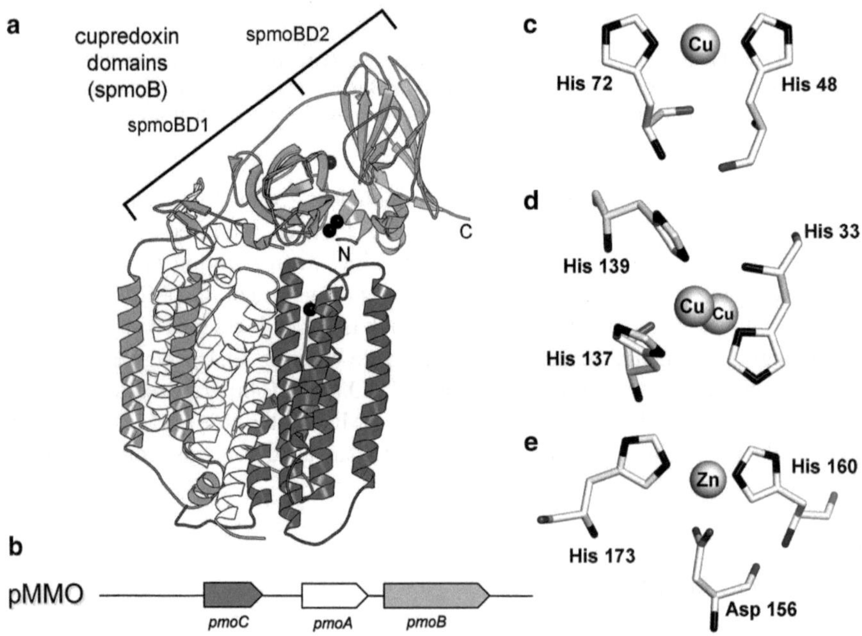

Figure 3 (**a**) The structure and subunit organization of a single pMMO αβγ protomer depicting the pmoA (white), pmoB (light gray), and pmoC (dark gray) subunits. The metal centers are shown as black spheres and from top down are the copper, dicopper, and zinc sites. (**b**) Organization of the pMMO operon. (**c**) Mononuclear copper site between the cupredoxin domains of pmoB. (**d**) Dinuclear copper site at the N-terminus of pmoB. (**e**) Zinc binding site in pmoC.

These residues are conserved across all known Proteobacterial methanotroph pmoB sequences [5]. Verrucomicrobia, which represent a distinct phylum of methanotrophs having pMMO-like sequences that are divergent from the Proteobacterial ones, do not conserve any of the ligands contributing to this site [5, 43]. It is unknown whether the Verrucomicrobia homologues oxidize methane or carry out an alternative function in these organisms. pmoB residues 1–32 are an N-terminal signal sequence that targets the protein to cytoplasmic membranes and they are cleaved off *in vivo* to facilitate assembly of the dinuclear site. The Cu−Cu distance was modeled at 2.5–2.6 Å based on X-ray absorption spectroscopy (XAS) data for the isolated protein [34, 44]. Similar Cu−Cu distances were observed in *M. trichosporium* OB3b and *Methylocystis* sp. strain M pMMO [39, 45]. At this resolution, finer details about the ligand geometry and identities of the exogenous terminal and bridging ligands could not be determined. Subsequent low resolution structures of pMMO from *M. trichosporium* OB3b (3.9 Å) and *Methylocystis* sp. strain M (2.68 Å) suffer from a similar problem that make it very difficult to definitively build an accurate representation of this dimetallic site [39, 45]. In the *Methylocystis* structure, two of the three centers are modeled with only one copper, suggesting how labile the metal ion may be at this site.

At the interface between the N- and C-terminal cupredoxin domains of the pmoB subunit of *M. capsulatus* (Bath), a mononuclear copper-binding site is coordinated by His48 and His72 (Figure 3c) [30]. Among homologous Proteobacterial and Verrucomicrobia pMMO proteins and the closely related ammonia monooxygenases, sequence alignments indicate His48 is not conserved and is replaced with either Asn, Gln, Thr, or Val [5]. *M. trichosporium* OB3b and *Methylocystis* sp. strain M pMMO substitute His48 for an asparagine and show no copper binding at this position in their structures [39, 45].

A third metal binding site was identified in the transmembrane region of the pmoA subunit facing the large inner channel of the pMMO trimer (Figure 3e). In the *M. capsulatus* (Bath) and *Methylocystis* sp. strain M pMMO structures, zinc, required for crystallization, binds the enzyme and is coordinated to Asp156, His160, and His173. The *M. trichosporium* OB3b pMMO structure, which did not require zinc for crystallization, has copper bound at this position, but it should be noted that this metal was added during purification. Although the residues at this site appear to be conserved across all species, there are lingering questions as to whether this site is biologically significant [5].

Lastly, in the initial pMMO structure from *M. capsulatus* (Bath), a fourth putative site unoccupied by metal ions was identified at the interface between the pmoA and pmoC subunits in the transmembrane region based on the localization of several hydrophilic His, Met, Glu, and Asp residues [30]. This region of the protein was hypothesized to house proposed trinuclear copper and diiron catalytic sites [12, 46]. A higher resolution *Methylosinus* sp. strain M pMMO structure later revealed that some of the transmembrane helixes in this region of the protein were mistraced in the original *M. capsulatus* (Bath) pMMO structure [39]. The resulting re-traced structure shows this fourth putative site to be non-existent.

2.3 *Identifying the Active Site*

2.3.1 Proposed Active Sites

Over the years, it has been a challenge to identify the location and metal content of the enzyme active site. Protein purifications from various laboratories yielding variable levels of copper atoms per protomer with and without 1–2 iron atoms produced spectroscopic data from which several different active site models were proposed [14, 33–39]. Early electron paramagnetic resonance (EPR) spectroscopic investigations of *M. capsulatus* (Bath) membrane fractions and purified protein containing 15–20 coppers per protomer yielded type II copper spectra with a hyperfine splitting pattern and an isotropic signal at $g = 2.06$ that were interpreted to suggest the existence of catalytic and electron transfer trinuclear copper centers with similar configurations to those of multicopper oxidases like laccase, ascorbate oxidase, and ceruloplasmin [47–50]. A similar type II Cu(II) EPR signal was observed in preparations from *M. trichosporium* Ob3b [51, 52].

Preparations of pMMO from *M. capsulatus* (Bath) and membrane fractions of *Methylomicrobium albus* BG8 from different laboratories, however, exhibited spectra more typical of type II Cu(II), which could be interpreted as a mononuclear copper site having a square planar composition and a coordination sphere comprising mostly histidines [33, 38, 53–55]. These findings, in addition to the purification and spectroscopic characterization of active pMMO with 2–3 copper atoms per protomer, called into question the plausibility of a trinuclear copper site [34, 44]. The first crystal structures of pMMO suggested that mononuclear and dinuclear copper centers should be considered carefully as the locus of catalytic activity, given their similarity to type II and type III copper centers in amino acid/peptide monooxygenases and catechol oxidase, respectively [40, 49].

Lastly, several laboratories, after accounting for a common cytochrome contamination, proposed pMMO to be an Fe–Cu enzyme, because pMMO preparations having higher iron content seemed to display better activity [12, 14, 38, 51, 56, 57]. Mössbauer spectroscopic study of purified pMMO fractions from *M. capsulatus* (Bath) containing significant levels of non-heme iron suggested the presence of a carboxylate-bridged diiron center similar to that in sMMO [57]. The three conserved residues that constituted the mononuclear zinc/copper site were proposed as the location for this iron center, despite the paucity of potential amino acid ligands, conserved or otherwise, needed to support a dinuclear site at this position.

The foregoing discrepancies between competing laboratories may have their origin in one or more of the following: differences in protein preparation leading to the loss of labile copper and/or iron; possible co-purification of pMMO with methanobactin, which would increase the amount of copper in the samples; contamination by the sMMO hydroxylase component (MMOH), which would add diiron centers; or the presence of bacteriohemerythrin, a hemerythrin-like protein expressed at high copper levels that is proposed to deliver O_2 to pMMO, which would also elevate iron levels [12, 27, 35, 46, 58].

2.3.2 Evidence for a Dicopper Active Site

Several lines of evidence indicate the dicopper site to be the locus of methane oxidation in pMMO. Metal reconstitution studies of apo-pMMO from *M. capsulatus* (Bath) revealed that the addition of 2–3 copper ions per protomer returned 90 % of the original activity whereas the addition of iron had no effect on turnover [39, 59]. Adding more than 2–3 equivalents of copper per apo-pMMO protomer inhibited the enzyme [60], possibly by hydrogen peroxide formation because the addition of catalase minimized this effect [61].

Studies of the *M. capsulatus* (Bath) pmoB cupredoxin domains provide the most significant line of evidence pointing toward the dicopper site as the one responsible for methane hydroxylation [60]. A recombinant soluble construct, termed spmoB, was created by replacing the two transmembrane helices anchoring the pmoB subunit to the membrane with a soluble synthetic linker that fused the two pmoB cytosolic domains (spmoBD1 residues 33–172 and spmoBD2 residues 265–414) (Figure 3a). After refolding the protein from inclusion bodies and reconstituting, spmoB bound ~3 copper ions and oxidized methane with an activity of 203.1 ± 20.2 nmol min^{-1}μmol^{-1} compared to that of the full length protein, 2290 ± 60 nmol min^{-1} μmol^{-1}. This result indicated that the cytosolic domains, and not the transmembrane region, are responsible for the hydroxylation chemistry. A His48Asn variant that removed the mononuclear copper site at the interface between the cupredoxin domains afforded a construct that bound ~2 coppers and displayed 10 % activity compared to that of the original spmoB. This decrease was attributed to a misfolding of the two domains, which share over ~1400 $Å^2$ of buried surface area. Conversely, a His137, 139Ala spmoB double variant targeting the dicopper site bound ~1 copper ion and all activity was abolished. XAS studies detailing the coordination environment around the copper centers of spmoB indicated three- to four-coordinate Cu(I) and an O/N rich environment with average Cu–O/N distances of 1.95 Å and a Cu–Cu interaction at 2.53 Å. These values are nearly identical to those obtained from fits to XAS data of purified and copper-reconstituted full-length pMMO.

The low activity of spmoB compared to the full length enzyme, while a concern, can be attributed to several factors. Not all of the protein may have been refolded properly, and delivery of electrons to the metal center may not be as efficient in the absence of the transmembrane domains. The dicopper site in the spmoB construct is highly solvent-exposed. It is conceivable that the transmembrane domains plus the lipid membrane play a significant role in coordinating reactivity of the reduced metal center with gaseous substrates so as to prevent buffer components from adventitiously quenching the reaction. In short, all signs point toward a dinuclear copper site as the likely pMMO active site. Although the location of the active site has become clarified, a detailed picture of the resting dimetallic center structure with the appropriate copper-coordinating protein- and solvent-derived ligands and their geometries remain to be determined. The square-planer 3–4 coordinate geometry indicated by EPR and XAS studies suggest that each copper at this site requires an additional 1–2 ligands [34, 53, 60], which presumably are solvent-derived oxo, hydroxo, and/or aqua species.

Beyond the residues that coordinate directly to the dicopper center, strictly conserved amino acids in the pmoB subunit in the second coordination sphere that could participate in O_2 and CH_4 activation process are scarce. Acidic residues at positions 339 and 35 in *M. capsulatus* (Bath) pMMO are intriguing candidates for further investigation into pMMO metal center assembly and proton transfer events, given their close proximity to the dicopper site (Figure 4a), but they are not conserved in related methanotrophs.

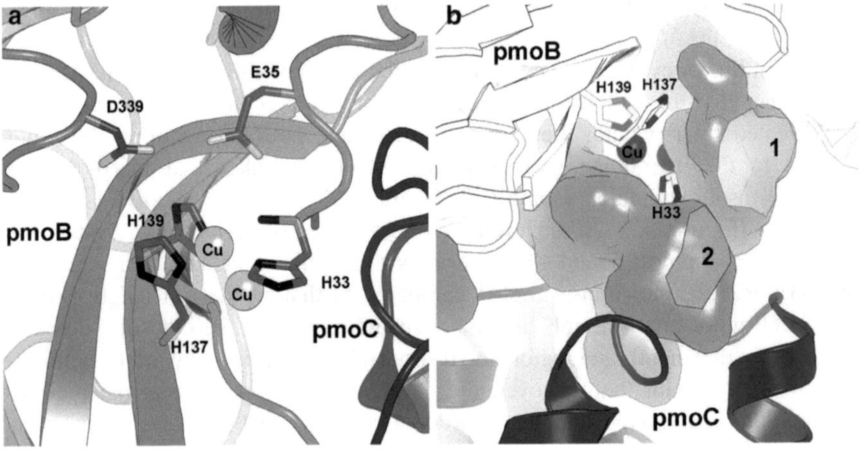

Figure 4 (**a**) Local environment surrounding the pMMO dicopper center. (**b**) Putative substrate-binding cavities (1 and 2) adjacent to the pMMO dicopper center. The pmoB and pmoC subunits are depicted as white and gray ribbons, respectively.

The absence of any conserved residues beyond those that coordinate to the metals stands in contrast to the situation for sMMO and related bacterial multicomponent monooxygenases (BMMs), where highly conserved residues in the second and third coordination spheres are essential for assembly of the hydroxylase diiron center, proton transfer (via conserved Thr213 and Asn214 residues), O_2 and methane access to the diiron center, as well as electron transfer.

2.4 *Substrate Access and Product Egress from the Dicopper Site*

2.4.1 Access to the Substrate-Binding Pocket

The pMMO dicopper site is located about 6 Å from the protein surface, almost at the interface between the transmembrane and cytosolic domains. Surface calculations show cavities on both faces of the dicopper center providing access to solvent through small openings in the protein surface (Figure 4b). Cavity 1 is shallow (5–6 Å deep), lined with polar residues, and relatively solvent-exposed. Cavity 2 lies between the first pmoB cupredoxin domain and pmoC at the transmembrane interface. This cavity is longer (~14 Å), narrower, and quite hydrophobic. These features are conserved in three pMMO crystal structures, but the variable dicopper center geometries observed at low resolution make it difficult to visualize the exact morphologies of these pockets. The narrow substrate profile of pMMO, comprising short-chain alkanes of up to five carbons in length, and enantioselective formation of *R*-2-alcohols tends to support the latter sterically restricted cavity as the substrate-binding pocket [62–64]. It is unclear whether the wider more polar cavity has any function, although one could speculate that it might support the binding of membrane associated quinones and help to facilitate direct electron transfer from the reductant to the dicopper center.

2.4.2 Substrate and Product Channeling

The requirement of a 300 kDa protein for activating such small substrates suggests that the large scaffold may play a significant role in channeling and coordinating the movement of molecules to and from the active site. Additional "holes" noted in the exterior of the pMMO structure at the membrane interface may serve this purpose by controlling hydrocarbon and O_2 access to the dicopper center [41], possibly by using a scheme similar to that observed in the sMMO and its homologue, toluene/o-xylene monooxygenase (ToMO), where a series of hydrophobic cavities facilitate dioxygen movement through the protein [65]. Dioxygen delivery may also be conferred by bacteriohemerythrins [66]. Like pMMO, expression of this O_2-binding protein is up-regulated at high copper-to-biomass ratios. For *M. capsulatus* (Bath), enhanced hydroxylation activity occurs with the addition of bacteriohemerythrin to pMMO-enriched membranes, suggesting this component to be important for delivering dioxygen to the intracytoplasmic membranes.

A cryo-electron microscopy structure of pMMO detailing how MDH docks on top of the pmoB cupredoxin domains suggests the two proteins participate in substrate channeling, possibly by using the central cavity as a mode for product egress [41, 42]. Some biochemical evidence for channeling may come from work using the suicide substrate acetylene, which, when activated by the enzyme, covalently attaches to nearby amino acids by a mechanism that has yet to be fully resolved. Before the pMMO crystal structures were obtained, mass spectrometric analysis of ^{14}C-acetylene-labeled pMMO identified modified sites in the pmoA and pmoB subunits and suggested early on that the active site did not reside in the pmoB cupredoxin domain but instead in the other subunits [16, 38, 67]. Similar results were obtained with AMO, in which His191 in the AmoA subunit (the analog to Tyr186 in the *M. capsulatus* (Bath) pmoA subunit) was distinctly labeled [68–70]. This residue is ~14–19 Å removed from the pmoB dicopper center and sits close to the central cavity of the pMMO trimer. The modification of residues distant from the dicopper site suggests the activated acetylene compound migrates through the protein before it covalently attaches and that internal cavities may play a specific role in coordinating substrate and product movement.

2.4.3 Electron Sources

The source of electrons for the pMMO system is currently unknown. Membrane-bound quinols associated with the electron transport chain have long been assumed to be the pMMO reductants [33, 34, 38, 51]. This assignment is consistent with steady state experiments on isolated membranes or purified protein, which have successfully employed duroquinol as the reductant. Other laboratories have noted NADH-dependent activity [35], although it is likely that there was co-purification of pMMO with a type 2 NADH:quinone oxidoreductase contaminant [14].

Recent co-purification of pMMO with MDH led to a reconsideration of a hypothesis in which electrons from the oxidation of methanol are recycled into the oxidation of methane [42, 71]. It is possible that the reduced pyrroloquinoline quinone ($PQQH_2$) cofactor generated by oxidation of methanol to formaldehyde in MDH donates its electrons back to pMMO as the last step in its catalytic cycle. Such an event could obviate any need to shuttle NADH generated by downstream reactions through additional redox proteins to supply pMMO with reductant and facilitate faster growth. Further work is required to definitively establish the electron source.

2.5 *Mechanism*

A major goal for researchers working on pMMO is to provide a detailed understanding of its mechanism, and especially how pMMO is unique compared to other hydrocarbon-oxidizing dinuclear copper proteins like tyrosinase and catechol oxidase. The problems associated with the purification and stability of pMMO, as well as the controversy surrounding the identification of the active site, have slowed progress, especially when compared to our understanding of sMMO.

Complicating matters further is the difficulty in studying membrane proteins like pMMO with rapid kinetic and spectroscopic methodologies in order to observe key intermediates in the O_2 and C–H bond activation processes. Although information is scarce, some significant progress has been made recently toward this goal. Here we examine current knowledge of the O_2 and C–H bond activation steps in pMMO as well as additional insight gained by related work from model compounds and computational studies.

2.5.1 Spectroscopic Identification of an Oxygen Intermediate

The first, and currently only, observed spectroscopically detected intermediate in pMMO was identified by addition of anoxic hydrogen peroxide to ascorbate-reduced *M. capsulatus* (Bath) pMMO, which over 12 h produced a stable optical feature at 345 nm with $\varepsilon = 10{,}000\ cm^{-1}\ M^{-1}$ [72]. This same optical spectrum was observed in spmoB samples, but here it appeared almost immediately after reacting the reduced protein with either O_2 or H_2O_2. It is not understood why pMMO reacts differently than spmoB toward O_2 and H_2O_2, but greater substrate accessibility to the active site in spmoB was suggested to play a role. As expected, the spmoB His48Asn variant behaved like the wild-type spmoB and the His137, His139 double variant exhibited no reactivity toward dioxygen or hydrogen peroxide, again identifying the dicopper site as the essential catalytic center. The addition of methane to this stable copper-oxygen complex in pMMO and spmoB results in a disappearance of the spectral feature, demonstrating that this oxygen-derived intermediate is catalytically competent and possibly the one responsible for C–H bond activation. The optical spectroscopic properties of the pMMO oxygen-derived species have been observed in type III copper systems like tyrosinase, hemocyanin, and multicopper oxidases, and these are characteristic of either a μ-η^2:η^2-peroxo Cu^{II}_2 or a hydroxo-bridged Cu^{II}_2 intermediate (Table 1) [73–76].

Table 1 Spectroscopic Parameters of Dicopper-Oxygen Complexes.

	Optical			
	λ_{max} (nm)	ε ($M^{-1}\ cm^{-1}$)	Cu–Cu (Å)	Cu^{II}_2/O_2 Species
pMMO, spmoB [5]	345	10,000		
Oxyhemocyanin [74]	350; 580	20,000; 1000	3.6	μ- η^2:η^2
Oxytyrosinase [74]	350; 580	18,000; 1000	3.4	μ- η^2:η^2
Met-hemocyanin [74]			3.1	di-μ-hydroxo
Multicopper oxidases [73]	330	5000	3.7	μ-hydroxo

The dinuclear metal centers in these related type III copper proteins are coordinated by 5–6 histidines, have much longer Cu–Cu distances of 3.1–3.6 Å, and appear more constrained by the surrounding protein scaffold. By contrast, the bidendate coordination of one copper in pMMO by the N-terminal histidine may offer greater flexibility and allow the metal center to adopt novel geometries that have yet to be identified. The exact structure of the pMMO intermediate remains uncertain, however, and further structural and biochemical characterization of this species is crucial to further our understanding of the enzyme chemistry. Unfortunately, the application of resonance Raman spectroscopy is limited by the presence of a persistent cytochrome contaminant in purifications of the native protein, and high concentrations of refolded spmoB have been difficult to achieve.

2.5.2 Computational Studies and Comparisons to Copper Model Compounds

Computational studies examining the reactivity of O_2 and CH_4 with the different copper sites in pMMO and the characterization of copper model complexes have provided valuable insights into the pMMO mechanism [46, 77–79].

In one particular set of calculations, a mixed-valent di-μ-oxo-$Cu^{II}Cu^{III}$ species was found to be more reactive toward methane than either a mononuclear Cu^{III}-oxo or a di-μ-oxo Cu_2^{III} intermediate [77, 78]. Such a reactive di-μ-oxo-$Cu^{II}Cu^{III}$ center could be generated in the enzyme if an electron were injected into either a μ-η^2:η^2-peroxo Cu_2^{II} or a di-μ-oxo Cu_2^{III} precursor, although the latter species has yet to be identified in biological systems [4, 80] (Figure 5). This electron could originate from a metal center, an exogenous reductant, or a protein residue. Methane oxidation by the reactive di-μ-oxo-$Cu^{II}Cu^{III}$ center was calculated to occur through a concerted, non-radical oxygen insertion mechanism (Figure 6).

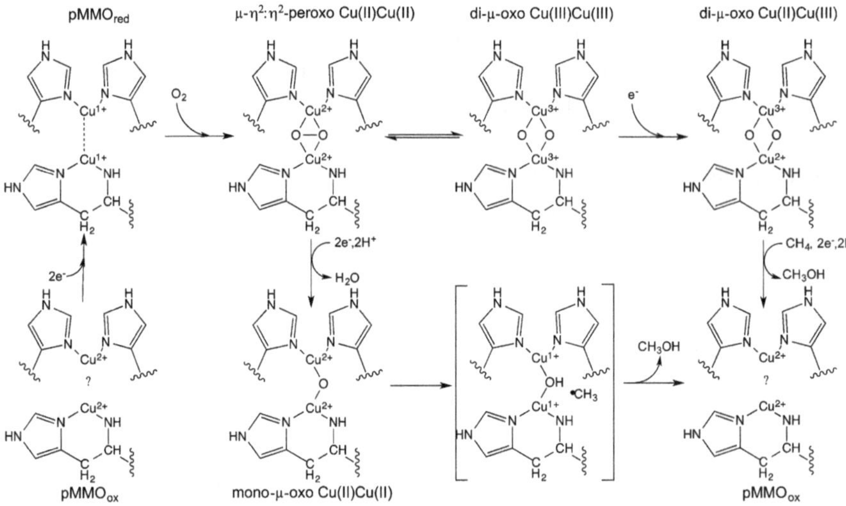

Figure 5 Proposed mechanisms for O_2 activation by pMMO. Figure adapted from [4].

A possibly significant development in the pMMO field was the identification of a copper zeolite, Cu-ZSM-5, that readily oxidizes methane to methanol at 100 °C [81]. Mechanistic characterization of this complex revealed that O_2 binding first results in a μ-η^2:η^2-peroxo Cu_2^{II} complex followed by subsequent conversion to a reactive bent mono-μ-oxo-Cu_2^{II} species [82–84]. Electrons for the process are proposed to come from spectator Cu^I ions in the zeolite. DFT calculations on this compound suggested the C–H activation step occurs at the bridging oxygen, resulting in a transient $[Cu–OH–Cu]^{2+}$ intermediate and methyl radical that rapidly recombines with the copper-bound hydroxyl radical to form methanol (Figures 5 and 6). Until further characterization of pMMO is carried out, it is unknown which, if any, of these mechanistic possibilities accurately represent the O_2 activation events and the essential catalytic intermediate in the enzyme.

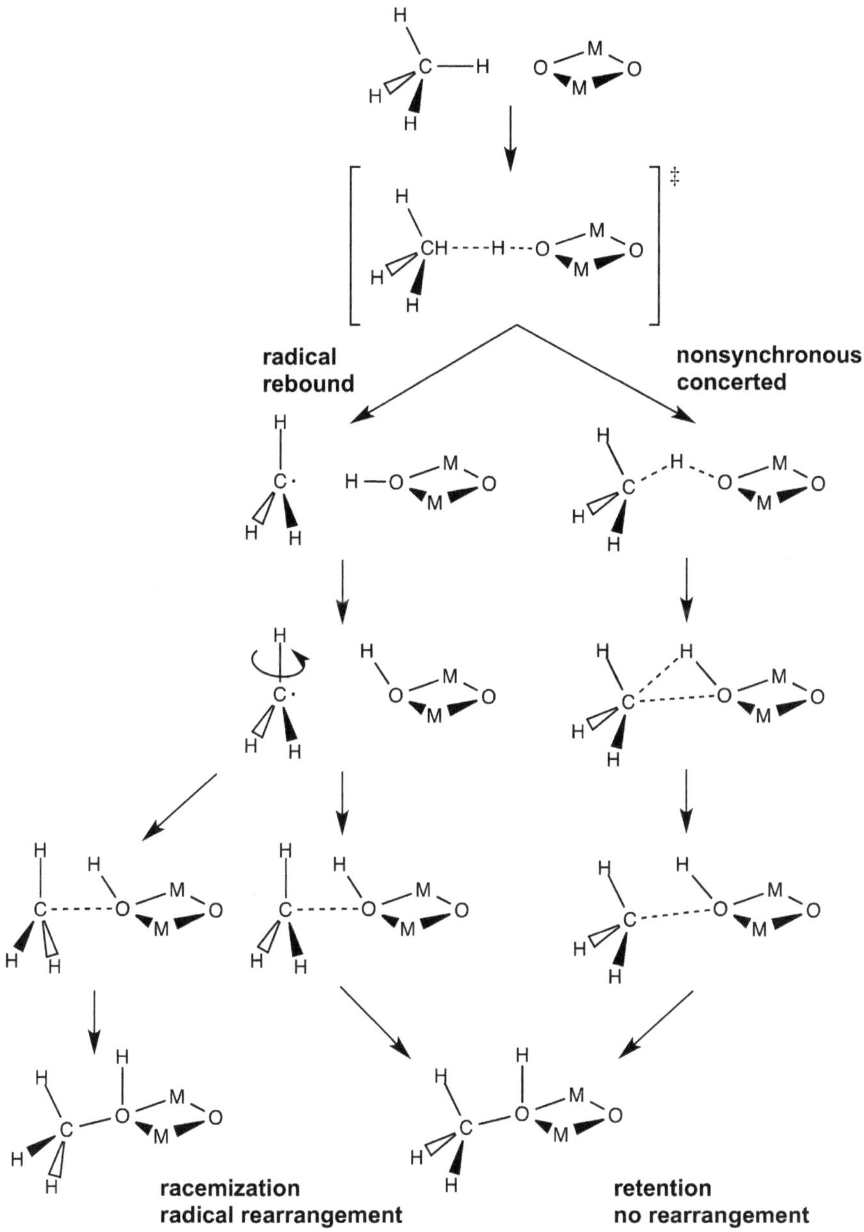

Figure 6 Radical and concerted mechanisms of C–H bond activation. M represents the reactive metal in either pMMO or sMMO.

2.5.3 Mechanism of C–H Bond Breaking

The C–H bond activation steps in pMMO and sMMO have been examined by using substrate probes to differentiate between cationic, alkyl radical, or concerted oxygen insertion mechanisms (Figure 6). These probes generally fall into two classes, chiral substrates and radical clock probes, the latter of which are too large to be accommodated by the pMMO active site. Reactions of pMMO with the (S) or (R)-[1-$^3H_1$,$^2H_1$] ethane and [2,2-$^2H_2$] butane assessed the reactivity of pMMO toward a chiral methyl group [27, 85, 86]. An inversion of configuration suggested rotation about the C–C bond of a radical or cationic ethyl intermediate. The calculated lifetime of such an intermediate, given the barrier of C–C bond rotation in an ethyl radical (0.15 kcal mol^{-1} at 30 °C), is ~180 fs [85]. Reactions with *M. capsulatus* (Bath) pMMO revealed complete retention of configuration, suggesting a concerted mechanism in which an oxygen atom is inserted between the C–H bond of a pentacoordinated species having either a C–O or C–Cu bond. A radical or cationic rebound mechanism exhibiting complete retention would have to occur faster than 10 fs in order to avoid detectable C–C bond rotation. This latter scenario is not feasible because it would require that radical capture be faster than the calculated rate constant for the decay of the transition state (6.6×10^{12} s^{-1} at 45 °C). However, given that pMMO has a narrow substrate-binding pocket, a mechanism invoking a bound radical with highly restricted rotation cannot be ruled out.

2.6 *Unresolved Questions*

Despite the progress made in the last 15 years, there are numerous unresolved questions concerning the structure and mechanism of pMMO. There is a great need for higher resolution structural data to further define the geometry and the solvent contributions to the resting dicopper active site. A better starting place for thinking about the pMMO mechanism is required, especially for those pursuing synthetic catalysts, model complexes, and computational studies.

Similarly, improvements in purification of both native pMMO and the spmoB construct are needed to facilitate more extensive mechanistic studies using time-resolved spectroscopic methods. Further insight into the structure of the long-lived peroxo species and discovery of new intermediates along the reaction pathway will provide a stronger framework for zeroing in on a mechanism for O_2 and C–H bond activation by this unique dicopper center. Finally, it is unknown how the requisite substrates are assembled at the metal center with appropriate timing in order to achieve efficient coupling of the different chemical processes. Resolving this question requires better understanding of the role of the large protein scaffold into which the dicopper unit is embedded and how both substrates and products are channeled to and from the active site in a controlled fashion. Much important work remains to be done.

3 Soluble Methane Monooxygenase

3.1 Genetics and System Components

3.1.1 Soluble Methane Monooxygenase

Soluble methane monooxygenase is a three-component enzyme system requiring a 251-kDa hydroxylase MMOH (*mmoXYZ*), a 38-kDa reductase MMOR (*mmoC*), and a 16-kDa regulatory or effector protein MMOB (*mmoB*) for optimal activity (Figure 7) [8, 10]. MMOH takes the form of an $(\alpha\beta\gamma)_2$ heterodimer and houses the carboxylate-bridged diiron active site ~12–14 Å below the surface of the hydroxylase α-subunit. MMOR, a [2Fe-2S]- and FAD-containing reductase, supplies electrons to the hydroxylase by consuming NADH. MMOB, a cofactorless protein, up-regulates activity by binding to the hydroxylase α-subunit near the diiron center to induce changes in the protein architecture that efficiently couple NADH consumption with substrate oxidation.

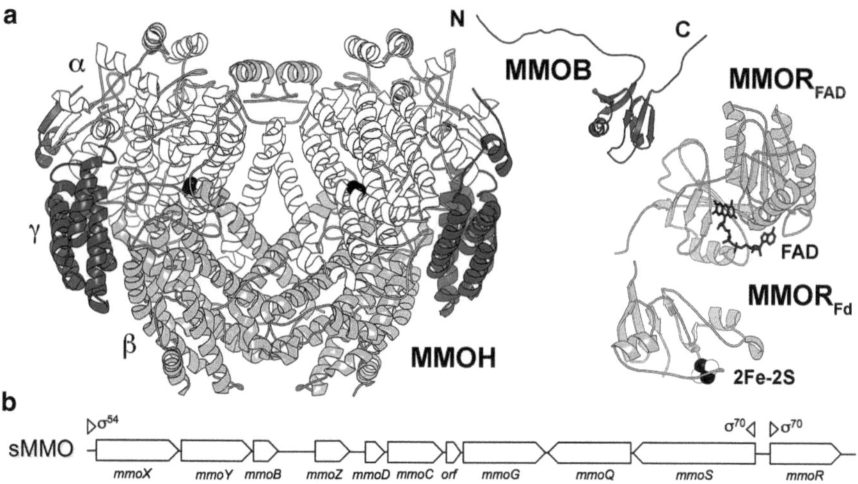

Figure 7 (**a**) sMMO components. (**b**) sMMO operon from *M. capsulatus* (Bath).

The sMMO operon from several methanotrophs has been sequenced and contains additional genes that are hypothesized to encode for proteins important for controlling expression and assembly of the system (Figure 7b) [87–90]. In the middle of the catalytic gene sequences is *mmoD* (formerly *orfY*), which encodes for a protein of unknown function (MMOD) and inhibits sMMO activity by competing with MMOB for the same binding site on MMOH [91, 92]. Comparisons to DmpK (also known as PHK) in related phenol hydroxylase systems have suggested that MMOD could play a role in the metal center assembly of MMOH, although this hypothesis has yet to be proved. MMOD has also been proposed to serve as a transcriptional regulator of the sMMO operon [93]. Downstream of the sMMO genes are *mmoQ*, *mmoS*, *mmoR* and *mmoG*. The *mmoR* gene encodes for a σ54-dependent transcriptional activator thought to control the copper-dependent expression of sMMO. MMOQ and MMOS, which show significant homology with two-component sensor-regulator systems, have been proposed to activate the σ54-dependent transcriptional activator, but exactly how these proteins sense the cell environment and interact has yet to be established [87, 94, 95]. Lastly, MMOG is a putative GroEL-like protein-folding chaperone that may facilitate proper assembly of MMOH and the transcriptional activator [87, 89, 96]. The organization and composition of the sMMO operon varies among the different methanotrophs.

Given the complex structure of MMOH and the extra genes required for its assembly, it is not surprising that no one has yet generated a robust heterologous expression system for the hydroxylase despite numerous attempts and claims. Consequently, almost all studies on MMOH have been conducted on native protein purified from either *Methylococcus capsulatus* (Bath), a thermophile with an optimal growth temperature at 45 °C, or *Methylosinus trichosporium* OB3b, a mesophile with optimal growth at 30 °C. The hydroxylase components from other BMMs, such as toluene 4-monooxygenase, (T4MO), toluene/o-xylene monoxygenase (ToMO), and PH (phenol hydroxylase), have been recombinantly expressed in *E. coli* [97–99]. These alternate systems allow conserved residues and structural features within these proteins to be probed by mutagenesis. Thus, some of the ideas about sMMO function presented below rely on analysis of and comparisons to related BMM systems.

3.1.2 Related Bacterial Multicomponent Monooxygenases and Substrate Specificities

sMMO is the most well characterized member of a larger family of bacterial multicomponent monooxygenases that can be subcategorized into sMMO/alkane monooxygenases (MMOs), four-component alkene/aromatic monooxygenases (TMOs), phenol hydroxylases, alkene monooxygenases (AMOs), tetrahydrofuran monooxygenases (THFMOs), and hyperthermophilic aromatic/alkene monooxygenases [100, 101]. Enhanced environmental screening and genomic sequencing have expanded the diversity and scope of these subfamilies since the genetics were last reviewed [101]. All of these systems utilize the same protein components, a hydroxylase, a regulatory protein, and a reductase. Members of the four-component toluene monooxygenases require both a flavin and Rieske protein to shuttle electrons to the hydroxylase. These different systems, which evolved to hydroxylate or epoxidize specific hydrocarbons for efficient carbon and energy assimilation in lower metabolic pathways, have relatively flexible substrate binding pockets that accommodate numerous alternate substrates.

The hydroxylase α-subunits are the most conserved among the different classes and share ~21–35 % sequence identity to MMOH and related alkane monooxygenases. Even though the residues that contribute to the assembly of the diiron center are conserved and the active site structures of related family members are similar, if not identical, to that of MMOH [102–104], none of these other systems can hydroxylate methane. sMMO is a special member of this enzyme class. Even the 65 % identical butane monoxygenase cannot activate methane [105], suggesting there are highly relevant and fundamental structure function relationships between the metal center and protein scaffold that are essential for tuning reactivity toward small alkanes.

3.2 *Component Structures and Function*

3.2.1 Soluble Methane Monooxygenase Hydroxylase

The 251-kDa hydroxylase component is an $(\alpha\beta\gamma)_2$ heterodimer with a 2-fold axis of symmetry (Figure 7) [106, 107]. A large canyon is formed at the interface between the $\alpha\beta\gamma$ protomer in the middle of the molecule that is important for the docking of the other protein components. The diiron center is located in the α-subunit and housed within a four-helix bundle comprising helices B, C, E, and F, the latter two of which form a surface of the canyon that is critical for binding MMOB and undergo helical rearrangements upon protein component binding (Figure 8) [108]. The remainder of

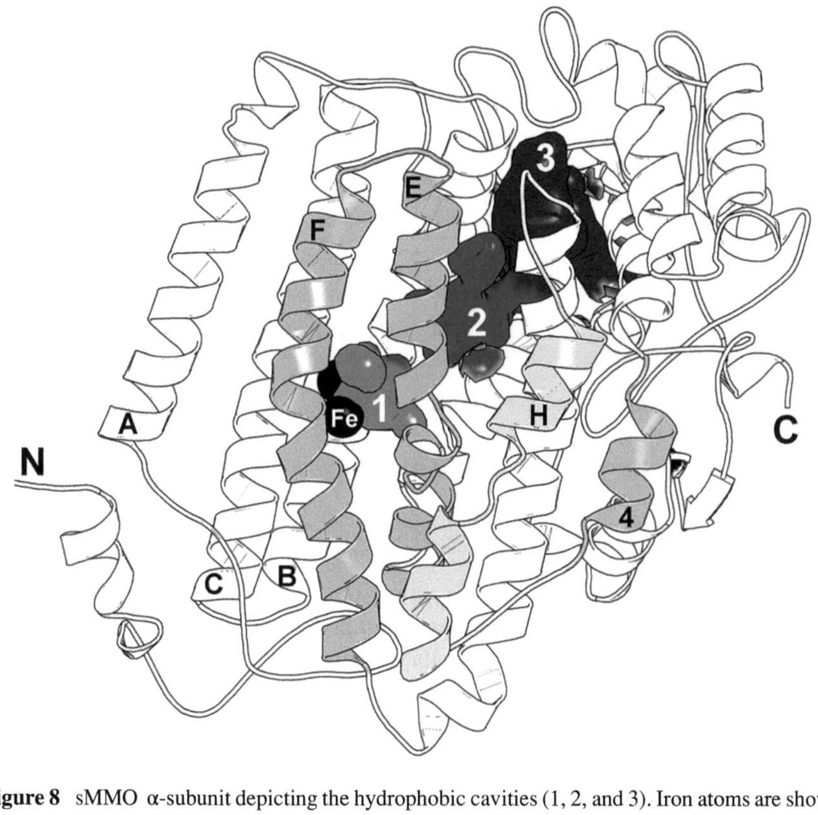

Figure 8 sMMO α-subunit depicting the hydrophobic cavities (1, 2, and 3). Iron atoms are shown as black spheres. Helices E, F, H, and 4, which are the primary MMOB docking surfaces on MMOH, are shaded light gray.

the protein, particularly the α-subunit, plays an important role in coordinating the efficient assembly and activation of the different substrates (electrons, protons, hydrocarbon, and dioxygen) at the active site using a series of hydrophobic cavities, hydrogen bonding networks, and electron transfer pathways (Figure 8) [109, 110].

At different times during the reaction cycle, MMOB and MMOR dock on to MMOH and induce changes in MMOH that are not only responsible for activating the diiron center, but may also help to coordinate the catalytic events [111]. The hydroxylase can accommodate the binding of two molecules of MMOB and MMOR, one on each face of the hydroxylase. Optimal activity in *M. capsulatus* (Bath) MMOH, however, is observed with a 1:2:0.5 H:B:R ratio, suggesting that MMOR binds transiently to the hydroxylase and services multiple diiron centers [111]. In the absence of MMOB, MMOH functions primarily as an NADH oxidase. More details about the specific effects of each component, the shuttling of substrates, and the coordination of catalytic events are discussed below.

3.2.2 The Reductase and Electron Transfer to the Hydroxylase

MMOR is a typical reductase with an N-terminal [2Fe-2S] ferredoxin ($MMOR_{Fd}$) domain and a C-terminal FAD-binding ($MMOR_{FAD}$) domain. NMR structures of the individual domains from *M. capsulatus* (Bath) are available [112, 113] (Figure 7). Electron transfer kinetics in the sMMO systems have been investigated in detail by using stopped-flow optical spectroscopy [114–117] (Figure 9).

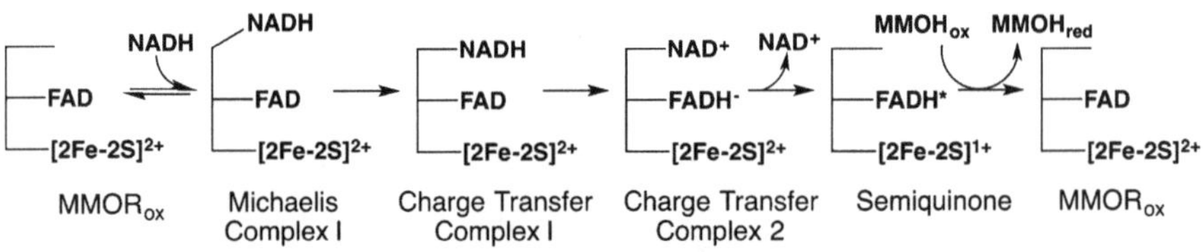

Figure 9 Electron transfer steps from MMOR to $MMOH_{ox}$. The figure was adapted from [8].

Briefly, after the binding of NADH to the $MMOR_{FAD}$ domain, the first electron transfer event produces the two-electron reduced FAD hydroquinone ($FADH^-$) and NAD^+. Following the release of NAD^+, a one-electron transfer step produces the flavin semiquinone ($FADH^{\bullet}$) and the reduced [2Fe-2S] intermediates. Two successive electron transfer steps ultimately yield the reduced hydroxylase, $MMOH_{red}$. The reduction potentials of the MMOR cofactors are not affected by the hydroxylase or the regulatory protein [117].

3.2.3 The Regulatory Protein and Interactions with the Hydroxylase

MMOB is organized into a structured 95 amino acid core with flexible N- and C-terminal tails of 35 and 11 amino acids, respectively [118, 119] (Figure 7). A newly available structure shows that it docks onto the MMOH α-subunit in the canyon region on top of helices E and F, which contribute glutamate and histidine ligands to the diiron center (Figure 10) [108]. The protein enhances the steady state reactivity of sMMO by 8–150 fold [108, 120], accelerates the rate of O_2 activation by the reduced metal center by 1000-fold [121], efficiently couples NADH consumption to substrate hydroxylation, alters the spectroscopic and redox properties of the diiron center, and influences the regio- and stereospecificity of the sMMO-catalyzed reaction [8, 10, 122]. Similar phenomena are observed in the other BMM systems [122–124]. A long-standing goal is to understand how the structural changes that shift the relative population of hydroxylase molecules into an active conformation account for the altered properties of the activated enzyme. The specifics of these changes in relation to sMMO activation are discussed in Section 3.4.2.

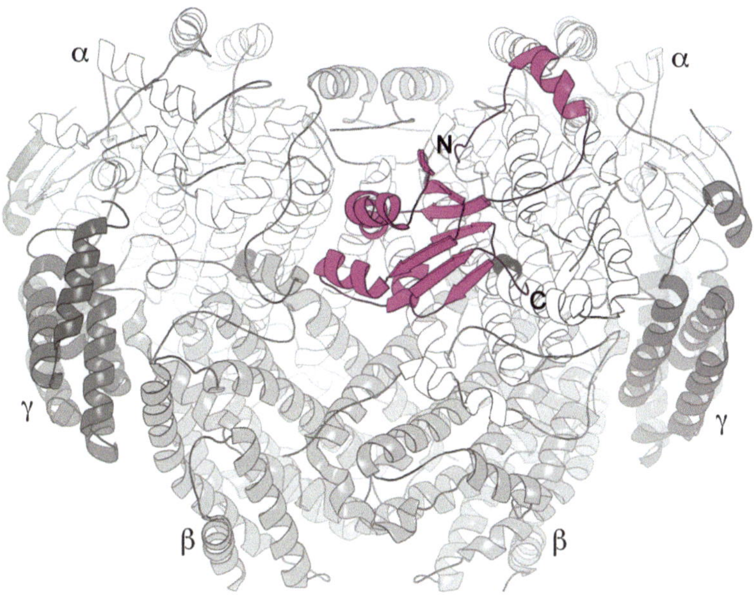

Figure 10 Structure of the MMOH-MMOB complex (PDB entry 4GAM). MMOB is colored magenta.

Also relevant to the activation process are contributions from the MMOB N-terminus that forms a ring-like structure when docked onto the surface of the MMOH α-subunit [108]. Removing residues from the N-terminus yields protein forms that have slightly diminished affinity for MMOH with significant loss of activity [108, 125–127]. Nevertheless, a N-terminal truncate of MMOB lacking the first 12 residues failed to alter the redox potentials of the hydroxylase diiron center [125]. The combined data indicate the MMOB N-terminus is essential for the regulatory protein to trigger important structural changes in MMOH required for activity. Similarly, MMOB C-terminal deletions have measurable effects on reaction rates and the efficient formation of sMMO intermediates [128].

3.3 The Diiron Center

The MMOH carboxylate-bridged diiron center has been the subject of numerous spectroscopic and structural investigations to ascertain how this metal unit activates dioxygen and hydrocarbons. These experiments, conducted in the presence and absence of the other protein components, substrates, products, and other chemical probes, have provided a considerable amount of information, not all of which is directly relevant to the focus of this chapter on catalysis. As the different mechanistic aspects are discussed, it is important to remember that it was the X-ray crystal structure of MMOH in the resting diiron(III) state ($MMOH_{ox}$), solved in the *absence* of MMOB, that provided the foundation for interpreting much of the data relevant to the sMMO catalytic cycle (Figure 11).

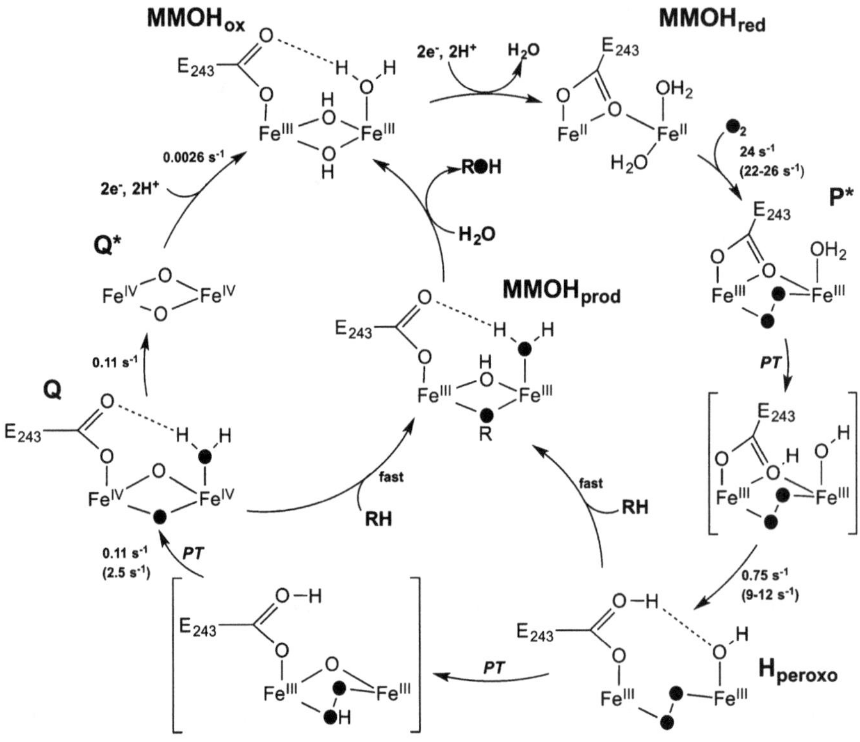

Figure 11 sMMO reaction cycle and intermediates. Proton transfer steps (PT) and rate constants for the *M. capsulatus* Bath and (*M. trichosporium* OB3b) systems are provided. P*, H_{peroxo} = diiron(III) peroxo intermediates; Q, Q* = diiron(IV) intermediates.

The oxidized ($MMOH_{ox}$) and reduced ($MMOH_{red}$) forms of MMOH are the most stable and therefore have received the most structural and spectroscopic scrutiny. The one-electron reduced mixed-valent $Fe^{II}Fe^{III}$ diiron center ($MMOH_{mv}$), while stable, is not an intermediate in the cycle. $MMOH_{mv}$, however, is EPR-active and has been useful for interrogating aspects of the sMMO system.

3.3.1 The Oxidized Hydroxylase

MMOH$_{ox}$ has two high-spin Fe^{III} ions that are coordinated by four glutamates and two histidines (Figure 12a) [107, 129, 130]. The iron atoms lie ~3.0–3.1 Å apart and are bridged by two solvent derived hydroxide ligands, which antiferromagnetically couple the iron ions and lead to a diamagnetic ground state with a characteristic Mössbauer spectrum [131–134]. Glu243 shows the most positional variability in the different X-ray structures of MMOH$_{ox}$. It forms hydrogen bonds to either the terminal water on Fe1 or the bridging hydroxides [130].

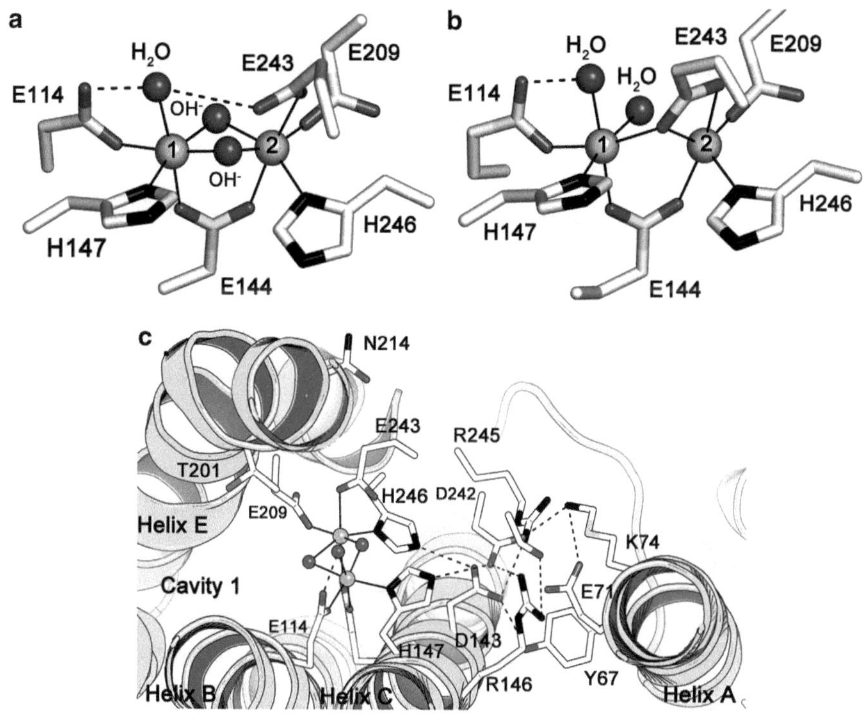

Figure 12 Structures of the (**a**) oxidized (PDB entry 1MTY) and (**b**) reduced (PDB entry 1FYZ) MMOH diiron centers in the absence of MMOB. (**c**) Conserved hydrogen bonding network behind the diiron center leading toward the canyon surface (PDB entry 1MTY). The nitrogen and oxygen atoms are colored black and gray, respectively.

The hydrophobic substrate-binding pocket lies distal to the coordinating histidines in front of the bridging hydroxide. Behind the diiron center, leading 10 Å from the histidine residues to helix A on the canyon surface of the hydroxylase, is an extensive hydrogen bonding network that is strictly conserved among the different BMMs (Figure 12c). The network may assist in the assembly of the diiron center. Functional roles, such as involvement in electron transfer, have yet to be established.

3.3.2 The Reduced Hydroxylase

Upon reduction of MMOH to the diiron(II) state ($MMOH_{red}$), both hydroxides are protonated to generate water, one of which is displaced when Glu243 undergoes a carboxylate shift to bind in a bridging, bidentate chelating mode to Fe2 (Figure 12b) [129, 130]. The remaining water stays coordinated to Fe1 and binds weakly to Fe2, based on the longer Fe–O distance. As a result of these changes, the Fe–Fe distance increases to 3.3–3.4 Å [135], and a possible coordination site opens up on Fe2. $MMOH_{red}$ is a weakly ferromagnetically coupled high-spin system with a characteristic EPR signal at $g = 16$ [132, 134, 136–138]. This EPR signal has been valuable for tracking the reaction of $MMOH_{red}$ with O_2.

Additional conformational changes of interest occur near the active site when $MMOH_{ox}$ is reduced in the absence of MMOB. Asn214, which sits above the diiron center on helix E, undergoes a rotomer shift from the protein surface to the interior (Figure 13) [130].

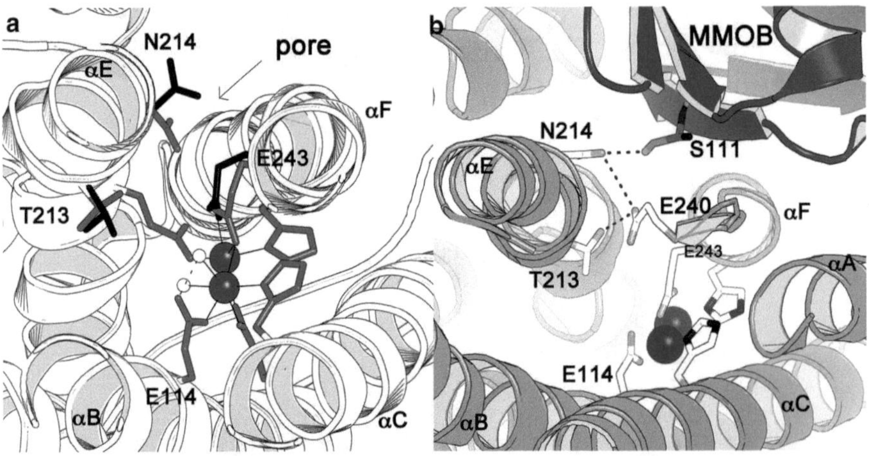

Figure 13 (**a**) The effects of redox state on the configuration of Thr213, Asn214, and Glu243 in MMOH. The side chains in $MMOH_{ox}$ and $MMOH_{red}$ are denoted as black and gray sticks, respectively. (**b**) Hydrogen bonding pattern between Thr213, Asn214, and Glu240 when MMOB is bound to the hydroxylase surface. The nitrogen and oxygen atoms are colored black and gray, respectively.

This shift most likely occurs because reduction breaks water-mediated hydrogen bonding interactions between Glu243 and Asn214. Such a mechanism suggests how movement of residues at the diiron center can trigger conformational changes on the protein surface. The rotomer change in Asn214 leads to the formation of a small pore from the surface to the active site that has been proposed as pathway by which hydrocarbons, dioxygen or protons can access the metal center or by which hydrophilic products like water and alcohols may leave [109, 110]. The structure of the MMOH-MMOB complex, however, reveals that MMOB docks to the hydroxylase on top of the pore and forms a hydrogen bond directly to Asn214 via its conserved residue, Ser111 (Figure 13b) [108]. A similar interaction occurs in structures of related BMM complexes [103, 104]. The exact function of Asn214 in sMMO and other BMM homologues is under investigation.

3.4 *Protein Component Complexes*

Understanding the interactions of MMOR and MMOB with the hydroxylase is essential for unraveling how sMMO efficiently performs its challenging catalytic task. Complexes formed between the different protein components play crucial roles in orchestrating the electron transfer, oxygen-activation, and hydrocarbon-activation events at the MMOH diiron center. Broadly stated, MMOR and MMOB can be viewed as having two very general regulatory functions, to control the movement of substrates to the active site and to alter the hydroxylase structure in such a way as to make the diiron center more reactive. How the components accomplish these tasks remains a focus of intense investigation.

3.4.1 Reductase Binding and Effects on the Hydroxylase

Initial chemical cross-linking studies between *M. trichosporium* (OB3b) MMOR and MMOH indicated that the reductase binds to the β-subunit [120]. More recent cross-linking studies on *M. capsulatus* (Bath) using just the $MMOR_{Fd}$ domain suggest that the primary interaction is actually with the MMOH α-subunit [139]. Spectroscopic studies of the MMOR-MMOH complex have shown that the reductase does not significantly influence properties of the hydroxylase diiron center [8], however, both MMOR and MMOB alter the redox potentials of MMOH [125, 131, 140–143]. Although there is some debate in the literature over their exact values, the general trends reflecting the effects of the different components on the MMOH redox potentials are consistent. Binding of MMOB decreases the potentials of the diiron center by ~100–200 mV, making it harder to reduce, whereas the docking of MMOR restores the potentials to the level observed before MMOB binding to MMOH to favor reduction. These changes are consistent with recent results indicating that MMOB displaces MMOR when it binds to MMOH [144].

Important questions concerning electron transfer in the sMMO system are where MMOR docks to MMOH and what is the electron transfer pathway through the hydroxylase to the diiron active site. An application of Marcus Theory to the electron transfer rates between $MMOR_{Fd}$ and MMOH estimates that the [2Fe-2S] cluster rests ~11–14 Å from the dinuclear iron center [114]. This distance is consistent with theories of efficient electron transfer rates in biological systems that generally position electron donors and acceptors within 14 Å of each other [145]. It should be noted that MMOB and MMOR do not form complexes with one another. Although it was previously suggested that they do not compete for the same binding site on MMOH [111, 120], the most recent studies indicate otherwise. In particular, they demonstrate that the MMOR Fd [Fe_2S_2] cluster docks near MMOH residues N214 and E240, the latter of which undergoes a conformational change that can facilitate the shuttling of both protons and electrons to the diiron center [144, 204]. MMOB, however, has a higher affinity for $MMOH_{red}$ than $MMOH_{ox}$ and can displace MMOR from the reduced hydroxylase. Consequently, MMOB can regulate the electron transfer process and possibly serve to prevent the quenching of the activated oxygen intermediates by MMOR.

3.4.2 Hydroxylase Activation by the Regulatory Protein

Before its crystal structure was determined, extensive spectroscopic interrogation of the MMOH-MMOB complex indicated that MMOB affects the electronic properties of the diiron center but alters its structure little. Correlating specific electronic changes with concrete structural modifications has been difficult, especially when trying to ascertain the structures of the different intermediates in the catalytic cycle. XAS studies of oxidized and reduced MMOH from *Methylococcus capsulatus* (Bath) and *Methylosinus trichosporium* (OB3b), which offer the most direct method for probing changes in coordination geometry, ligand type and number, and metal-ligand distances in the absence of a crystal structure, indicated that MMOB had no remarkable affect on the spectra or fitting parameters of $MMOH_{ox}$ or $MMOH_{red}$ [135, 146]. The only notable difference was that the Debye-Waller factors, a measure of thermal order for a fit atom, decrease [147]. This finding suggested that binding of the regulatory protein to the hydroxylase serves to limit the motion of the metal coordinating ligands.

EPR spectroscopic investigations of $MMOH_{ox}$ that was cryo-reduced at 77 K by γ-radiation to produce an EPR active mixed-valent $Fe^{II}Fe^{III}$ dimetallic center that maintains the structure of $MMOH_{ox}$ revealed that the addition of MMOB has no observable affect on the oxidized diiron(III) cluster [148]. In the presence of products like methanol and phenol, however, spectral changes in γ-irradiated $MMOH_{ox}$ occur when MMOB is added and indicated that MMOB influences the manner by which exogenous ligands bind to the metal center. Consistent with this notion is the observation that MMOB promotes DMSO and glycerol coordination to $MMOH_{ox}$, whereas in the absence of MMOB, the binding of these compounds is undetectable. MMOB also induces a spectral change in the $g = 1.84$ EPR signal of chemically reduced $MMOH_{mv}$, indicating that MMOB can alter the magnetic coupling of the metal center [120]. For $MMOH_{red}$, the characteristic $g = 16$ EPR signal of fully reduced hydroxylase is not altered significantly in the presence of MMOB [120, 132, 133, 136, 137].

Further investigation of $MMOH_{red}$ by circular dichroism (CD) and magnetic circular dichroism (MCD) spectroscopy demonstrated that MMOB influences the structure of the fully reduced diiron center by altering the ligand field environment of only one of the iron atoms [149, 150]. The ligand field environment of the second iron atom only changes when MMOB is bound to MMOH in the presence of substrates and inhibitors. Although it is unknown which specific iron atom experiences these different effects, the data appear consistent with EPR results suggesting that MMOB helps to generate open coordination sites at the diiron center for dioxygen activation [149, 150]. Rapid freeze quench (RFQ) EPR spectroscopic experiments reveal that MMOB increases the rate of O_2 association with the enzyme by 1000-fold, which corroborates this assessment and demonstrates how MMOB can either accelerate the formation of intermediates or facilitate dioxygen access to the active site [121]. The structural changes observed in the MMOH-MMOB complex tend to support the latter function [108]. Recent double electron-electron resonance spectroscopic studies indicate that the redox state of the diiron center strongly effects the conformation of the MMOB N-terminus on the surface of

MMOH. In particular, binding to the reduced protein better orders the flexible N-terminal tail of MMOB, allowing Tyr8 to interact with MMOH and exert its allosteric effects that allow passage of substrate to the cavity at the active site diiron center [144].

A phenomenon commonly observed in the steady state turnover of sMMO is a change in the product distribution when MMOB is added to the reconstituted system [151]. For example, without MMOB the hydroxylation of propane, butane, and nitrobenzene by MMOH yields 61 % 1-propanol, 6 % 1-butanol, and 10 % *p*-nitrophenol. With MMOB, yields of 93 % 1-propanol, 56 % 1-butanol, and 90 % *p*-nitrophenol are achieved. The general trend for the many substrates that have been tested with sMMO is that MMOB shifts regiospecificity such that mostly primary alcohols and para-substituted aromatic alcohols are formed. In other words, MMOB changes the morphology of the active site such that the primary carbon on alkanes and the C4 position on mono-substituted aromatics are preferentially presented to the activated oxygen species at the diiron center. Similar changes in regiospecificity are observed in the toluene monooxygenase systems when the regulatory protein is added [152], suggesting the general structural effect is conserved among BMMs. Initial insight into the regulatory protein-dependent regiospecificity changes was provided by the structure of a 6-bromohexanol-soaked MMOH crystal in which an unexpected α- to π-helix transition in MMOH helix E elongated the active site pocket and increased its overall volume [109]. The structures of the MMOH-MMOB, PHH-PHM, and T4MOH-T4MOD complexes later revealed a similar structural change, in addition to new ones that would explain some, but not all, of the observations described above.

3.4.3 Structures of Regulatory Protein-Hydroxylase Complexes

The long sought structure of the MMOH-MMOB complex was solved to 2.9 Å resolution and published in 2013 (Figure 10) [108]. The structure not only confirmed many previously generated ideas about the complex based on the available biochemical information but also revealed intriguing new structural changes that stand in contrast to the structures of related BMM hydroxylase-regulatory protein complexes. The folded MMOB core docks onto MMOH α-subunit helices A, E, and F in the canyon of the $\alpha_2\beta_2$ interface while the long N-terminal tail forms a ring-shaped structure on the α-subunit surface over helices H and 4 (Figure 8). MMOB-induced conformational changes in the hydroxylase are observed primarily in helices E, F, H, and 4. As observed in the 6-bromohexanol-soaked structure of MMOH, residues 212–216 on helix E undergo a small α- to π-helix transition, resulting in conserved Thr213 shifting away from the active site toward the interface between helices E and F (Figure 14). Asn214 moves toward the hydroxylase surface to hydrogen bond with Ser111 of MMOB, and Leu216 slides into the back of the active site cavity 1. The net effect is a lengthening of the substrate binding pocket. Glu209, a ligand to Fe2, is shifted slightly away from its original position in the MMOH structure, but the coordination geometry to the metal center appears unchanged.

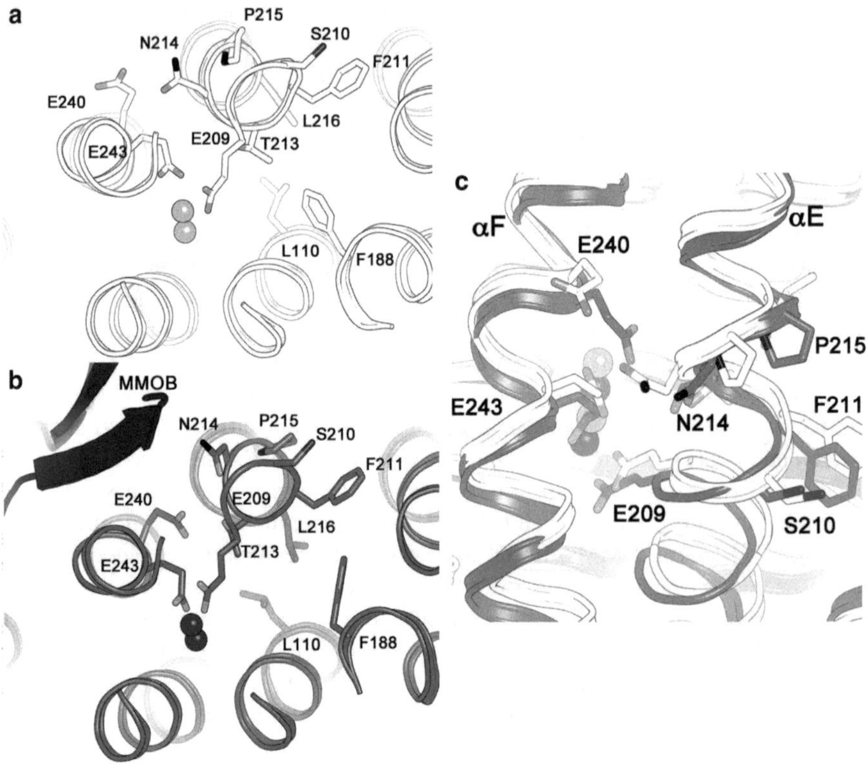

Figure 14 MMOB-induced conformational changes in α-subunit helices E and F. (**a**) $MMOH_{ox}$, (**b**) MMOH-B complex. (**c**) Top down view comparing the helical changes in the $MMOH_{ox}$ (light gray) and MMOH-B (dark gray) structures.

On helix F, residues 240–244 wind more tightly, resulting in the rotation of Glu240 toward the interface between helices E and F (Figure 14). This configuration is stabilized by hydrogen bonding interactions between Glu240 and Thr213 on MMOH and Ser111 on MMOB (Figure 13b). Concomitant with these movements in helices E and F, Glu243 undergoes a carboxylate shift (Figure 15). The positional adjustments in helices H and 4 most likely stem from a repacking of the interface between helices E and F and an optimization of interactions with the MMOB N-terminus. These latter adjustments do not appear to directly influence the active site or the various cavities within the hydroxylase, but may have other, yet undetermined roles. Similar re-arrangements of helices E and F occur in the phenol hydroxylase and toluene 4-monooxygenase systems [103, 104].

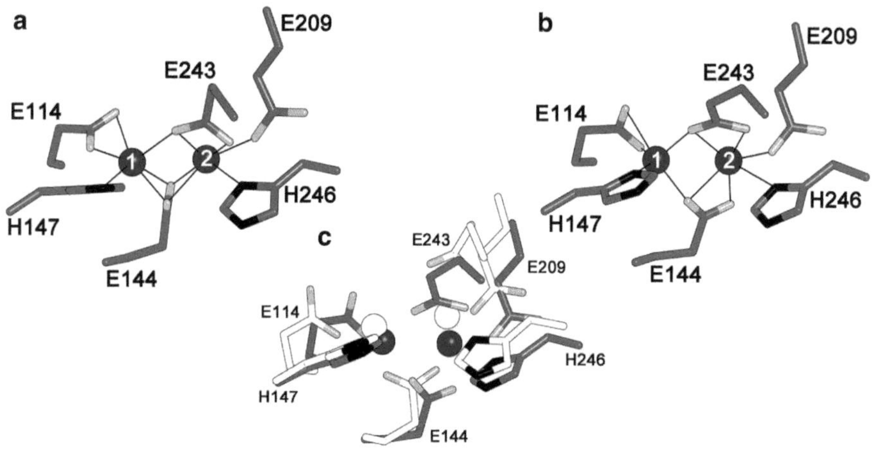

Figure 15 Structure of the MMOH-B diiron active site from two different protomers (**a**, **b**). (**c**) Overlay of the $MMOH_{ox}$ (white) and MMOH-B (gray) diiron centers.

3.4.4 The Activated Soluble Methane Monooxygenase Diiron Center

From the MMOH-MMOB structure, it is clear that the changes in the conformations and positions of helixes E and F propagate toward the diiron center and affect the coordination environment. The "observed" changes, however, must be interpreted with a degree of caution. First, the limited 2.9-Å resolution of the structure does not allow for clear visualization of solvent-derived ligands and will affect the accuracy by which it is possible to model the positions of the amino acid ligands. Second, although the oxidized form of the MMOH-MMOB complex was crystallized, it is conceivable that the X-ray beam led to cryo-reduction of the metal center by one or two electrons to form the mixed-valent or fully reduced state, respectively. As a result, the oxidation state of the diiron center in the MMOH-MMOB structure is uncertain. The metal center has geometric features that are similar to those of both $MMOH_{ox}$ and $MMOH_{red}$ (Figure 15). The carboxylate shift in Glu243 is similar to the one in $MMOH_{red}$ but the average Fe–Fe distance in each of the four metal centers found in the asymmetric unit of the crystal is 3.1 Å, a value closer to the 3.0–3.1 Å distance observed in $MMOH_{ox}$ as opposed to the 3.3–3.4 Å distance in $MMOH_{red}$ and $MMOH_{mv}$ [130].

Although it is a low-resolution structure, some of the iron ligands in the MMOH-MMOB complex seem to adopt new geometries. Glu114 appears to assume a bidentate chelating mode of coordination to Fe1. Glu144 also adjusts its coordination to the iron atoms. In one active site, the side chain rotates by 90° such that the two carboxylate oxygen atoms are orthogonal to the Fe–Fe vector (Figure 15a). In another active site, one oxygen atom bridges both irons and Fe2 is coordinated in a bidentate chelating fashion (Figure 15b). These new configurations for Glu114 and Glu144 may be in response to gross lateral movement of helices E and F that force Glu209, Glu243, His246, and the iron atoms to translate toward the C-terminus of helix F (Figure 15c and 16).

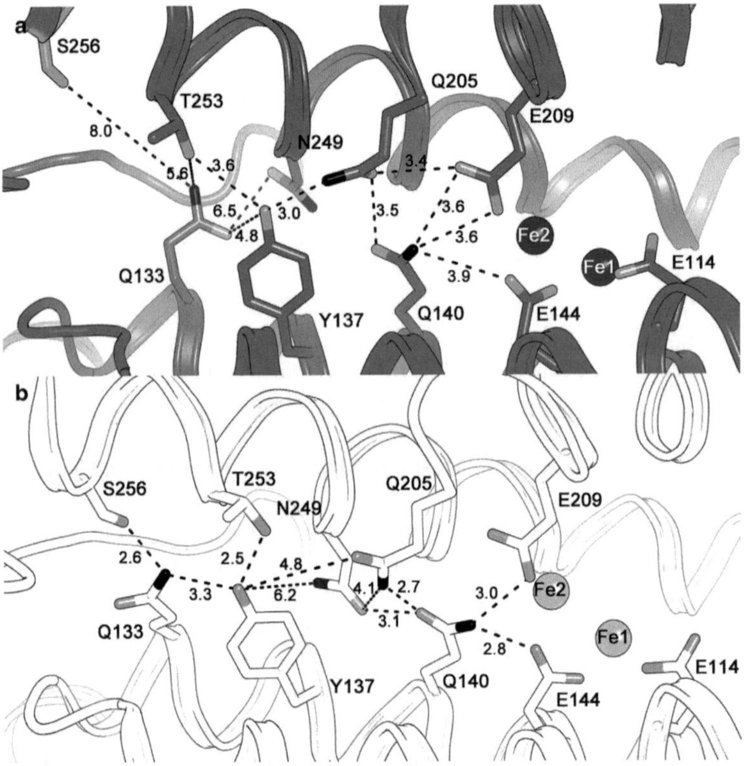

Figure 16 Second and third coordination sphere hydrogen-bonding patterns near Fe2 and Glu209 in (**a**) the MMOH-B complex and in (**b**) $MMOH_{ox}$.

An overlay of the $MMOH_{ox}$ and MMOH-MMOB structures shows the positions of Fe1 and Fe2 to shift by 0.7 Å and 1.3 Å, respectively, to accommodate the helical motions and new positions of some of the amino acid ligands. Because helices B and C do not move much, it is possible that the side-chains of Glu114, Glu144, and His147 on these helices are forced to adjust accordingly to maintain interactions with the iron atoms (Figures 15 and 16). To support this lateral movement along the axis of the four-helix bundle, the hydrogen bonding patterns of the second coordination sphere near Glu209 and Fe2 rearrange. Most notably, Gln140 is no longer in contact with Glu209 and Glu144 (Figure 16).

3.4.5 Comparisons to Toluene Monooxygenases and Phenyl Hydroxylase

One important question in the BMM field is, "Why do the other BMMs not hydroxylate methane?" In the absence of the regulatory protein, the structures of the oxidized and reduced ToMOH and T4MOH diiron centers are nearly identical to those in MMOH [102, 104, 153]. The most glaring difference between sMMO and the TMOs is the occurrence of a long, 30-Å channel in the α-subunit that is required to transport aromatic substrates to the diiron center. This channel contrasts with the cavities in MMOH that follow a similar, but not identical, route through the α-subunit. It is unlikely that the channel affects hydroxylation chemistry, although the greater solvent access in TMOs was a concern because high-valent intermediates could be quenched if the activated metal center were not protected [102].

The global structures of the PHH-PHM and T4MOH-T4MOD complexes are similar to that of MMOH-MMOB with the regulatory protein binding over α-subunit helices A, E, and F in the hydroxylase canyon region. Despite having a much shorter N-terminus, the regulatory protein in these related BMM systems induces nearly identical conformation changes in helices E and F near the diiron center, demonstrating a conserved method for "activating" the hydroxylases. The one surprise from these structures is that the helical changes collapse the substrate channel to create a single pocket near the active site, while simultaneously opening up access to other α-subunit cavities [104]. These cavities were later identified as a major route for O_2 access to the metal center in ToMO [65].

The electron density around the diiron center of the 2.3 Å PHH-PHM structure is difficult to interpret and there are questions about the oxidation state of the metal center and overall occupancy of PHM in the crystal structure. For T4MOH-D, structures of the oxidized and reduced diiron centers look similar to the ones found in $MMOH_{ox}$ and $MMOH_{red,}$ (Figures 12 and 17). Although it is difficult to make informative comparisons to the metal center geometry observed in the low resolution structure of the MMOH-B complex at this time, a significant difference between the structures of TMO and sMMO hydroxylase-regulatory protein complexes lies in the second coordination sphere.

Near Fe1 in T4MOH, Glu104 hydrogen bonds with Gln141, which may serve to constrain the position of this iron ligand, as does a water molecule in MMOH (Figure 17) [203]. If Glu114 does in fact alter its configuration in the MMOH-MMOB complex, this difference could be one contributing factor. On the opposite side of the active site near Fe2, the hydrogen bonding patterns are unaltered in the T4MOH and T4MOH-D structures, suggesting a rigid assembly of the T4MOH metal center. By contrast, this network reconfigures in MMOH when the regulatory protein is bound (Figure 16). A relaxing of hydrogen bonding interactions between Gln140 and the coordinating Glu209 and Glu144 ligands possibly affords the sMMO diiron center more conformational freedom. It is tempting to speculate that the additional conformational constraints in other BMM systems prevent them from achieving the high-valent Q intermediate observed sMMO responsible for activating high energy C–H bonds in alkanes.

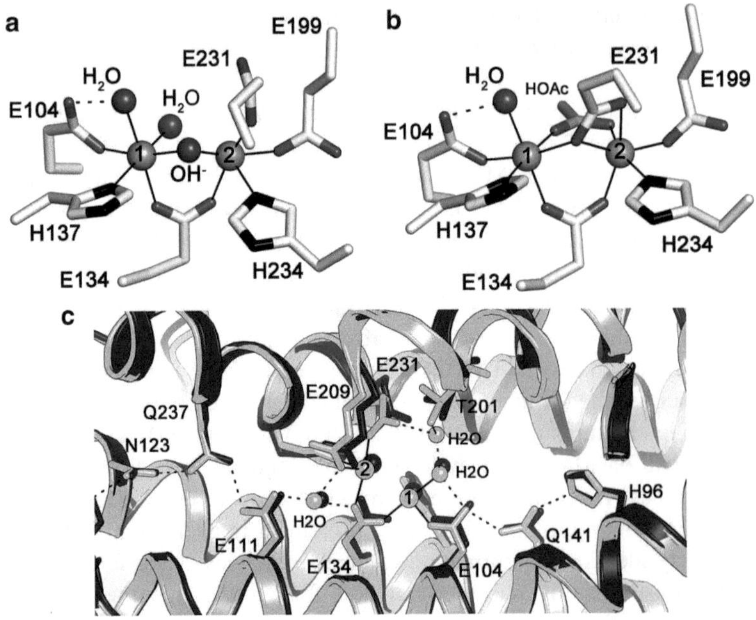

Figure 17 Structures of the (**a**) oxidized (PDB entry 3DHH) and (**b**) reduced (PDB entry 3DHI) T4MOH-D diiron centers. (**c**) Overlay of the second and third coordination sphere hydrogen-bonding patterns near Fe1 and Fe2 in the oxidized T4MOH (dark gray) and the T4MOH-D (light gray) structures.

3.5 Substrate Access to the Catalytic Diiron Center

A challenge for all BMM systems is to deliver substrates to the diiron center with the appropriate timing to support catalysis (Figure 11). The structure of the hydroxylase has suggested defined routes for O_2, hydrocarbon, electrons, and protons to the active site. The timing of delivery of these substrates is critical, not only for maximum catalytic efficiency, but also to avoid unwanted events such as quenching-activated oxygen intermediates. Such quenching leads to the uncoupling of electron consumption from productive utilization of oxygen to convert methane to methanol. Otherwise, MMOH can become an NADH oxidase. In the next section we describe pathways for substrate access and product egress during the catalytic cycle.

3.5.1 Cavities for O_2 and Hydrocarbons

Several hydrophobic cavities have been identified in the MMOH α-subunit, three of which, (cavities 1, 2, and 3) bind xenon, halogenated alkanes, and alcohols of up to eight carbon atoms in length (Figure 8) [109, 110, 154]. The entrance to each cavity is gated by a set of hydrophobic residues that adopt alternate rotomer conformations in MMOH crystal structures with different product alcohols bound. Cavities 1 and 2 are discontinuous and separated by Leu110 (not shown) and Phe188 (Figure 18a). Cavities 2 and 3 are continuous, but somewhat restricted near Val105, Phe109 and Leu289. The opening and closing of these gates have been proposed as a means for controlling substrate entrance to and/or product egress from the active site pocket (cavity 1) during catalysis [109, 155]. Structures of $MMOH_{red}$ show no changes in cavity structure or positioning of the amino acid gates. Binding of MMOB to the hydroxylase, and subsequent changes in helices E and F, however, force a repositioning of the Leu110 and Phe188 side chains, connecting cavities 1 and 2 such that a pathway extends from the protein surface through the α-subunit to the active site (Figure 18b and 18c). Thus, one of the major functions of MMOB is to facilitate the creation of a freely diffusible space for methane and dioxygen to access the diiron center [108].

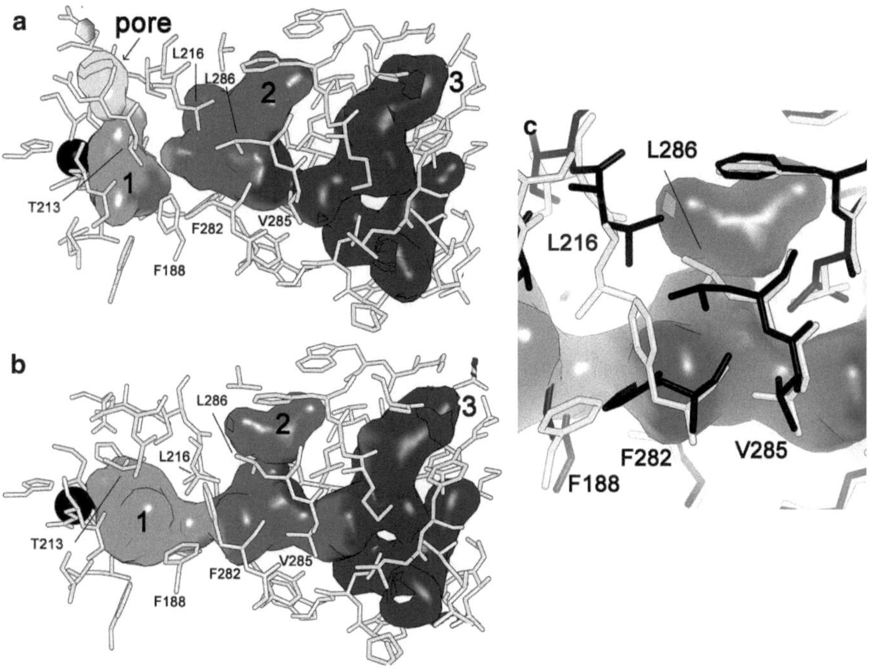

Figure 18 Cavities (1, 2, 3) in the (**a**) MMOH and (**b**) MMOH-B structures. Labeled residues exhibit the largest positional changes. (**c**) Overlay of the MMOH (black) and MMOH-B (light gray) structures depicting positional differences at the interface between the cavities. Cavity surfaces correspond to those in the MMOH-B complex. Val105, Phe109, and Leu289 at the interface between cavities 2 and 3 are not visible in this orientation of the structure. The figure was adapted from [108].

The best evidence that these cavities are functional comes from studies of the TMO system. The structure of the T4MOH-D complex reveals a similar series of cavities through the hydroxylase α-subunit once the regulatory protein binds and collapses the aromatic substrate channel leading to the active site [104]. Taking advantage of a Thr201Ser mutation in ToMOH that allows observation of the reaction of O_2 with the reduced diiron center, optical stopped-flow spectroscopic investigations of ToMOH variants that restrict or enhance O_2 passage through the cavities revealed that the hydroxylase uses this ~40 Å route as a primary pathway for controlling O_2 diffusion to the iron center [65]. The inability to express sMMOH in a heterologous host has thus far prevented similar experiments being conducted with this enzyme.

In addition to the cavities, a hydrophilic pore formed between MMOH helices E and F involving Asn214, Thr213, and Glu240 was hypothesized to provide the most direct route for the passage of small gaseous substrates or product alcohols to and from the diiron center [109]. Redox dependent shifts in Asn214 rearrange the orientation of amino acid side chains comprising the pore. Because the regulatory protein docks on top of these helices in the different BMM hydroxylase-effector complexes, the pathway leading from the diiron center to the surface through this pore is blocked. Biochemical data showing that MMOB bound to the hydroxylase surface significantly decreases the rate of adventitious iron loss from reduced MMOH, further supports the notion that MMOB can restrict the passage of molecules to and from the diiron center through the pore [92].

Occlusion of the pore by MMOB as well as its hydrophilic character suggest that the pore is unlikely to provide the route of hydrophobic gases to the active site during the catalytic cycle, but dissociation of MMOB could generate a route for hydrophilic methanol and water release at the conclusion of the reaction cycle [108]. The importance of the pore for product release is supported by steady-state and pre-steady state studies of MMOH using an MMOB Asn107Gly/Ser109Ala/Ser110Ala/Thr111A quadruple variant that effectively sculpted the MMOB surface oriented over the hydroxylase diiron center [156]. When monitoring reactions with large substrates like nitrobenzene, this quadruple variant generated a two-fold increase in both steady-state turnover and the rate of product release from the diiron center following oxidation. The rates of methane hydroxylation and product release were unaffected. Work with the quad mutant suggests that product alcohols can escape the enzyme through a pathway at the MMOH-B interface. A major conclusion from all of this work is that the regulatory protein exerts significant control over the movement of compounds into and out of the active site at different stages of the catalytic cycle.

3.5.2 Proton Delivery

The delivery of protons is required for reduction of the dimetallic center and the O_2 activation steps of the sMMO catalytic cycle [157–159]. Solvent is considered the primary source of the protons, but their pathway to the diiron center is largely unknown. The structural changes in MMOH helices E and F near the pore suggest a mechanism for proton delivery whereby movement of Glu240 from the hydroxylase surface toward the interior initiates opening of the pore and formation of a hydrogen bonding network at the MMOH-MMOB interface using conserved Asn214, Thr213, and Ser111 on MMOB (Figures 13 and 14). Because these residues are unlikely to directly ferry a proton, we propose that water or hydronium ions, such as the one bridging Gln230 (Glu240 in MMOH) and the terminal water coordinated to Fe1 in the T4MOH-D complex, may be important participants in this relay process (Figure 17c).

Unfortunately, little work has been done to investigate the proposed proton transfer pathway except for studies on the role of the conserved threonine in TMOs. Mutagenesis of this residue in T4MOH and ToMOH revealed that substitution of either hydrophobic or hydrophilic residues at this position yields a less active enzyme with efficient coupling but altered substrate regiospecificity [159, 160]. The T4MOH Thr201Ala variant resulted in significant release of hydrogen peroxide from the enzyme following reaction with $T4MOHD_{red}$ with O_2, suggesting this threonine may help to stabilize the appropriate iron-bound activated oxygen intermediates [161]. Further study in ToMOH indicated that Thr201 mutants affect the formation and decay kinetics of a peroxodiiron(III) intermediate, and that this residue plays a major role in the proton transfer steps required to generate this species [159, 162]. If the route for proton relay involves Glu240, Asn214, and Thr213, initializing this pathway relay for the controlled delivery of protons through the pore may be one of the primary functions of regulatory proteins in the BMM family.

3.6 Dimetallic Activation of O_2 and Methane

Efforts to unravel the sMMO mechanism have focused on how MMOH activates dioxygen, the identity of intermediates that react with hydrocarbons, and how C–H bonds are broken. A detailed understanding of the structures and properties of the different intermediates has been of interest as many laboratories aim to generate synthetic hydrocarbon oxidation catalysts based on a dimetallic iron unit. Mechanistic studies have identified several intermediate species during rapid kinetic experiments, some of which have overlapping properties that are difficult to deconvolute [9]. The fact that more than one intermediate can react with hydrocarbon substrates at one point in time led to confusion about the nature of the C–H bond activation and oxygen insertion process [11]. Among the laboratories [9, 10] focused on establishing a mechanism for the *M. capsulatus* (Bath) *(Mc)* and *M. trichosporium* Ob3b (*Mt*) systems, there is general consensus as to the nature of the major intermediates in the sMMO reaction cycle (Figure 11). Some disagreement still exists as to what their exact structures are and how one might transition to the next. In this section we first discuss single-turnover kinetic studies aimed at identifying the transient oxygen intermediates and their reactivity with substrates, followed by studies focusing on the C–H bond-breaking steps.

3.6.1 Reaction of O_2 with the Reduced Hydroxylase

The transfer of two electrons and two protons to the diiron(III) center of $MMOH_{ox}$ initiates the sMMO catalytic cycle by generating fully reduced $MMOH_{red}$. Its subsequent reaction with O_2 is most efficient when two equivalents of MMOB are present, suggesting MMOB-induced conformational changes gate O_2 access to the activated metal center [121]. O_2-promoted decay of $MMOH_{red,}$ which has been monitored at 4 °C in *M. capsulatus* (Bath) by RFQ Mössbauer and in *M. trichosporium* OB3b by EPR spectroscopy, was measured to be 24 s^{-1} at pH 7.0 and 22 s^{-1} at pH 7.7, respectively [141, 163]. The intermediate that forms from the decay of $MMOH_{red}$ is P* (vide infra), but the absence of any observed pH or O_2 concentration dependence for the decay of *Mt* $MMOH_{red}$ suggested that a rapid irreversible O_2 binding event occurs before P* in which O_2 is bound to the hydroxylase but not to the diiron center. This intermediate, termed O, alternatively was interpreted to represent a tight but reversible O_2 Michaelis complex to $MMOH_{red}$. However, steady state experiments on *Mc* MMOH using an oxygen electrode failed to provide any evidence for such a complex [164].

3.6.2 Peroxo Intermediates

After reaction with O_2, peroxodiiron(III) intermediates are formed. Mössbauer data at 4 K on rapid freeze-quenched samples of $MMOH_{red}$ mixed with O_2 reveal a quadrupole doublet with $\delta = 0.66$ mm s^{-1} and $\Delta E_Q = 1.51$ mm s^{-1} [165, 166]. These parameters are indicative of an antiferromagnetically coupled, high-spin diiron(III) center in which each Fe^{III} ion has a similar coordination geometry. UV-vis stopped-flow spectroscopy revealed optical features at 420 nm and 720 nm in *Mc* MMOH and 700 nm in *Mt* MMOH [141, 166, 167]. Although the decay constants of this intermediate measured by RFQ Mössbauer and stopped-flow UV-vis spectroscopy were similar in *Mc* MMOH (~0.36 s^{-1} at 4 °C), the measured rate constants obtained by these methods for the formation of the peroxo intermediate were significantly different, namely, ~1 s^{-1} by stopped-flow and ~25 s^{-1} by Mössbauer [167]. This discrepancy suggested the existence of two peroxodiiron(III) species, P* and H_{peroxo}, having similar Mössbauer properties but different, overlapping optical signals. A re-examination of the intermediates in *Mt* MMOH by stopped-flow UV-vis spectroscopy required an additional intermediate preceding H_{peroxo} to fit the data. This intermediate, P*, which was presumed to have no optical features at 420 nm, formed with a rate constant of 22–26 s^{-1} at 4 °C followed by H_{peroxo} at ~9–12 s^{-1} [168, 169]. This rate was consistent with earlier observations, establishing P* as the first intermediate resulting from $MMOH_{red}$ reacting with O_2. More recently, in a MMOB His33Ala variant in *Mt* MMOH, the conversion of P* to P was retarded, facilitating the trapping of larger quantities for characterization [168]. This intermediate had Mössbauer parameters indicative of a diiron(II) center, no significant optical features, and no $g = 16$ EPR signal that is characteristic of the diiron(II) center in $MMOH_{red}$. In contrast, subsequent UV-vis stopped flow studies of *Mc* MMOH in the presence of methane, to eliminate underlying optical signals of intermediate Q, suggest that P* is an optically active diiron(III) center with absorption maxima at 420 nm ($\varepsilon = 3500\,M^{-1}\,cm^{-1}$) and 720 nm ($\varepsilon = 1250\,M^{-1}\,cm^{-1}$) that overlap with the ensuing H_{peroxo} and Q intermediates [205]. It is unclear, based on the latter two investigations, whether or not P* in *Mc* MMOH is the same intermediate as that observed in *Mt* MMOH.

Table 2 Spectroscopic Parameters of Peroxodiiron(III) Intermediates.

	Optical		Mössbauer		
	λ_{max} (nm)	ε ($M^{-1}\,cm^{-1}$)	δ (mm s^{-1})	ΔE_Q (mm s^{-1})	Peroxide Binding Mode
MMOH H_{peroxo} (*Mc*)	420; 720	3880; 1350[a]	0.66	1.51[b]	
MMOH H_{peroxo} (*Mt*)	725	2500[c]	0.67	1.51[d]	
MMOH P* (*Mc*)	420; 720	3500; 1250[a]			
$ToMOH_{peroxo}$			0.54	0.67[e]	
ToMOH $Thr201Ser_{peroxo}$	675	1500[f]	0.67	1.51[f]	
RNR-R2 Asp84Glu	700	1500[g]	0.63	1.58[g]	μ-1,2[h]
Δ9-desaturase	700	1200[i]	0.68; 0.64	1.90; 1.06[i]	μ-1,2[j]
Frog M ferritin	650[k]		0.62	1.08[k]	μ-1,2[l]
hDOHH[m]	630	2800	0.55; 0.58	1.16; 0.88	μ-1,2
$[Fe_2(\mu\text{-}1,2\text{-}O_2)\text{-}(\mu\text{-}O_2CCH_2Ph)_2\text{-}\{HB(pz')_3\}_2]$[n]	694	2650	0.66	1.40	gauche μ-1,2
$[Fe_2(\mu\text{-}OH)(\mu\text{-}1,2\text{-}O_2)\text{-}(6\text{-}Me_2\text{-}BPP)_2]^{+}$[o]	644	3000	0.50	1.31	*cis*-μ-1,2
$[Fe_2(\mu\text{-}O)(\mu\text{-}1,2\text{-}O_2)\text{-}(6\text{-}Me_3\text{-}TPA)_2]^{2+}$[p]	494; 648	1100;1200	0.54	1.68	*cis*-μ-1,2

[a][157]
[b][141]
[c][158]
[d][166]
[e][173]
[f][162]
[g][190], RNR-R2 = ribonucleotide reductase R2 subunit, Asp84Glu mutation.
[h][191]
[i][192]
[j][193]
[k][194]
[l][195]
[m][196], hDOHH = human deoxyhypusine hydroxylase.
[n][197], pz' = 3,5-bis(isopropyl)-pyrazolyl.
[o][198], BPP = *N,N*-bis(2-pyridylmethyl)-3-aminopropionate.
[p][199], TPA = tris(2-pyridylmethyl)amine.

Assuming P* to be a diiron(III) complex, the observation of optical bands is key, for it allows for more informed comparisons to peroxodiiron(III) species in well characterized model complexes and enzymes (Table 2). Based on similarities in their optical and Mössbauer properties, P* and H_{peroxo} are expected to have a μ-1,2-peroxo diiron(III) core but probably differ in the configuration of surrounding ligands. Possible peroxo structures are presented in Figure 19a and take into consideration the different binding modes of Glu-243 (**1–8**). The gauche μ-1,2 configuration of the $[Fe_2(\mu\text{-}1,2\text{-}O_2)(\mu\text{-}O_2CCH_2Ph)_2\{HB(pz')_3\}_2]$ model complex is the most interesting model because its Mössbauer parameters are nearly identical to those of P* (**3** or **4**) [170]. Computational work also seems to favor this binding mode as opposed to *cis*-μ-1,2 and μ- η^2:η^2 geometries (**1**, **2**, and **7**) [171, 172]. Interestingly, the optical and Mössbauer properties of the ToMOH peroxo intermediate are quite different from those of the MMOH peroxo intermediate, suggesting that the structure of the ToMOH peroxo species (possibly **5** or **6**) is the point of departure in their mechanisms [172–174].

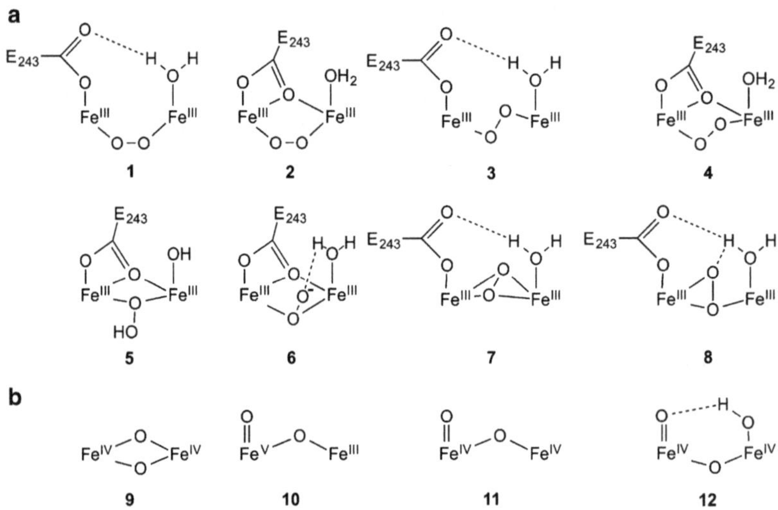

Figure 19 (**a**) Proposed structures of peroxodiiron(III) intermediates. (**b**) Possible structures for intermediate Q. The figure was adapted from [9].

The conversion of P* to H_{peroxo} is pH-dependent and has a kinetic solvent isotope effect (KSIE) in D_2O of k_H/k_D=2.0 for *Mc* MMOH and k_H/k_D=1.3 for *Mt* MMOH [157, 158]. Fits to the data from both enzyme systems are consistent with the kinetic model 1 in Figure 20. Given the similar spectroscopic parameters of these intermediates, the major structural difference between P* and H_{peroxo} is probably a proton transfer event that does not significantly perturb the diiron(III)-oxygen core (Figure 11). Because the spectroscopic properties of peroxo- and hydroperoxo-diiron(III) species are expected to be significantly different [175], Glu243 was proposed to be the recipient of this proton [9]. Further work is required to confirm the structures of the P* and H_{peroxo} intermediates, which are important unsolved questions.

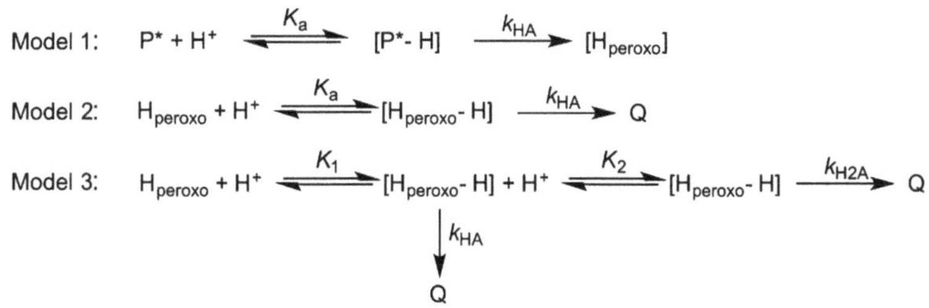

Figure 20 Kinetic models for the sMMO proton transfer steps. The figure was adapted from [9].

3.6.3 Intermediates Q and Q*

Electron-rich substrates, like propylene and diethylether, react with H_{peroxo} to form oxidized products and $MMOH_{ox}$ via a two-electron transfer mechanism [167, 176, 177] (Figure 11). In the absence of such substrates, homolytic cleavage of the H_{peroxo} O–O bond leads to formation of intermediate Q, the species responsible for activating the C–H bond in methane. Q is an antiferromagnetically coupled diiron(IV) intermediate with a bright yellow color and absorption maxima at 330 nm and 420 nm (Table 3) [141, 163, 165, 178]. The measured rate constants for Q formation, 0.36 s^{-1} and 2.5 s^{-1} at 4 °C in *Mc* and *Mt* MMOH, respectively, coincide with the rate of H_{peroxo} decay [157, 169]. Mössbauer spectral studies of Q from *Mt* MMOH show a single quadrupole doublet, indicating that both Fe(IV) atoms have similar ligand environments [178]. This information, coupled with an XAS spectrum of *Mt* Q fit with a short Fe–Fe distance of 2.46 Å and Fe–O/N bond lengths of 1.77 Å and 2.05 Å, lead to the interpretation of Q as being a di(μ-oxo)diiron(IV) species (Figure 19b, **9**) [166].

In the presence of methane, Q converts to $MMOH_{prod}$ (or $MMOH_{ox}$) with rate constants of 3.9 s^{-1} and 2.9 s^{-1} at 4 °C for *Mc* and *Mt* MMOH, respectively [128, 157, 169]. In the absence of substrates, Q decays to $MMOH_{ox}$ by acquiring two electrons and two protons by a relatively uncharacterized process. An intermediate in this decay pathway, Q*, was identified by UV-vis stopped-flow spectroscopy to have an absorption maximum at 420 nm, like Q, but a unique shoulder at 455 nm that is unaffected by substrate concentration [157]. The formation and decay rates for Q* were measured at 4 °C in *M. capsulatus* (Bath) MMOH to be 0.01 s^{-1} and 0.0028 s^{-1}, respectively. The reason for the formation of such an intermediate is unclear, but, the collapse of Q to a non-reactive thermodynamically

Table 3 Spectroscopic Parameters of Intermediate Q and High-Valent Non-heme Diiron Complexes.

	Optical		Mössbauer		
	λ_{max} (nm)	ε (M^{-1} cm^{-1})	δ (mm s^{-1})	ΔE_Q (mm s^{-1})	Fe–Fe (Å)
MMOH Q (*Mc*)	420	8415[a]	0.21; 0.14	0.68; 0.55[b]	
MMOH Q (*Mt*)	330; 420	7500; 7500[c]	0.17	0.53[d]	2.46[e]
MMOH Q* (*Mc*)[f]	420 (455 shoulder)				
$[Fe_2(\mu\text{-}O)_2(L)_2]^{3+}$	366; 616	7900; 5200[g]	0.48; 0.08	1.6; 0.5[h]	2.68[i]
$[Fe_2(\mu\text{-}O)_2(L)_2]^{4+}$[j]	485; 875	9800; 2200	−0.04	2.09	2.73

[a][157]
[b][141]
[c][163]
[d][178]
[e][166]
[f][157]
[g][200], L = 5-Me_3-TPA; TPA = tris(2-pyridylmethyl)amine.
[h][201], L = 6-Me_3-TPA.
[i][202], L = 5-Et_3-TPA.
[j][183], L = tris((4-methoxy-3,5-dimethylpyrid-2-yl)d_2-methyl)amine.

more stable intermediate such as Q* in the absence of substrate may protect the active site from radical initiated damage by the more reactive Q species.

The conversion H_{peroxo} to Q is pH-dependent and exhibits KSIEs of $k_H/k_D = 1.8$ for *Mc* MMOH and $k_H/k_D = 1.4$ for *Mt* MMOH, making Q formation a proton-dependent step [157, 158]. Q to Q* conversion, on the other hand, is pH-independent. Kinetic data for the pH dependent conversion of H_{peroxo} to Q are best fit in *Mt* MMOH by using single ionizing system and a pK_a of 7.6 (Figure 20, kinetic model 2). For this same step, *Mc* MMOH requires a doubly ionizing system with pK_a values of 7.8 and 7.2, suggesting a second proton transfer event during this transformation (Figure 20, kinetic model 3). Accounting for all the available kinetic data, proton-assisted heterolytic cleavage of the O–O bond best describes the possible series of transitions from P* to Q in *Mc* sMMO (Figure 11) [9, 157].

The discovery of Q* has added to a growing list of evidence suggesting the structure of Q needs further evaluation. Unlike the Mössbauer spectrum of *Mt* Q in which one quadrupole doublet was observed, early analysis of *Mc* Q samples showed two quadrupole doublets of equal intensity [141, 165], suggesting the presence of a second similar species that may be Q*. The long lifetime of Q* is a concern with regard to the preparation of samples for RFQ Mössbauer and XAS. For example, in the 186 seconds age time of the optical stopped flow experiments performed in the absence of substrate to examine the kinetics of the different Q intermediates, it was estimated that 19 % and 60 % of the species present were Q and Q*, respectively [157]. It is conceivable that fits to XAS data on *Mt* MMOH modeled Q* or an average of the two diiron(IV) intermediates as opposed to Q alone. Adding to the concern, many DFT calculations fail to reproduce the short Fe–Fe distance observed by XAS [179, 180]. To generate this intermediate, protein scaffold may compress the iron atoms in order to achieve favorable energetic conversion from H_{peroxo} to Q, as suggested by QM/MM studies [180].

Several synthetic catalysts that achieve a high-valent diiron species capable of oxygen insertion between C–H bonds offer some insight into alternative possibilities for the structure of Q, in which the oxidizing potency of the complex is concentrated at a single $Fe^{IV}{=}O$ unit (Figure 20). As an example, a $[Fe_2(H_2Hbamb)_2(N\text{-melm})_2]$ complex oxidized cyclohexane to cyclohexanol via an $Fe^{II}Fe^{IV}$ intermediate with a terminal $Fe^{IV}{=}O$ species [181]. An analogous $Fe^{III}Fe^{V}$ center was suggested for Q in MMOH (**10**). A valence-delocalized $[OH{-}Fe^{III}{-}O{-}Fe^{IV}{=}O]^{2+}$ unit complexed to tris((4-methoxy-3,5-dimethyl-pyrid-2-yl)d_2-methyl)amine oxidized C–H bonds a million times faster than a valence-delocalized "diamond core" analogue, $[Fe^{3.5}(\mu\text{-}O)_2Fe^{3.5}]^{3+}$ [182]. In one instance, oxidation by a complex with a $[Fe^{IV}(\mu\text{-}O)_2Fe^{IV}L^2]^{4+}$ core was 100 times slower than the corresponding mononuclear Fe(IV)=O complex using the same ligand [183]. These findings suggest that a terminal Fe=O unit (**11** and **12**) may be more effective than a di(μ-oxo)diiron(IV) unit for C–H bond activation.

3.6.4 The Product-Bound Hydroxylase

The oxidation of substrates by Q leads to the product-bound state, $MMOH_{prod}$. This intermediate has only been kinetically observed for oxidation of nitrobenzene to *p*-nitrophenol. Here, product release from the hydrophobic active site promotes formation of the optically detectable *p*-nitrophenolate ion in solution [141, 163]. For this substrate, the rate of steady-state turnover is similar to the rate of product release, indicating product release to be the rate-limiting step in the reaction, at least when nitrobenzene is the substrate. Product displacement may be accomplished either by ligand exchange with solvent or by reduction of the metal center to reinitiate the catalytic cycle. Further studies of product release with other substrates more similar to methane would be valuable.

X-ray structures of $MMOH_{ox}$ crystals soaked with halogenated and non-halogenated alcohols of 1–6 carbons in length reveal that these molecules can readily displace a hydroxide ion and bind to the bridging position between the iron atoms, leaving the surrounding ligand environment unperturbed [109, 154]. The product-bound structures are consistent with EPR, ENDOR, UV-vis, and resonance Raman spectroscopic characterization of $MMOH_{prod}$ in the absence of MMOB [148, 184, 185]. The addition of MMOB influences exogenous ligand binding to the diiron center, as indicated by EPR [148]. It is conceivable that the structure of the diiron center in the MMOH-MMOB product complex is different from the one observed in MMOH alone.

3.6.5 C–H Bond Activation by Different Intermediates

Like the O_2-activation steps, C–H bond activation by sMMO has also received significant attention to determine the mechanism by which the different intermediates activate hydrocarbons. Mechanistic chiral alkane and radical clock substrate probes that rearrange to specific products when a particular hydrocarbon intermediate is generated have been employed to qualitatively differentiate between radical, cationic, and concerted mechanisms (Figure 6). The results of these experiments were not straightforward and have been the subject of considerable debate, reviewed thoroughly elsewhere [11]. Computational methods have added additional insight and, in some cases, explain some of the more paradoxical results.

Small chiral alkanes that closely mimic the natural sMMO substrate were used as reporters for the H-atom abstraction mechanism by intermediate Q [186, 187]. In principle, a concerted mechanism would produce complete retention or inversion of stereochemistry whereas radical and cationic intermediates would predict full racemization (Figure 7). For *M. capsulatus* (Bath) and *M. trichosporium* OB3b sMMO, hydroxylation of (S) or (R)-[1-$^3H_1$,$^2H_1$] ethane yielded a 70:30 retention: inversion of configuration ratio. Reactions with [2-^{3}H] butane exhibited ~90 % retention. These results could be interpreted as evidence for a radical intermediate, but to achieve 70 % retention of configuration requires the ethyl radical to recombine with the iron-bound hydroxyl radical at a rebound rate of 1×10^{13} s^{-1} [188]. This large value excludes the possibility of a discrete radical intermediate, which would racemize in less than 1×10^{-10} s. DFT calculations provided some clarity to this discrepancy, demonstrating that sMMO may have a "bound-radical" intermediate with restricted rotation owing to weak interactions with the diiron center [189].

Radical clock substrate probes offered an additional way to probe the C–H activation process. For these substrates, H-atom or hydride abstraction by sMMO leads to a radical or cationic intermediate on a constrained hydrocarbon substrate that subsequently rearranges to produce a lower energy, less constrained product alcohol. An example of this rearrangement is shown in Figure 21. Some probes detect only radical products, whereas others, in theory, elegantly differentiate between either a cationic or radical rearrangement [11].

Because the rate constant for the rearrangement of these substrates is known, it is possible to estimate the lifetimes of transient species. Analysis of the reactions with the various probes shows the predominant products to be unrearranged hydroxylated compounds. The minor products, however, were derived from *both* radical and cationic intermediates. These contradictory findings, while perplexing, suggested that sMMO utilizes more than one oxidizing species to activate C–H bonds. Stopped-flow spectroscopic experiments monitoring the reaction of H_{peroxo} with propylene provided the first direct evidence that this Q precursor is reactive toward electron-rich substrates [167]. Further investigation of H_{peroxo} reactions with ethers demonstrated that this intermediate reacts with more electron-rich substrates faster than Q, via a two-electron transfer mechanism as opposed to single-electron transfer steps by Q [176] (Figure 22).

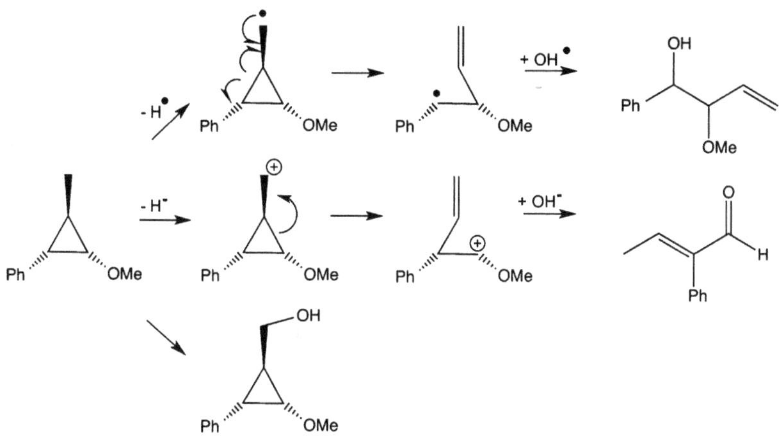

Figure 21 Example of a radical clock substrate (*trans*,*trans*-2-methoxy-3-phenylcyclopropyl)-methane used on sMMO, that can differentiate between radical and cationic based hydroxylation. The figure was adapted from [8].

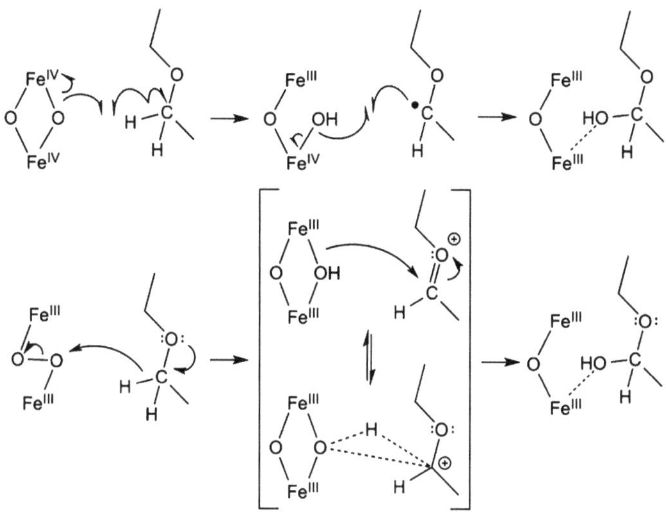

Figure 22 Proposed mechanism of C–H bond activation of diethylether by intermediate Q (**top**) and H_{peroxo} (**bottom**). The figure was adapted from [176].

A concern stemming from the work with radical clock substrates was the low percentage of re-rearranged radical clock-derived products [11]. If Q were to react with hydrocarbons via a traditional radical-based mechanism, a higher yield of radical-derived products was expected. Like the chiral alkane results, DFT calculations suggest the low yields of radical-derived products are explained by a mechanism using "bound radical" with restricted rotation (Figure 6) [189]. Calculations examining a concerted mechanism indicated this pathway to be less favorable.

4 Concluding Remarks and Future Directions

In the last twenty-five years, investigations of the pMMO and sMMO systems, including the organisms that utilize these enzymes, have considerably improved our understanding of the biological transformation of methane to methanol at dinuclear copper and iron centers. As the general picture of the O_2 and C–H activation processes becomes clear, attention now turns toward pursuing a more detailed mechanistic understanding of the hydroxylation chemistry. In particular, it is important to identify the structures of the activated dioxygen intermediates and to sort out the intricacies of the electron and proton transfer events critical for the conversion of one intermediate to the next for methane oxidation.

The controlled delivery of the requisite protons, electrons, oxygen, and hydrocarbon substrates at the appropriate time in the catalytic cycle is essential for efficient hydroxylation chemistry. Failure to regulate when the different substrates arrive at the active site would lead to uncoupling, quenching the activated oxygen intermediates and forming hydrogen peroxide and/or water.

How sMMO controls the timing and assembly of substrates at its diiron center is now reasonably well understood, but this information is lacking for the pMMO system. With sMMO, having multiple protein components has made the process challenging, and information about the electron transfer steps from MMOR to MMOH is incomplete. Although we are much closer to understanding the intracies of these enzymes, how the sMMO and pMMO protein scaffolds and metal centers work in concert remains a continuing goal as we try to reveal how biology masterfully tunes the reactivity of these unique dinuclear metal centers to carry out their challenging hydroxylation chemistry.

Abbreviations and Definitions

AMO	ammonia monooxygenase
BMMs	bacterial multicomponent monooxygenases
CD	circular dichroism
DFT	density functional theory

EM	electron microscopy
ENDOR	electron nuclear double resonance spectroscopy
EPR	electron paramagnetic resonance spectroscopy
FAD	flavin adenine dinucleotide
Fd	ferredoxin
H_4HBamb	2,3-bis(2-hydroxybenzamido)-dimethylbutane
H_{peroxo}	sMMO peroxo intermediate
KSIE	kinetic solvent isotope effect
Mc	*Methylococcus capsulatus* (Bath)
MCD	magnetic circular dichroism spectroscopy
melm	*N*-methylimizadole
Mt	*Methylosinus trichosporium* OB3b
MDH	methanol dehydrogenase
MMO	methane monooxygenase
MMOB	soluble methane monoxygenase regulatory protein
MMOD	soluble methane monoxygenase component D
MMOH	soluble methane monoxygenase hydroxylase
$MMOH_{ox}$	oxidized soluble methane monoxygenase hydroxylase
$MMOH_{prod}$	product bound soluble methane monoxygenase hydroxylase
$MMOH_{red}$	reduced soluble methane monoxygenase hydroxylase
MMOH-B	soluble methane monoxygenase hydroxylase-regulatory protein complex
MMOG	soluble methane monoxygenase GroEL-like chaperone
MMOR	soluble methane monoxygenase reductase
NADH	nicotinamide adenine dinucleotide (reduced)
P*	sMMO peroxo intermediate
PH	phenol hydroxylase
PHH	hydroxylase component of phenol hydroxylase
PHM	regulatory protein component of phenol hydroxylase
PHH-M	phenol hydroxylase regulatory protein complex
pMMO	particulate methane monoxygenase
pmoA	26-kDa subunit of pMMO
pmoB	45-kDa subunit of pMMO
pmoC	23-kDa subunit of pMMO
pz'	3,5-bis(isopropyl)-pyrazolyl
Q	sMMO intermediate
Q*	sMMO intermediate (activated)
QM/MM	quantum mechanics/molecular mechanics
sMMO	soluble methane monoxygenase
spmoB	soluble domains of the pmoB subunit
spmoBD1	soluble domain 1 of the pmoB subunit
spmoBD2	soluble domain 2 of the pmoB subunit
T4MO	toluene 4-monooxygenase
T4MOD	toluene 4-monooxygenase regulatory protein

T4MOH	toluene 4-monooxygenase hydroxylase
T4MOH-D	toluene 4-monooxygenase hydroxylase-regulatory protein complex
THFMO	tetrahydrofuran monooxygenase
TMO	toluene monooxygenase family
ToMO	toluene/o-xylene monoxygenase
ToMOD	toluene/o-xylene monooxygenase regulatory component
ToMOH	toluene/o-xylene monooxygenase hydroxylase component
ToMOH-D	toluene/o-xylene monooxygenase hydroxylase-regulatory protein complex
TPA	tris(2-pyridylmethyl)amine
XAS	X-ray absorption spectroscopy
ZSM-5	zeolite socony mobil-5

Acknowledgments We thank Michael McCormick for providing Pymol files for Figure 17. This work was supported by National Institute of General Medical Sciences (GM-032134 to S.J.L.).

References

1. US Environmental Protection Agency, in "Methane and Nitrous Oxide Emissions From Natural Sources", Ed US Environmental Protection Agency, Washington DC, USA, 2010.
2. R. S. Hanson, T. E. Hanson, *Microbiol. Rev.* **1996**, *60*, 439–471.
3. H. Jiang, Y. Chen, P. X. Jiang, C. Zhang, T. J. Smith, J. C. Murrell, X. H. Xing, *Biochem. Eng. J.* **2010**, *49*, 277–288.
4. R. Balasubramanian, A. C. Rosenzweig, *Acc. Chem. Res.* **2007**, *40*, 573–580.
5. M. A. Culpepper, A. C. Rosenzweig, *Crit. Rev. Biochem. Mol.* **2012**, *47*, 483–492.
6. A. S. Hakemian, A. C. Rosenzweig, *Annu. Rev. Biochem.* **2007**, *76*, 223–241.
7. R. L. Lieberman, A. C. Rosenzweig, *Crit. Rev. Biochem. Mol.* **2004**, *39*, 147–164.
8. M. Merkx, D. A. Kopp, M. H. Sazinsky, J. L. Blazyk, J. Müller, S. J. Lippard, *Angew. Chem. Int. Edit.* **2001**, *40*, 2782–2807.
9. C. E. Tinberg, S. J. Lippard, *Acc. Chem. Res.* **2011**, *44*, 280–288.
10. B. J. Wallar, J. D. Lipscomb, *Chem. Rev.* **1996**, *96*, 2625–2657.
11. M.-H. Baik, M. Newcomb, R. A. Friesner, S. J. Lippard, *Chem. Rev.* **2003**, *103*, 2385–2419.
12. J. D. Semrau, A. A. DiSpirito, S. Yoon, *FEMS Microbiol. Rev.* **2010**, *34*, 496–531.
13. H. Ali, J. C. Murrell, *Microbiology* **2009**, *155*, 761–771.
14. D. W. Choi, R. C. Kunz, E. S. Boyd, J. D. Semrau, W. E. Antholine, J. I. Han, J. A. Zahn, J. M. Boyd, A. M. de la Mora, A. A. DiSpirito, *J. Bacteriol.* **2003**, *185*, 5755–5764.
15. S. H. Stanley, S. D. Prior, D. J. Leak, H. Dalton, *Biotechnol. Lett.* **1983**, *5*, 487–492.
16. S. D. Prior, H. Dalton, *J. Gen. Microbiol.* **1985**, *131*, 155–163.
17. R. Balasubramanian, A. C. Rosenzweig, *Curr. Opin. Chem. Biol.* **2008**, *12*, 245–249.
18. M. W. Fitch, D. W. Graham, R. G. Arnold, S. K. Agarwal, P. Phelps, G. E. Speitel, G. Georgiou, *Appl. Environ. Microbiol.* **1993**, *59*, 2771–2776.
19. A. A. DiSpirito, J. A. Zahn, D. W. Graham, H. J. Kim, C. K. Larive, T. S. Derrick, C. D. Cox, A. Taylor, *J. Bacteriol.* **1998**, *180*, 3606–3613.
20. C. M. Tellez, K. P. Gaus, D. W. Graham, R. G. Arnold, R. Z. Guzman, *Appl. Environ. Microbiol.* **1998**, *64*, 1115–1122.
21. J. Colby, D. I. Stirling, H. Dalton, *Biochem. J.* **1977**, *165*, 395–402.

22. I. J. Higgins, D. J. Best, R. C. Hammond, *Nature* **1980**, *286*, 561–564.
23. J. Green, H. Dalton, *J. Biol. Chem.* **1989**, *264*, 17698–17703.
24. B. G. Fox, J. G. Borneman, L. P. Wackett, J. D. Lipscomb, *Biochemistry* **1990**, *29*, 6419–6427.
25. K. K. Andersson, W. A. Froland, S. K. Lee, J. D. Lipscomb, *New J. Chem.* **1991**, *15*, 411–415.
26. M. J. Rataj, J. E. Kauth, M. I. Donnelly, *J. Biol. Chem.* **1991**, *266*, 18684–18690.
27. S. I. Chan, K. H. C. Chen, S. S. F. Yu, C. L. Chen, S. S. J. Kuo, *Biochemistry* **2004**, *43*, 4421–4430.
28. K. Burrows, A. Cornish, D. Scott, I. J. Higgins, *J. Gen. Microbiol.* **1984**, *130*, 327–333.
29. D. D. Smith, H. Dalton, *Eur. J. Biochem.* **1989**, *182*, 667–671.
30. R. L. Lieberman, A. C. Rosenzweig, *Nature* **2005**, *434*, 177–182.
31. S. Stolyar, A. M. Costello, T. L. Peeples, M. E. Lidstrom, *Microbiology* **1999**, *145*, 1235–1244.
32. D. J. Arp, L. A. Sayavedra-Soto, N. G. Hommes, *Arch. Microbiol.* **2002**, *178*, 250–255.
33. P. Basu, B. Katterle, K. K. Andersson, H. Dalton, *Biochem. J.* **2003**, *369*, 417–427.
34. R. L. Lieberman, D. B. Shrestha, P. E. Doan, B. M. Hoffman, T. L. Stemmler, A. C. Rosenzweig, *Proc. Natl. Acad. Sci. USA* **2003**, *100*, 3820–3825.
35. H. H. T. Nguyen, S. J. Elliott, J. H. K. Yip, S. I. Chan, *J. Biol. Chem.* **1998**, *273*, 7957–7966.
36. S. S. F. Yu, K. H. C. Chen, M. Y. H. Tseng, Y. S. Wang, C. F. Tseng, Y. J. Chen, D. S. Huang, S. I. Chan, *J. Bacteriol.* **2003**, *185*, 5915–5924.
37. A. Miyaji, T. Kamachi, I. Okura, *Biotechnol. Lett.* **2002**, *24*, 1883–1887.
38. J. A. Zahn, A. A. DiSpirito, *J. Bacteriol.* **1996**, *178*, 1018–1029.
39. S. M. Smith, S. Rawat, J. Telser, B. M. Hoffman, T. L. Stemmler, A. C. Rosenzweig, *Biochemistry* **2011**, *50*, 10231–10240.
40. R. L. Lieberman, A. C. Rosenzweig, *Dalton Trans.* **2005**, 3390–3396.
41. A. Kitmitto, N. Myronova, P. Basu, H. Dalton, *Biochemistry* **2005**, *44*, 10954–10965.
42. N. Myronova, A. Kitmitto, R. F. Collins, A. Miyaji, H. Dalton, *Biochemistry* **2006**, *45*, 11905–11914.
43. H. J. Op den Camp, T. Islam, M. B. Stott, H. R. Harhangi, A. Hynes, S. Schouten, M. S. M. Jetten, N. K. Birkeland, A. Pol, P. F. Dunfield, *Environ. Microbiol. Rep.* **2009**, *1*, 293–306.
44. R. L. Lieberman, K. C. Kondapalli, D. B. Shrestha, A. S. Hakemian, S. M. Smith, J. Telser, J. Kuzelka, R. Gupta, A. S. Borovik, S. J. Lippard, B. M. Hoffman, A. C. Rosenzweig, T. L. Stemmler, *Inorg. Chem.* **2006**, *45*, 8372–8381.
45. A. S. Hakemian, K. C. Kondapalli, J. Telser, B. M. Hoffman, T. L. Stemmler, A. C. Rosenzweig, *Biochemistry* **2008**, *47*, 6793–6801.
46. S. I. Chan, S. S. F. Yu, *Acc. Chem. Res.* **2008**, *41*, 969–979.
47. H. H. T. Nguyen, A. K. Shiemke, S. J. Jacobs, B. J. Hales, M. E. Lidstrom, S. I. Chan, *J. Biol. Chem.* **1994**, *269*, 14995–15005.
48. H. H. T. Nguyen, K. H. Nakagawa, B. Hedman, S. J. Elliott, M. E. Lidstrom, K. O. Hodgson, S. I. Chan, *J. Am Chem. Soc.* **1996**, *118*, 12766–12776.
49. A. C. Rosenzweig, M. H. Sazinsky, *Curr. Opin. Struc. Biol.* **2006**, *16*, 729–735.
50. H. H. A. T. Nguyen, M. Zhu, S. J. Elliott, K. H. Nakagawa, B. Hedman, A. M. Costello, T. L. Peeples, B. Wilkinson, H. Morimoto, P. G. Williams, H. G. Floss, M. E. Lidstrom, K. O. Hodgson, S. I. Chan, *Microbial Growth on C(1) Compounds* **1996**, 150–158.
51. M. Takeguchi, K. Miyakawa, I. Okura, *J. Mol. Catal. A-Chem.* **1998**, *132*, 145–153.
52. M. Takeguchi, I. Okura, *Catal. Surv. Jpn.* **2000**, *4*, 51–63.
53. H. Yuan, M. L. P. Collins, W. E. Antholine, *J. Am Chem. Soc.* **1997**, *119*, 5073–5074.
54. H. Yuan, M. L. P. Collins, W. E. Antholine, *Biophys. J.* **1999**, *76*, 2223–2229.
55. S. S. Lemos, M. L. P. Collins, S. S. Eaton, G. R. Eaton, W. E. Antholine, *Biophys. J.* **2000**, *79*, 1085–1094.
56. L. V. Tumanova, I. A. Tukhvatullin, D. S. Burbaev, R. I. Gvozdev, K. K. Andersson, *Russ. J. Bioinorg. Chem.* **2005**, *34*, 177–185.

57. M. Martinho, D. W. Choi, A. A. DiSpirito, W. E. Antholine, J. D. Semrau, E. Münck, *J. Am Chem. Soc.* **2007**, *129*, 15783–15785.
58. O. A. Karlsen, J. R. Lillehaug, H. B. Jensen, *Mol. Microbiol.* **2008**, *70*, 15–26.
59. S. M. Smith, R. Balasubramanian, A. C. Rosenzweig, *Meth. Enzymol.* **2011**, *495*, 195–210.
60. R. Balasubramanian, S. M. Smith, S. Rawat, L. A. Yatsunyk, T. L. Stemmler, A. C. Rosenzweig, *Nature* **2010**, *465*, 115–U131.
61. A. Miyaji, T. Kamachi, I. Okura, T. Baba, *J. Mol. Catal. B-Enzym.* **2010**, *62*, 121–121.
62. J. C. Murrell, T. J. Smith, in *Handbook of Hydrocarbon and Lipid Microbiology*, Ed K. N. Timmis, Springer, Heidelberg, Germany, 2010, Vol. 1, pp. 1045–1055.
63. A. Miyaji, T. Miyoshi, K. Motokura, T. Baba, *Biotechnol. Lett.* **2011**, *33*, 2241–2246.
64. S. J. Elliott, M. Zhu, L. Tso, H. H. T. Nguyen, J. H. K. Yip, S. I. Chan, *J. Am Chem. Soc.* **1997**, *119*, 9949–9955.
65. W. J. Song, G. Gucinski, M. H. Sazinsky, S. J. Lippard, *Proc. Natl. Acad. Sci. USA* **2011**, *108*, 14795–14800.
66. K. H. C. Chen, H. H. Wu, S. F. Ke, Y. T. Rao, C. M. Tu, Y. P. Chen, K. H. Kuei, Y. S. Chen, V. C. C. Wang, W. C. Kao, S. I. Chan, *J. Inorg. Biochem.* **2012**, *111*, 10–17.
67. S. A. Cook, A. K. Shiemke, *Arch. Biochem. Biophys.* **2002**, *398*, 32–40.
68. M. R. Hyman, P. M. Wood, *Biochem. J.* **1985**, *227*, 719–725.
69. M. R. Hyman, D. J. Arp, *J. Biol. Chem.* **1992**, *267*, 1534–1545.
70. S. Gilch, M. Vogel, M. W. Lorenz, O. Meyer, I. Schmidt, *Microbiology* **2009**, *155*, 279–284.
71. G. M. Tonge, D. E. F. Harrison, C. J. Knowles, I. Higgins, *FEBS Lett.* **1975**, *58*, 293–299.
72. M. A. Culpepper, G. E. Cutsail, B. M. Hoffman, A. C. Rosenzweig, *J. Am Chem. Soc.* **2012**, *134*, 7640–7643.
73. E. I. Solomon, U. M. Sundaram, T. E. Machonkin, *Chem. Rev.* **1996**, *96*, 2563–2605.
74. E. I. Solomon, J. W. Ginsbach, D. E. Heppner, M. T. Kieber-Emmons, C. H. Kjaergaard, P. J. Smeets, L. Tian, J. S. Woertink, *Faraday Discuss.* **2011**, *148*, 11–39.
75. C. R. Andrew, K. P. McKillop, A. G. Sykes, *Biochim. Biophys. Acta* **1993**, *1163*, 17–25.
76. T. Zalateva, L. Santagostini, L. Bubacco, L. Casella, B. Salvato, M. Beltramini, *J. Inorg. Biochem.* **1998**, *72*, 211–215.
77. Y. Shiota, K. Yoshizawa, *Inorg. Chem.* **2009**, *48*, 838–845.
78. K. Yoshizawa, Y. Shiota, *J. Am. Chem. Soc.* **2006**, *128*, 9873–9881.
79. P. P. Y. Chen, S. I. Chan, *J. Inorg. Biochem.* **2006**, *100*, 801–809.
80. R. A. Himes, K. D. Karlin, *Curr. Opin. Chem. Biol.* **2009**, *13*, 119–131.
81. M. H. Groothaert, P. J. Smeets, B. F. Sels, P. A. Jacombs, R. A. Schoonheydt, *J. Am. Chem. Soc.* **2005**, *127*, 1394–1395.
82. P. J. Smeets, R. G. Hadt, J. S. Woertink, P. Vanelderen, R. A. Schoonheydt, B. F. Sels, E. I. Solomon, *J. Am. Chem. Soc.* **2010**, *132*, 14736–14738.
83. J. S. Woertink, P. J. Smeets, M. H. Groothaert, M. A. Vance, B. F. Sels, R. A. Schoonheydt, E. I. Solomon, *Proc. Natl. Acad. Sci. USA* **2009**, *106*, 18908–18913.
84. P. Vanelderen, R. G. Hadt, P. J. Smeets, E. I. Solomon, R. A. Schoonheydt, B. F. Sels, *J. Catal.* **2011**, *284*, 157–164.
85. B. Wilkinson, M. Zhu, N. D. Priestley, H. H. T. Nguyen, H. Morimoto, P. G. Williams, S. I. Chan, H. G. Floss, *J. Am. Chem. Soc.* **1996**, *118*, 921–922.
86. S. S. F. Yu, L. Y. Wu, K. H. C. Chen, W. I. Luo, D. S. Huang, S. I. Chan, *J. Biol. Chem.* **2003**, *278*, 40658–40669.
87. R. Csaki, L. Bodrossy, J. Klem, J. C. Murrell, K. L. Kovacs, *Microbiology* **2003**, *149*, 1785–1795.
88. T. Nakamura, T. Hoaki, S. Hanada, A. Maruyama, Y. Kamagata, H. Fuse, *FEMS Microbiol. Lett.* **2007**, *277*, 157–164.
89. G. P. Stafford, J. Scanlan, I. R. McDonald, J. C. Murrell, *Microbiology* **2003**, *149*, 1771–1784.
90. A. R. Theisen, M. H. Ali, S. Radajewski, M. G. Dumont, P. F. Dunfield, I. R. McDonald, S. N. Dedysh, C. B. Miguez, J. C. Murrell, *Mol. Microbiol.* **2005**, *58*, 682–692.

91. M. Merkx, S. J. Lippard, *J. Biol. Chem.* **2002**, *277*, 5858–5865.
92. M. H. Sazinsky, M. Merkx, E. Cadieux, S. Y. Tang, S. J. Lippard, *Biochemistry* **2004**, *43*, 16263–16276.
93. V. Izzo, G. Leo, R. Scognamiglio, L. Troncone, L. Birolo, A. Di Donato, *Arch. Biochem. Biophys.* **2011**, *505*, 48–59.
94. U. E. Ukaegbu, S. Henery, A. C. Rosenzweig, *Biochemistry* **2006**, *45*, 10191–10198.
95. U. E. Ukaegbu, A. C. Rosenzweig, *Biochemistry* **2009**, *48*, 2207–2215.
96. J. Scanlan, M. G. Dumont, J. C. Murrell, *FEMS Microbiol. Lett.* **2009**, *301*, 181–187.
97. V. Cafaro, V. Izzo, R. Scognamiglio, E. Notomista, P. Capasso, A. Casbarra, P. Pucci, A. Di Donato, *Appl. Environ. Microbiol.* **2004**, *70*, 2211–2219.
98. J. D. Pikus, J. M. Studts, C. Achim, K. E. Kauffmann, E. Münck, R. J. Steffan, K. McClay, B. G. Fox, *Biochemistry* **1996**, *35*, 9106–9119.
99. V. Cafaro, R. Scognamiglio, A. Viggiani, V. Izzo, I. Passaro, E. Notomista, F. D. Piaz, A. Amoresano, A. Casbarra, P. Pucci, A. Di Donato, *Eur. J. Biochem.* **2002**, *269*, 5689–5699.
100. J. G. Leahy, P. J. Batchelor, S. M. Morcomb, *FEMS Microbiol. Rev.* **2003**, *27*, 449–479.
101. E. Notomista, A. Lahm, A. Di Donato, A. Tramontano, *J. Mol. Evol.* **2003**, *56*, 435–445.
102. M. H. Sazinsky, J. Bard, A. Di Donato, S. J. Lippard, *J. Biol. Chem.* **2004**, *279*, 30600–30610.
103. M. H. Sazinsky, P. W. Dunten, M. S. McCormick, A. DiDonato, S. J. Lippard, *Biochemistry* **2006**, *45*, 15392–15404.
104. L. J. Bailey, J. G. Mccoy, G. N. Phillips, B. G. Fox, *Proc. Natl. Acad. Sci. USA* **2008**, *105*, 19194–19198.
105. B. L. Dubbels, L. A. Sayavedra-Soto, D. J. Arp, *Microbiology* **2007**, *153*, 1808–1816.
106. N. Elango, R. Radhakrishnan, W. A. Froland, B. J. Wallar, C. A. Earhart, J. D. Lipscomb, D. H. Ohlendorf, *Protein Sci.* **1997**, *6*, 556–568.
107. A. C. Rosenzweig, C. A. Frederick, S. J. Lippard, P. Nordlund, *Nature* **1993**, *366*, 537–543.
108. S. J. Lee, M. S. McCormick, S. J. Lippard, U. S. Cho, *Nature* **2013**, *494*, 380–384.
109. M. H. Sazinsky, S. J. Lippard, *J. Am. Chem. Soc.* **2005**, *127*, 5814–5825.
110. D. A. Whittington, A. C. Rosenzweig, C. A. Frederick, S. J. Lippard, *Biochemistry* **2001**, *40*, 3476–3482.
111. G. T. Gassner, S. J. Lippard, *Biochemistry* **1999**, *38*, 12768–12785.
112. L. L. Chatwood, J. Müller, J. D. Gross, G. Wagner, S. J. Lippard, *Biochemistry* **2004**, *43*, 11983–11991.
113. J. Müller, A. A. Lugovskoy, G. Wagner, S. J. Lippard, *Biochemistry* **2002**, *41*, 42–51.
114. J. L. Blazyk, G. T. Gassner, S. J. Lippard, *J. Am. Chem. Soc.* **2005**, *127*, 17364–17376.
115. J. L. Blazyk, S. J. Lippard, *Biochemistry* **2002**, *41*, 15780–15794.
116. J. L. Blazyk, S. J. Lippard, *J. Biol. Chem.* **2004**, *279*, 5630–5640.
117. D. A. Kopp, G. T. Gassner, J. L. Blazyk, S. J. Lippard, *Biochemistry* **2001**, *40*, 14932–14941.
118. K. J. Walters, G. T. Gassner, S. J. Lippard, G. Wagner, *Proc. Natl. Acad. Sci. USA* **1999**, *96*, 7877–7882.
119. S.-L. Chang, B. J. Wallar, J. D. Lipscomb, K. H. Mayo, *Biochemistry* **1999**, *38*, 5799–5812.
120. B. G. Fox, Y. Liu, J. E. Dege, J. D. Lipscomb, *J. Biol. Chem.* **1991**, *266*, 540–550.
121. Y. Liu, J. C. Nesheim, S. K. Lee, J. D. Lipscomb, *J. Biol. Chem.* **1995**, *270*, 24662–24665.
122. M. H. Sazinsky, S. J. Lippard, *Accounts Chem. Res.* **2006**, *39*, 558–566.
123. K. H. Mitchell, J. M. Studts, B. G. Fox, *Biochemistry* **2002**, *41*, 3176–3188.
124. E. Cadieux, V. Vrajmasu, C. Achim, J. Powlowski, E. Münck, *Biochemistry* **2002**, *41*, 10680–10691.
125. J. Kazlauskaite, H. A. O. Hill, P. C. Wilkins, H. Dalton, *Eur. J. Biochem.* **1996**, *241*, 552–556.
126. S. Chang, B. J. Wallar, J. D. Lipcomb, J. D. Mayo, *Biochemistry* **2001**, *40*, 9539–9551.
127. H. Brandstetter, D. A. Whittington, S. J. Lippard, C. A. Frederick, *Chem. Biol.* **1999**, *6*, 441–449.
128. J. Y. Zhang, J. D. Lipscomb, *Biochemistry* **2006**, *45*, 1459–1469.
129. A. C. Rosenzweig, P. Nordlund, P. M. Takahara, C. A. Frederick, S. J. Lippard, *Chem. Biol.* **1995**, *2*, 632.

130. D. A. Whittington, S. J. Lippard, *J. Am. Chem. Soc.* **2001**, *123*, 827–838.
131. M. P. Woodland, D. S. Patil, R. Cammack, H. Dalton, *Biochim. Biophys. Acta.* **1986**, *873*, 237–242.
132. J. G. Dewitt, J. G. Bentsen, A. C. Rosenzweig, B. Hedman, J. Green, S. Pilkington, G. C. Papaefthymiou, H. Dalton, K. O. Hodgson, S. J. Lippard, *J. Am. Chem. Soc.* **1991**, *113*, 9219–9235.
133. B. G. Fox, J. D. Lipscomb, *Biochem. Biophys. Res. Comm.* **1988**, *154*, 165–170.
134. B. G. Fox, K. K. Surerus, E. Münck, J. D. Lipscomb, *J. Biol. Chem.* **1988**, *263*, 10553–10556.
135. J. G. Dewitt, A. C. Rosenzweig, A. Salifoglou, B. Hedman, S. J. Lippard, K. O. Hodgson, *Inorg. Chem.* **1995**, *34*, 2505–2515.
136. B. G. Fox, M. P. Hendrich, K. K. Surerus, K. K. Andersson, W. A. Froland, J. D. Lipscomb, E. Münck, *J. Am Chem. Soc.* **1993**, *115*, 3688–3701.
137. M. P. Hendrich, E. Münck, B. G. Fox, J. D. Lipscomb, *J. Am. Chem. Soc.* **1990**, *112*, 5861–5865.
138. B. M. Hoffman, B. E. Sturgeon, P. E. Doan, V. J. DeRose, K. E. Liu, S. J. Lippard, *J. Am. Chem. Soc.* **1994**, *116*, 6023–6024.
139. D. A. Kopp, E. A. Berg, C. E. Costello, S. J. Lippard, *J. Biol. Chem.* **2003**, *278*, 20939–20945.
140. K. E. Liu, S. J. Lippard, *J. Biol. Chem.* **1991**, *266*, 12836–12839.
141. K. E. Liu, A. M. Valentine, D. L. Wang, B. H. Huynh, D. E. Edmondson, A. Salifoglou, S. J. Lippard, *J. Am. Chem. Soc.* **1995**, *117*, 10174–10185.
142. Y. Liu, J. C. Nesheim, K. E. Paulsen, M. T. Stankovich, J. D. Lipscomb, *Biochemistry* **1997**, *36*, 5223–5233.
143. K. E. Paulsen, Y. Liu, B. G. Fox, J. D. Lipscomb, E. Münck, M. T. Stankovich, *Biochemistry* **1994**, *33*, 713–722.
144. W. Wang, S. J. Lippard, *J. Am. Chem. Soc.* **2014**, *136*, 2244–2247.
145. C. C. Page, C. C. Moser, P. L. Dutton, *Curr. Opin. Chem. Biol.* **2003**, *7*, 551–556.
146. L. Shu, Y. Lui, J. D. Lipscomb, L. J. Que, *J. Biol. Inorg. Chem.* **1996**, *1*, 297–304.
147. D. Jackson Rudd, M. H. Sazinsky, M. Merkx, S. J. Lippard, B. Hedman, K. O. Hodgson, *Inorg. Chem.* **2004**, *43*, 4579–4589.
148. R. Davydov, A. M. Valentine, S. Komar-Panicucci, B. M. Hoffman, S. J. Lippard, *Biochemistry* **1999**, *38*, 4188–4197.
149. S. C. Pulver, W. A. Froland, J. D. Lipscomb, E. I. Solomon, *J. Am. Chem. Soc.* **1997**, *119*, 387–395.
150. N. Mitic, J. K. Schwartz, B. J. Brazeau, J. D. Lipscomb, E. I. Solomon, *Biochemistry* **2008**, *47*, 8386–8397.
151. W. A. Froland, K. K. Andersson, S. K. Lee, Y. Liu, J. D. Lipscomb, *J. Biol. Chem.* **1992**, *267*, 17588–17597.
152. J. D. Pikus, J. M. Studts, K. McClay, R. J. Steffan, B. G. Fox, *Biochemistry* **1997**, *36*, 9283–9289.
153. M. S. McCormick, M. H. Sazinsky, K. L. Condon, S. J. Lippard, *J. Am. Chem. Soc.* **2006**, *128*, 15108–15110.
154. D. A. Whittington, M. H. Sazinsky, S. J. Lippard, *J. Am. Chem. Soc.* **2001**, *123*, 1794–1795.
155. A. C. Rosenzweig, H. Brandstetter, D. A. Whittington, P. Nordlund, S. J. Lippard, C. A. Frederick, *Protein Struct. Funct. Genet.* **1997**, *29*, 141–152.
156. B. J. Wallar, J. D. Lipscomb, *Biochemistry* **2001**, *40*, 2220–2233.
157. C. E. Tinberg, S. J. Lippard, *Biochemistry* **2009**, *48*, 12145–12158.
158. S. Y. Lee, J. D. Lipscomb, *Biochemistry* **1999**, *38*, 4423–4432.
159. W. J. Song, M. S. McCormick, R. K. Behan, M. H. Sazinsky, W. Jiang, J. Lin, C. Krebs, S. J. Lippard, *J. Am. Chem. Soc.* **2010**, *132*, 13582–13585.
160. J. D. Pikus, K. H. Mitchell, J. M. Studts, K. McClay, R. J. Steffan, B. G. Fox, *Biochemistry* **2000**, *39*, 791–799.
161. N. L. Elsen, L. J. Bailey, A. D. Hauser, B. G. Fox, *Biochemistry* **2009**, *48*, 3838–3846.

162. W. J. Song, R. K. Behan, S. G. Naik, B. H. Huynh, S. J. Lippard, *J. Am. Chem. Soc.* **2009**, *131*, 6074–6075.
163. S. K. Lee, J. C. Nesheim, J. D. Lipscomb, *J. Biol. Chem.* **1993**, *268*, 21569–21577.
164. S. S. Stahl, W. A. Francisco, M. Merkx, J. P. Klinman, S. J. Lippard, *J. Biol. Chem.* **2001**, *276*, 4549–4553.
165. K. E. Liu, D. Wang, B. H. Huynh, D. E. Edmondson, A. Salifoglou, S. J. Lippard, *J. Am. Chem. Soc.* **1994**, *116*, 7465–7466.
166. L. J. Shu, J. C. Nesheim, K. Kauffmann, E. Münck, J. D. Lipscomb, L. Que, *Science* **1997**, *275*, 515–518.
167. A. M. Valentine, S. S. Stahl, S. J. Lippard, *J. Am. Chem. Soc.* **1999**, *121*, 3876–3887.
168. R. Banerjee, K. K. Meier, E. Münck, J. D. Lipcomb, *Biochemistry* **2013**, *52*, 4331–4342.
169. B. J. Brazeau, J. D. Lipcomb, *Biochemistry* **2000**, *39*, 13503–13515.
170. D. Jahng, A. K. Sun, C. S. Kim, T. K. Wood, *J. Cell. Biochem.* **1995**, *59*, 44–44.
171. W. G. Han, L. Noodleman, *Inorg. Chem.* **2008**, *47*, 2975–2986.
172. A. D. Bochevarov, J. N. Li, W. J. Song, R. A. Friesner, S. J. Lippard, *J. Am. Chem. Soc.* **2011**, *133*, 7384–7397.
173. L. J. Murray, R. Garcia-Serres, S. Naik, B. H. Huynh, S. J. Lippard, *J. Am. Chem. Soc.* **2006**, *128*, 7458–7459.
174. L. J. Murray, S. J. Lippard, *Accounts Chem. Res.* **2007**, *40*, 466–474.
175. K. P. Jensen, C. B. Bell, III., M. D. Clay, E. I. Solomon, *J. Am. Chem. Soc.* **2009**, *131*, 12155–12171.
176. L. G. Beauvais, S. J. Lippard, *J. Am. Chem. Soc.* **2005**, *127*, 7370–7378.
177. C. E. Tinberg, S. J. Lippard, *Biochemistry* **2010**, *49*, 7902–7912.
178. S.-K. Lee, B. G. Fox, W. A. Froland, J. D. Lipcomb, E. Münck, *J. Am. Chem. Soc.* **1993**, *115*, 6450–6451.
179. B. F. Gherman, M. H. Baik, S. J. Lippard, R. A. Friesner, *J. Am. Chem. Soc.* **2004**, *126*, 2978–2990.
180. D. Rinalde, D. M. Phillip, S. J. Lippard, R. A. Friesner, *J. Am. Chem. Soc.* **2007**, *129*, 3135–3147.
181. G. T. Rowe, E. V. Rybak-Akimova, J. P. Caradonna, *Inorg. Chem.* **2007**, *46*, 10594–10606.
182. G. Xue, R. De Hont, E. Münck, L. Que, *Nature Chem.* **2010**, *2*, 400–405.
183. G. Q. Xue, D. Wang, R. De Hont, A. T. Fiedler, X. P. Shan, E. Münck, L. Que, *Proc. Natl. Acad. Sci. USA* **2007**, *104*, 20713–20718.
184. S. K. Smoukov, D. A. Kopp, A. M. Valentine, R. Davydov, S. J. Lippard, B. M. Hoffman, *J. Am. Chem. Soc.* **2002**, *124*, 2657–2663.
185. K. K. Andersson, T. E. Elgren, L. Que, J. D. Lipcomb, *J. Am. Chem. Soc.* **1992**, *114*, 8711–8713.
186. N. D. Priestley, H. G. Floss, W. A. Froland, J. D. Lipcomb, P. G. Williams, H. Morimoto, *J. Am. Chem. Soc.* **1992**, *114*, 7561–7562.
187. A. M. Valentine, B. Wilkinson, K. E. Liu, S. KomarPanicucci, N. D. Priestley, P. G. Williams, H. Morimoto, H. G. Floss, S. J. Lippard, *J. Am. Chem. Soc.* **1997**, *119*, 1818–1827.
188. T. J. Sears, P. M. Johnson, P. Jin, S. Oatis, *J. Chem. Phys.* **1996**, *104*, 781–792.
189. B. F. Gherman, B. D. Dunietz, D. A. Whittington, S. J. Lippard, R. A. Friesner, *J. Am. Chem. Soc.* **2001**, *123*, 3836–3837.
190. J. M. Bollinger, Jr., C. Krebs, A. Vicol, S. Chen, B. A. Ley, D. E. Edmondson, B. H. Huynh, *J. Am. Chem. Soc.* **1998**, *120*, 1094–1095.
191. P. Moënne-Loccoz, J. Baldwin, B. A. Ley, T. M. Loehr, J. M. Bollinger, Jr., *Biochemistry* **1998**, *37*, 14659–14663.
192. J. A. Broadwater, C. Achim, E. Münck, B. G. Fox, *Biochemistry* **1999**, *38*, 12197–12204.
193. J. A. Broadwater, J. Y. Ai, T. M. Loehr, J. Sanders-Loehr, B. G. Fox, *Biochemistry* **1998**, *37*, 14664–14671.
194. A. S. Pereira, W. Small, C. Krebs, P. Tavares, D. E. Edmondson, E. C. Theil, B. H. Huynh, *Biochemistry* **1998**, *37*, 9871–9876.

195. P. Moënne-Loccoz, C. Krebs, K. Herlihy, D. E. Edmondson, E. C. Theil, B. H. Huynh, T. M. Loehr, *Biochemistry* **1999**, *38*, 5290–5295.
196. V. V. Vu, J. P. Emerson, M. Martinho, Y. S. Kim, E. Münck, M. H. Park, L. J. Que, *Proc. Natl. Acad. Sci. USA* **2009**, *106*, 14814–14819.
197. K. Kim, S. J. Lippard, *J. Am. Chem. Soc.* **1996**, *118*, 4914–4915.
198. X. Zhang, H. Furutachi, S. Fujinami, S. Nagatomo, Y. Maeda, Y. Watanabe, T. Kitagawa, M. Suzuki, *J. Am. Chem. Soc.* **2005**, *127*, 826–827.
199. Y. Dong, Y. Zhang, L. Shu, E. C. Wilkinson, L. Que, Jr., K. Kauffmann, E. Münck, *J. Am. Chem. Soc.* **1997**, *119*, 12683–12684.
200. Y. Dong, H. Fujii, M. P. Hendrich, R. A. Leising, G. Pan, C. R. Randall, E. C. Wilkinson, Y. Zang, L. Que, Jr., *J. Am. Chem. Soc.* **1995**, *117*, 2778–2792.
201. Y. Dong, L. Que, Jr., K. Kauffmann, E. Münck, *J. Am. Chem. Soc.* **1995**, *117*, 11377–11378.
202. H.-F. Hsu, Y. Dong, L. Shu, V. G. Young, Jr., L. Que, Jr., *J. Am. Chem. Soc.* **1999**, *121*, 5230–5237.
203. B. D. Dunietz, M. D. Beachy, Y. Cao, D. A. Whittington, S. J. Lippard, R. A. Friesner, *J. Am. Chem. Soc.* **2000**, *122*, 2828–2839.
204. W. Wang, R. E. Iacob, R. P. Luoh, J. R. Engen, S. J. Lippard, *J. Am. Chem. Soc.* **2014**, *136*, 9754–9762.
205. C. E. Tinberg, S. J. Lippard, *Biochemistry* **2010**, *49*, 7902–7912.

Chapter 7
Metal Enzymes in "Impossible" Microorganisms Catalyzing the Anaerobic Oxidation of Ammonium and Methane

Joachim Reimann, Mike S.M. Jetten, and Jan T. Keltjens

Contents

J. Reimann • M.S.M. Jetten (✉) • J.T. Keltjens (✉)
Department of Microbiology, Institute of Wetland and Water Research (IWWR), Radboud University of Nijmegen, Heyendaalseweg 135, 6525AJ Nijmegen, The Netherlands
e-mail: j.reimann@science.ru.nl; m.jetten@science.ru.nl; j.keltjens@science.ru.nl

P.M.H. Kroneck, M.E. Sosa Torres (eds.), *Sustaining Life on Planet Earth: Metalloenzymes Mastering Dioxygen and Other Chewy Gases*, Metal Ions in Life Sciences 15, DOI 10.1007/978-3-319-12415-5_7

Abstract Ammonium and methane are inert molecules and dedicated enzymes are required to break up the N-H and C-H bonds. Until recently, only aerobic microorganisms were known to grow by the oxidation of ammonium or methane. Apart from respiration, oxygen was specifically utilized to activate the inert substrates. The presumed obligatory need for oxygen may have resisted the search for microorganisms that are capable of the anaerobic oxidation of ammonium and of methane. However extremely slowly growing, these "impossible" organisms exist and they found other means to tackle ammonium and methane. Anaerobic ammonium-oxidizing (anammox) bacteria use the oxidative power of nitric oxide (NO) by forging this molecule to ammonium, thereby making hydrazine (N_2H_4). Nitrite-dependent anaerobic methane oxidizers (N-DAMO) again take advantage of NO, but now apparently disproportionating the compound into dinitrogen and dioxygen gas. This intracellularly produced dioxygen enables N-DAMO bacteria to adopt an aerobic mechanism for methane oxidation.

Although our understanding is only emerging how hydrazine synthase and the NO dismutase act, it seems clear that reactions fully rely on metal-based catalyses known from other enzymes. Metal-dependent conversions not only hold for these key enzymes, but for most other reactions in the central catabolic pathways, again supported by well-studied enzymes from model organisms, but adapted to own specific needs. Remarkably, those accessory catabolic enzymes are not unique for anammox bacteria and N-DAMO. Close homologs are found in protein databases where those homologs derive from (partly) known, but in most cases unknown species that together comprise an only poorly comprehended microbial world.

Keywords anaerobic methane oxidation • anammox • denitrification • hydrazine oxidation • hydrazine synthesis • methanol dehydrogenase • nitric oxide • octaheme proteins

Please cite as: *Met. Ions Life Sci.* 15 (2015) 257–313

1 Introduction

Ammonia and methane are rather inert molecules, which is due to the high bond dissociation energy of the N-H (~400 kJ · mole^{-1}) and C-H (429 kJ · mole^{-1}) bonds. Aerobic ammonium-oxidizing and methane-oxidizing microorganisms are able to break up these bonds using the oxidative power of oxygen. By the action of two evolutionary highly related membrane-bound iron/copper enzymes, ammonia monooxygenase (AMO) and particulate methane monooxygenase (pMMO), ammonia and methane are activated to more accessible hydroxylamine (NH_2OH) and methanol (CH_3OH), respectively [1–3]. Besides pMMO, aerobic methane-oxidizing bacteria, the methanotrophs, can dispose of the unrelated soluble iron-only methane monooxygenase (sMMO) for methane activation (see Chapter 6 of this volume). Aerobic ammonium-oxidizing bacteria (AOB) and aerobic methanotrophs have been investigated for more than a century. All species studied until lately were members of the domain of Bacteria, but also ammonium-oxidizing Archaea (AOA) are known by now [4]. Although AOA typically thrive at very low oxygen concentrations, they too depend on oxygen to convert ammonium into hydroxylamine [5, 6].

The common notion that oxygen would be indispensable for ammonium and methane conversion may have blindfolded the search for microorganisms that would make a living of the oxidation of these substrates with electron acceptors other than O_2. Today we know that these "impossible" microorganisms do exist and in this chapter we will go into the biochemical basis of two groups of bacteria that drive ammonium and methane conversion with nitrite as the oxidant: The anaerobic ammonium-oxidizing (anammox) bacteria and the nitrite/nitrate-dependent anaerobic methane oxidizers (N-DAMO). Characteristics of these organisms are extreme slow growth rates, and the requirement of dedicated enrichment techniques and molecular detection tools [7, 8], factors that may have impeded an earlier identification.

Anammox bacteria have been first described only 20 years ago [9]. The active search for these organisms was spurred by the observation that ammonium disappeared in laboratory wastewater treatment reactors in the apparent absence of O_2 and as the result of a biological agent [9]. Following their first enrichment [10], anammox bacteria have been detected in a multitude of species and subspecies by now [11, 12]. They have been found worldwide and occur in essentially each anaerobic environment where fixed nitrogen (ammonium, nitrate, nitrite) is lost as dinitrogen gas (N_2). In many of these ecosystems, anammox bacteria play a key role in nitrogen cycle processes. Apart from their biogeochemical relevance, anammox bacteria offer us an economically attractive and environmentally friendly alternative to current wastewater treatment technology [13].

Anaerobic oxidation of methane is performed by two completely different groups of microorganisms: Archaea and members of the NC10 phylum of Bacteria. Archaeal methane oxidizers are found in deep-sea methane seeps where they couple, either or not in symbiotic interaction, methane oxidation with the reduction of sulfate [14]. Next to this, an archaeal enrichment culture designated *Candidatus* 'Methanoperedens nitroreducens' was recently described that was able of methane oxidation at the expense of nitrate [15]. Intriguingly, the process took place in concurrence with anammox bacteria. Even so intriguingly, the original NC10 enrichment culture grown on methane and nitrate harbored a significant population of Archaea that phylogenetically affiliated those of the sulfate- and nitrate-reducing systems [16]. The archaeal species disappeared completely when the NC10 culture was shifted from nitrate to nitrite [17]. Although many details still have to be resolved, it seems clear that methane oxidation by Archaea is achieved by the reversal of the methane formation pathway [14], which is described in Chapter 6 of Vol. 14 of this series [138]. NC10 bacteria, represented by *Candidatus* 'Methanomirabilis oxyfera' [18], employ another strategy to tackle methane. These bacteria produce oxygen themselves, allowing the monooxygenation of methane into methanol that is subsequently oxidized to CO_2, as is carried out by aerobic methanotrophs [19].

Considering the inert nature of ammonium and methane, the primary question was how both substrates are activated by anammox and N-DAMO bacteria with nitrite (or nitrate) as oxidant. Whereas we are only at the beginning of a fundamental understanding of the molecular mechanisms underlying the anammox and N-DAMO processes, it has become clear that these microorganisms evolved novel biochemical systems to catalyze these reactions. This biochemistry is fully metal ion-based. However, not only the specific activation reactions are supported by metal chemistry. Most of the other pathway key reactions rely, in some way or another, on metal ions (Figure 1 in Section 2.1 and Figure 2 in Section 2.2). Here, anammox and N-DAMO bacteria take advantage of enzyme systems already known from other organisms, which have been adapted to own metabolic requirements by subtle structural modifications.

2 Pathways of Nitrite-Driven Anaerobic Oxidation of Ammonium and Methane

Anammox bacteria are autotrophic microorganisms, i.e., they are able to use CO_2 as the sole carbon source [10]. The energy for growth is obtained from the oxidation of ammonium with nitrite at which N_2 is produced as the end product (eq. 1). Principally, also ammonium oxidation coupled to nitrate reduction would provide sufficient energy to sustain growth (eq. 2), but this process does not comply with the biochemistry as far as we know it and highly enriched laboratory cultures are obtained only with nitrite.

$$NH_4^+ + NO_2^- \rightarrow N_2 + 2\,H_2O \qquad \left(\Delta G^{0\prime} = -357\,\text{kJ} \cdot \text{mol}^{-1} N_2\right) \quad (1)$$

$$5\,NH_4^+ + 3\,NO_3^- \rightarrow 4\,N_2 + 9\,H_2O + 2\,H^+ \qquad \left(\Delta G^{0\prime} = -452\,\text{kJ} \cdot \text{mol}^{-1} N_2\right) \quad (2)$$

$$3\ CH_4 + 8\ NO_2^- + 8\ H^+ \rightarrow 3\ CO_2 + 4\ N_2 + 10\ H_2O \qquad \left(\Delta G^{0\prime} = -696\,kJ \cdot mol^{-1} N_2\right) \quad (3)$$

$$5\ CH_4 + 8\ NO_3^- + 8\ H^+ \rightarrow 5\ CO_2 + 4\ N_2 + 14\ H_2O \qquad \left(\Delta G^{0\prime} = -613\,kJ \cdot mol^{-1} N_2\right) \quad (4)$$

Like anammox bacteria, *M. oxyfera* is an autotroph. With methane as electron donor, nitrite is reduced in catabolism to N_2, as in anammox bacteria (eq. 3). Theoretically, nitrate might act as an electron acceptor (eq. 4), but its use resists a proper redox balance (see Section 2.2). In laboratory practice, enrichment of *M. oxyfera* is only achieved with nitrite [17, 18].

Before discussing the individual biochemical reactions, we will first give an overview of how these steps are connected in the central catabolic pathways of the anammox and N-DAMO bacteria.

2.1 The Anammox Pathway

The crucial question at the start of anammox biochemistry was how ammonium could be activated in the apparent absence of oxygen. The clue to its understanding came from the fortuitous observation that metabolism by whole cells of ammonium and hydroxylamine, an artificial substrate, was accompanied by the transient accumulation of hydrazine (N_2H_4) [20]. Subsequent whole cell and biochemical studies established the role of hydrazine as an intermediate, also under physiological conditions [21, 22]. These studies identified a second intermediate proposed earlier, NO [21], while their results disfavored a role as such of hydroxylamine, as postulated initially [20]. Our current understanding is that anammox catabolism essentially is comprised of three consecutive reactions with two intermediates: (1) the one-electron reduction of nitrite to NO catalyzed by a nitrite reductase (Nir), (2) the condensation of ammonium and NO together with the input of three electrons leading to hydrazine, and (3) the oxidation of hydrazine to N_2, which generates four electrons that drive steps (1) and (2) in a cyclic way (Figure 1).

In this scheme (Figure 1), hydrazine synthesis would be catalyzed by a biochemical novelty, hydrazine synthase (HZS) [21]. Hydrazine oxidation would be brought about by hydrazine dehydrogenase/oxidase (HDH/HZO) representing a specific variant of the octaheme hydroxylamine oxidoreductase (HAO), a key enzyme in AOB (see Section 3.3.1). These expectations have been confirmed by the isolation and (initial) characterization of anammox enzymes [22]. Anammox metabolism as depicted in Figure 1 proceeds with a cyclic electron flow. This cyclic flow would take place such that oxidation and reduction reactions performed by membrane-bound

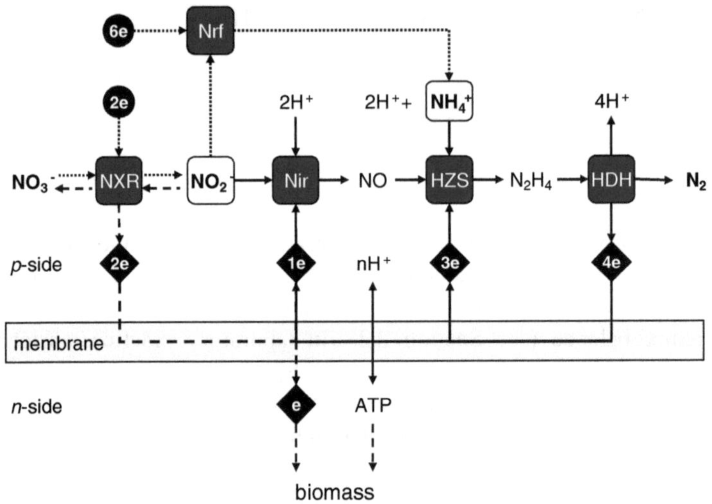

Figure 1 Scheme of the anaerobic ammonium oxidation (anammox) process with nitric oxide (NO) and hydrazine (N_2H_4) as intermediates and dinitrogen (N_2) as the end product. Primary substrates, ammonium (NH_4^+) and nitrite (NO_2^-), are indicated by the rounded squares. Reactions catalyzed by metal-ion-dependent enzymes are highlighted by dark grey rounded squares. Electron transfer processes within the cell are represented by black diamonds. Proton-motive force-driven ATP synthesis is schematically shown by proton translocation towards the positive (*p*-) side of the membrane system and ATP generation at the negative (*n*-) side. The *p*- and *n*-sides are the anammoxosome intracellular organelle and the cell cytoplasm, respectively. Electrons used for biosynthetic reduction reactions are replenished to the anammox cyclic electron flow by the oxidation of nitrite (dashed lines). "Disguised denitrification" by the reduction of nitrate or nitrite to ammonium at the expense of external organic or inorganic electron donors (black circles) is indicated by dotted lines. Abbreviations: HDH, hydrazine dehydrogenase; HZS, hydrazine synthase; Nir, nitrite reductase; Nrf, ammonium-forming nitrite reductase; NXR, nitrite:nitrate oxidoreductase.

processes are connected with the pumping of protons across these membranes, thus sustaining a proton-motive force to drive ATP synthesis. In anammox bacteria, respiratory processes would reside at a large intracellular cell organelle, termed the anammoxosome [23]. Present evidence supports such chemiosmotic mechanism in all major respects. Moreover, ongoing research underpins the central role of the anammoxosome in energy metabolism [24]. However, the detailed discussion of anammox energy metabolism is out of scope of the present chapter and we would like to refer interested readers to recent reviews that have been published on this matter [25, 26].

Reductive catabolic reactions in CO_2 fixation draw from the cyclic electron flow and electrons have to be replenished. The longstanding observation that autotrophic growth of anammox bacteria is always associated with the oxidation of nitrite to nitrate indicates that the electrons derive from nitrite [7]. The question how these electrons enter the cycle has not been resolved, considering that nitrite is a relative poor electron donor (E_0' NO_3^-/NO_2^- = + 0.43 V). Nevertheless, anammox cells contain high amounts of the iron-molybdoprotein nitrite:nitrate oxidoreductase (NXR), which is related to nitrate reductase found in denitrifiers (see Section 3.4.1). Indeed, NXR is capable of nitrate reduction [27] implying that anammox bacteria also use nitrate as terminal electron acceptor (eq. 2). Under chemolithoautotrophic conditions, i.e., with ammonium as the only electron donor, this is impossible. The reason for this is that in order to be reduced to nitrite, nitrate reduction needs electrons that drain the cyclic electron flow without replenishment. However, when grown in the presence of one of many different external inorganic or organic electron donors nitrate can be reduced to nitrite to support the anammox process (Figure 1) [28]. Under such conditions anammox bacteria may present themselves "disguised" as denitrifiers [28]. Now, half of the nitrite formed from nitrate is reduced to ammonia, which combines with the other half of nitrite in the anammox process. The apparent result is the complete reduction of nitrate into N_2, as is done by true denitrifiers. The reduction of nitrite into ammonia requires an ammonium-forming nitrite reductase as described in Chapter 9 of ref. [138]. The nature of this nitrite reductase in anammox bacteria will be the topic of Section 3.4.2.

2.2 *The Pathway of Nitrite-Driven Methane Oxidation*

An answer how *M. oxyfera* could couple the anaerobic oxidation of methane with the reduction to nitrite to produce CO_2 and N_2 (eq. 3) came from genome sequencing, whole-cell transcriptomic and proteomic analyses, and complementary physiological experiments using stable isotopes [18]. Initially, genome sequencing results were puzzling. *M. oxyfera* lacks the N_2-forming key enzyme in the denitrification pathway, nitrous oxide (N_2O) reductase (N_2OR) (see Chapter 8 of ref. [138]). At the same time, the full complement of the aerobic methane oxidation pathway was present in the genome of *M. oxyfera*. These contradictory findings could be resolved by the observation that the organism produced oxygen derived from nitrite during its metabolism [18]. The metabolic model that is consistent with all findings thus far is shown in Figure 2.

The central proposition of the model is that *M. oxyfera* harbors a nitric oxide reductase- (NOR) like enzyme that does not reduce NO into N_2O, as common NORs do (see Chapter 4), but that disproportionates NO into N_2 and O_2. The nature of this NO dismutase (NOD) and other NOR-like proteins found in the genome of *M. oxyfera* is the subject of Sections 4.1.2, 4.1.3, and 4.1.4. The production of O_2 serves the activation of methane into methanol that is oxidized to CO_2 thereby generating the electrons for methane monooxygenation and nitrite reduction (Section 4.2). One may note from Figure 2 that the substrate and electron flows are not balanced. The conversion of 8 molecules of nitrite requires the input of eight

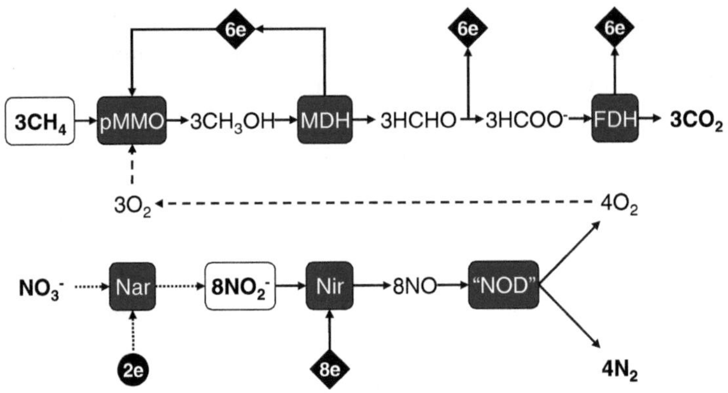

Figure 2 Scheme of the nitrite-dependent process of anaerobic methane oxidation (N-DAMO). Primary substrates, nitrite (NO_2^-) and methane (CH_4), are indicated by the rounded squares. Reactions catalyzed by metal enzymes are highlighted by dark grey squares. One may note that nitrite-dependent methane oxidation according to eq. (3) gives a surplus of one O_2 and four electrons that are converted into two water molecules. Nitrate-dependent methane oxidation (eq. 4) is only feasible with an additional external electron donor (black circle; see text). Abbreviations: FDH, formate dehydrogenase; MDH, methanol dehydrogenase; Nar, nitrate reductase; Nir, nitrite reductase; NOD, nitric oxide dismutase; pMMO, particulate methane monooxygenase.

electrons, at which four O_2 and four N_2 molecules are produced. Three O_2 molecules are required for methane activation, whilst a net 12 electrons are released in the oxidation pathway. The residual O_2 molecule and four electrons are respired into water [29]. By the same argumentation it is seen that nitrate-dependent methane oxidation by *M. oxyfera* alone (eq. 4) is not feasible for two reasons. At one hand, the conversion of eight molecules of NO_3^- would generate only four O_2 molecules, which does not sustain the monoxygenation of five CH_4 molecules. On the other hand, the oxidation of five molecules of methane generates an insufficient amount of electrons (20) for the reduction of eight molecules of nitrate into NO (24). Therefore, *M. oxyfera* has to rely on a (methane-oxidizing) nitrate-reducing partner that supplies N-DAMO with a certain amount of nitrite (or electron donor) to balance its substrate and electron flows. In the original N-DAMO enrichment culture on methane and nitrate [16] mentioned in the Introduction, this nitrite might have been supplied by the archaeal members of the culture.

3 Enzymes in Anammox Metabolism

3.1 Nitrite Reduction to Nitric Oxide

Even though it is well established that NO is an intermediate in anammox metabolism [22], it is less clear which enzyme catalyzes the reduction of nitrite into NO. In nature, two different and well investigated metalloenzymes are widely available that catalyze the reduction: The iron- and heme *d*-containing cytochrome cd_1 (NirS), and the copper-containing NirK [30] (eq. 5).

$$NO_2^- + 2\,H^+ + \mathbf{e}^- \rightarrow NO + H_2O \quad (E_0{}' = +0.38\,V) \qquad (5)$$

The genome of *Kuenenia stuttgartiensis*, the organism that has been used for most studies, contains a gene cluster (kuste4136-4140) encoding the full complement of a functional NirS [21, 22]. In the kuste4136 gene product all structural and functional amino acids are completely conserved with respect to NirS having a known crystal structure [31]. Kust4137 codes for a *c*-type heme (NirC) that might act as the one-electron shuttle in nitrite reduction. The gene products of kuste4138-4140 share significant sequence identities with NirN, NirJ, and NirF, respectively, known to be involved in the assembly of the heme *d* prosthetic group and further maturation of NirS [32]. However, NirS (kuste4136) and NirC (kuste4137) seem to be hardly expressed in the transcriptome and the whole-cell proteome [21, 22]. In striking contrast, NirS is among the highest expressed proteins in the marine anammox species *Scalindua profunda* [33]. Unlike the two species mentioned, anammox strain KSU-1 [34] and *Jettenia asiatica* [35] lack cyt cd_1 nitrite reductase, but they encode and express the copper-containing NirK. More surprising is the absence of both NirS and NirK in the genomes of *Brocadia* species ([36]; M. Oshiki and D. Speth, personal communication), even though cell free systems of *Brocadia sinica* display nitrite-reducing and NO-producing activity that easily supports metabolic requirements. Thus, *Brocadia* and possibly also other anammox species must have a nitrite reductase different from known NirS and NirK.

3.2 Hydrazine Synthesis

Hydrazine synthesis is the exemplary invention of anammox bacteria. The presence of hydrazine synthase enables anammox bacteria to convert ammonium anaerobically, thus providing for these organisms a specific and important niche in microbial ecosystems. Besides this, hydrazine synthase is the second enzyme, next to NOR (see Chapter 4 of this Volume), that is able to forge an N-N bond. Direct purification from *K. stuttgartiensis* identified HZS as the heterotrimeric gene products of kuste2859-2861 [22], as predicted before [21]. In this organism, HZS comprises no less than ~20 % of the cellular proteins. The kuste2859-2861 genes form part of a larger gene cluster (kuste2854-2861) of which the other genes may encode additional protein components that are needed for HZS expression and activity (Figure 3a). Kuste2857 that codes for a sigma 54-type transcriptional regulator is conspicuous in the expression regulation respect.

HZS catalyzes hydrazine synthesis from ammonium (or ammonia) and NO with the input of three electrons (eq. 6). Experiments with purified enzyme confirmed this reaction [22]. However, *in vivo* rates were exceedingly low (20 $nmol \cdot h^{-1} \cdot mg^{-1}$ protein) and enzyme activity needed the presence of one of

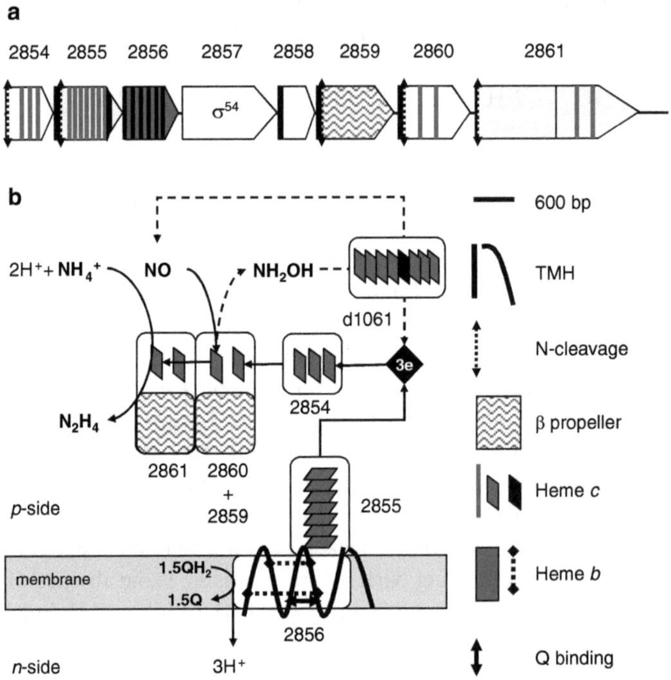

Figure 3 The hydrazine synthase system in *Kuenenia stuttgartiensis*. (**a**) Gene cluster organization. Gene lengths and the position of structural motifs are drawn to scale (bp, base pairs). Gene and protein identifiers (kuste numbers) are placed on top of the genes and next to their gene products. (**b**) Proposed functional organization of the gene products. Putative heme *b*- and menaquinone (Q)-binding sites in kuste2856 were suggested by sequence comparison with the highly homologous γ subunit (FdnI, b_{556}) of menaquinone-dependent formate dehydrogenase (FdhN) from *Escherichia coli* [40]. Possible recycling of spilled hydroxylamine during hydrazine synthesis by the action of the hydroxylamine oxidoreductase-like kustd1061 is indicated by dashed lines. Structural motifs are specified in the figure. Abbreviations: N-cleavage, N-terminal signal sequence cleavage site; TMH, transmembrane-spanning helix.

the HAO-like proteins (kustc1061) described next to artificially pull the reaction by removing the product (hydrazine) and push it by providing electrons via bovine heart cytochrome *c*. Still, *in vitro* rates were only about 1 % of *in vivo* rates. This loss in activity was already obtained by merely breaking the cells, indicating the disruption of a tightly coupled enzyme system and the requirement of other physiological relevant electron and substrate carriers. Nevertheless, HZS would be a very slow enzyme, also when fully active, possibly explaining the slow growth rate of anammox bacteria.

$$NO + NH_3 + 3\ H^+ + 3\ \mathbf{e}^- \rightarrow N_2H_4 + H_2O \quad (E_0{}' = +0.06\,V) \tag{6}$$

$$NH_2OH \rightarrow NO + 3\,H^+ + 3\,e^- \qquad (E_0{}' = -0.03\,V) \tag{7}$$

$$N_2H_4 \rightarrow N_2 + 4\,H^+ + 4\,e^- \qquad (E_0{}' = -0.75\,V) \tag{8}$$

$$NH_2OH + NH_3 \rightarrow N_2H_4 + H_2O \quad \left(\Delta G^{0\prime} = -38\,kJ \cdot reaction^{-1}\right) \tag{9}$$

Isolated HZS also catalyzes the oxidation of hydroxylamine (~1 $\mu mol \cdot min^{-1} \cdot mg^{-1}$ protein; eq. 7) and of its own product, hydrazine (34 $nmol \cdot min^{-1} \cdot mg^{-1}$; eq. 8) with rates that are one to five orders of magnitude higher than physiological hydrazine synthesis. At this stage, our understanding of the molecular mechanism underlying hydrazine synthesis is hidden in a cloud of questions. The initial biochemical characterization of the enzyme [22] and structural properties predicted from the kuste2859-2861 gene products [12, 26] might allow a first glance into this cloud (Figure 3b).

All three kuste2859-2861 genes are preceded by N-terminal signal sequences and lack any other coding sequences for transmembrane helices (TMHs), indicating the export of processed proteins across a cell membrane. In agreement herewith, antibodies raised against specific peptides localize HZS exclusively in the anammoxosome [24]. Sequence analysis suggests kuste2859 to be composed entirely of β propeller strands. In *S. profunda*, the kuste2859 and kuste2860 homologs are fused (scal00025) [33]. Also kuste2861 consists mainly of β propeller sheets, notably in its N-terminal part. In its C-terminus, two heme *c* binding motifs (CXXCH) are present. Kuste2860 contains two heme *c* binding motifs as well and sequence comparison affiliates this protein with two well-characterized enzymes with completely different functions: Cytochrome *c* peroxidase [37] and the MauG protein [38]. Cytochrome *c* peroxidase mediates the reduction of hydrogen peroxide; MauG catalyzes the intricate insertion of oxygen atoms into a precursor of tryptophan tryptophylquinone, the cofactor of methylamine dehydrogenase. Cyt *c* peroxidase and MauG harbor two *c*-type hemes each, one low-spin heme acting in electron transfer and one high-spin heme representing the catalytic center. These high-spin hemes are known to bind NO [39]. By homology, kuste2860 is inferred to be an NO-binding catalytic subunit. Binding of NO together with the addition of three electrons as proposed in Figure 3, would reduce NO to hydroxylamine (reversed reaction 7). The observed hydroxylamine oxidizing activity of HZS is consistent with this proposal. Next, the condensation of NH_2OH and ammonia together with the withdrawal of water would produce hydrazine (eq. 9). Although hydrazine synthesis in this way is exergonic *per se*, it will not be an easy reaction, taking into account the high dissociation energy of the N-H bond. To overcome the energy barrier, one might envisage the presence in HZS of a second catalytic site in which a strong oxidizing group binds ammonia to weaken the N-H bonds.

As a part of the kuste2854-2861 gene cluster, kuste2854 encodes an exported soluble triheme cyt *c* protein, which would make it an excellent three-electron carrier in hydrazine synthesis (Figure 3b) [26]. Kuste2855 codes for a protein with seven *c*-type hemes that, after export out of cytoplasm remains bound to the

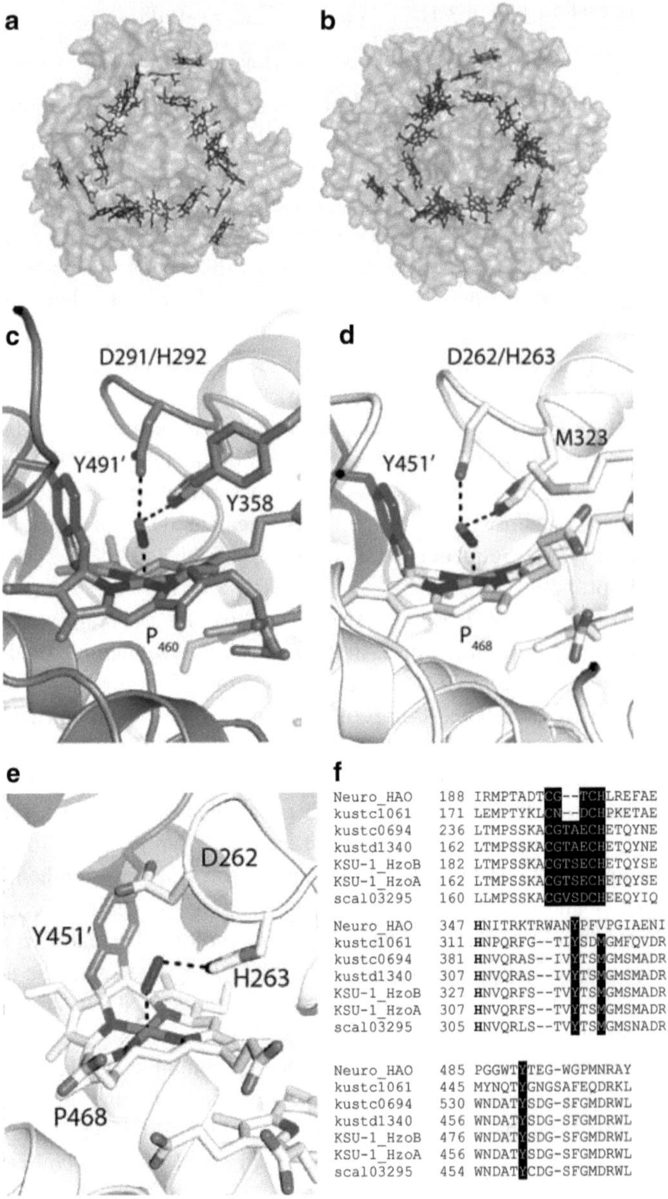

Figure 4 Structures of the homotrimeric octaheme proteins hydroxylamine oxidoreductase from *Nitrosomonas europaea* (NeHAO) and hydroxylamine oxidase kustc1061 from *Kuenenia stuttgartiensis*. Panels (**a**) and (**b**) show the conserved spatial arrangement of the 24 *c*-type hemes in NeHAO and kustc1061, respectively, as seen from the bottom of the enzymes. Panels (**c**) and (**d**) display the conserved active site configuration at heme-4 of NeHAO and kustc1061, respectively, in crystals soaked with hydroxylamine. One may note the tyrosine (Y491' and Y451') from a neighbouring subunit that makes two covalent bonds with the heme-4 in both enzymes. NeHAO (**c**) harbors an additional tyrosine (Y358) near the active site that is replaced at this

(anammoxosome) membrane by one TMH in its C-terminus. The membrane-bound kuste2856 shares significant sequence identity to the heme *b*-containing γ subunit (FdnI, cyt b_{556}) of menaquinone (MQ)-dependent formate dehydrogenase (FDH-N, FdnGHI) [26]. Sequence identity includes the conservation of the topology of the TMHs and of amino acids that are involved in binding of the two heme *b* groups and of menaquinone in the FDH-N crystal structure [40]. Herewith, kuste2855, perhaps in concert with kust2856, could act as a quinol:cyt *c* oxidoreductase at which the soluble kuste2854 shuttles the electrons to HZS (Figure 3b). The absence of this membrane system in experiments with purified HZS could be an explanation for its observed low activity [26].

3.3 *Anammox and Its Multiple Hydroxylamine Oxidoreductase-Like Proteins*

3.3.1 A Collection of Hydroxylamine Oxidoreductase-Like Proteins

The identification of hydrazine as an intermediate in the anammox process immediately suggested a candidate enzyme that could oxidize the compound into N_2 (eq. 8) [21]. It has been known for a long time that hydrazine can be oxidized efficiently by hydroxylamine oxidoreductase (HAO) from aerobic ammonium oxidizers [41]. In these organisms, the physiological reaction of HAO is the four-electron oxidation of hydroxylamine into the endproduct nitrite (eq. 10) [42].

$$NH_2OH + H_2O \rightarrow NO_2^- + 5\,H^+ + 4\,\mathbf{e}^- \quad (E_0{}' = +0.065\,V) \qquad (10)$$

HAO from *Nitrosomonas europaea* (NeHAO) has been well characterized over many years and its crystal structure is known [43, 44]. NeHAO is a homotrimeric protein in which each subunit binds eight *c*-type hemes (Figure 4a). Seven of these hemes are low-spin by proximal and distal histidine ligations. Together, they represent an electron-wiring circuit to the subunit's exit and to a next subunit. The catalytic center is heme-4 that binds and converts hydroxylamine at the distal position (Figure 4c). A remarkable property of NeHAO is the presence of two covalent

Figure 4 (continued) position by a methionine (M323) in kustc1061 (**d**). (**e**) Crystals of kustc1061 soaked in hydrazine reveal the presence of a diazene (NH=NH)-like molecule bound to the heme-4 iron. Multiple amino acid sequence alignments (**f**) of NeHAO, kustc1061 and the hydrazine dehydrogenase/oxidase (HDH/HZO) octaheme proteins from *K. stuttgartiensis* (kustc0694, kustd1340), anammox strain KSU-1 (HzoB, HzoA) and *Scalindua profunda* (scal03295) highlight the typical CX_4CH heme-3 binding motif in the HDHs (top), the active site tyrosine in NeHAO that is replaced by a methionine by a two-amino acid contraction in the anammox proteins (middle), and the conserved tyrosine that is covalently bound to the catalytic heme-4 in these octaheme proteins (bottom). Structural figures were prepared in PyMOL [48] using the following protein database (PDB ID) files: (a) 4N4N, (b) 4N4J, (c) 4N4O, (d) 4N4K, and (e) 4N4L [44].

bonds between a tyrosine from the next subunit and the active-site heme-4, constituting the P460 prosthetic group. The group is named after its specific absorption maximum around 460 nm in the fully reduced protein. Through these bonds, all three subunits are covalently linked as a 200 kDa complex. For as-yet not fully understood reasons, the P_{460}-type heme favors substrate oxidation reactions [45]. In contrast, other related multiheme proteins that lack the tyrosine heme-4 crosslink are equipped for reductive substrate conversions.

Rather than one, no less than ten different HAO paralogs are detected in the genome of *K. stuttgartiensis*, six of which being highly expressed at the transcriptome and proteome levels (Figure 5) [21, 22]. The collection of HAO-like proteins is highly conserved in other genomes of anammox bacteria as well. The straightforward questions then were, which one(s) would be the physiological hydrazine dehydrogenase(s), what would the others do, and how would the different HAO paralogs be tuned to a presumably specific function? Sequence comparison and phylogenetic analysis allow their classification into a number of subgroups (Figure 5) [12, 26, 45]. Four of the proteins (kustc0694, kustd1340, kustc1061, and kusta0043) possess the tyrosine in their C-terminus that makes the covalent bonding in HAO. Six of these lack the particular tyrosine, at which kustc0458, kuste4574, and kuste2479 cluster as one subgroup sharing ~50 % sequence identity to each other. The residual three, lowly expressed ones (kustd2021, kuste2435, and kuste2457) display less than 30 % sequence identities, both to each other and to the seven ones already mentioned. Direct evidence for the function of some HAO-like paralogs has come from direct purification and subsequent characterization.

3.3.2 Hydroxylamine Oxidation to Nitric Oxide

The best understood HAO-like protein from anammox bacteria, both regarding its structure and function, is kustc1061 from *K. stuttgartiensis* [44]. The protein has also been purified from anammox strain KSU-1 [46] and from *Brocadia anammoxidans* [47]. In agreement with what is seen in its protein sequence (Figure 5), kustc1061 is an octaheme protein in which the (three) subunits are covalently bound via the conserved tyrosine to the catalytic heme-4, like in NeHAO [44]. Despite an only 30 % sequence identity, kustc1061 and NeHAO share a highly similar overall architecture (Figure 4). The similarity encompasses the spatial position and orientation of the eight heme groups, and the structure of the P_{460} catalytic heme-4. Such heme arrangement was also found in other multiheme proteins, like cyt *c* nitrite reductase (see Chapter 9 of ref. [138]). Still, there appears to be one significant difference between kustc1061 and NeHAO, which affects the functionality of both enzymes. NeHAO contains a specific tyrosine near the catalytic site, which is absent in kustc1061 (Figures 4c and d).

Both kustc1061 and NeHAO catalyze the oxidation of hydroxylamine, but with different outcomes. While kustc1061 produces NO in a three-electron

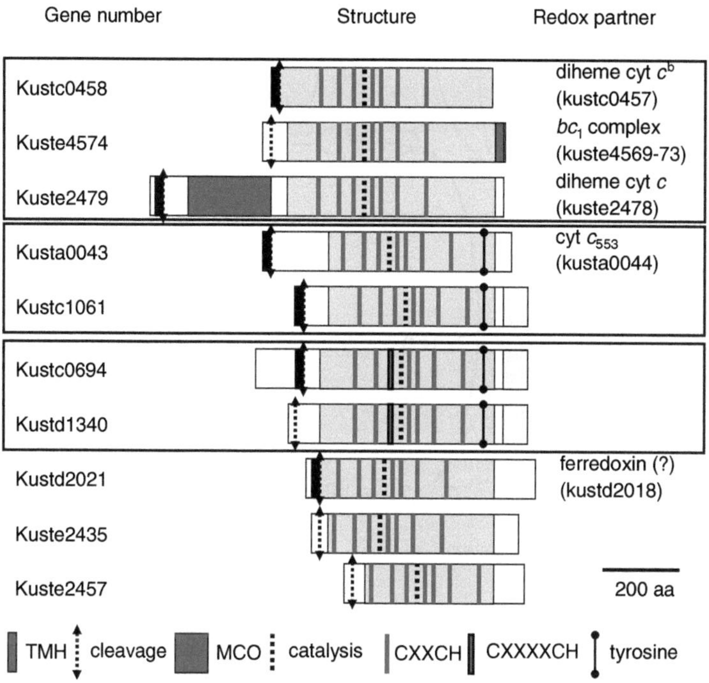

Figure 5 A family of ten hydroxylamine oxidoreductase (HAO)-like proteins in the genome of *Kuenenia stuttgartiensis* [12, 26]. Lengths of the polypeptides are drawn to scale (aa, amino acids). Homologous stretches that are rich in *c*-type hemes are vertically aligned. HAO-like proteins showing high sequence identities, suggestive of a same function, are boxed. Structural motifs are specified in the figure with the following abbreviations: TMH, transmembrane-spanning helix; cleavage, N-terminal signal sequence cleavage site; MCO, multicopper oxidase; catalysis, catalytic *c*-type heme; CXXH, heme *c*-binding motif; CXXXXCH, divergent heme *c*-binding motif; tyrosine, tyrosine involved in covalent bonding with the catalytic heme.

oxidation (eq. 7), the reaction product of NeHAO is nitrite (eq. 10). Hydroxylamine reacts spontaneously with as-isolated fully oxidized kustc1061, at which the enzyme undergoes a two-electron reduction and hydroxylamine gets oxidized to the HNO nitrosyl state. In crystals both of kustc1061 and HAO soaked with hydroxylamine, the nitrosyl is bound to the iron of the P_{460} catalytic heme [44] (Figures 4c and d). The subsequent one-electron oxidation of HNO yields NO, which is released in case of kustc1061. The sequence of events suggests the catalytic mechanism shown in Figure 6. In the reaction catalyzed by NeHAO, a water molecule has to be added to NO to produce nitrite after one more one-electron oxidation. The tyrosine next to the NeHAO catalytic site, which is absent in kustc1061, is perfectly positioned to mediate such water addition [44].

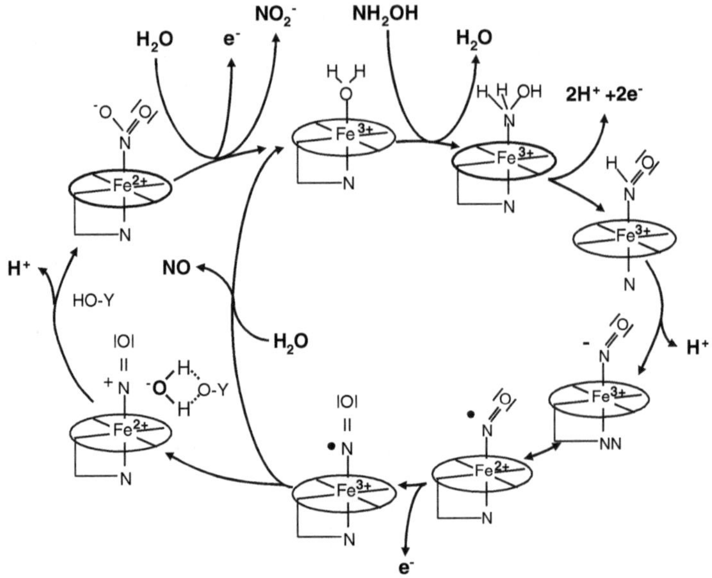

Figure 6 Reaction mechanism of hydroxylamine oxidation to nitrite by hydroxylamine oxidoreductase (HAO) (outer cycle) and to nitric oxide (NO) by kustc1061 (inner cycle). The scheme combines models proposed for the mechanism of kustc1061 [44] and of HAO ([49] and references therein) assisted by a specific tyrosine near the HAO active site for the addition of water.

Besides hydroxylamine conversion, kustc1061 catalyzes the four-electron oxidation of hydrazine (eq. 8), albeit with 50-fold lower catalytic efficiency (k_{cat}/K_m) [44]. Analogously to hydroxylamine, hydrazine added to the oxidized protein results in a diazene-like (NH=NH) molecule bound to the P_{460} iron [44] (Figure 4e). Even though hydrazine is not the physiological substrate of kustc1061, this observation might be of relevance in understanding the function of the genuine hydrazine dehydrogenase described in the next section.

It is understood by now that hydroxylamine is not an intermediate and the function of kustc1061 as a specific hydroxylamine oxidase is not easy to reconcile with anammox central metabolism as we know it. However, hydroxylamine could be an intermediate in hydrazine synthesis (see previous Section 3.2; Figure 3) and, as suggested before [26], failure to complete the difficult condensation reaction with ammonia might result in the spillage of hydroxylamine. Now, kustc1061 would expertly serve the regeneration of NO and three electrons for a next synthesis attempt. In this respect, it may be interesting to note that kustc1061 homologs in other anammox genomes are clustered with the HZS system [33–36].

3.3.3 Hydrazine Oxidation and Nitrogen Formation

Direct purification from *K. stuttgartiensis* identified the genuine N_2-forming hydrazine dehydrogenase/oxidase (HDH/HZO) as kustc0694 [22]. The homologous enzyme from anammox strain KSU-1 had been described before [50], but its function was not fully assured by then. Kustc0694 is a member of the HAO-like family introduced above (Figure 5). The purified protein is 100 amino acids shorter than annotated in the genome database because of an incorrect assignment of the start codon and N-terminal signal peptide cleavage [22]. This cleavage is most likely associated with the export of kustc0694 into the anammoxosome [24]. Kustc0694 is over 95 % identical, both at the nucleotide and amino acid levels, to kustd1340, most likely representing a second HDH in *K. stuttgartiensis* and other anammox species.

Like kustc1061 and NeHAO, kustc0694 is a homotrimeric protein carrying eight *c*-type hemes per subunit and the three subunits are covalently linked to each other [22, 50]. A peculiar feature of kustc0694 and its close homologs is the divergent CX_4CH binding motif for heme-3 (Figures 4f and 5). The rational for this is not understood. The increase in length between both cysteines that bind the heme *c* most likely affects the position of this heme with respect to neighboring hemes, in its turn affecting electron transfer between the hemes. Upon reduction of the oxidized protein, the characteristic P_{460} absorption maximum is observed [50]. This observation, the covalent binding of the subunits, and the presence in the kustc0694 sequence of the tyrosine that makes two bonds to heme-4 in kustc1061 and NeHAO (Figure 4) are highly suggestive that heme-4 is the catalytic site also in kustc0694.

Kustc0694 catalyzes with high V_{max} and low K_m the four-electron oxidation of hydrazine to N_2 according to equation 8 [22]. Hydroxylamine is not a substrate, but rather a strong competitive inhibitor with respect to hydrazine. Also NO, a free intermediate in anammox metabolism, competitively inhibits hydrazine oxidation with low K_i. Incubation of kustc1061 with hydrazine results in the two-electron-oxidized diazene bound to the P_{460} heme (Figure 4e). Although this remains to be shown, the same could hold for kustc0694. The subsequent two-electron oxidation of diazene would then make N_2. The passage of two times two electrons through an electron wire composed of a series of one-electron (heme) carriers will need a fine-timing of the electron circuit. Similarly, the P_{460} catalytic center has to be fine-tuned for hydrazine turnover. How these aspects are resolved is a matter of ongoing research, which will include the resolution of the kustc0694 crystal structure. Future investigations have to address one more important issue, namely how the low-redox potential electrons derived from hydrazine oxidation (eq. 8) are utilized to build up a proton-motive force.

3.3.4 What About the Other Hydroxylamine Oxidoreductase-Like Proteins?

Kustc0458, kuste4574, and kuste2479 share significant sequence identities, especially regarding the presence and the position in their protein sequences of eight *c*-type hemes (Figure 5). This might suggest that these three proteins have a similar function. All three lack the tyrosine in the C-terminus, which is responsible for the binding with a catalytic heme, indicative of reductive transformations [45].

Kustc0458 has been purified from K. *stuttgartiensis* and its crystal structure has been resolved as a heterododecameric ($\alpha_6\beta_6$) complex composed of a dimer of kustc0458 octaheme trimers at which each kustc0458 subunit is associated with the diheme kustc0457 [51]. Altogether, the complex harbors no less than sixty *c*-type hemes. Again, the catalytic center is heme-4, but this heme is not covalently bound to a next subunit. Correspondingly, the kustc0457-0458 complex does not catalyze the oxidation of nitrogenous substrates and the only activity that could be measured was the reduction of nitrite, presumably to ammonium [51]. However, this activity was obtained with reduced methyl viologen as a powerful reductant. Since reduced viologen dyes also support reductive reactions by NeHAO [52] and kustc1061 [44] artificially, the question remains if nitrite reduction to ammonium represents the physiological reaction of kustc0457-0458. An answer to this question may be found by the isolation and identification of the physiological electron carrier of the complex. Here, an attractive candidate would be the gene product that is directly linked to the kustc0457 and kustc0458 genes, namely kustc0456. Sequence analysis identifies kustc0456 as a novel type-1 copper-containing cupredoxin [26]. Cupredoxins serve as electron carriers in many respiratory and photosynthetic systems [53], including NO-forming nitrite reductase NirS [54].

Kustc0458 shows 47 % sequence identity to kuste4574 (Figure 5). The identity comprises the conservation in the alignment of the eight heme-*c* binding sites including six or seven histidines that make up the distal ligands to the electron-wiring *c*-type hemes, leaving heme-4 as the catalytic site also in kuste4574. As already mentioned, kuste4574 lacks the tyrosine that is involved in the covalent bonding of heme-4 in NeHAO, kustc0694, and kustc1061, indicative of a reductive catalysis. Importantly, kuste4574 forms part of a gene cluster (kuste4569-74) that codes for one of three bc_1 complexes in *K. stuttgartiensis* (Figure 7) and this bc_1 complex could be most remarkable by its composition.

Like canonical bc_1/b_6f complexes, the kuste4569-74 genes code for a 2Fe-2S Rieske subunit (kuste4569), a heme *b*- and quinone-binding part (kustc4571-72) and a *c*-type heme (kuste4573). However, detailed sequence analyses affiliate this anammox system, as well as the alternative system encoded by kustd1480-85, with a novel type of *bc* complexes [55, 56]. The Rieske subunit could have aberrant redox properties [55], the heme *b*- and quinone-binding part is split across two proteins similar to photosynthetic b_6f systems, and the *c*-type heme is a hexaheme protein. Besides HAO-like kuste4574, kuste4569-74 gene cluster codes for one more polypeptide, an FAD-binding NAD(P)H oxidoreductase (kuste4570). If assembled as one whole, the question is what the complex would do, and an answer can only be a speculative one: The simultaneous coupling by a bifurcation mechanism of (mena)quinol oxidation to the reduction of both a nitrogenous substrate and $NAD(P)^+$ [26]. If the reduction encompasses the conversion of nitrite into NO, this would answer two open issues at the same time, the nature of the elusive nitrite reductase and the way NAD(P)H is generated for biosynthetic purposes.

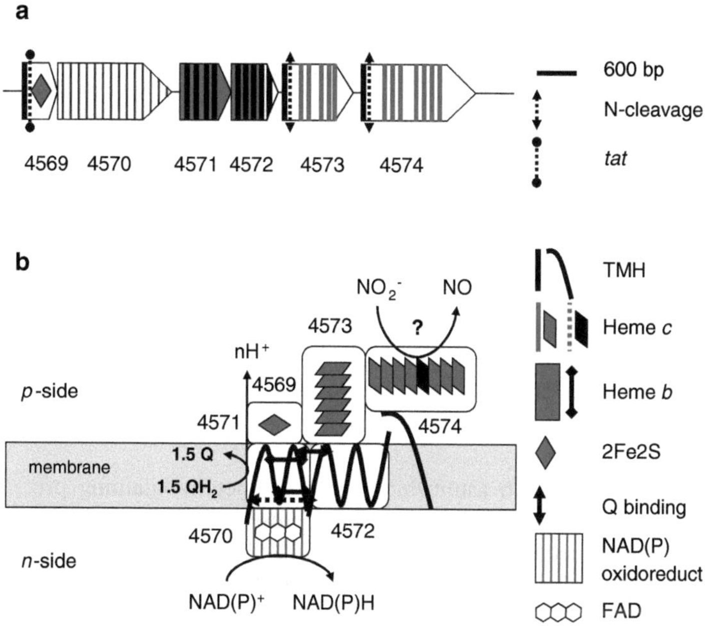

Figure 7 Gene cluster organization (**a**) and possible functional organization of its gene products (**b**) of a novel cytochrome *bc* complex found in anammox bacteria. The complex consists of Rieske-type 2Fe2S iron sulfur protein, a membrane-bound heme *b* part split across two proteins and a *c*-type cytochrome [12, 26] together with a HAO-like octaheme protein and a FAD-dependent NAD(P)H oxidoreductase. Heme *b*- and quinone (Q)-binding sites were inferred from the conservation of amino acids in crystal structures of bc_1 and b_6f complexes implicated in the binding of these compounds [57]. The anammox *bc* complex has been suggested to couple menaquinol oxidation to the reduction of nitrite and $NAD(P)^+$ by a bifurcation mechanism [26]. Numbers in (a) and (b) refer to the gene identifiers (kuste) of *Kuenenia stuttgartiensis*. Gene lengths and the position of structural motifs are drawn to scale (bp, base pairs). Structural motifs are specified in the figure. Abbreviations are the following: N-cleavage, N-terminal signal sequence cleavage site; *tat*, twin-arginine translocation signal; TMH, transmembrane-spanning helix.

Kuste2479 is another interesting HAO-like protein that is well conserved amongst anammox bacteria. While its C-terminal part with its eight *c*-type hemes shows significant (>50 %) sequence identity to kustc0458 and kuste4574, the N-terminal sequence is comprised of a leader sequence for protein export and an amino acid sequence that maps to the broad family of copper-containing laccases [58], and to two-domain small laccases more specifically [59]. The neighboring gene kuste2479 codes for a diheme protein in the N-terminus and a linocyn in the C-terminus. Linocyns assemble into isohedral nanocompartments that enclose oxidative-stress response proteins [60]. After export out of the cell, the assembly might act in the removal of reactive oxygen or nitrosative species. The presence and identification of such nanobodies in anammox cultures remains to be established, which will not be an easy task since kuste2479 and kuste2478 are only lowly expressed.

3.4 The Oxidation of Nitrite and Reduction of Nitrate

3.4.1 The Nitrite:Nitrate Oxidoreductase System

In anammox metabolism, the nitrite:nitrate oxidoreductase (NXR) serves two opposite goals, (1) the oxidation of nitrite to replenish the cyclic electron flow upon draining due to the loss of intermediates (NO, hydrazine) or electrons for biosynthetic activities, and (2) the reduction of nitrate to nitrite during growth in the presence of external electron donors other than ammonium (eq. 11) (Figure 1). In *K. stuttgartiensis*, structural genes (*NxrABC*) coding for the NXR protein form part of a large gene cluster (kustd1699-1713), which is highly conserved among anammox genomes.

$$NO_2^- + H_2O \rightleftharpoons NO_3^- + 2\,H^+ + 2\,\mathbf{e}^- \quad (E_0' = +0.43\,V) \qquad (11)$$

The cluster encodes an astonishing range of metal-containing proteins (Figure 8a). The *NxrABC* structural genes are represented by kustd1700, kustd1703, and kustd1704, respectively. Herein, kustd1700 codes for a protein (1148 amino acids) that is preceded by a twin-arginine translocation (*tat*) signal for protein export and that belongs to the broad class of molybdopterin (MPT) enzymes. Members of this class catalyze a variety of oxidation and reduction reactions of N-, S- or carbon-substrates (see [61] for a review). Amongst many others, the family includes nitrate reductase (NarGHI), formate dehydrogenase (see Section 4.2.4), acetylene hydratase (see Chapter 2 of ref. [138]), and dimethylsulfoxide (DMSO) reductase and dimethylsulfide dehydrogenase (see Chapter 11 of ref. [138]). Quite a few MPT proteins have been well characterized by the resolution of their atomic structures [61]. Their common property is the presence of a catalytic molybdenum atom that is coordinated by four sulfur atoms derived from two antiparallel-oriented dithiolene-containing pterins that are esterified with guanine dinucleotide (Mo-*bis*PGD) (Figure 8b). The pterin occurs in two conformers, as a tricyclic (pyranopterin) and a bicyclic structure with an open furan (molybdopterin). A fifth ligand to molybdenum is provided by a serine, cysteine, selenocysteine, or a mono-/bidentate oxo/hydroxyl group (D222 in NarG) [62]. In the N-terminal part, a conserved set of cysteines is usually present to bind a 4Fe4S cluster (FS0). While sequence analysis unambiguously identifies kustc1700 as a molybdopterin protein, sequence identity with MPT enzymes of known function, including NarG nitrate reductase, is moderate (10–25 %). By and large, kustd1700 harbors the series of amino acids that have been identified in other MPT proteins with the binding of the Mo-*bis*PGD cofactors, but the full conservation with any known protein is not seen. This would imply that Mo and its cofactors are bound such as to facilitate its specific catalytic function, the oxidation of nitrite to nitrate. In agreement herewith, sequence identity is highest (~60 %) with putative *Nxr* genes found in the genomes of the nitrite oxidizers *Candidatus* 'Nitrospira defluvii' [63] and *Nitrospina gracilis* [64]. Despite the overall divergence, kustd1700

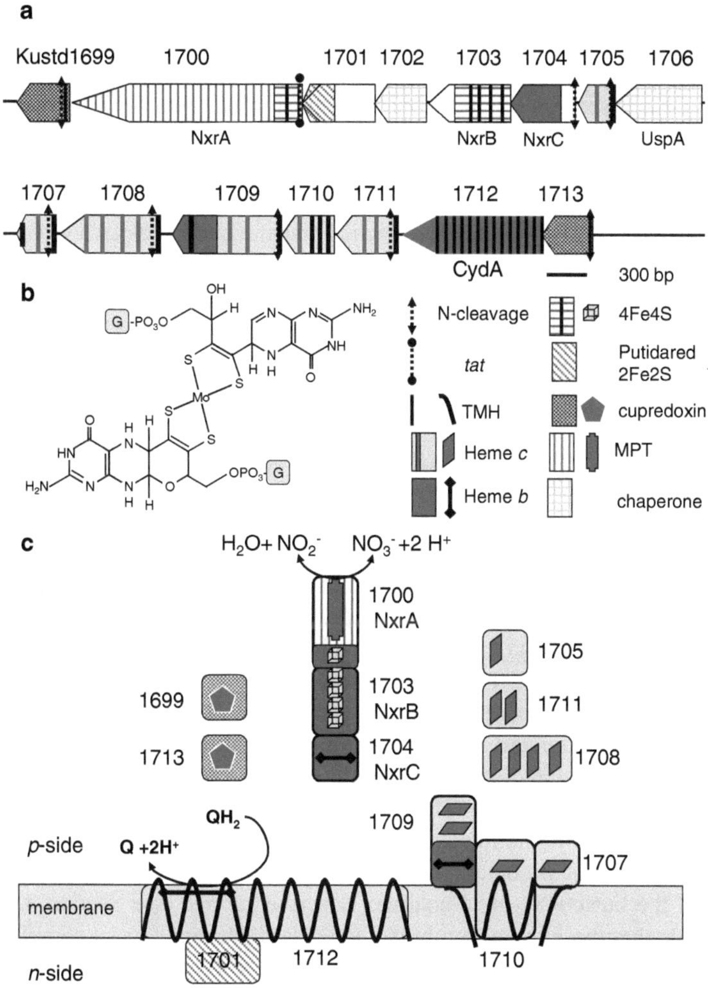

Figure 8 Gene cluster organization (**a**) and possible functional organization (**c**) of the nitrite: nitrate oxidoreductase (NXR) system in *Kuenenia stuttgartiensis*. Numbers in (a) and (c) refer to the gene identifiers (kustd) of *Kuenenia stuttgartiensis*; gene lengths and the position of structural motifs in (a) are drawn to scale (bp, base pairs). Structural motifs are specified in the figure. The system is suggested to be comprised of the catalytic NxrABC protein, membrane-bound protein complexes, including a protein (kustd1712) partly related to subunit I (cydA) of the cytochrome *bd* quinol oxidase (CydAB) [71] and several heme-*c* type as well as novel cupredoxin-type [26] soluble electron carriers. NxrA houses a Mo-*bis*(pyrano guanine dinucleotide) cofactor common to all molybdopterin enzymes of which the chemical structure is shown in (**b**). In this structure the open furan form of the pterin moiety is displayed above and the tricyclic pterin below the molybdenum that is coordinated by the dithiolene groups; G, guanine monophosphate. Putative heme *b*- and quinone (Q)-binding sites were inferred from the CydAB sequence [71]. Abbreviations: N-cleavage, N-terminal signal sequence cleavage site; *tat*, twin-arginine translocation signal; TMH, transmembrane-spanning helix; putidared, putidaredoxin-like 2Fe2S iron sulfur cluster; MPT, Mo-*bis*(pyrano guanine dinucleotide) cofactor; UspA, universal stress protein A.

contains the aspartate (D222) that binds Mo in NarG. Also the set of FS0-binding cysteines is found, albeit with one remarkable difference. In NarG the iron sulfur cluster is coordinated by a $HX_3CX_3CX_{34}C$ sequence [62, 65], whereas a $CX_3DX_3CX_{39}C$ sequence is found at this position in kustd1700 and homologous nitrifier sequences. The difference in coordination likely will affect the redox potential of the FS0 iron-sulfur cluster in its role to transfer electrons to or from the molybdopterin catalytic site.

Like in many MPT proteins [61], electron transfer encompasses a second subunit (NxrB, kustd1704) that contains the conserved set of cysteines to accommodate 4 additional iron sulfur clusters (FS1-4), in which FS1-3 represent 4Fe4S clusters and FS4 is a 3Fe4S moiety. Unlike archetypal MPT proteins [61], the third subunit (NxrC, kustd1704) is not a membrane anchor. Rather, kustd1704 is a soluble protein with an N-terminal signal sequence for protein export (Figure 8). Importantly, NxrC shares highest sequence identity with the heme *b*-containing C subunits of a subset of MPT proteins, as there are dimethylsulfide dehydrogenase [66], selenate reductase [67], chlorate reductase [68], and ethylbenzene dehydrogenase [69]. Sequence identity includes the amino acids that coordinate heme *b* in the crystal structure of ethylbenzene dehydrogenase [69].

Besides the three structural subunits, the kustd1699-1713 gene cluster encodes a series of other proteins that apparently play a role in the functioning of the NXR system. The gene product of kustd1702 is related to the NarJ and TorD proteins that act as chaperones in MPT protein maturation [70]. A chaperone role is also anticipated for the universal stress protein A (UspA)-related kustd1706. Kustd1701 is characterized by four C-terminal cysteines with the typical spacing of adrenodoxin/putidaredoxin-like 2Fe2S clusters. Kustd1707, kustd1710, and kustd1709 have one or more TMHs at which the former two carry one *c*-type heme each with an orientation as proposed in Figure 8c. Kustd1709 is remarkable by its fusion of heme *b*-containing kustd1704 and kustc0457, found as the diheme subunit of the kustc0457-0458 complex described above (see Section 3.3.4). It is conceivable that these three membrane-associated proteins form a complex, possibly associated with one or more of the remaining polypeptide components of the kustd1699-1713 gene cluster. Herein, kustd1712 would be membrane-bound by no less than 15 predicted TMHs. The first N-terminal half of kustd1712 displays significant sequence identity with the quinone- and heme *b*-binding subunit (CydA) of cytochrome *bd* oxidase (CydAB) [71]. Kustd1705, kustd1708, and kustd1711 again are preceded by N-terminal cleavage sites and possess CXXCH motifs for the binding of one, four, and two *c*-type hemes, respectively. In front and at the end of the kustd1699-1713 cluster, two genes are found that encode exported proteins with type-1 cupredoxin features that are homologous to kustc0456 encountered before (Section 3.3.4). Together, these proteins provide an array of mobile and membrane-bound carriers that somehow are involved in electron and possibly proton-translocating processes related with nitrite oxidation and nitrate reduction. However, the specific role of each of them remains to be established.

3.4.2 Nitrite Reduction to Ammonium

In the presence of inorganic electron donors such as Fe(II) compounds, Mn^{2+} and possibly dihydrogen, or organic compounds, most notably formate, acetate or propionate, anammox bacteria may adopt a lifestyle "disguised" as denitrifier [28]. Different species may even have specialized in the use of certain electron donors, providing them a competitive advantage. For instance, *Candidatus* 'Brocadia fulgida' was specifically enriched in a medium that contained ammonium and nitrite supplemented with a low (1 mM) concentration of acetate [28]. In contrast, *Candidatus* 'Anammoxoglobus propionicus' came to the fore under the same enrichment conditions and with the same inoculum, except that propionate substituted for acetate [28]. As outlined in Section 2.1, during "disguised" denitrification half of nitrite is reduced to ammonium, which combines with the other half to produce dinitrogen gas in the anammox process (Figure 1). Consequently, all N_2 is produced from nitrite or nitrate, as is also the case with "true" denitrification. The key enzyme in "disguised" denitrification would be the nitrite reductase making ammonium. This is also the key enzyme in the so-called DNRA bacteria that make a living from the dissimilatory nitrite/nitrate reduction to ammonium.

$$NO_2^- + 8\,H^+ + 6\,\mathbf{e}^- \rightarrow NH_4{}^+ + 2\,H_2O \quad (E_0{}' = +0.43\,V) \tag{12}$$

DNRA bacteria occur in broad and likely only partly appreciated diversity of species in many ecosystems where they make a short circuit in the N-cycle [72]. The organisms use nitrite as terminal electron acceptor, which is reduced by the input of six electrons to ammonium without the release of intermediates (eq. 12). This type of nitrite respiration is mediated by the NrfHA/NrfABCD systems that are discussed in Chapter 9 of ref. [138]. Briefly, NrfH is a membrane-bound tetraheme *c* protein belonging to the NapC/NirT family of menaquinol oxidases [45, 73]. NrfH accepts the electrons from menaquinol and transfers these to its associated catalytic part, NrfA. In enteric bacteria the alternative, partly soluble NrfABCD system is found in which the soluble pentaheme cytochrome *c* NrfB mediates electron transfer between the catalytic NrfA (alternatively named cytochrome c_{552} nitrite reductase) and the membrane-associated iron-sulfur protein NrfC and NrfD, which is an assumed quinol dehydrogenase of the NrfD/PsrC family [73]. Thus, NrfA represents the catalytic enzyme in both systems. NrfA is a homodimeric pentaheme cytochrome *c* protein (unrelated to NrfB) localized in the periplasm. The five *c*-type hemes (1–5) of NrfA fully superimpose hemes 4–8 of the HAO-like proteins shown in Figure 4. Heme-1 in NrfA represents the catalytic site and it is the equivalent of heme-4 in the HAOs. A distinctive feature of this heme-1 is the presence of a lysine as a proximal ligand, which is signified by the CXXCK motif in the protein sequence. The presence of the lysine seems to be a governing factor in the reactivity of NrfA, next to a conserved triad of amino acids near the active site (R114, Y218, H277; numbering according to the *Wolinella succinogenes* structure, PDB ID 1FS7) [74].

The nature of a NrfA-like protein in anammox bacteria has been elusive for quite some time. The reason for this is that no pentaheme protein could be found in sequenced genomes with the diagnostic CXXCK motif and the RYH triad. Members of the *Campylobacteriaceae* such as *Campolybacter jejuni* [75] and *Anaeromyxobacter dehalogenans* [76] that are known for their DNRA activity lack the particular pentaheme proteins as well. However, cytochrome c_{552} nitrite reductases have been annotated in their genomes, yet bearing the common CXXCH binding motif for heme-1. Deletion and expression analyses of the alternative *Campolybacter jejuni* pentaheme protein demonstrated that it is a nitrite reductase making ammonium [77]. The alternative NrfAs remain to be characterized regarding the nature of the proximal ligand to heme-1. It is possible that this ligand is a lysine derived from another part of the protein sequence, rather than a histidine. Furthermore, detailed sequence analyses affiliated the *Campylobacteriaceae* and *A. dehalogenans* proteins with a broad class of NrfAs that covered 18 clades (A-R), three of which (J-L) being characterized by a CXXCH motif for heme-1 [78]. Importantly, close homologs of one of the three clades (K) are found in the genomes of anammox strain KSU-1 (KSU1_B0055) [34], of *Brocadia* species [36] and of *Jettenia* ([35]; D. Speth and M. Oshiki, personal communication). In these genomes, *nrfA* genes are localized next to the ones coding for tetraheme NrfH proteins. However, the genome of *K. stuttgartiensis*, which is able of ammonium formation from nitrite [28], lacks the particular homolog, as does *Scalindua*.

Besides pentaheme nitrite reductases, two different octaheme proteins are known that reduce nitrite to ammonium with high catalytic efficiencies, both of which crystal structures have been resolved: A homohexameric enzyme from *Thioalkalivibrio* species [79] and a homodimeric enzyme from *Shewanella oneidensis* [80]. Despite a limited amino acid sequence identity, the spatial arrangement of the eight hemes is highly similar in both types. In fact, heme arrangements are conserved with respect the ones of the octaheme HAO-like proteins described above (Figure 4) and of pentaheme NrfA. Like in NrfA, the catalytic hemes are proximally ligated by a lysine, albeit in different fashions. In the *S. oneidensis* enzyme, the catalytic heme-2 receives the coordinating ε-amine from a lysine in a distal part of the amino acid sequence [80], whereas the catalytic heme-4 in the *Thioalkalivibrio* proteins is lysine-coordinated by the CXXCK heme-4 binding sequence [79]. Notwithstanding the extant collection of octaheme proteins (Figure 5), homologs of those octaheme nitrite reductases are absent in genomes of anammox bacteria. Still, this does not rule out that one of the anammox HAOs fulfils a physiological function of an ammonium-forming nitrite reductase. In this respect, it is interesting to note that kustd2021 homologs (Figure 5) are consistently present in genomes that lack the alternative (CXXCH) NrfA, whereas genomes of anammox species that possess such NrfA are devoid of kustd2021 homologs.

4 Enzymes in Nitrite-Driven Methane Oxidation

M. oxyfera derives its energy for growth from the oxidation of methane coupled to the reduction of nitrite or nitrate (eqs 3 and 4). Presently, none but one of the enzymes involved in these processes has been purified and characterized. Progress in the field is severely hampered by the extremely slow growth rate and growth yield of the organism, which is enriched as tight aggregates that withstand a convenient cell breaking and protein extraction. Consequently, our knowledge on *M. oxyfera* metabolism mainly relies on what its genome tells us and what we have learned from detailed functional and structural studies on homologous enzymes from methanotrophic, methylotrophic, and denitrifying model organisms. A striking aspect of the genome of *M. oxyfera* is the redundancy of many of its catabolic enzymes [18]. In the genome, the inventory is found for two different nitrate reductases, multiple heme-copper oxidases including Nor-like proteins, three different methanol dehydrogenases, two different pathways for formaldehyde-formate interconversion and three different formate/formyl dehydrogenase systems. The rationale behind this redundancy is unclear and an answer may come only from future physiological, biochemical and perhaps genetic research.

4.1 The Nitrite Reduction Route

4.1.1 Nitrite and Nitrate Reduction to Nitric Oxide

The genome of *M. oxyfera* harbors the inventory for two different nitrate reductases, the membrane-bound NarGHI system briefly mentioned in Section 3.4.1, which is localized in the cytoplasm, and the periplasmic NapAB system [30, 81]. Both systems are well studied in denitrifiers [81], which includes the elucidation of several crystal structures [62, 65, 82]. Both systems are only lowly expressed at the transcriptome and proteome levels in *M. oxyfera*, which makes sense in enrichment cultures grown on nitrite.

In the genome of *M. oxyfera*, the NarGHI system is encoded on the DAMO_0779-0774 gene cluster (Figure 9a). Structural genes are represented by DAMO_0778 (*narG*), DAMO_0776 (*narH*) and DAMO_0774 (*narI*). The genes are interspersed by genes coding for a hemerythrin-like protein (DAMO_0777) and the common NarJ (DAMO_0775) with a chaperone function [70]. At the end of the cluster, DAMO_ 0779 codes for a *c*-type triheme protein that could be bound to the membrane by one TMH in its C-terminus. DAMO_ 0779 shows ~40 % sequence identity to proteins found in the genomes of the nitrite-oxidizing *Nitrospina* (CCQ90255) and of *Sinorhizobium fredii* (CCE99719) where it has been annotated as subunit II of cytochrome *cb* oxidase, and to the *K. stuttgartiensis* kustc0457 diheme protein described before (Section 3.3.4). Quite remarkably, NarGHI from *M. oxyfera* display highest sequence identities (~80 %) to the corresponding polypeptides from the aerobic nitrite oxidizers *Nitrococcus mobilis* and *Nitrobacter hamburgensis* that will employ their NarGHI for nitrite oxidation. Sequence identity with the NarG (45 %), NarH (51 %), and NarI (31.5 %) subunits of the

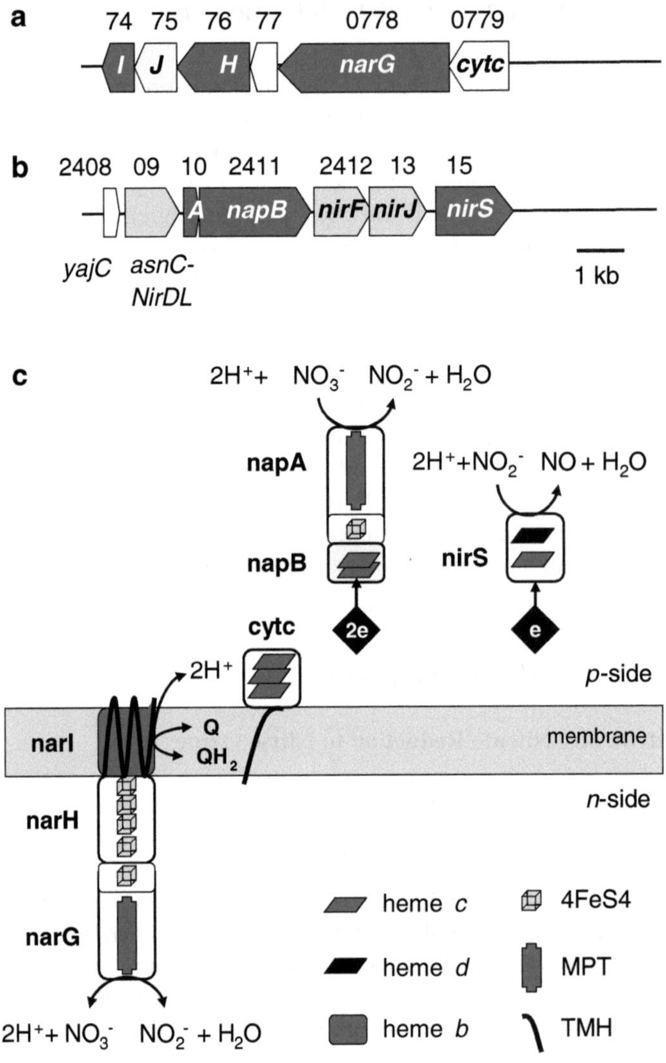

Figure 9 Gene cluster organization of (**a**) the cytoplasmic dissimilatory nitrate reductase (Nar), and (**b**) periplasmic dissimilatory nitrate reductase (Nap) and nitrite reductase (Nir) systems in *Methylomirabilis oxyfera*. Gene lengths and intergenic regions are drawn to scale. Genes coding for catalytic proteins are shaded dark grey (white lettering) and for accessory proteins light grey; see text for the function of the accessory proteins. Gene orientation is indicated by the arrow heads and numbers refer the gene identifiers (DAMO_). The localization of the enzymes systems and their putative functional organization are displayed in (**c**). Herein the *p*- and *n*-sides refer to the positive (periplasmic) and negative (cytoplasmic) sides of the cell. As yet unknown carriers that transfer electrons for nitrate and nitrite reduction are indicated by black diamonds. Structural motifs are specified in the Figure with the following abbreviations: MPT, Mo-*bis*(pyrano guanine dinucleotide) cofactor; TMH, transmembrane-spanning helix.

structurally resolved nitrate reductase from *Escherichia coli* [62, 65] is less, but all structural features are conserved in *M. oxyfera* NarGHI. Conserved features comprise domain organization of the three subunits, amino acids that bind the Mo-*bis*PGD prosthetic group and the 4Fe4S cluster in NarG, amino acids in NarH that coordinate the four iron-sulfur clusters, and those in NarI that are related with the binding of two heme *b* molecules and menaquinone. Unlike many other NarG subunits, DAMO_0778 is not preceded by a *tat* signal for protein export, but in other systems the N-terminus is protected from cleavage to leave the nitrate reductase in the cytosol [83].

Contrary to Nar, the Nap system is localized in the periplasm (Figure 9c). In its essence, the Nap system is composed of the heterodimeric NapAB in which NapA is a molybdopterin protein bearing a 4Fe4S cluster in its N-terminal region, as is the case for NarG and other MPT proteins. NapB is a *c*-type diheme protein. NapAB is well defined by the elucidation of the crystal structures from four different microbial species [82]. The catalytic molybdenum is coordinated by six sulfur atoms, four of which originate from the two dithiolene molybdopterin cofactors that both are present as tricyclic (pyranopterin) ring structures (Figure 8b). The *M. oxyfera* genes coding for NapAB (DAMO_2411 and DAMO_2410, respectively) form part of a larger gene cluster (DAMO_2408-2415) that includes the inventory for the nitrite reductase described next (Figure 9b). In DAMO_2411 and DAMO_2410, all amino acids implicated in NapAB crystal structures with the binding and structuring of the Mo-*bis*PGD, iron-sulfur and heme *c* cofactors are conserved. The identity suggests functionality of NapAB from *M. oyfera*, if expressed. In agreement with a periplasmic localization, DAMO-2410 contains a *tat* signal for protein export. Remarkably, the DAMO_2408-2415 cluster lacks any genes that in other Nap systems code for membrane-bound components, such as NapD, NapF, NapG, and NapH, that provide the electrons for nitrate reduction by the oxidation of menaquinol [84]. Since genes coding for such membrane-bound Nap components are absent in other parts of the genome as well, reduction of nitrate by *M. oxyfera* Nap might rely on a reduced cytochrome *c* pool.

The reduction of nitrite to NO in *M. oxyfera* is catalyzed by cytochrome cd_1 (NirS) of which structural and accessory genes are contained within the DAMO_2408-2415 cluster (Figure 9b). In this cluster, DAMO_2409 codes for a gene product comprising a fusion between a AsnC-family transcriptional regulator, and NirD/G and NirL/H involved in the biosynthesis of heme *d*. DAMO_2412 and DAMO_2413 code for two more heme *d* biosynthesis proteins, NirF and NirJ, respectively. The catalytic nitrite-reducing enzyme, NirS is derived from the DAMO_2415 gene. The DAMO_2415 amino acid sequence shows a remarkable high identity (61 %) to the one of *K. stuttgartiensis* NirS (kuste4136; see Section 3.1), but the *M. oxyfera* sequence is also highly similar to NirS from *Pseudomonas aeruginosa* NirS (53 % identity) and from *Paracoccus pantotrophus* NirS (53 % identity), both having known crystal structures [85]. Indeed, the identity encompasses the conservation in DAMO_2415 of all structurally and functionally relevant amino acids found in the crystal structures. In agreement with the central role in metabolism, DAMO_2415 is one of the highest expressed genes in the genome and one of the most abundant proteins in the proteome [18]. Translated DAMO_2415 is preceded by an N-terminal signal sequence and antibodies raised against specific peptide sequences confirm its localization in the periplasm [86]. Protein translocation might be supported by the product of the DAMO_2408 gene in front of the DAMO_2408-2415 cluster (Figure 9b), which codes for YajC, one of the components of the Sec-YidC-YajC machinery for protein export and membrane protein insertion [87]. However, the DAMO_2408-2415 cluster again lacks one conspicuous candidate: The gene encoding NirC, a *c*-type cytochrome that provides the electron for nitrite reduction.

Besides the dissimilatory NirS, *M. oxyfera* contains the gene complement for an assimilatory nitrite reductase (NasBD). NasBD is composed of the NasD small subunit carrying a 2Fe2S Rieske-type iron-sulfur cluster and the NasB large subunit with a siroheme as its catalytic cofactor. The enzyme catalyzes the NAD(P)H-dependent reduction of nitrite to ammonium, similar to NrfA (see Section 3.4.2) [88]. In the genome of *M. oxyfera*, *nasB* (DAMO_0865; *nirB*) and *nasD* (DAMO_0864; *nirD*) belong to a larger gene cluster (DAMO_0859-0867) that covers a catalytic and regulatory inventory for nitrogen assimilation, such as NAD(P)H-dependent glutamine synthetase (GOGAT) and the NR_I-NR_{II} (*glnLG*; *ntrBC*) two-component regulatory system. However, genes coding for assimilatory nitrate reductase (NasAB) that are often found in *Nas* operons [89] are absent both in the DAMO_0859-0867 gene cluster as well as in other parts of the genome.

4.1.2 Nitric Oxide Reductase-like Proteins in *M. oxyfera*

Free radical nitric oxide is a highly reactive compound and nature has invented a plethora of enzymes that are capable to reduce NO to the less toxic N_2O (eq. 13). Denitrifying Bacteria and Archaea contain membrane-bound respiratory nitric oxide reductases (NOR). Interestingly, such NORs are also found in some non-denitrifying pathogenic bacteria, presumably to protect the cell against NO released by the host's immune system [90].

$$2\,NO + 2\,H^+ + 2\,e^- \rightarrow H_2O + N_2O \quad (E_0' = +1.18\,V) \qquad (13)$$

Initially two NOR subclasses were proposed based on amino acid sequences and type of electron donor [91], cytochrome *c*-dependent NOR (cNOR) that receives electrons from soluble protein electron donors (either cytochromes *c* or pseudoazurin) and quinol-dependent NOR (qNOR). Three-dimensional structures at atomic resolution were obtained for both types [92] and helped to establish the molecular details of this class of enzymes. An additional hybrid NOR, termed qCu_ANOR, has been described for *Bacillus azotoformans* that oxidizes both soluble cytochromes and membrane-embedded quinols [93]. The recent sequencing of the genome of *B. azotoformans* identified this qCu_ANOR as a sNOR [94], one of several novel NOR types [95] (see next). In the first genome survey of *M. oxyfera* three homologous nitric oxide reductase-encoding genes were detected, all three proteins belonging to the group of qNORs. However, as outlined below two of these proteins likely have another function, whereas two more NORs can be added to the *M. oxyfera* repertoire due to erroneous annotation resulting from at least some confusion in the genomic NOR scenery.

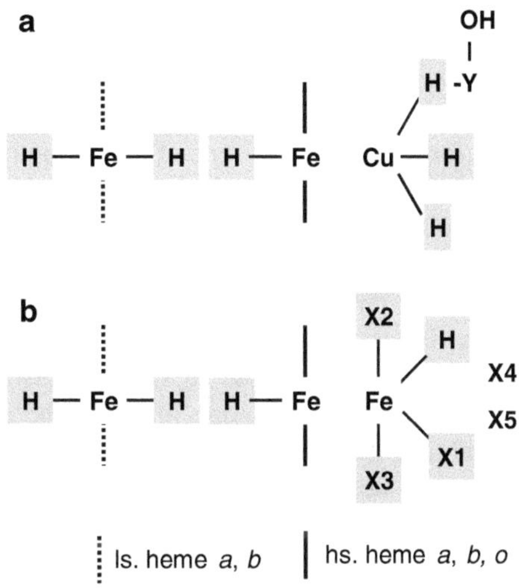

c

	X1	X2	X3	X4	X5
cNOR	H	H	E	E	E
qNOR	H	H	E	E	E
sNOR	H	H	N	Y	X
gNOR	D	H	D	L	E
NOD	H	N	Y	E	Q

Figure 10 Schematic representation of the catalytic site of members of the heme-copper oxidase superfamily represented by (**a**) oxidases and (**b**) dissimilatory nitric oxide reductases (NORs). The catalytic site comprises a low-spin (ls) heme for electron transfer and a binuclear catalytic part composed of a high-spin (hs) heme with either a copper ion (Cu_B) as in oxidases or an iron ion (Fe_B) as is the case with NORs. The different amino acids (H, X_1-X3) that coordinate these moieties are highlighted by light grey squares. X4 and X5 are second shell amino acids within the Fe_B coordination sphere at which the X5 (glutamate) in cytochrome *c*-dependent NOR (cNOR) has been implicated in proton transfer to the active site. X5 in sNORs is variable. (**c**) Amino acid variations (X1-X5) that are found in established and putative novel types of NORs are specified in the table.

Early sequence analyses [96] unambiguously identified NORs as members of the heme-copper oxidase (HCO) superfamily that comprises terminal oxidases such as the mitochondrial cytochrome *c* oxidase or the bacterial *cbb*$_3$ oxidase (see also Chapter 4). All members of the superfamily share sequence similarities in the catalytic subunit I. This subunit I is membrane-bound by 12 or 14 TMHs. In all known heme-copper oxidases, subunit I harbors a six-coordinated low-spin heme (*a* or *b*-type) that functions as electron mediator to the active site (Figure 10).

The active site, where substrate binds and is converted is a unique dinuclear center formed by a five-coordinated high-spin heme (*a*, *b* or *o*-type) and a non-heme metal. In oxygen-reducing HCO, the non-heme metal is a copper ion (Cu_B) able to cycle between Cu^+ and Cu^{2+}. All examined NORs harbor an iron ion at this position (Fe_B), which together with the heme-Fe forms a dinuclear iron site (Figure 10b), suitable to accommodate two NO molecules simultaneously and catalyze the reductive conversion to N_2O. Although it was found that the crystallized qNOR from *Geobacillus stearothermophilus* coordinated a zinc ion as the non-heme metal [92a] (Figure 11), it is regarded as an artifact based on the loss of catalytic activity when iron is substituted for zinc. How the type of metal contributes to the specific catalytic abilities resulting in NO or O_2 reduction is not understood. Interestingly in this respect is that some NO and O_2 reductases exhibit cross reactivity towards O_2 and NO, respectively [97].

The three metallic cofactors (two heme Fe and one Cu/Fe) in subunit I are ligated by six histidine residues (Figure 10). For a long time, these six histidines were thought to be invariant and diagnostic for all HCOs. Two histidines coordinate the electron-transferring low-spin heme iron, one histidine the active site high-spin heme iron and the remaining three are non-heme metal-coordinating. A crucial active-site residue, which is absent in NORs, is a conserved tyrosine that is covalently bound to one of the Cu_B-ligating histidines and acts as a temporary electron and proton donor during O_2 reduction [98]. This cross-linked tyrosine is critical to distinguish between NO- and O_2-reducing HCOs. In contrast, a characteristic feature for cNOR and qNOR enzymes is a conserved glutamate, which is part of the Fe_B ligation sphere providing a distorted trigonal bipyramidal coordination geometry to the iron ion. Yet, a detailed survey by Hemp and Gennis [95] of sequences that had been automatically annotated in Archaeal genomes as HCOs, highly suggests that the NOR picture is far from complete and more diffuse. For example, members of a family, referred to as sNORs [95], are phylogenetically closest related to O_2-reducing B-type oxidases and share a considerable sequence similarity to members of this subfamily. However, they lack the cross-linked tyrosine, and unlike cNORs and qNORs the Fe_B-ligating glutamate is substituted by an asparagine. The particular features are retrieved in the above-mentioned qCu_ANOR from *B. azotoformans* [94] and they are also found in HCO-like proteins from other denitrifying bacteria, including *M. oxyfera*.

4.1.3 The Making of Oxygen

As shown by genome analysis and activity experiments, O_2 is not only an essential but also a measureable intermediate in the nitrite-dependent anaerobic oxidation of methane in *M. oxyfera* [18]. Since external supply of oxygen was excluded, Ettwig et al. [18] postulated a mechanism by which NO serves as the source of intracellularly produced O_2. In the proposed scheme two molecules of NO are disproportionated into N_2 and O_2 (eq. 14).

$$2\,\mathrm{NO} \rightarrow \mathrm{O_2} + \mathrm{N_2} \quad \left(\Delta G^{0\prime} = -173\,\mathrm{kJ \cdot mol^{-1} O_2}\right) \tag{14}$$

This reaction is thermodynamically highly exergonic but kinetically challenging. The conversion of NO into O_2 and N_2 involves the breaking of two N-O double bonds together with the formation of an N-N triple bond and an O-O double bond and is thus assumed to proceed with low turnover rates. This chemistry has not been observed in biological systems before. Therefore, a candidate "nitric oxide dismutase" (NOD) that could catalyze this reaction could not be identified by homology.

Besides the primary question, which enzyme would catalyze NO dismutation, even puzzling was that the genome of *M. oxyfera* coded for three qNOR homologs [18]. Two of these (DAMO_2437 and DAMO_2434) were among the most abundantly expressed proteins; the third one (DAMO_1889) was hardly expressed. The presence in the genome and expressions of these qNOR-like proteins opposed the observation that the microorganism produced N_2O at trace levels only, while the gene coding for the only known enzyme capable of N_2O reduction, N_2OR, could not be detected. Furthermore, the detailed inspection of the DAMO_2437 and DAMO_2434 sequences exhibited amino acid substitutions at positions that were recognized in canonical qNOR as being essential for electron transfer and reductive conversion of NO [99].

Crystallization of the *Geobacillus stearothermophilus* qNOR in the presence of the quinol analogue 2-heptyl hydroxyquinoline N-oxide (HQNO) permitted the identification of a specific binding site for (mena)quinol (Figure 11b) [92a]. Part of this binding pocket is provided by several semi-conserved hydrophobic residues that form a binding groove for the hydrocarbon tail of the quinol moiety. Specific interactions with functional groups of the HQNO core are formed by a set of highly conserved amino acids. His328 and Asp746 (*Geobacillus* numbering) form H-bonds with the NO and OH groups, respectively. In DAMO_2437 and DAMO_2434 both of these residues are substituted mostly by functionally dissimilar amino acids (Figure 11c) and it is thus conceivable that quinol binding and consequently electron donation is impaired in these proteins. Since alternative electron entry sites are absent (cytochrome *c* as in cNOR and Cu_A as in sNOR) in DAMO_2437 and DAMO_2434, electron-consuming NO conversion seems unlikely. Additional features of DAMO_2437 and DAMO_2434 are a slightly altered active site configuration (Figure 11e). One of the three Fe_B coordinating His residues (His560 DAMO_2437 numbering) is exchanged for an asparagine. Next, the highly conserved second coordination shell Glu581, which was implicated in proton transfer to the active site in cNOR [100] is replaced by a glutamine. Although variations in the amino acid composition of the catalytic site in NOR seem to be more widespread than anticipated, the particular deviations observed in DAMO_2437 and DAMO_2434 clearly disfavor a role as NOR. The deviations very well might facilitate another reaction, NO disproportionation.

Importantly, the specific substitutions are conserved in some other proteins in genome databases, namely close homologs of DAMO_2437 and DAMO_2434 in

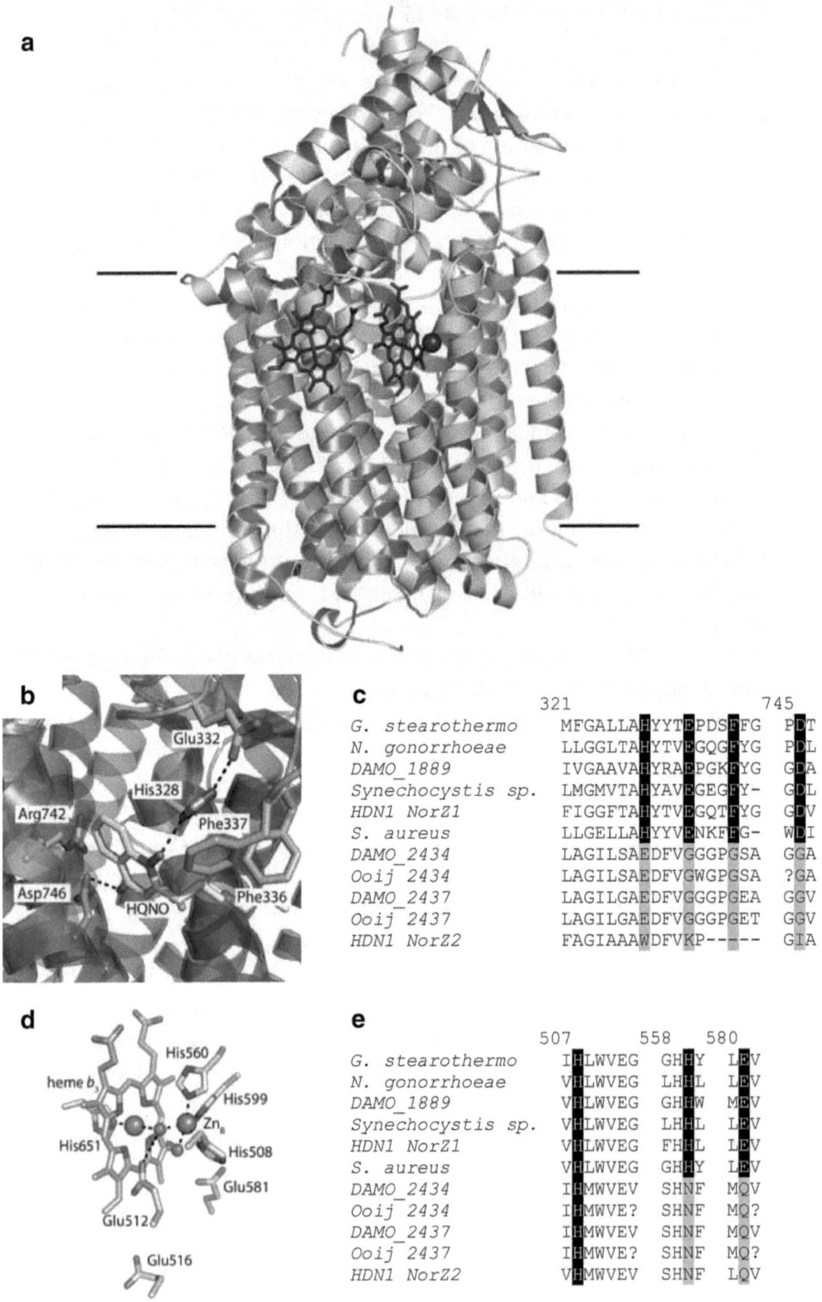

Figure 11 Overall structure of quinol-dependent nitric oxide reductase (qNOR) from *Geobacillus stearothermophilus* (**a**), its quinol-binding site (**b**) and active site configuration (**d**). Specific differences in quinol-binding (**c**) and catalytic sites (**e**) between qNORs and putative NO dismutases (NOD) are highlighted in the multiple amino acid sequence alignments. (**a**) Ribbon representation of the qNOR overall structure and the position of the active site. Horizontal lines bound the membrane lipid bilayer. (**b**) Quinol binding site harboring the quinol analog 2-heptyl

the NC10 enrichment culture strain Ooij and a putative NOR from the γ proteobacterial strain HdN1 (Figures 11c and e) [99]. HdN1 is capable of growth on C6-C20 alkanes [101]. Growth on hexadecane proceeds with oxygen, nitrate or nitrite as electron acceptors but not with N_2O. However, N_2O does serve as the oxidant with hexadecanol or hexadecanoic acid as substrates, compounds that do not require the oxidative activation of a C-H bond. Even though the genome of HdN1 lacks any genes coding for the glycyl-radical-catalyzed activation of alkanes, two or three monooxygenase-encoding genes are present. Taken together, these observations suggest that alkane activation in HdN1 and methane activation in *M. oxyfera* might proceed in similar ways, supported by oxygen that is made via nitrate and nitrite by a NOR-like NO dismutase.

4.1.4 Nitric Oxide and Oxygen Reduction

Unlike DAMO_2437 and DAMO_2434, the amino acid sequence of DAMO_1889 complies with a functional qNOR (Figures 11c and e). As already mentioned, the gene is hardly expressed, which accords the insignificant N_2O production in *M. oxyfera* cultures growing on methane and nitrite [18]. Nevertheless, the situation regarding NO reduction could be more complicated at the genomic and physiological levels.

The nitrite-dependent conversion of methane (eq. 3) is associated with a theoretical surplus of one molecule of O_2 and four electrons. Whole cells are capable of O_2 reduction at rates that exceed the methane-converting metabolic activity [29]. Indeed, the genome of *M. oxyfera* seemed to harbor no less than three different HCO-type oxidases, next to a plant-like alternative oxidase [18, 29]. Supported by substrate-specific O_2 reduction rates and expression data, it was argued [29] that the residual O_2 would be reduced to water by DAMO_1118-19. This protein complex had been annotated as a bo_3-type oxidase belonging to the class A2 of heme-copper oxidases. A reevaluation now of the protein sequence reveals that DAMO_1118-19 is not an O_2 reductase but rather a member of a divergent type of gNORs. Similarly, DAMO_0801-02, which was being annotated as ba_3 oxidase, appears to belong to

Figure 11 (continued) hydroxyquinoline *N*-oxide (HQNO). The Zn atom seen in (**d**) most likely represents a crystallization artifact and would be replaced by an Fe atom in the catalytically active enzyme. qNORS from different microbial species with protein identifiers in parentheses are the following: *G. stearothermo, G. stearothermophilus* (3YAG); *N. gonorrhoeae, Neisseria gonorrhoeae* 1291(EEH62456); DAMO_1889, qNOR from *Methylomirabilis oxyfera*; *Synochocystis* sp. (BAA18795); HDN1 NorZ1, γ Proteobacterium strain HDN1 (CBL45628); *S. aureus, Staphylococcus aureus* (EGL94648). Putative NODs: DAMO_2434, Ooij 2434, DAMO_2437, Ooij 2437, proteins from the *M. oxyfera* enrichment cultures "Twente" (DAMO) and "Ooij" [18]; HDN1 NorZ2 (CBL43845). Structural figures (PDB ID: 3AYG; [92a]) were prepared in PyMOL [48].

the new NO-reductase type of sNORs. The only oxidase that stands a proper annotation, namely as an aa_3-type oxidase, is DAMO_1162-65. The aa_3-type oxidase, which is the most efficient proton-pumping terminal oxidase but which has an only low affinity for oxygen, tends to be expressed under high-oxygen conditions. A role of the DAMO_1162-65 aa_3-type oxidase as being responsible for O_2 reduction could be ruled out on the basis of too low activities and too low expression values indeed [29]. Hence, it is still an open issue, which enzyme from *M. oxyfera* acts in O_2 removal. Considering that oxygen reductase activity with turnover numbers up to 8.4 μmol $O_2 \cdot min^{-1} \cdot mg^{-1}$ protein has been demonstrated for c- and qNOR [97, 102], one of the NO reductases, perhaps even gNOR DAMO_1118-19, might moonlight as an O_2 reductase. And taking the slow metabolic rates of *M. oxyfera* (1–2 nmol $CH_4 \cdot min^{-1} \cdot mg^{-1}$ protein) into account, an even minute amount of NOR would sustain oxygen reduction.

However lowly expressed under standard growth, the question remains what the redundant set of known qNOR (DAMO_1889), and new gNOR (DAMO_1118-19) and sNOR (DAMO_0801-02) would do. The set may provide the organism with a means to cope with nitrosative stress invoked by other members of a denitrifying community. Next, NO is an essential intermediary substrate of nitrite-dependent methane oxidation and the tight balancing of its production and subsequent (quite slow) conversion is necessary to avoid harmful concentrations. The balance could be easily disturbed by the uptake of external electron donors more oxidized than methane, such as formate, immediately resulting in excess nitrite reduction to NO. Under such conditions, reductive conversion of NO into N_2O is advantageous over its disproportionation into N_2 and O_2, which is expected to occur very slowly thereby requiring a vast amount of enzyme for appreciable activity.

4.2 *The Methane Oxidation Pathway*

4.2.1 The Activation of Methane into Methanol

In aerobic methanotrophs, inert methane is activated to methanol by the insertion of an oxygen atom from O_2. The other oxygen atom is reduced to water (eq. 15). Methane activation is catalyzed by two different methane monooxygenases (MMO), a particulate, membrane-bound (pMMO) and a soluble (sMMO) form. *M. oxyfera* contains only pMMO. Although the enzyme has already been discussed in detail in Chapter 6 of this Volume, we will recall some of its main structural properties for the sake of comparison.

$$CH_4 + O_2 + 2\ H^+ + 2\ \mathbf{e}^- \rightarrow CH_3OH + 2\ H_2O \quad (E_0{}' = +1.46\,V) \qquad (15)$$

Supported by the resolution of the crystal structures from three different methanotrophs [103], the common view is that pMMO is composed of $\alpha_3\beta_3\gamma_3$ trimers with three copies each of the pmoA, pmoB, and pmoC subunits ordered in a cylindrical way within the cell membrane. PmoA is primarily composed of 6–7 TMHs; three conserved periplasmic aromatic amino acids (F50, W51, W54; numbering according to the *Methylococcus capsulatus* structure) are proposed to line a methane substrate channel [2]. Also pmoC is membrane-bound by 5 TMHs. PmoB is structured by two cupredoxin domains located in the N-and C-terminal parts. Both domains are separated by two TMHs, but they are positioned next to each other by a linker loop; a third TMH is N-terminally cleaved. The cupredoxin domains face the periplasm and represent the only soluble part of the enzyme complex. Sandwiched between the two cupredoxin domains, three histidines (H33, H137, H139) ligate a dicopper center in the *M. capsulatus* protein [103], but the *Methylocystis* structure is more ambiguous, showing both one and two copper atoms at this side [103]. By stripping the two THMs from pmoB and connecting the cupredoxin domains by a short linker such that the overall structure was maintained, a copper-containing recombinant protein was made that was catalytically active by itself, both with methane and the abiological substrate propylene [104]. This clever approach clearly shows that pmoB is the catalytic site. Also the other subunits may bind copper and/or zinc atoms, but there is at least some disagreement concerning the relevance of these metals; they could have been the artificial result of enzyme purification and crystallization methods [2].

The genome of *M. oxyfera* harbors one pMMO, the subunits of which are encoded by DAMO_2451 (*pmoC1*), DAMO_2450 (*pmoA*), and DAMO_2448 (*pmoB*). Consistent with their central role in metabolism, the three genes are among the highest expressed, both at the transcriptome and proteome levels [18]. Immunogold labeling of affinity-purified antibodies raised against the soluble part of pmoB localize it at the periplasmic side of the cell membrane as expected [86]. DAMO_2339 codes for a second *pmoC*, which is identical to DAMO_2451. Downstream of DAMO_2448, DAMO_2446 has been annotated as pmoD1 of which a second copy (pmoD2; 84 % sequence identity) is found as DAMO_2340. PmoD is predicted to be an exported protein that remains membrane-bound by one TMH in the C-terminus. Homologs of this protein are found in genomes of many methanotrophic and non-methanotrophic species. Contrary to *M. oxyfera*, those *pmoD* homologs are never linked to genes encoding pMMOs and it is questionable whether pmoD forms part of the pMMO complex in *M. oxyfera*. At first sight, sequence identities of pmoA, pmoB, and pmoC from *M. oxyfera* are not very high (~50 %) with respect to the ones in pMMOs with known crystal structures and phylogenetic analyses place all three *M. oxyfera* subunits in a separate clade [18]. Nevertheless, subunit sequences of verified pMMOs are also quite divergent. In fact, a closer comparison reveals the complete conservation in *M. oxyfera* pMMO of all secondary structural elements (TMHs and their topology, β barrel strands in the cupredoxin domains) with respect to those with known crystal structures. Moreover, all amino acids that have been implicated in the binding of metals, no matter their eventual biochemical relevance, are present in pmoA, pmoB, and pmoC from *M. oxyfera*. These similarities suggest a same overall architecture of *M. oxyfera* pMMO and a same function, namely the oxygen-dependent activation of methane (eq. 15). In agreement herewith, in experiments with whole cells the role of oxygen was substantiated in propylene epoxidation, a widely applied artificial reaction to follow pMMO activity [18].

$$CH_4 + NO \rightarrow CH_3OH + 0.5\ N_2 \quad \left(\Delta G^{0\prime} = -211\,\text{kJ} \cdot \text{reaction}^{-1}\right) \tag{16}$$

$$CH_4 + N_2O \rightarrow CH_3OH + N_2 \quad \left(\Delta G^{0\prime} = -223\,\text{kJ} \cdot \text{reaction}^{-1}\right) \tag{17}$$

Structural considerations and the observed role for oxygen in propylene epoxidation do not preclude an alternative view for methane activation by *M. oxyfera*, namely by using the oxidative power of NO (or N_2O) (eqs 16 and 17). Although thermodynamically highly feasible, the reactions are incompatible with the physiology of *M. oxyfera*. Apart from the mechanistic problem how dinitrogen gas would be produced in case of activation by one NO, it can be inferred that substantial amounts of N_2O should be formed to maintain the redox balance. Since no electrons are required for methane activation (Figure 2), the only way to get rid of the extra electrons would be nitrite reduction to the level of N_2O. In active metabolizing cells, the gas can only be detected at trace levels [18].

4.2.2 Methanol Oxidation

Microorganisms have two different types of enzymes at their disposal to oxidize methanol, the product of methane activation. Gram-positive bacteria utilize methanol dehydrogenases that take $NAD(P)^+$ as the electron acceptor and that are localized in the cytoplasm [105]. In Gram-negative methylo- or methanotrophs, a periplasmic MDH is present, which belongs to the broad class of pyrroloquinoline quinone (PQQ)-containing proteins (quinoproteins) [106]. The particular quinoproteins are structured by eight sets of four-stranded, antiparallel W-shaped β sheets that tightly pack the protein backbone into eight propeller-like peptide sequences according to a pseudo eight-fold symmetry axis (Figure 12). The catalytic cofactor, PQQ, is located central to the propellers, buried deeply inside a hydrophobic pocket. PQQ is kept in position by a series of in-plane hydrogen bonds to the side chains and carbonyl groups of a conserved set of amino acids.

Different types of quinoproteins catalyze the oxidation of a wide range of alcohol and aldehyde substrates and each type is more or less tuned for a specific substrate. The oxidation occurs at the PQQ cofactor and involves a base-catalyzed proton abstraction in concert with the transfer of a hydride from the substrate to PQQ (Figure 13) [107]. By accepting the hydride, PQQ gets reduced to $PQQH_2$. After the subsequent transfer of two electrons to an associated electron carrier, a *c*-type heme protein, $PQQH_2$ is re-oxidized and the enzyme is prepared for a next reaction cycle. In most quinoproteins studied to date, a calcium ion in close vicinity to PQQ assists in hydride abstraction. By acting as a Lewis-acid, the C5-O5 bond in PQQ is polarized, which facilitates the nucleophilic addition of the hydride to C5 (Figure 13). The Ca^{2+} ion is kept in the right configuration by the coordination to two highly conserved amino acids, a glutamate and an asparagine (Figures 12c and d). Proton abstraction is assisted by a conserved aspartate.

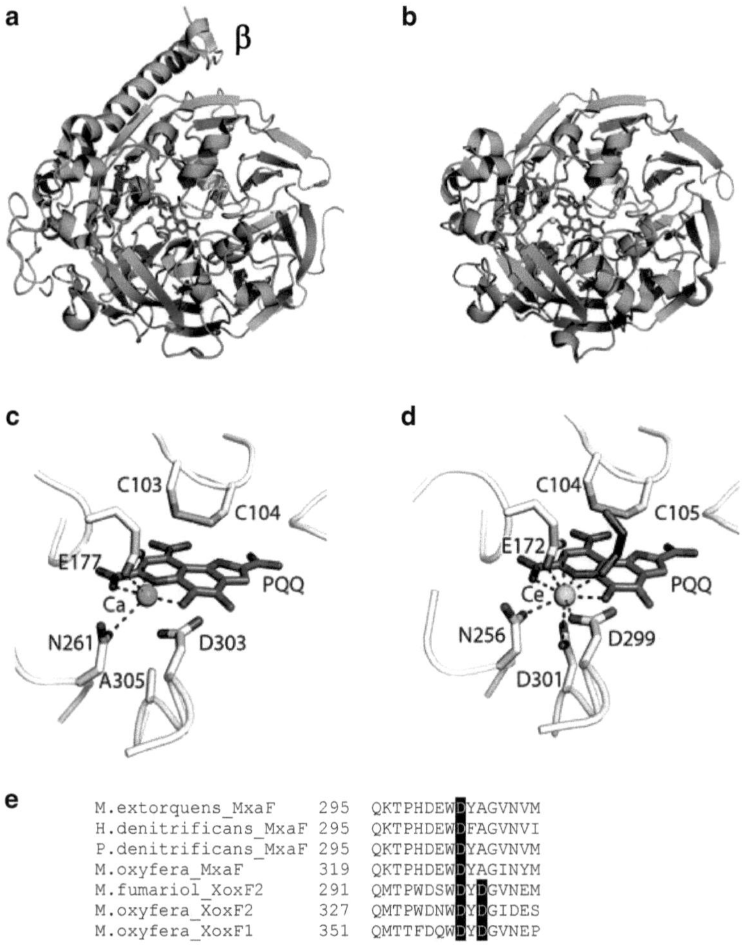

Figure 12 Structures of MxaFI- and XoxF-type methanol dehydrogenases (MDH). A bottom view of the MxaFI-type MDH from *Methylobacterium extorquens* AM1 (PDB ID: 1W6S; [109 g]) composed of the large (MxaF) subunit with the PQQ catalytic site in the center and its small (MxaI, β subunit) is shown in panel (**a**). Panel (**c**) presents a detailed view of the catalytic site in which a Ca ion is coordinated by PQQ. A bottom view of the XoxF-type MDH from *Methylacidiphilum fumariolicum* SolV (PDB ID: 4MAE) [114] and details of its catalytic site are presented in panels (**b**) and (**d**), respectively. Note that PQQ now is associated with a Ce ion that is additionally coordinated by an aspartate (D301). This aspartate residue is also present in two XoxF-MDHs from *Methylomirabilis oxyfera* (**e**). In the sequence alignment (**e**), proteins with their (protein) database identifiers in parentheses are the following: M.extorquens_MxaF (1 W61); H.denitrificans_MxaF (2D0V); P.denitrificans_MxaF (1LRW); M.oxyfera_MxaF (DAMO_0112); M.fumariol_XoxF2 (4MAE); M.oxyfera_XoxF1 (DAMO_0124); M.oxyfera_XoxF2 (DAMO_0134). Structural figures were prepared by PyMOL [48].

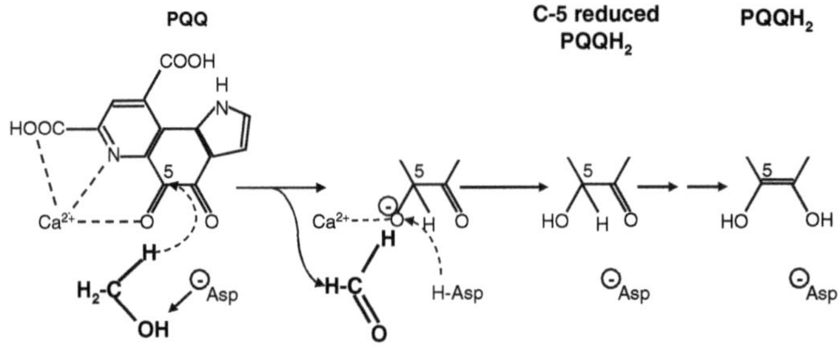

Figure 13 Catalytic mechanism of PQQ-dependent methanol dehydrogenase. The scheme was adapted from [108].

$$CH_3OH \rightarrow HCHO + 2\,H^+ + 2\,e^- \quad (E_0{}' = -0.182\,V) \qquad (18)$$

MDH catalyzes the oxidation of methanol to formaldehyde according to equation (18). The genome of *M. oxyfera* codes for three MDH quinoprotein systems (MDH1-3) that are uniquely located on one large cluster (DAMO_0112-0138) (Figure 14). In the MDH1 system (DAMO_0112-122), DAMO_0112 (*mxaF*) and DAMO_0115 (*mxaI*) encode the large (α) and small (β) subunits, respectively, of the canonical heterodimeric ($\alpha_2\beta_2$) MxaFI-type of MDH. This type of MDH that carries calcium as a PQQ-associated metal cofactor is well understood by the resolution of the crystal structures from seven different microbial species [109]. Besides these structural genes, the MDH1 system comprises a range of other genes that are commonly associated with the functionality of MxaFI-MDH and that are ordered as also seen in *M. oxyfera* [110]. DAMO_0114 codes for the cognate electron carrier, cytochrome c_L (MxaG) and DAMO_0113 for a periplasmic solute-binding protein (MxaJ) with an as-yet unknown function. DAMO_0116-0121 find their homologs (MxaRSACKLD) for instance in *Methylobacterium extorquens*, a methylotrophic model organism, where these gene products have been implemented with Ca^{2+} insertion [110]. In the latter organism, MxaE (DAMO_0122 in *M. oxyfera*) is additionally required for MxaFI maturation. Quite astonishingly, *M. oxyfera* lacks the gene inventory (*pqqABCDEF*) [111] for PQQ biosynthesis [112]. This would make the organism dependent on other microorganisms in its environment that supply *M. oxyfera* with a crucial cofactor of one of its key enzymes.

The MDH2 and MDH3 systems code for divergent types of MDH quinoproteins, termed XoxF-MDHs, which display sequence identities of less than 50 % with MxaF large subunits. XoxF proteins are widely found in genomes of methylotrophs and methanotrophs, but also in genomes of organisms that are as yet unknown for such lifestyles. Their function has been enigmatic for quite some time [19]. The recent purification and characterization of XoxF proteins from different species [113], and the elucidation of the atomic structure of one of these [114] solved the enigma. Just like MxaFI-MDH, XoxF is a methanol-oxidizing quinoprotein. In fact, large subunit structures are fully superimposable (Figures 12a and b). However,

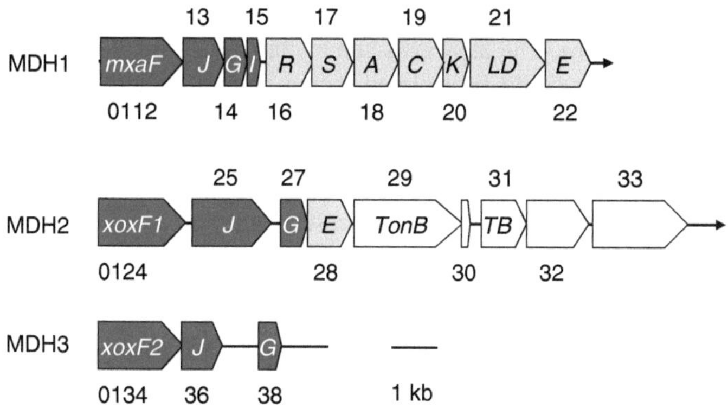

Figure 14 Genomic organization of three methanol dehydrogenase (MDH) systems within the DAMO_0112-038 gene cluster from *Methylomirabilis oxyfera*. Structural genes are shaded dark grey with white lettering and accessory genes are in light grey. Gene lengths and intergenic regions are drawn to scale. TonB, TB, TonB-dependent uptake system.

there are two major differences: XoxF-MDH is a homodimeric protein (α_2) composed of a large PQQ-containing subunit only and PQQ is associated with a rare earth element (REE) instead of Ca^{2+} (Figure 12d). REEs, such as cerium or lanthanum, are stronger Lewis acids, lending XoxF-MDHs a catalytic efficiency that is superior to their MxaFI-MDH counterparts [114, 115]. Because radii of REE ions are bigger than of Ca^{2+}, their proper accommodation requires a specific adaptation of the amino acid sequence. This adaptation is found in a specific aspartate residue that provides an additional ligand and that is located two positions downstream of the catalytic aspartate (Figure 12e) [114, 115].

The various features can be discerned in the further inspection of the MDH2 (DAMO_0124-0133) and MDH3 (DAMO_0134-0138) gene systems from *M. oxyfera* (Figure 14). Herein, DAMO_0124 and DAMO_0134 code for XoxF large subunit homologs [116]. Their amino acid sequences show the conservation of the catalytic aspartate and of the amino acids involved in PQQ and metal association, including the aspartate related with REE coordination (Figure 12e). Still, DAMO_0124 and DAMO_0134 share only 43 % sequence identity and phylogenetic analysis addresses these proteins to two different types of XoxF-MDHs, namely XoxF1 and XoxF2, respectively. (One may note that no less than five types of XoxF proteins can be distinguished [19]). The MDH1 and MDH2 systems lack genes that code for small subunits, but genes encoding cytochrome c_L (*xoxG*) and *xoxJ* paralogs are present (Figure 14). Next, the MDH2 and MDH3 sub-clusters lack any genes involved in Ca^{2+} insertion, which is common in XoxF systems [115]. Quite unusually, the MDH2 sub-cluster harbors a second *mxaE* copy. Two genes in the latter sub-cluster code for TonB-dependent uptake proteins. These uptake proteins are consistently detected in genomes of microorganisms that contain XoxF-type MDHs and they might play a role in the acquirement of poorly soluble REEs [115].

The straightforward question is which of the three MDHs is functionally expressed in *M. oxyfera*. The answer is somewhat surprising. The only methanol-oxidizing enzyme that could be isolated was a heterodimer ($\alpha_2\beta_2$) composed of XoxF1 (DAMO_0124) large subunits and MxaI (DAMO_0115) small subunits [116]. Although the diagnostic REE-binding aspartate is apparent in the DAMO_0124 amino acid sequence, it is unknown whether the protein does harbor a REE indeed. It is conceivable that in the presence of more available Ca^{2+} the latter atom is bound, its coordination being assisted by the small subunit. Such assistance would allocate the subunit a function.

4.2.3 Formaldehyde Oxidation

Out of the many ways to oxidize formaldehyde biochemically, *M. oxyfera* potentially uses two (Figure 15). In the first pathway, 5,6,7,8-tetrahydromethanopterin (H_4MPT) acts as the one-carbon unit (C1) carrier. Enzymes of this pathway are highest expressed at the transcriptional and protein levels [18] indicating that major part of the formaldehyde flux passes this branch. The second, only lowly expressed pathway is an analogous one with 5,6,7,8-tetrahydrofolate (H_4F) as C1-carrier serving biosynthetic reactions (Figure 15c).

The H_4MPT-dependent pathway is found in many methylo- and methanotrophs [19], and also in methylotrophic methanogens. Hydrogenotrophic methanogens that produce methane by the reduction of CO_2 or by the disproportionation of formate use the route in the reversed order (see Chapter 6 of ref. [138]). In *M. oxyfera*, all enzymes of the H_4MPT pathway are encoded on a large gene cluster (DAMO_0454-0473) (Figure 15a) highly resembling the organization in methylotrophs [117]. Besides H_4MPT-dependent metabolic enzymes, the *M. oxyfera* cluster comprises genes that encode one of three formate/formyl oxidation systems described hereafter (see Section 4.2.4) as well as a conserved set of genes that have been implicated in H_4MPT biosynthesis [117]. Next, the DAMO_0454-0473 cluster comprises various conserved genes that are absent in methylotrophic C1-transfer clusters, possibly involved in the as-yet elusive synthesis of the second C1-carrier in the route, methanofuran (MFR).

In the H_4MPT route, formaldehyde oxidation starts with the binding of the substrate to the C1 carrier to make 5,10-methylene- H_4MPT (Figure 15b). This reaction is catalyzed by the formaldehyde activating enzyme (Fae). 5,10-methylene-H_4MPT oxidation by methylene-H_4MPT dehydrogenase (Mtd) generates 5,10-methenyl-H_4MPT, which is converted to 5-formyl-H_4MPT by methenyl-H_4MPT cyclohydrolase (Mch). Fae (DAMO_0454) from *M. oxyfera* displays 58 % sequence identity to Fae from *M. extorquens* [118] and Mch (DAMO_0461) is 40–44 % identical to Mch enzymes from different sources with known crystal structures ([119]; PDB ID: 4FIO). In agreement with the anticipated function, sequence identities include the presence of critical amino acids for the binding of substrates and for catalysis. However, the role of the Mtd protein (DAMO_0455) could be more ambiguous. Whereas Mtd proteins from

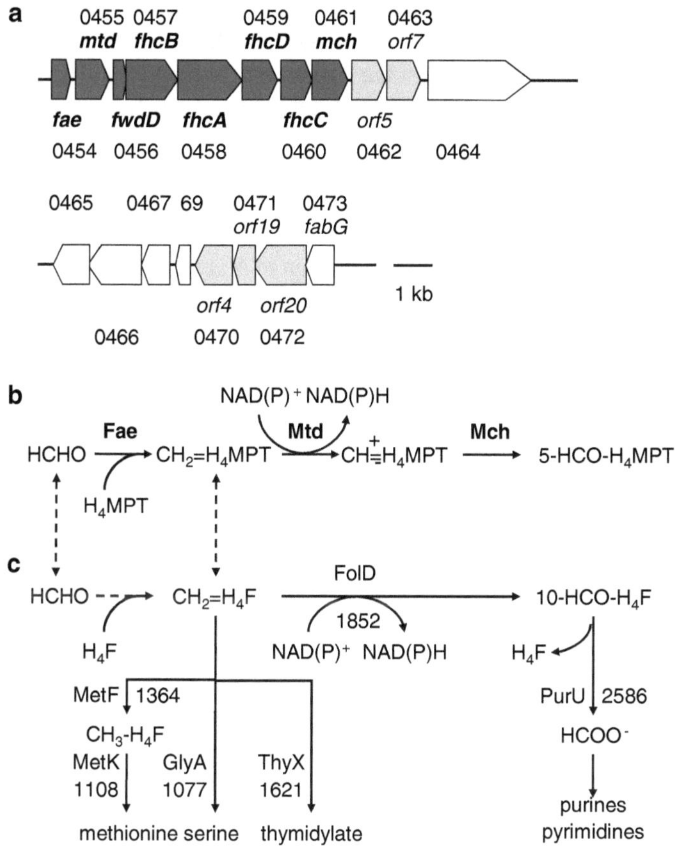

Figure 15 Tetrahydromethanopterin (H_4MPT)- and tetrahydrofolate (H_4F)-dependent formaldehyde and formate conversion systems in *Methylomirabilis oxyfera*. (**a**) Gene arrangement of the H_4MPT-dependent route on the DAMO_0454-0473 gene cluster. Gene lengths and intergenic regions are drawn to scale. Structural genes are shaded dark grey. A conserved set of genes also found in genomes of other methylotrophs that are suggested to be involved in the biosynthesis of H_4MPT [117] is shaded light grey. Herein, *orf5* codes for an α-L-glutamate ligase (RimK family), *orf7* for an ATP: dephospho-CoA triphosphoribosyl transferase (CitG), *orf4* for β-ribofuranosylaminobenzene 5′-phosphate synthase, *orf19* for a conserved protein (DUF447), and orf20 for a dihydropteroate synthase-related protein; FabG, 3-oxoacyl-(acyl-carrier protein) reductase. Other abbreviations are as specified in the text. (**b**) Pathway of H_4MPT-dependent formaldehyde oxidation. (**c**) Pathway of formaldehyde conversion to formate and its linkage with biosynthetic reactions using H_4F as the C1 carrier. Protein codes are indicated together with protein identifiers (DAMO_). The connection between H_4MPT (b)- and H_4F (c)-dependent pathways is indicated by dashed lines.

methanogenic and sulfate-reducing Archaea that use coenzyme F_{420} as electron acceptor are already quite divergent from $NAD(P)^+$-dependent Mtd proteins from methylotrophs, the latter also distribute in no less than five different phylogenetic clades (MtdA-E) [19]. These Mtd types markedly differ in substrate affinities.

The MtdB enzyme is specific for 5,10-methylene-H_4MPT in combination with both NAD^+ for energy generation and $NADP^+$ for biosynthesis [120]. The MtdA protein exclusively relies on $NADP^+$ but it can use 5,10-methylene-H_4MPT and 5,10-methylene-H_4F to variable degrees [121]. MtdC proteins are even more promiscuous in these respects [122]. The Mtd from *M. oxyfera* takes a unique intermediary position in between the MtdB and MtdA clades [19]. The sequence comparison between the *M. oxyfera* protein and MtdA from *M. extorquens* having a known crystal structure [123] reveals the nearly complete conservation of $NADP^+$-binding amino acids, but amino acids in the H_4MPT/H_4F-binding domain in *M. extorquens* MtdA are less well conserved in the *M. oxyfera* protein sequence. Apparently, *M. oxyfera* Mtd evolved such as to optimally sustain catabolism, i.e., substrate oxidation to CO_2, and C1 anabolism through H_4F-dependent reactions (Figures 15b and c).

The H_4F-dependent counterpart in between the formaldehyde and formate oxidation states is the bifunctional $NAD(P)^+$-dependent 5,10-methylene-H_4F:5,10-methenyl-H_4F cyclohydrolase (FolD, DAMO_1852) that generates 10-formyl-H_4F (Figure 15c). Substrates might be supplied to this enzyme in connection with Mtd activity just described or by the non-enzymic binding of formaldehyde to H_4F generating 5,10-methylene-H_4F. The various H_4folate derivatives are substrates for amino acid (serine, methionine) and purine (thymidylate) syntheses that depend on C1 compounds. In *M. oxyfera*, all enzymes involved use H_4F as the C1 carrier, underlining the central role of this cofactor in biosynthetic processes. Formate for purine synthesis can be formed by 10-formyl-H_4F lyase (PurU, DAMO_2586) that hydrolyzes 10-formyl-H_4F into formate and H_4F. However, there may be more means to make formate.

4.2.4 Formate Oxidation

Formate is a powerful electron donor that plays a key role in the energy metabolism of many respiratory and fermentative microorganisms. Nature has evolved an amazing variation in the way formate is oxidized to CO_2 by formate dehydrogenases (eq. 19), first and for all regarding their structural and functional organization [124]. The common property is the presence of a pterin cofactor (*bis*PGD) in the catalytic FDH subunit [125]. In FDHs, the cofactor is coordinated to either molybdenum (Mo-*bis*PGD) or tungsten (W-*bis*PGD) at which the metal additionally binds either cysteine or selenocysteine.

$$\mathrm{HCOO^- + H^+ \rightarrow CO_2 + 2\,H^+ + 2\,\mathbf{e}^-} \qquad (\mathrm{E_0}' = -0.44\,\mathrm{V}) \qquad (19)$$

$$\text{formyl-MFR} + \mathrm{H_2O \rightarrow CO_2 + 2\,H^+ + 2\,\mathbf{e}^-} + \text{MFR} \qquad (\mathrm{E_0}' = -0.51\,\mathrm{V}) \qquad (20)$$

$$\text{5-formyl-}\mathrm{H_4}\text{MPT} + \text{MFR} \rightarrow \mathrm{H_4}\text{MPT} + \text{formyl-MFR} \qquad \left(\Delta G^{0\prime} = -5\,\mathrm{kJ}\cdot\text{reaction}^{-1}\right) \qquad (21)$$

$$\text{formyl-MFR} + H_2O \rightarrow \text{MFR} + HCOO^- \quad \left(\Delta G^{0\prime} = -15\,\text{kJ} \cdot \text{reaction}^{-1}\right) \quad (22)$$

In methanogenic and sulfate-reducing Archaea, the formyl group bound to methanofuran is oxidized to CO_2 by formyl-MFR dehydrogenase (fMFR-DH) (eq. 20; see also Chapter 6 of ref. [138]). As is the case with FDHs, the catalytic subunit is a MPT protein carrying either molybdenum (Fmd systems) or tungsten (Fwd systems) [126]. Again, these metals are coordinated to either cysteine or selenocysteine.

M. oxyfera has the disposal of three different formate/formyl oxidation systems (Figure 16). The first one (FwdDFhcBADC; DAMO_0456-0460) is coded as part of the formaldehyde-oxidizing gene cluster (DAMO_0454-0473) (Figure 15). Herein, *M. oxyfera* FhcBADC resembles the four-subunit formyl transferase/hydrolase (Fhc) from aerobic methylotrophs that is associated in these organisms with the C1 cluster as well [127]. Indeed, *M. oxyfera* FhcD (DAMO_0459) shows 48 % sequence identity to FhcD from *M. extorquens* that has been identified as 5-formyl-H_4MPT: methanofuran formyltransferase catalyzing reaction (21) [128]. In methanogenic and sulfate-reducing Archaea, this formyltransferase is present as a separate enzyme. Remarkably, the sequence identity of DAMO_0459 is even higher (53 %) with the formyltransferase from the distantly related methanogen *Methanopyrus kandleri*, having a known crystal structure [129], than with *M. extorquens* FhcD. *M. oxyfera* FhcA (DAMO_0458) is homologous both to *M. extorquens* FhcA and to the FwdA subunit of archaeal Fwd systems. This polypeptide is a member of the superfamily of metallo-dependent amidohydrolases and likely catalyzes formyl-MFR hydrolysis to formate (eq. 22). Formation of formate by the *M. extorquens* FhcCDAB complex is consistent with such a role [127]. *M. oxyfera* FhcC (DAMO_0460) has its homologs in FhcC from methylotrophs and in archaeal FwdC/FmdC subunits. FhcB from *M. oxyfera* (DAMO_0457) is an interesting case.

While the sequence identity with FhcB from *M. extorquens* is only low (22 %), the identity with cysteine- or selenocysteine-containing FwdB subunits [130] is much higher, especially with respect to *M. kandleri* (~46 %). In Archaea, FwdB mostly represents the MPT-catalytic subunit [130, 131]. FhcB from *M. oxyfera* shares with the *M. kandleri* FwdB subunits (MK0259 and MK1527) the presence in the N-terminus of four cysteines that commonly coordinate the 4Fe4S cluster in MPT catalytic subunits. Next, three cysteines are conserved in DAMO_0457 with respect to MK0259, two of which are substituted in MK1527 by a selenocysteine and a serine. These similarities highly suggest a role of the DAMO_0456-0460 system as a formyltransferase/cysteine-containing formylmethanofuran dehydrogenase. In agreement with the proposal, DAMO_0456-0460 possesses one more component, FwdD (DAMO_0456), which is unique for a non-Archaeon. Thus far, the FwdD subunit has only been found in archaeal tungsten- or molybdenum-containing (here: FmdE) fMFR-DHs. *M. extorquens* FhcB lacks the afore-mentioned cysteines and the Fhc complex is devoid of an FwdD-like subunit [127]. These differences would make the complex from aerobic methylotrophs inactive in formyl oxidation, restricting its function to a formyltransferase/formate-generating hydrolase [127]. Irrespective of its novelty, the FwdDFhcBADC system from *M. oxyfera* leaves

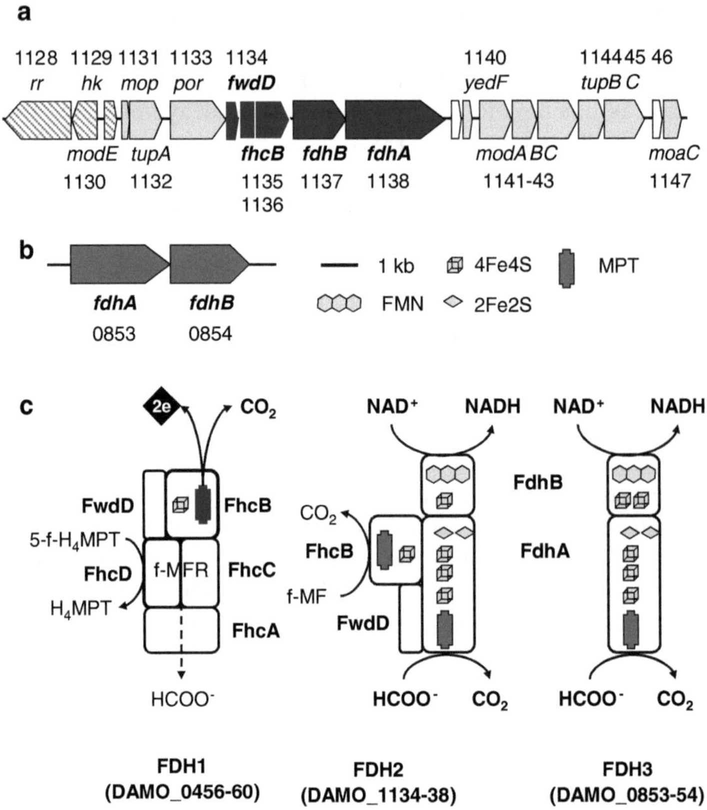

Figure 16 Three formyl/formate dehydrogenase (FDH) systems in *Methylomirabilis oxyfera*. The gene cluster organization of FDH 2 (DAMO_1128-1148) and FDH3 (DAMO_0853-0854) are shown in panels (**a**) and (**b**), respectively. Gene lengths and intergenic regions are drawn to scale. Structural genes are shaded dark grey and accessory genes are in light grey. Regulatory genes are hatched in diagonals. Genes coding for FDH1 form part of the H_4MPT-dependent formaldehyde oxidation gene cluster shown in Figure 15a. Panel (**c**) illustrates a model for the functional organization of FDHI, FDH2, and FHD3. Structural motifs are specified in the figure. Abbreviations and further explanations: rr, response regulator; hk, histidine kinase; modE, molybdenum uptake regulator; modABC, molybdenum uptake system; tupABC, tungsten uptake system; mop, molybdenum-pterin binding protein; por, porin; yedF, selenium metabolism protein; moaC, molybdopterin biosynthesis protein; MPT, *bis*(pyrano guanine dinucleotide); f-MFR, formylmethanofuran.

us with one issue: Where do the electrons from formyl/formate oxidation (eq. 20) go to? All components are predicted to be cytoplasmic soluble proteins and the encoding genes are not linked to ones (FwdFG) that code for iron-sulfur electron transfer proteins in archaeal Fwd systems [130, 131]. Unfortunately, the genome of *M. oxyfera* does not give an immediate clue to the issue.

Next to the previous formyl-MFR-dependent formyl/formate dehydrogenase system, which is highest expressed at the transcriptional and protein levels [18], the genome of *M. oxyfera* has the inventory of two alternative and much lower expressed systems (Figure 16a, b). Both harbor a canonical formate dehydrogenase catalytic subunit (FdhA). The first alternative (DAMO_1134-1138) seems to represent a hybrid between fMFR-DH, common formate dehydrogenase and NADH-dehydrogenase (complex I, Nuo) elements. The DAMO_1134-1138 genes form part of a larger gene cluster (DAMO_1128-1147) (Figure 16a). The cluster encodes more components required for formyl/formate/dehydrogenase functionality, such as molybdate and tungstate ABC-type transporters, a porin that might facilitate uptake of these metals, and selenocysteine and molybdopterin (MoaC) biosynthetic genes. A two-component regulatory system and ModE-type transcriptional regulator in front of the gene cluster could be involved in expression control.

Regarding the DAMO_1134-1138 formyl/formate dehydrogenase, DAMO_1134 is a FwD paralog (53 % sequence identity to DAMO_0456). DAMO_1135 and DAMO_1136 have been annotated as two genes, but the DAMO_1135 stop codon most likely represents a selenocysteine codon. Together the translated DAMO_1135-36 sequences fully cover DAMO_0457 FhcB and the FwdB subunits of the *M. kandleri* cysteine- (MK0259) and selenocysteine (MK1527) Fwds. Most notably, the presumed end of DAMO_1135 matches the position (amino acid 120) of selenocysteine in MK1527. These properties suggest a role of DAMO_1134-36 as a selenocysteine-containing fMFR-DH. In DAMO_1138, all features are conserved that have been identified in the crystal structure of *E. coli* formate dehydrogenase H (FDH-H) [132] with the binding of the molybdo-*bis*PGD prosthetic group and the preceding 4Fe4S cluster. Sequence conservation of DAMO_1138 with respect to tungsten-containing FDH from *Desulfovibrio gigas*, another enzyme of which the atomic structure has been resolved [133], is much lower (22 % identity) than with *E. coli* FDH-H (36 %). This is indicative that DAMO_1138 binds Mo at its catalytic metal. However, whereas FDH-H is a selenocysteine protein, in DAMO_1138 a cysteine is present at the equivalent position. Still, there is one more major difference. DAMO_1138 has an N-terminal extension that shows significant sequence identity with the NuoG subunit of bacterial complex I. The identity includes the presence of the set of cysteines that coordinate two 2Fe2S and two 4Fe4S clusters in NuoG [134] (Figure 16c). Otherwise, the extension is not unique for *M. oxyfera* and it is also seen in formate dehydrogenases from sulfate reducers and homoacetogens. DAMO_1137 is homologous to the beta subunit of formate dehydrogenase (FdhB). Similarly to NuoF in complex I [134], FdhB is the $NAD(P)^+$- and flavin mononucleotide (FMN)-binding subunit of FDH. Taken together, the analyses suggest that DAMO_1134-1138 is a bifunctional enzyme, capable of formylmethanofuran and of formate oxidation (eqs 19 and 20) using $NAD(P)^+$ as the electron acceptor.

The third formate dehydrogenase is the simplest one consisting of two subunits only (DAMO_0853-0854) (Figure 16). The DAMO_0853 sequence is 39.5 % identical to the DAMO_1138 FdhA subunit. This identity includes a NuoG-like N-terminal part. Similarly, in its C-terminal part DAMO_0854 displays 34 % sequence identity to DAMO_1137 FdhB. Remarkably, DAMO_0854 (710 amino acids) is much longer than DAMO_1137 (472 amino acids). The difference concerns the presence in DAMO_0854 of the N-terminal extension that is related to the NuoE subunit [134] of the bacterial (NuoA-N) complex I. Again, the data are in agreement with a function of DAMO_0853-0854 as a $NAD(P)^+$-dependent FDH.

All this brings us to the role of formate in the metabolism of *M. oxyfera*, which could be a complex one like in methylotrophs [135]. It is known that cells of *M. oxyfera* instantaneously use exogenous formate as the electron donor for nitrite reduction and methane activation [18]. Thus, *M. oxyfera* is capable of formate uptake. Within the cells, formate can be the product of formyl-H_4F lyase (PurU) activity. Next, formate could be made by methanol dehydrogenase. MDHs not only catalyze the oxidation of methanol to formaldehyde, but also its subsequent oxidation to formate, and formaldehyde-oxidizing activity is particularly high in REE-containing XoxF MDHs [114]. Thirdly, formate might be produced by the above FwdDFhcBADC system (DAMO_0456-0460) similar to the *M. extorquens* Fhc complex (eq. 22). Extrusion of excess formate or import from the environment might serve the balancing of electron fluxes in oxidative and reductive metabolic processes. Obviously, formate movements would require an appropriate transporter, which is available in the genome as a FocA/NirC-type formate/nitrite transporter (DAMO_2284).

5 General Conclusions

For a long time, it was believed that the anaerobic oxidation of ammonium and of methane would be impossible. The reason for this was that enzymes, that were capable to break up the N-H bond of ammonium and the C-H bond of methane, depended on oxygen. From this chapter, it may have become clear that anaerobic ammonium and methane oxidizers do exist and that their existence went with the development of new enzyme systems that might involve subtle and elegant variations of metal-based catalyses that we already know from other, even completely different enzymic reactions. In anammox bacteria, ammonium is activated into hydrazine using the oxidative power of NO (see Section 3.2). In the N-DAMO process by *M. oxyfera*, a modification of a common NOR might have turned this enzyme into a machinery with another functionality, a nitric oxide dismutase generating N_2 and O_2 (see Section 4.1.3), the latter compound enabling *M. oxyfera* to utilize the same pathway for methane oxidation as aerobic methanotrophs do (see Section 4.2). While many details regarding their catalytic mechanisms still remain to be elucidated, the novel hydrazine synthase and NO dismutase are extremely slow enzymes. Obviously, the slow rate of these key enzymes immediately affects the growth rates of the microorganisms, which is very low as well. These slow growth rates, requiring dedicated enrichment techniques, may have been another reason why anammox and N-DAMO bacteria have evaded our attention. Although slow, the presence of the novel ammonium- and methane-activating systems provides the microorganisms with a unique niche in anaerobic ecosystems, which is a central one in case of anammox bacteria [8, 11, 12].

Genome sequencing of anammox bacteria and of *M. oxyfera*, supported by a beginning of the functional characterization of enzymes involved, has been surprising and rewarding. A first surprise was the genetic redundancy of many enzymes that are related with central catabolic pathways [21, 26]. For instance, anammox bacteria harbor a core set of no less than eight to ten different HAO-like enzymes that are tuned for hydrazine and hydroxylamine oxidation, possibly for nitrite reduction and still to be discovered other functions by subtle changes around their heme *c* catalytic sites (see Section 3.3). In *M. oxyfera* five different NOR-like proteins might catalyze NO dismutation and NO reduction, again brought about by specific amino acid substitutions near the active sites (see Sections 4.1.2, 4.1.3, and 4.1.4). Besides these, anammox bacteria and *M. oxyfera* are likely to be goldmines for divergent catalytic proteins and respiratory complexes, often representing new combinations of systems that have been well-studied in current model organisms. The putative novel *bc* complexes, new complex I-formate dehydrogenase combinations and NXR systems in anammox bacteria [21, 26], and the REE-containing MDH (see Section 4.2.2.) and formate dehydrogenases (see Section 4.2.4) in *M. oxyfera* are just a few examples. However, apart from enzymes that lend anammox bacteria and *M. oxyfera* their specific ecological position, such as hydrazine synthase and hydrazine dehydrogenase, many proteins are not unique for these microorganisms. Close homologs are readily found in protein databases and in genomes assembled by metagenomic approaches, covering a variety of poorly studied or as-yet unknown environmental species.

In the end, anammox bacteria and *M. oxyfera* may stand model for a general principle: There are no such things as "impossible" microorganisms. Each biogeochemical process that is associated with a negative Gibbs free energy change finds a microbe that can thrive on the reaction. Time, a plethora of protein templates, metal-based catalytic mechanisms, and genetic toolboxes will allow the accommodation of metabolic pathways to specific needs. Their uncovering may take time as well, especially with respect to slow-growing species that are likely more the rule than the exception in natural systems.

Abbreviations and Definitions

$\Delta G^{0\prime}$ $E_0{}'$	Gibbs free energy changes and standard midpoint redox potentials, both at pH 7, were taken or calculated from data presented in [136, 137]
AMO	ammonium monooxygenase
anammox	anaerobic ammonium oxidation
AOA	ammonium-oxidizing Archaea
AOB	ammonium-oxidizing Bacteria
ATP	adenosine 5′-triphosphate
*bis*PGD	*bis*(pyrano guanine dinucleotide)

C1	one-carbon compound
cNOR	cytochrome *c*-dependent nitric oxide reductase
Cyt *c*	cytochrome *c*
DMSO	dimethylsulfoxide
DNRA	dissimilatory nitrite/ nitrate reduction forming ammonium
FAD	flavin adenine dinucleotide
Fae	formaldehyde activating enzyme
Fhc	formyl transferase/hydrolase
FDH	formate dehydrogenase
FMD	molybdenum-containing formylmethanofuran dehydrogenase
fMFR	formylmethanofuran
fMFR-DH	formylmethanofuran dehydrogenase
Fwd	tungsten-containing formylmethanofuran dehydrogenase
H_4F	5,6,7,8-tetrahydrofolate
H_4MPT	5,6,7,8-tetrahydromethanopterin
HAO	hydroxylamine oxidoreductase
HCO	heme-copper oxidase
HDH/HZO	hydrazine dehydrogenase/oxidase
HQNO	2-heptyl hydroxyquinoline N-oxide
HZS	hydrazine synthase
MCO	multicopper oxidase
MDH	methanol dehydrogenase
methanogen	Archaeon that grows by the formation of methane
methanotroph	organism capable of (aerobic) growth on methane
methylotroph	organism capable of growth on methylated compound(s)
MFR	methanofuran
MMO	methane monooxygenase
MPT	molybdopterin
$MQ(H_2)$	(reduced) menaquinone
Mtd	methylenetetrahydromethanopterin dehydrogenase
MxaFI	heterotetrameric calcium-containing methanol dehydrogenase
NAD(P)H	nicotinamide adenine dinucleotide (phosphate) reduced
N-cycle	biogeochemical nitrogen cycle
N-DAMO	nitrite-dependent anaerobic methane oxidation
N_2OR	nitrous oxide (N_2O) reductase
Nap	periplasmic dissimilatory nitrate reductase
Nar	cytoplasmic dissimilatory nitrate reductase
Nas	assimilatory nitrate reductase
NeHAO	*Nitrosomonas europaea* hydroxylamine oxidoreductase
Nir	nitrite reductase
Nrf	ammonium-forming nitrite reductase
NOD	nitric oxide dismutase
NOR	nitric oxide reductase
NXR	nitrite:nitrate oxidoreductase

pMMO	particulate membrane-bound methane monooxygenase
PQQ	pyrroloquinoline quinone
qNOR	quinol-dependent nitric oxide reductase
quinoprotein	protein containing PQQ at its active site
REE	rare earth element
sMMO	soluble methane monooxygenase
tat	twin-arginine translocation signal
TMH	transmembrane-spanning helix
XoxF	homodimeric rare earth element-containing methanol dehydrogenase

Acknowledgments The contribution by Joachim Reimann was supported by the Netherlands Organization for Scientific Research (ALW grant number 822.02.005). The work by Mike Jetten is facilitated by two European Union Advanced Research Grants (ERC 232937, ERC 339880) and by the Dutch Government "Zwaartekrachtsubsidie" (Gravitation Grant) to the Soehngen Institute for Anaerobic Microbiology (SIAM 024 002 002).

References

1. D. J. Arp, L. Y. Stein, *Crit. Rev. Biochem. Mol. Biol.* **2003**, *38*, 471–495.
2. M. A. Culpepper, A. C. Rosenzweig, *Crit. Rev. Biochem Mol. Biol.* **2012**, *47*, 483–492.
3. T. J. Lawton, J. Ham, T. Sun, A. C. Rosenzweig, *Proteins*, **2014**, *82*, 2263–2267.
4. R. Hatzenpichler, *Appl. Environ. Microbiol.* **2012,** *78*, 7501–7510.
5. W. Martens-Habbena, P. M. Berube, H. Urakawa, J. R. de la Torre, D. A. Stahl, *Nature* **2009**, *461*, 976–979.
6. N. Vajrala, W. Martens-Habbena, L. A. Sayavedra-Soto, A. Schauer, P. J. Bottomley, D. A. Stahl, D. J. Arp, *Proc. Natl. Acad. Sci. USA* **2013**, *110*, 1006–1011.
7. M. Strous, J. J. Heijnen, J. G. Kuenen, M. S. M. Jetten, *Appl. Microbiol. Biotechnol.* **1998**, *50*, 589–596.
8. M. C. Schmid, B. Maas, A. Dapena, K. van de Pas-Schoonen, J. van de Vossenberg, B. Kartal, L. van Niftrik, I. Schmidt, I. Cirpus, J. G. Kuenen, M. Wagner, J. S. Sinninghe Damsté, M. Kuypers, N. P. Revsbech, R. Mendez, M. S. M. Jetten, M. Strous, *Appl. Environ. Microbiol.* **2005**, *71*, 1677–1784.
9. (a) A. A. van de Graaf, A. Mulder, P. de Bruijn, M. S. M. Jetten, L. A. Robertson, J. G. Kuenen, *Appl. Environ. Microbiol.* **1995**, *61*, 1246–1251. (b) A. A. van de Graaf, P. de Bruijn, L. A. Robertson, M. S. M. Jetten, J. G. Kuenen, *Microbiology* **1996**, *142*, 2187–2196.
10. M. Strous, J. A. Fuerst, E. H. Kramer, S. Logemann, G. Muyzer, K.T. van de Pas-Schoonen, R. Webb, J. G. Kuenen, M. S. M. Jetten, *Nature* **1999**, *400*, 446–449.
11. H. J. M. Op den Camp, B. Kartal, D. Guven, L. A. van Niftrik, H. C. Haaijer, W. R. van der Star, K. T. van de Pas-Schoonen, A. Cabezas, Z. Ying, M. C. Schmid, M. M. Kuypers, J. van de Vossenberg, H. R. Harhangi, C. Picioreanu, M. C. van Loosdrecht, J. G. Kuenen, M. Strous, M. S. M. Jetten, *Biochem. Soc. Trans.* **2006**, *34*, 174–178.
12. B. Kartal, L. van Niftrik, J. T. Keltjens, H. J. M. Op den Camp, M. S. M. Jetten, *Adv. Microb. Physiol.* **2012**, *60*, 211–262.
13. B. Kartal, J. G. Kuenen, M. C. van Loosdrecht, *Science* **2010**, *328*, 702–703.

14. (a) K. Knittel, A. Boetius, *Annu. Rev. Microbiol.* **2009**, *63*, 311–334. (b) R. K. Thauer, *Curr. Opin. Microbiol.* **2011**, *14*, 292–299. (c) J. Milucka, T. G. Ferdelman, L. Polerecky, D. Franzke, G. Wegener, M. Schmid, I. Lieberwirth, M. Wagner, F. Widdel, M. M. Kuypers, *Nature* **2012**, *491*, 541–546. (d) T. Holler, G. Wegener, H. Niemann, C. Deusner, T. G. Ferdelman, A. Boetius, B. Brunner, F. Widdel, *Proc. Natl. Acad. Sci. USA* **2011**, *108*, E1484–1490.
15. M. F. Haroon, S. Hu, Y. Shi, M. Imelfort, J. Keller, P. Hugenholtz, Z. Yuan, G. W. Tyson, *Nature* **2013**, *500*, 567–570.
16. A. A. Raghoebarsing, A. Pol, K. T. van de Pas-Schoonen, A. J. Smolders, K. F. Ettwig, W. I. Rijpstra, S. Schouten, J. S. Damsté, H. J. M. Op den Camp, M. S. M. Jetten, M. Strous, *Nature* **2006**, *440*, 918–921.
17. K. F. Ettwig, S. Shima, K. T. van de Pas-Schoonen, J. Kahnt, M. H. Medema, H. J. M. Op den Camp, M. S. M. Jetten, M. Strous, *Environ. Microbiol.* **2008**, *10*, 3164–3173.
18. K. F. Ettwig, M. K. Butler, D. Le Paslier, E. Pelletier, S. Mangenot, M. M. Kuypers, F. Schreiber, B. E. Dutilh, J. Zedelius, D. de Beer, J. Gloerich, H. J. Wessels, T. van Alen, F. Luesken, M. L. Wu, K. T. van de Pas-Schoonen, H. J. M. Op den Camp, E. M. Janssen-Megens, K. J. Francoijs, H. Stunnenberg, J. Weissenbach, M. S. M. Jetten, M. Strous, *Nature* **2010**, *464*, 543–548.
19. (a) L. Chistoserdova, M. G. Kalyuzhnaya, M. E. Lidstrom, *Annu. Rev. Microbiol.* **2009**, *63*, 477–499. (b) L. Chistoserdova, *Environ. Microbiol.* **2011**, *13*, 2603–2622.
20. (a) A. A. van de Graaf, P. de Bruijn, L. A. Robertson, M. S. M. Jetten, J. G. Kuenen, *Microbiology* **1997,** *143*, 2415–21. (b) J. Schalk, H. Oustad, J. G. Kuenen, M. S. M Jetten, *FEMS Microbiol. Lett.* **1998**, *158*, 61–67.
21. M. Strous, E. Pelletier, S. Mangenot, T. Rattei, A. Lehner, M. W. Taylor, M. Horn, H. Daims, D. Bartol-Mavel, P. Wincker, V. Barbe, N. Fonknechten, D. Vallenet, B. Segurens, C. Schenowitz-Truong, C. Médigue, A. Collingro, B. Snel, B. E. Dutilh, H. J. M. Op den Camp, C. van der Drift, I. Cirpus, K. T. van de Pas-Schoonen, H. R. Harhangi, L. van Niftrik, M. Schmid, J. T. Keltjens, J. van de Vossenberg, B. Kartal, H. Meier, D. Frishman, M. A. Huynen, H. W. Mewes, J. Weissenbach, M. S. M. Jetten, M. Wagner, D. Le Paslier, *Nature* **2006,** *440*, 790–794.
22. B. Kartal, W. J. Maalcke, N. M. de Almeida, I. Cirpus, J. Gloerich, W. Geerts, H. J. M. Op den Camp, H. R. Harhangi, E. M. Janssen-Megens, K. J. Francoijs, H. G. Stunnenberg, J. T. Keltjens, M. S. M. Jetten, M. Strous, *Nature* **2011**, 479,127–130.
23. (a) M. R. Lindsay, R. I. Webb, M. Strous, M. S. M. Jetten, M. K. Butler, R. J. Forde, J. A. Fuerst, *Arch. Microbiol.* **2001**, *175*, 413–429. (b) L. A. van Niftrik, J. A. Fuerst, J. S. Sinninghe Damsté, J. G. Kuenen, M. S. M. Jetten, M. Strous, *FEMS Microbiol. Lett.* **2004**, *233*, 7–13. (c) L. van Niftrik, W. J. Geerts, E. G. van Donselaar, B. M. Humbel, A. Yakushevska, A. J. Verkleij, M. S. M. Jetten, M. Strous, *J. Struct. Biol.* **2008**, *161*, 401–410. (d) L. van Niftrik, W. J. Geerts, E. G. van Donselaar, B. M. Humbel, R. I. Webb, J. A. Fuerst, A. J. Verkleij, M. S. M. Jetten, M. Strous, *J. Bacteriol.* **2008**, *190*, 708–717.
24. (a) L. van Niftrik, M. van Helden, S. Kirchen, E. G. van Donselaar, H. R. Harhangi, R. I. Webb, J. A. Fuerst, H. J. M. Op den Camp, M. S. M. Jetten, M. Strous, *Mol. Microbiol.* **2010**, *77*, 701–715. (b) R. Karlsson, A. Karlsson, O. Bäckman, B. R. Johansson, S. Hulth, *FEMS Microbiol. Lett.* **2009**, *297*, 87–94. (c) R. Karlsson, A. Karlsson, O. Bäckman, B. R. Johansson, S. Hulth, *FEMS Microbiol. Lett.* **2014**, *354*, 10–18.
25. L. van Niftrik, M. S. M. Jetten, *Microbiol. Mol. Biol. Rev.* **2012**, *76*, 585–596.
26. B. Kartal, N. M. de Almeida, W. J. Maalcke, H. J. M. Op den Camp, M. S. M. Jetten, J. T. Keltjens, *FEMS Microbiol. Rev.* **2013**, *37*, 428–461.
27. N. M. de Almeida, W. J. Maalcke, J. T. Keltjens, M. S. M. Jetten, B. Kartal B, *Biochem. Soc. Trans.* **2011**, *39*, 303–308.
28. (a) B. Kartal, J. Rattray, L. A. van Niftrik, J.van de Vossenberg, M. C. Schmid, R. I. Webb, S. Schouten, J. A. Fuerst, J.S. Sinninghe Damsté, M. S. M. Jetten, M. Strous, *Syst. Appl. Microbiol.* **2007**, *30*, 39–49. (b) B. Kartal, M. M. M. Kuypers, G. Lavik, J. Schalk, H. J. M. Op

den Camp, M. S. M. Jetten, M. Strous, *Environ. Microbiol.* **2007**, *9*, 635–642. (c) B. Kartal, L. van Niftrik, J. Rattray, J. L.van de Vossenberg, M. C. Schmid, J. Sinninghe Damsté, M. S. M. Jetten, M. Strous, *FEMS Microbiol. Ecol.* **2008**, *63*, 46–55.

29. M. L. Wu, S. de Vries S, T. A. van Alen, M. K. Butler, H. J. M. Op den Camp, J. T. Keltjens, M. S. M. Jetten, M. Strous, *Microbiology* **2011**, *157*, 890–898.
30. (a) S. J. Ferguson, V. Fülöp, *Subcell. Biochem.* **2000**, *35*, 519–540. (b) O. Einsle, P. M. H. Kroneck, *Biol. Chem.* **2004,** *385*, 875–883. (c) P. Tavares, A. S. Pereira, J. J. Moura, I. Moura, *J. Inorg. Biochem.* **2006**, *100*, 2087–2100. (d) I. S. MacPherson, M. E. Murphy, *Cell. Mol. Life Sci.* **2007**, *64*, 2887–2899. (e) S. Rinaldo A. Arcovito, G. Giardina, N. Castiglione, M. Brunori, F. Cutruzzolà, *Biochem. Soc. Trans.* **2008**, *36*, 1155–1159. (f) A. C. Merkle, N. Lehnert, *Dalton Trans.* **2012**, *41*, 3355–3368.
31. (a) P. A. Williams, V. Fülöp, E. F. Garman, N. F. Saunders, S. J. Ferguson, J. Hajdu, *Nature* **1997**, 389, 406–412. (b) D. Nurizzo, M. C. Silvestrini, M. Mathieu, F. Cutruzzolà, D. Bourgeois, V. Fülöp V, J. Hajdu, M. Brunori, M. Tegoni, C. Cambillau, *Structure* **1997**, *5*, 1157–1171. (c) D. Nurizzo, F. Cutruzzolà, M. Arese, D. Bourgeois, M. Brunori, C. Cambillau, M. Tegoni, *Biochemistry* **1998**, *37*, 13987–13996.
32. (a) S. Kawasaki, H. Arai, T. Kodama, Y. Igarashi, *J. Bacteriol.* **1997**, *179*, 235–242. (b) S. Bali, A. D. Lawrence, S. A. Lobo, L. M. Saraiva, B. T. Golding, D. J. Palmer, M. J. Howard, S. J. Ferguson, M. J. Warren, *Proc. Natl. Acad. Sci. USA* **2011**, *108*, 18260–18265.
33. J. van de Vossenberg, D. Woebken, M. J. Maalcke, H. J. Wessels, B. E. Dutilh, B. Kartal, E. M. Janssen-Megens, G. Roeselers, J. Yan, D. Speth, J. Gloerich, W. Geerts, E. van der Biezen, W. Pluk, K. J. Francoijs, L. Russ, P. Lam, S. A. Malfatti, S. G. Tringe, S. C. Haaijer, H. J. M. Op den Camp, H. G. Stunnenberg, R. Amann, M. M. Kuypers, M. S. M. Jetten, *Environ Microbiol.* **2013**, *15*, 1275–1289.
34. D. Hira, H. Toh, C. T. Migita, H. Okubo, T. Nishiyama, M. Hattori, K. Furukawa, T. Fujii, *FEBS Lett.* **2012**, *586*, 1658–1663.
35. Z. Hu D. R. Speth, K. J. Francoijs, Z. X. Quan, M. S. M. Jetten, *Front. Microbiol.* **2012**, *3*, 366.
36. F. Gori, S. G. Tringe, B. Kartal, E. Marchiori M. S. M. Jetten, *Biochem. Soc. Trans.* **2011**, *39*, 1799–1804.
37. (a) G. W. Pettigrew, A. Echalier, S. R. Pauleta, *J. Inorg. Biochem.* **2006**, *100*, 551–567. (b) J. M. Atack, D. J. Kelly, *Adv. Microb. Physiol.* **2007**, *52*, 73–106.
38. (a) C. M. Wilmot, V. L. Davidson, *Curr. Opin. Chem. Biol.* **2009**, *13*, 469–474. (b) C. M. Wilmot, E. T. Yukl, *Dalton. Trans.* **2013**, **42**, 3127–3135. (c) S. Shin S, V. L. Davidson, *Arch. Biochem. Biophys.* **2014**, *544C*, 112–118.
39. (a) S. L. Edwards, T. L. Poulos, *J. Biol. Chem.* **1990**, *265*, 2588–2595. (b) E. T. Yukl, B. R. Goblirsch, V. L. Davidson, C. M. Wilmot, *Biochemistry* **2011**, *50*, 2931–2938.
40. M. Jormakka, S. Törnroth, B. Byrne, S. Iwata, *Science* **2002**, *295*, 1863–1868.
41. (a) A. B. Hooper, A. Nason, *J. Biol. Chem.* **1965**, *240*, 4044–4057. (b) K. R. Terry, A. B. Hooper, *Biochemistry* **1981**, *20*, 7026–7032.
42. A. B. Hooper, T. Vannelli, D. J. Bergmann, D. M. Arciero, *Antonie Van Leeuwenhoek* **1997**, *71*, 59–67.
43. (a) N. Igarashi, H. Moriyama, T. Fujiwara, Y. Fukumori, N. Tanaka, *Nat. Struct. Biol.* **1997**, *4*, 276–284. (b) P. E. Cedervall, A. B. Hooper, C. M. Wilmot, *Acta Crystallogr. Sect F Struct. Biol. Cryst. Commun.* **2009**, *65*, 1296–1298. (c) P. Cedervall, A. B. Hooper, C. M. Wilmot, *Biochemistry* **2013**, *52*, 6211–6218.
44. W. J. Maalcke, A. Dietl, S. J. Marritt, J. N. Butt, M. S. M. Jetten, J. T. Keltjens, T. R. Barends, B. Kartal, *J. Biol. Chem.* **2014**, *289*, 1228–1242.
45. (a) M. G. Klotz, M. C. Schmid, M. Strous, H. J. M. op den Camp, M. S. M. Jetten, A. B. Hooper, *Environ. Microbiol.* **2008**, *10*, 3150–3163. (b) J. Simon, M. G. Klotz, *Biochim. Biophys. Acta* **2013**, *1827*, 114–135.
46. M. Shimamura, T. Nishiyama, K. Shinya, Y. Kawahara, K. Furukawa, T. Fujii, *J. Biosci. Bioeng.* **2008**, *105*, 243–248.

47. J. Schalk, S. de Vries, J. G. Kuenen, M. S. M. Jetten, *Biochemistry* **2000**, *39*, 5405–5412.
48. W. L. DeLano, *The PyMOL Molecular Graphics System*, DeLano Scientific, San Carlos, CA, USA, 2002, http://www.pymol.org.
49. M. L. Fernández, D. A. Estrin, S. E. Bari, *J. Inorg. Biochem.* **2008**, *102*, 1523–1530.
50. M. Shimamura, T. Nishiyama, H. Shigetomo, T. Toyomoto, Y. Kawahara, K. Furukawa, T. Fujii, *Appl. Environ. Microbiol.* **2007**, *73*, 1065–1072.
51. W. J. Maalcke, *Multiheme Protein Complexes of Anaerobic Ammonium-Oxidizing Bacteria*, PhD Thesis, Radboud University of Nijmegen (The Netherlands), 2012.
52. (a) J. Kostera, J. McGarry, A. A. Pacheco, *Biochemistry* **2010**, *49*, 8546–8553. (b) A. A. Pacheco, J. McGarry, J. Kostera, A. Corona, *Methods Enzymol.* **2011**, *486*, 447–463.
53. (a) H. B. Gray, B. G. Malmström, R. J. P. Williams, *J. Biol. Inorg. Chem.* **2000**, *5*, 551–559. (b) H. B. Gray, J. R. Winkler, *Biochim. Biophys. Acta* **2010**, *1797*, 1563–1572. (c) M. Choi, V. L. Davidson, *Metallomics* **2011**, *3*, 140–151.
54. (a) I. V. Pearson, M. D. Page, R. J. van Spanning, S. J. Ferguson, *J. Bacteriol.* **2003**, *185*, 6308–6315. (b) K. A. Sam, S. A. Fairhurst, R. N. Thorneley, J. W. Allen, S. J. Ferguson, *J. Biol. Chem.* **2008**, *283*, 12555–12563.
55. F. ten Brink, B. Schoepp-Cothenet, R. van Lis, W. Nitschke, F. Baymann, *Biochim. Biophys. Acta* **2013**, *1827*, 1392–1406.
56. D. V. Dibrova, D. A. Cherepanov, M. Y. Galperin, V. P. Skulachev, A. Y. Mulkidjanian, *Biochim. Biophys. Acta* **2013**, *1827*, 1407–1427.
57. (a) X. Gao, X. Wen, L. Esser, B. Quinn, L. Yu, C. A. Yu, D. Xia, *Biochemistry* **2003**, *42*, 9067–9080. (b) X. Gao, X. Wen, C. Yu, L. Esser, S. Tsao, B. Quinn, L. Zhang, L. Yu, D. Xia, *Biochemistry* **2002**, *41*, 11692–11702. (c) H. Palsdottir, C. G. Lojero, B. L. Trumpower, C. Hunte, *J. Biol. Chem.* **2003**, *278*, 31303–31311. (d) L. Esser, B. Quinn, Y. F. Li, M. Zhang, M. Elberry, L. Yu, C. A. Yu, D. Xia, *J. Mol. Biol.* **2004**, *341*, 281–302. (e) L. Esser, M. Elberry, F. Zhou, C. A. Yu, L. Yu, D. Xia, *J. Biol. Chem.* **2008**, *283*, 2846–2857. (f) S. R. Solmaz, C. Hunte, *J. Biol. Chem.* **2008**, *283*, 17542–17549. (g) J. Yan, G. Kurisu, W. A. Cramer, *Proc. Natl. Acad. Sci. USA* **2006**, *103*, 69–74. (h) E. Yamashita, H. Zhang, W. A. Cramer, *J. Mol. Biol.* **2007**, *370*, 39–52.
58. (a) H. Claus, *Arch. Microbiol.* **2003**, *179*, 145–150. (b) P. Giardina, V. Faraco, C. Pezzella, A. Piscitelli, S. Vanhulle, G. Sannia, *Cell. Mol. Life Sci.* **2010**, *67*, 369–385. (c) P. Di Gennaro, A. Bargna, G. Sello, *Appl. Microbiol. Biotechnol.* **2011**, *90*, 1817–1827.
59. (a) M. C. Machczynski, E. Vijgenboom, B. Samyn, G. W. Canters, *Protein Sci* **2004**, *13*, 2388–2397. (b) T. Skálová, J. Dohnálek, L. H. Østergaard, P. R. Østergaard, P. Kolenko, J. Dusková, A. Stepánková, J. Hasek, *J. Mol. Biol.* **2009**, *385*, 1165–1178.
60. M. Sutter, D. Boehringer, S. Gutmann, S. Günther, D. Prangishvili, M. J. Loessner, K. O. Stetter, E. Weber-Ban, N. Ban, *Nat. Struct. Mol. Biol.* **2008**, *15*, 939–947.
61. R. A. Rothery, G. J. Workun, J. H. Weiner, *Biochim. Biophys. Acta* **2008**, *1778*, 1897–1929.
62. M. Jormakka, D. Richardson, B. Byrne, S. Iwata, *Structure* **2004**, *12*, 95–104.
63. S. Lücker, M. Wagner, F. Maixner, E. Pelletier, H. Koch, B. Vacherie, T. Rattei,J. S. Damsté, E. Spieck, D. Le Paslier, H. Daims, *Proc. Natl. Acad. Sci. USA* **2010**, 107, 13479–13484.
64. S. Lücker, B. Nowka, T. Rattei, E. Spieck, H. Daims, *Front. Microbiol.* **2013**, *4*, 27.
65. M. G. Bertero, R. A. Rothery, M. Palak, C. Hou, D. Lim, F. Blasco, J. H. Weiner, N. C. Strynadka, *Nature Struct. Mol. Biol.* **2003**, *10*, 681–687.
66. C. A. McDevitt, G. R. Hanson, C. J. Noble, M. R. Cheesman, A. G. McEwan, *Biochemistry* **2002**, *41*, 15234–15244.
67. (a) I. Schröder, S. Rech, T. Krafft, J. M. Macy, *J. Biol. Chem.* **1997**, *272*, 23765–23768. (b) E. C. Lowe, S. Bydder, R. S. Hartshorne, H. L. Tape, E. J. Dridge, C. M. Debieux, K. Paszkiewicz, I. Singleton, R. J. Lewis, J. M. Santini, D. J. Richardson, C. S. Butler, *J. Biol. Chem.* **2010**, 285, 18433–18442.
68. H. D. Thorell, K. Stenklo, J. Karlsson, T. Nilsson, *Appl. Environ. Microbiol.* **2003**, *69*, 5585–5592.
69. D. P. Kloer, C. Hagel, J. Heider, G. E. Schulz, *Structure* **2006**, *14*, 1377–1388.

70. (a) P. Lanciano, A. Vergnes, S. Grimaldi, B. Guigliarelli, A. Magalon, *J. Biol. Chem.* **2007**, *282*, 17468–17474. (b) S. Zakian, D. Lafitte, A. Vergnes, C. Pimentel, C. Sebban-Kreuzer, R. Toci, J. B. Claude, F. Guerlesquin, A. Magalon, *FEBS J.* **2010**, *277*, 1886–1895.
71. (a) V. B. Borisov, R. B. Gennis, J. Hemp, M. I. Verkhovsky, *Biochim. Biophys. Acta* **2011**, *1807*, 1398–1413. (b) A. Giuffrè, V. B. Borisov, M. Arese, P. Sarti, E. Forte, *Biochim. Biophys. Acta* **2014**, *1837*, 1178–1187.
72. S. B. Mohan, M. Schmid, M. S. M. Jetten, J. Cole, *FEMS Microbiol. Ecol.* **2004**, *49*, 433–443.
73. (a) J. Simon, M. Kern, *Biochem. Soc. Trans.* **2008**, *36*, 1011–1016. (b) M. Kern, J. Simon, *Biochim. Biophys. Acta* **2009**, *1787*, 646–656.
74. (a) O. Einsle, A. Messerschmidt, P. Stach, G. P. Bourenkov, H. D. Bartunik, R. Huber, P. M. H. Kroneck, *Nature* **1999**, *400*, 476–480. (b) O. Einsle, A. Messerschmidt, R. Huber, P. M. H. Kroneck, F. Neese, *J. Am. Chem. Soc.* **2002**, *124*, 11737–11745.
75. M. J. Sellars, S. J. Hall, D. J. Kelly, *J. Bacteriol.* **2002**, *184*, 4187–4196.
76. R. A. Sanford, J. R. Cole, J. M. Tiedje, *Appl. Environ. Microbiol.* **2002**, *68*, 893–900.
77. (a) M. S. Pittman, K. T. Elvers, L. Lee, M. A. Jones, R. K. Poole, S. F. Park, D. J. Kelly, *Mol. Microbiol.* **2007**, *63*, 575–590. (b) M. Kern, F. Eisel, J. Scheithauer, R. G. Kranz, J. Simon, *Mol. Microbiol.* **2010**, *75*, 122–137.
78. A. Welsh, J. Chee-Sanford, L. Connor, F. Löffler, R. Sanford, *Appl. Environ. Microbiol.* **2014**, *80*, 2110–2119.
79. (a) K. M. Polyakov, K. M. Boyko, T. V. Tikhonova, A. Slutsky, A. N. Antipov, R. A. Zvyagilskaya, A. N. Popov, G. P. Bourenkov, V. S. Lamzin, V. O. Popov, *J. Mol. Biol.* **2009**, *389*, 846–862. (b) A. A. Trofimov, K. M. Polyakov, K. M. Boyko, T. V. Tikhonova, T. N. Safonova, A. V. Tikhonov, A. N. Popov, V. O. Popov, *Acta Crystallogr. D Biol. Crystallogr.* **2010**, *66*, 1043–1047. (c) T. Tikhonova, A. Tikhonov, A. Trofimov, K. Polyakov, K. Boyko, E. Cherkashin, T. Rakitina, D. Sorokin, V. Popov, *FEBS J.* **2012**, *279*, 4052–4061. (d) T. V. Tikhonova, A. A. Trofimov, V. O. Popov, *Biochemistry (Moscow)* **2012**, *77*, 1129–1138.
80. (a) C. G. Mowat, E. Rothery, C. S. Miles, L. McIver, M. K. Doherty, K. Drewette, P. Taylor, M. D. Walkinshaw, S. K. Chapman, G. A. Reid, *Nature Struct. Mol. Biol.* **2004**, *11*, 1023–1024. (b) S. J. Atkinson, C. G. Mowat, G. A. Reid, S. K. Chapman, *FEBS Lett.* **2007**, *581*, 3805–3808.
81. (a) J. J. Moura, C. D. Brondino, J. Trincão, M. J. Romão, *J. Biol. Inorg. Chem.* **2004**, *9*, 791–799. (b) G. Fritz, O. Einsle, M. Rudolf, A. Schiffer, P. M. H. Kroneck, *J. Mol. Microbiol. Biotechnol.* **2005**, *10*, 223–233. (b) P. J. González, C. Correia, I. Moura, C. D. Brondino, J. J. Moura, *J. Inorg. Biochem.* **2006**, *100*, 1015–1023. (c) J. Simon, R. J. van Spanning, D. J. Richardson, *Biochim. Biophys. Acta* **2008**, *1777*, 1480–1490. (d) B. Kraft, M. Strous, H. E. Tegetmeyer, *J Biotechnol.* **2011**, *155*, 104–117. (e) C. Sparacino-Watkins, J. F. Stolz, P. Basu, *Chem. Soc. Rev.* **2014**, *43*, 676–706.
82. (a) J. M. Dias, M. E. Than, A. Humm, R. Huber, G. P. Bourenkov, H. D. Bartunik, S. Bursakov, J. Calvete, J. Caldeira, C. Carneiro, J. J. Moura, I. Moura, M. Romão, *Structure* **1999**, *7*, 65–79. (b) P. Arnoux, M. Sabaty, J. Alric, B. Frangioni, B. Guigliarelli, J. M. Adriano, D. Pignol, *Nat. Struct. Biol.* **2003**, *10*, 928–934. (c) B. J. Jepson, S. Mohan, T. A. Clarke, A. J. Gates, J. A. Cole, C. S. Butler, J. N. Butt, A. M. Hemmings, D. J. Richardson, *J. Biol. Chem.* **2007**, *282*, 6425–6437. (d) S. Najmudin, P. J. González, J. Trincão, C. Coelho, A. Mukhopadhyay, N. M. Cerqueira, C. C. Romão, I. Moura, J. J. Moura, C. D. Brondino, M. J. Romão, *J. Biol. Inorg. Chem.* **2008**, *13*, 737–753. (e) C. Coelho, P. J. González, J. G. Moura, I. Moura, J. Trincão, M. J. Romão, *J. Mol. Biol.* **2011**, *408*, 932–948.
83. B. Ize, S. J. Coulthurst, K. Hatzixanthis, I. Caldelari, G. Buchanan, E. C. Barclay, D. J. Richardson, T. Palmer, F. Sargent, *Microbiology* **2009**, *155*, 3992–4004.
84. (a) T. H. Brondijk, D. Fiegen, D. J. Richardson, J. A. Cole, *Mol. Microbiol.* **2002**, *44*, 245–255. (b) T. H. Brondijk, A. Nilavongse, N. Filenko, D. J. Richardson, J. A. Cole, *Biochem. J.* **2004**, *379*, 47–55. (c) A. Nilavongse, T. H. Brondijk, T. W. Overton, D. J.

Richardson, E. R. Leach, J. A. Cole, *Microbiology* **2006**, *152*, 3227–3237. (d) M. Kern, J. Simon, *Microbiology* **2009**, *155*, 2784–2794. (e) M. Kern, J. Simon, *Mol. Microbiol.* **2008**, *69*, 1137–1152.

85. (a) T. Sjögren, M. Svensson-Ek, J. Hajdu, P. Brzezinski, *Biochemistry* **2000**, *39*, 10967–10974. (b) A. Jafferji, J. W. Allen, S. J. Ferguson, V. Fulop, *J. Biol. Chem.* **2000**, *275*, 25089–25094.
86. M. L. Wu, T. A. van Alen, E. G. van Donselaar, M. Strous, M. S. M. Jetten, L. van Niftrik, *FEMS Microbiol. Lett.* **2012**, *334*, 49–56.
87. R. J. Schulze, J. Komar, M. Botte, W. J. Allen, S. Whitehouse, V. A. Gold, A. Lycklama, J. A. Nijeholt, K. Huard, I. Berger, C. Schaffitzel, I. Collinson, *Proc. Natl. Acad. Sci. USA* **2014**, *111*, 4844–4849.
88. V. M. Luque-Almagro, A. J. Gates, C. Moreno-Vivián, S. J. Ferguson, D. J. Richardson, M. D. Roldán, *Biochem. Soc. Trans.* **2011**, *39*, 1838–1843.
89. P. J. Simpson, D. J. Richardson, R. Codd, *Microbiology* **2010**, *156*, 302–312.
90. (a) L. Philippot, *Trends Microbiol.* **2005,** *13*, 191–192. (b) R. Cramm, R. A. Siddiqui, B. Friedrich, *J. Bacteriol.* **1997**, *179*, 6769–6777.
91. (a) W. G. Zumft, *J. Inorg. Biochem.* **2005**, *99*, 194–215. (b) J. Hendriks, A. Oubrie, J. Castresana, A. Urbani, S. Gemeinhardt, M. Saraste, *Biochim. Biophys. Acta* **2000**, *1459*, 266–273.
92. (a) Y. Matsumoto, T. Tosha, A. V. Pisliakov, T. Hino, H. Sugimoto, S. Nagano, Y. Sugita, Y. Shiro, *Nature Struct. Mol. Biol.* **2012**, *19*, 238–245. (b) T. Hino, Y. Matsumoto, S. Nagano, H. Sugimoto, Y. Fukumori, T. Murata, S. Iwata, *Science* **2010**, *330*, 1666–1670.
93. Suharti, M. J. F. Strampraad, I. Schröder, S. de Vries, *Biochemistry* **2001**, *40*, 2632–2639.
94. K. Heylen, J.T. Keltjens, *Front. Microbiol.* **2012**, *3*, 371.
95. J. Hemp, R. B. Gennis, *Results Probl. Cell Differ.* **2008**, *45*, 1–31.
96. (a) M. Saraste, J. Castresana, *FEBS Lett.* **1994**, *341*, 1–4. (b) J. van der Oost, A. P. N. de Boer, I.-W. L. Gier, W. G. Zumft, A. H. Stouthamer, R. J. M. van Spanning, *FEMS Microbial. Lett.* **1994**, *121*, 1–10.
97. (a) T. Fujiwara, Y. Fukumori, *J. Bacteriol*, **1996**. *178*, 1866–1871. (b) U. Flock, N. J. Watmough, P. Ädelroth, *Biochemistry* **2005**, *44*, 10711–10719. (c) G. Butland, S. Spiro, N. J. Watmough, D. J. Richardson, *J. Bacteriol.* **2001**, *183*, 189–199. (d) N. Sakurai, T. Sakurai, *Biochemistry* **1997**, *36*, 13809–13815. (e) A. Giuffrè, G. Stubauer, P. Sarti, M. Brunori, W. G. Zumft, G. Buse, T. Soulimane, *Proc. Natl. Acad. Sci. USA* **1999**, *96*, 14718–14723. (f) E. Forte, A. Urbani, M. Saraste, P. Sarti, M. Brunori, A. Giuffrè, *Eur. J. Biochem.* **2001**, *268*, 6486–6491.
98. (a) D. A. Proshlyakov, M. A. Pressler, C. DeMaso, J. F. Leykam, D. L. deWitt, G. T. Babcock, *Science* **2000**, *290*, 1588–1591. (b) R. B. Gennis, *Biochim. Biophys. Acta* **1998**, *1365*, 241–248. (c) G. T. Babcock, *Proc. Natl. Acad. Sci. USA* **1999**, *96*, 12971–12973 (d) M. R. A. Blomberg, P. E. M. Siegbahn, M. Wikström, *Inorg. Chem.* **2003**, *42*, 5231–5243. (e) M. Iwaki, A. Puustinen, M. Wikström, P. R. Rich, *Biochemistry* **2003**, *42*, 8809–8817.
99. K. F. Ettwig, D. R. Speth, J. Reimann, M. L. Wu, M. S. M. Jetten, J. T. Keltjens, *Front. Microbiol.* **2012**, *3*, 273.
100. (a) F. H. Thorndycroft, G. Butland, D. J. Richardson, N. J. Watmough, *Biochem. J.* **2007**, *401*, 111–119. (b) U. Flock, P. Lachmann, J. Reimann, N. J. Watmough, P. Ädelroth, *J. Inorg. Biochem.* **2009**, *103*, 845–850.
101. (a) P. Ehrenreich, A. Behrends, J. Harder, F. Widdel, *Arch. Microbiol.* **2000**, *173*, 58–64. (b) J. Zedelius, R. Rabus, O. Grundmann, I. Werner, D. Brodkorb, F. Schreiber, P. Ehrenreich, A. Behrends, H. Wilkes, M. Kube, R. Reinhardt, F. Widdel, *Environ. Microbiol. Rep.* **2010**, *3*, 125–135.
102. L. Salomonsson, J. Reimann, T. Tosha, N. Krause, N. Gonska, Y Shiro, P. Ädelroth, *Biochim. Biophys. Acta* **2012**, *1817*, 1914–1920.

103. (a) R. L. Lieberman, A. C. Rosenzweig, *Nature* **2005**, *434*, 177–182. (b) A. S. Hakemian, K. C. Kondapalli, J. Telser, B. M. Hoffman, T. L. Stemmler, A. C. Rosenzweig, *Biochemistry* **2008**, *47*, 6793–6801. (c) S. M. Smith, S. Rawat, J. Telser, B. M. Hoffman, T. L. Stemmler, A. C. Rosenzweig, *Biochemistry* **2011**, *50*, 10231–10340.
104. R. Balasubramanian, S. M. Smith, S. Rawat, L. A. Yatsunyk, T. L. Stemmler, A. C. Rosenzweig, *Nature* **2010**, *465*, 115–119.
105. (a) H. J. Hektor, H. Kloosterman, L. Dijkhuizen, *J. Mol. Catal. B* **2000**, *8*, 103–109. (b) H. J. Hektor, H. Kloosterman, L. Dijkhuizen, *J. Biol. Chem.* **2002**, *277*, 46966–46973.
106. (a) C. Anthony, M. Ghosh, *Prog. Biophys. Mol. Biol.* **1998**, *69*, 1–21. (b) C. Anthony, P. Williams, *Biochim. Biophys. Acta* **2003**, *1647*, 18–23. (c) C. Anthony, *Arch. Biochem. Biophys.* **2004**, *428*, 2–9.
107. (a) L. Masgrau, J. Basran, P. Hothi, M. J. Sutcliffe, N. S. Scrutton, *Arch. Biochem. Biophys.* **2004**, *428*, 41–51. (b) S. Y. Reddy, T. C. Bruice, *J. Am. Chem. Soc.* **2003**, *125*, 8141–8150. (c) V. L. Davidson, *Arch. Biochem. Biophys.* **2004**, *428*, 32–40. (d) S. Y. Reddy, T. C. Bruice, *Protein Sci.* **2004**, *13*, 1965–1978. (e) X. Zhang, S. Y. Reddy, T. C. Bruice, *Proc. Natl. Acad. Sci. USA* **2007**, *104*, 745–749.
108. (a) A. Oubrie, B. W. Dijkstra, *Protein Sci.* **2000**, *9*, 1265–1273. (b) A. Oubrie, H. J. Rozeboom, K. H. Kalk, E. G. Huizinga, B. W. Dijkstra, *J. Biol. Chem.* **2002**, *277*, 3727–3732.
109. (a) Z. X. Xia, W. W. Dai, J. P. Xiong, Z. P. Hao, V. L. Davidson, S. White, F. S. Mathews, *J. Biol. Chem.* **1992**, *267*, 22289–22297. (b) Z. Xia, W. Dai, Y. Zhang, S. A. White, G. D. Boyd, F. S. Mathews, *J. Mol. Biol.* **1996**, *259*, 480–501. (c) Z. X. Xia Y. N. He, W. W. Dai, S. A. White, G. D. Boyd, F. S. Mathews, *Biochemistry* **1999**, *38*, 1214–1220. (d) Z. X. Xia, W. W. Dai, Y. N. He, S. A. White, F. S. Mathews, V. L. Davidson, *J. Biol. Inorg. Chem.* **2003**, *8*, 843–854. (e) M. Ghosh, C. Anthony, K. Harlos, M. G. Goodwin, C. Blake, *Structure* **1995**, *3*, 177–187. (f) M. Ghosh, A. Avezoux, C. Anthony, K. Harlos, C. C. Blake, *Experientia* **1994**, *71*, 251–260. (g) P. A. Williams, L. Coates, F. Mohammed, R. Gill, P. T. Erskine, A. Coker, S. P. Wood, C. Anthony, J. B. Cooper, *Acta Crystallogr. D Biol. Crystallogr.* **2005**, *61*, 75–79. (h) J. M. Choi, H. G. Kim, J. S. Kim, H. S. Youn, S. H. Eom, S. L. Yu, S. W. Kim, S. H. Lee, *Acta Crystallogr. Sect. F Struct. Biol. Cryst. Commun.* **2011**, *67*, 513–516.
110. (a) D. J. Anderson, C. J. Morris, D. N. Nunn, C. Anthony, M. E. Lidstrom, *Gene* **1990**, *90*, 173–176. (b) C. J. Morris, Y. M. Kim, K. E. Perkins, M. E. Lidstrom, *J. Bacteriol.* **1995**, *177*, 6825–6831. (c) L. Chistoserdova, S. W. Chen, A. Lapidus, M. E. Lidstrom, *J. Bacteriol.* **2003**, *185*, 2980–2987.
111. (a) N. Goosen, R. G. M. Huinen, P. Vandeputte, *J. Bacteriol.* **1992**, *174*, 1426–1427. (b) S. Puehringer, M. Metlitzky, R. Schwarzenbacher, *BMC Biochemistry* **2008**, *9*, 8. (c) N. Gliese, V. Khodaverdi, H. Görisch, *Arch. Microbiol.* **2010**, *192*, 1–14.
112. M. L. Wu, K. F. Ettwig, M. S. M. Jetten, M. Strous, J. T. Keltjens, L. van Niftrik, *Biochem. Soc. Trans.* **2011**, *39*, 243–248.
113. (a) Y. Hibi, K. Asai, H. Arafuka, M. Hamajima, T. Iwama, K. Kawai, *J. Biosci. Bioeng.* **2011**, *111*, 547–549. (b) N. A. Fitriyanto, M. Fushimi, M. Matsunaga, A. Pertiwiningrum, T. Iwama, K. Kawai, *J. Biosci. Bioeng.* **2011**, *111*, 613–617. (c) T. Nakagawa, R. Mitsui, A. Tani, K. Sasa, S. Tashiro, T. Iwama, T. Hayakawa, K. Kawai, *PloS One* **2012**, *7*, e50480.
114. A. Pol, T. R. M. Barends, A. Dietl, A. F. Khadem, J. Eygensteyn, M. S. M. Jetten, H. J. M. op den Camp, *Environ. Microbiol.* **2014**, *16*, 255–264.
115. J. T. Keltjens, A. Pol, J. Reimann, H. J. M. op den Camp, *Appl. Microbiol. Biotechnol.* **2014**, *98*, 6163–6183.
116. M. L. Wu, H. J. C. T. Wessels, A. Pol, H. J. M. Op den Camp, M. S. M. Jetten, L. Van Niftrik, J. T. Keltjens, **2014**, *submitted for publication.*
117. (a) L. Chistoserdova, J. A. Vorholt, R. K. Thauer, M. E. Lidstrom, *Science* **1998**, *281*, 99–102. (b) J. A. Vorholt, L. Chistoserdova, S. M. Stolyar, R. K. Thauer, M. E. Lidstrom, *J. Bacteriol.* **1999**, *181*, 5750–5757. (c) L. Chistoserdova, J. A. Vorholt, M. E. Lidstrom, *Genome Biol.* **2005**, *6*, 208. (d) L. Chistoserdova, M. E. Rasche, M. E. Lidstrom, *J. Bacteriol.* **2005**, *187*, 2508–2512. (e) M. G. Kalyuzhnaya, N. Korotkova, G. Crowther, C. J. Marx,

M. E. Lidstrom, *J. Bacteriol.* **2005**, *187*, 4607–4614. (f) M. G. Kalyuzhnaya, L. Chistoserdova, *Methods Enzymol.* **2005**, *397*, 443–54. (g) M. G. Kalyuzhnaya, O. Nercessian, A. Lapidus, L. Chistoserdova, *Environ. Microbiol.* **2005**, *7*, 1909–1916. (h) M. G. Kalyuzhnaya, S. Bowerman, O. Nercessian, M. E. Lidstrom, L. Chistoserdova, *Appl. Environ. Microbiol.* **2005**, *71*, 8846–8854. (i) L. Chistoserdova, A. Lapidus, C. Han, L. Goodwin, L. Saunders, T. Brettin, R. Tapia, P. Gilna, S. Lucas, P. M. Richardson, M. E. Lidstrom, *J. Bacteriol.* **2007**, *189*, 4020–4027. (j) S. Vuilleumier, L. Chistoserdova, M. C. Lee, F. Bringel, A. Lajus, Y. Zhou, B. Gourion, V. Barbe, J. Chang, S. Cruveiller, C. Dossat, W. Gillett, C. Gruffaz, E. Haugen, E. Hourcade, R. Levy, S. Mangenot, E. Muller, T. Nadalig, M. Pagni, C. Penny, R. Peyraud, D. G. Robinson, D. Roche, Z. Rouy, C. Saenampechek, G. Salvignol, D. Vallenet, Z. Wu, C. J. Marx, J. A. Vorholt, M. V. Olson, R. Kaul, J. Weissenbach, C. Médigue, M. E. Lidstrom, *PLoS One* **2009**, *4*, e5584. (k) M. G. Kalyuzhnaya, D. A. Beck, D. Suciu, A. Pozhitkov, M. E. Lidstrom, L. Chistoserdova, *ISME J.* **2010**, *4*, 388–398. (l) E. L. Hendrickson, D. A. Beck, T. Wang, M. E. Lidstrom, M. Hackett, L. Chistoserdova, *J. Bacteriol.* **2010**, *192*, 4859–4867.

118. P. Acharya, M. Goenrich, C. H. Hagemeier, U. Demmer, J. A. Vorholt, R. K. Thauer, U. Ermler, *J. Biol. Chem.* **2005**, *280*, 13712–13719.
119. (a) W. Grabarse, M. Vaupel, J. A. Vorholt, S. Shima, R. K. Thauer, A. Wittershagen, G. Bourenkov, H. D. Bartunik, U. Ermler, *Structure* **1999**, *7*, 1257–1268. (b) V. Upadhyay, U. Demmer, E. Warkentin, J. Moll, S. Shima, U. Ermler, *Biochemistry* **2012**, *51*, 8435–8443.
120. C. H. Hagemeier, L. Chistoserdova, M. E. Lidstrom, R. K. Thauer, J. A. Vorholt, *Eur. J. Biochem.* **2000**, *267*, 3762–3769.
121. J. A. Vorholt, L. Chistoserdova, M. E. Lidstrom, R. K. Thauer, *J. Bacteriol.* **1998**, *180*, 5351–5356.
122. J. A. Vorholt, M. G. Kalyuzhnaya, C. H. Hagemeier, M. E. Lidstrom, *J. Bacteriol.* **2005**, *187*, 6069–6074.
123. U. Ermler, C. H. Hagemeier, A. Roth, U. Demmer, W. Grabarse, E. Warkentin, J. A. Vorholt, *Structure* **2002**, *10*, 1127–1137.
124. (a) M. Jormakka, B. Byrne, S. Iwata, *Curr. Opin. Struct. Biol.* **2003**, *13*, 418–423. (b) J. J. Moura, C. D. Brondino, J. Trincão, M. J. Romão, *J. Biol. Inorg. Chem.* **2004**, *9*, 791–799. (c) R. G. Efremov, L. A. Sazanov, *Biochim. Biophys. Acta.* **2012**, *1817*, 1785–1795.
125. C. S. Mota, M. G. Rivas, C. D. Brondino, I. Moura, J. J. Moura, P. J. González, N. M. Cerqueira, *J. Biol. Inorg. Chem.* **2011**, *16*, 1255–1268.
126. J. A. Vorholt, R. K. Thauer, *Met. Ions Biol. Syst.* **2002**, *39*, 571–619.
127. B. K. Pomper, O. Saurel, A. Milon, J. A. Vorholt, *FEBS Lett.* **2002**, *523*, 133–137.
128. B. K. Pomper, J. A. Vorholt, *Eur. J. Biochem.* **2001**, *268*, 4769–4775.
129. (a) U. Ermler, M. Merckel, R. K. Thauer, S. Shima, *Structure* **1997**, *5*, 635–646. (b) P. Acharya, E. Warkentin, U. Ermler, R. K. Thauer, S. Shima, *J. Mol. Biol.* **2006**, *357*, 870–879.
130. J. A. Vorholt, M. Vaupel, R. K. Thauer, *Mol. Microbiol.* **1997**, *23*, 1033–1042.
131. (a) A. Hochheimer, D. Linder, R. K. Thauer, R. Hedderich, *Eur. J. Biochem.* **1996**, *242*, 156–162. (b) A. Hochheimer, R. A. Schmitz, R. K. Thauer, R. Hedderich, *Eur. J. Biochem.* **1995**, *234*, 910–920.
132. (a) V. N. Gladyshev, J. C. Boyington, S. V. Khangulov, D. A. Grahame, T. C. Stadtman, P. D. Sun, *J. Biol. Chem.* **1996**, *271*, 8095–8100. (b) J. C. Boyington, V. N. Gladyshev, S. V. Khangulov, T. C. Stadtman, P. D. Sun, *Science* **1997**, *275*, 1305–1308.
133. H. Raaijmakers, S. Macieira, J. M. Dias, S. Teixeira, S. Bursakov, R. Huber, J. J. Moura, I. Moura, M. J. Romão, *Structure* **2002**, *10*, 1261–1272.
134. (a) R. G. Efremov, R. Baradaran, L. A. Sazanov, *Nature* **2010**, *465*, 441–445. (b) R. G. Efremov, L. A. Sazanov, *Curr. Opin. Struct. Biol.* **2011**, *21*, 532–540.
135. G. J. Crowther, G. Kosály, M. E. Lidstrom, *J. Bacteriol.* **2008**, *190*, 5057–5062.
136. (a) R. K. Thauer, K. Jungermann, K. Decker, *Bacteriol. Rev.* **1977**, *41*, 100–180.

137. (b) http://nshs-science.net/chemistry/common/pdf/R-standard_enthalpy_of_formation.pdf (c) http://www.nist.gov/data/PDFfiles/jpcrd6.pdf (d) http://fizyczna.chem.pg.gda.pl/files/2012/10/chf_epm_cr_00.pdf
138. *The Metal-Driven Biogeochemistry of Gaseous Compounds in the Environment*, Eds P.M.H. Kroneck, M.E. Sosa Torres; Vol. 14 of *Metal Ions in Life Sciences*, Eds A. Sigel, H. Sigel, R.K.O. Sigel; Springer Science+Business Media, Dordrecht, The Netherlands, 2014.

Index

P.M.H. Kroneck, M.E. Sosa Torres (eds.), *Sustaining Life on Planet Earth: Metalloenzymes Mastering Dioxygen and Other Chewy Gases*, Metal Ions in Life Sciences 15, DOI 10.1007/978-3-319-12415-5

M

N

Q

R

S

T

U

V

W

X